Discrete Variational Derivative Method

A Structure-Preserving
Numerical Method for
Partial Differential Equations

CHAPMAN & HALL/CRC
Numerical Analysis and Scientific Computing

Aims and scope:
Scientific computing and numerical analysis provide invaluable tools for the sciences and engineering. This series aims to capture new developments and summarize state-of-the-art methods over the whole spectrum of these fields. It will include a broad range of textbooks, monographs, and handbooks. Volumes in theory, including discretisation techniques, numerical algorithms, multiscale techniques, parallel and distributed algorithms, as well as applications of these methods in multi-disciplinary fields, are welcome. The inclusion of concrete real-world examples is highly encouraged. This series is meant to appeal to students and researchers in mathematics, engineering, and computational science.

Editors

Choi-Hong Lai
*School of Computing and
Mathematical Sciences
University of Greenwich*

Frédéric Magoulès
*Applied Mathematics and
Systems Laboratory
Ecole Centrale Paris*

Editorial Advisory Board

Mark Ainsworth
*Mathematics Department
Strathclyde University*

Todd Arbogast
*Institute for Computational
Engineering and Sciences
The University of Texas at Austin*

Craig C. Douglas
*Computer Science Department
University of Kentucky*

Ivan Graham
*Department of Mathematical Sciences
University of Bath*

Peter Jimack
*School of Computing
University of Leeds*

Takashi Kako
*Department of Computer Science
The University of Electro-Communications*

Peter Monk
*Department of Mathematical Sciences
University of Delaware*

Francois-Xavier Roux
ONERA

Arthur E.P. Veldman
*Institute of Mathematics and Computing Science
University of Groningen*

Proposals for the series should be submitted to one of the series editors above or directly to:
CRC Press, Taylor & Francis Group
4th, Floor, Albert House
1-4 Singer Street
London EC2A 4BQ
UK

Discrete Variational Derivative Method

A Structure-Preserving Numerical Method for Partial Differential Equations

Daisuke Furihata
Osaka University
Osaka, Japan

Takayasu Matsuo
University of Tokyo
Tokyo, Japan

CRC Press
Taylor & Francis Group
Boca Raton London New York

CRC Press is an imprint of the
Taylor & Francis Group, an **informa** business

A CHAPMAN & HALL BOOK

Chapman & Hall/CRC
Taylor & Francis Group
6000 Broken Sound Parkway NW, Suite 300
Boca Raton, FL 33487-2742

© 2011 by Taylor and Francis Group, LLC
Chapman & Hall/CRC is an imprint of Taylor & Francis Group, an Informa business

No claim to original U.S. Government works

Printed in the United States of America on acid-free paper
10 9 8 7 6 5 4 3 2 1

International Standard Book Number: 978-1-4200-9445-9 (Hardback)

This book contains information obtained from authentic and highly regarded sources. Reasonable efforts have been made to publish reliable data and information, but the author and publisher cannot assume responsibility for the validity of all materials or the consequences of their use. The authors and publishers have attempted to trace the copyright holders of all material reproduced in this publication and apologize to copyright holders if permission to publish in this form has not been obtained. If any copyright material has not been acknowledged please write and let us know so we may rectify in any future reprint.

Except as permitted under U.S. Copyright Law, no part of this book may be reprinted, reproduced, transmitted, or utilized in any form by any electronic, mechanical, or other means, now known or hereafter invented, including photocopying, microfilming, and recording, or in any information storage or retrieval system, without written permission from the publishers.

For permission to photocopy or use material electronically from this work, please access www.copyright.com (http://www.copyright.com/) or contact the Copyright Clearance Center, Inc. (CCC), 222 Rosewood Drive, Danvers, MA 01923, 978-750-8400. CCC is a not-for-profit organization that provides licenses and registration for a variety of users. For organizations that have been granted a photocopy license by the CCC, a separate system of payment has been arranged.

Trademark Notice: Product or corporate names may be trademarks or registered trademarks, and are used only for identification and explanation without intent to infringe.

Library of Congress Cataloging-in-Publication Data

Furihata, Daisuke.
　　Discrete variational derivative method : a structure-preserving numerical method for partial differential equations / Daisuke Furihata, Takayasu Matsuo.
　　　　p. cm. -- (Chapman and Hall/CRC numerical analysis and scientific computation series)
　　Summary: "Many important problems in engineering and science are modeled by nonlinear partial differential equations (PDEs). A new trend in PDEs, called structure-preserving numerical methods, has recently developed. This book is devoted to one such technique, called the discrete variational derivative method. First, the text introduces the key factors and the basic ideas of this method, followed by target problems solvable by the method. The second section describes the rigorous mathematics in detail along with relevant applications, which are illustrated by worked examples. It concludes with a comprehensive listing of essential references on structure-preserving algorithms for advanced readers"-- Provided by publisher.
　　Includes bibliographical references and index.
　　ISBN 978-1-4200-9445-9 (hardback)
　　1. Differential equations, Partial--Numerical solutions. 2. Nonlinear theories. 3. Engineering mathematics. I. Matsuo, Takayasu. II. Title. III. Series.

QA377.F87 2010
518'.64--dc22
　　　2010043677

Visit the Taylor & Francis Web site at
http://www.taylorandfrancis.com

and the CRC Press Web site at
http://www.crcpress.com

Contents

Preface		ix
1	**Introduction and Summary**	**1**
1.1	An Introductory Example: Spinodal Decomposition	1
1.2	History	10
1.3	Derivation of Dissipative or Conservative Schemes	12
	1.3.1 Procedure for First-Order Real-Valued PDEs	12
	1.3.2 Procedure for First-Order Complex-Valued PDEs	19
	1.3.3 Procedure for Systems of First-Order PDEs	24
	1.3.4 Procedure for Second-Order PDEs	27
1.4	Advanced Topics	34
	1.4.1 Design of Higher-Order Schemes	34
	1.4.2 Design of Linearly Implicit Schemes	40
	1.4.3 Further Remarks	47
2	**Target Partial Differential Equations**	**49**
2.1	Variational Derivatives	49
2.2	First-Order Real-Valued PDEs	52
2.3	First-Order Complex-Valued PDEs	58
2.4	Systems of First-Order PDEs	60
2.5	Second-Order PDEs	65
3	**Discrete Variational Derivative Method**	**69**
3.1	Discrete Symbols and Formulas	69
3.2	Procedure for First-Order Real-Valued PDEs	75
	3.2.1 Discrete Variational Derivative: Real-Valued Case	75
	3.2.2 Design of Schemes	80
	3.2.3 User's Choices	87
3.3	Procedure for First-Order Complex-Valued PDEs	93
	3.3.1 Discrete Variational Derivative: Complex-Valued Case	93
	3.3.2 Design of Schemes	96
3.4	Procedure for Systems of First-Order PDEs	101
	3.4.1 Design of Schemes	105
3.5	Procedure for Second-Order PDEs	110
	3.5.1 First Approach: Direct Variation	111
	3.5.2 Second Approach: System of PDEs	115
3.6	Preliminaries on Discrete Functional Analysis	119

	3.6.1 Discrete Function Spaces	119
	3.6.2 Discrete Inequalities	121
	3.6.3 Discrete Gronwall Lemma	126

4 Applications 129
- 4.1 Target PDEs 1 129
  - 4.1.1 Cahn–Hilliard Equation 129
  - 4.1.2 Allen–Cahn Equation 149
  - 4.1.3 Fisher–Kolmogorov Equation 153
- 4.2 Target PDEs 2 155
  - 4.2.1 Korteweg–de Vries Equation 157
  - 4.2.2 Zakharov–Kuznetsov Equation 159
- 4.3 Target PDEs 3 164
  - 4.3.1 Complex-Valued Ginzburg–Landau Equation 164
  - 4.3.2 Newell–Whitehead Equation 165
- 4.4 Target PDEs 4 167
  - 4.4.1 Nonlinear Schrödinger Equation 167
  - 4.4.2 Gross–Pitaevskii Equation 180
- 4.5 Target PDEs 5 182
  - 4.5.1 Zakharov Equations 183
- 4.6 Target PDEs 7 185
  - 4.6.1 Nonlinear Klein–Gordon Equation 185
  - 4.6.2 Shimoji–Kawai Equation 189
- 4.7 Other Equations 191
  - 4.7.1 Keller–Segel Equation 191
  - 4.7.2 Camassa–Holm Equation 195
  - 4.7.3 Benjamin–Bona–Mahony Equation 212
  - 4.7.4 Feng Equation 222

5 Advanced Topic I: Design of High-Order Schemes 227
- 5.1 Orders of Accuracy of Schemes 227
- 5.2 Spatially High-Order Schemes 229
  - 5.2.1 Discrete Symbols and Formulas 229
  - 5.2.2 Discrete Variational Derivative 231
  - 5.2.3 Design of Schemes 233
  - 5.2.4 Application Examples 238
- 5.3 Temporally High-Order Schemes: Composition Method . . . 247
- 5.4 Temporally High-Order Schemes: High-Order Discrete Variational Derivatives 248
  - 5.4.1 Discrete Symbols 249
  - 5.4.2 Central Idea for High-Order Discrete Derivative . . . 250
  - 5.4.3 Temporally High-Order Discrete Variational Derivative and Design of Schemes 251

6 Advanced Topic II: Design of Linearly Implicit Schemes 271
- 6.1 Basic Idea for Constructing Linearly Implicit Schemes 271
- 6.2 Multiple-Points Discrete Variational Derivative 274
  - 6.2.1 For Real-Valued PDEs 274
  - 6.2.2 For Complex-Valued PDEs 275
- 6.3 Design of Schemes 277
  - 6.3.1 For Real-Valued PDEs 277
  - 6.3.2 For Complex-Valued PDEs 279
- 6.4 Applications 280
  - 6.4.1 Cahn–Hilliard Equation 280
  - 6.4.2 Odd-Order Nonlinear Schrödinger Equation 283
  - 6.4.3 Ginzburg–Landau Equation 283
  - 6.4.4 Zakharov Equations 284
  - 6.4.5 Newell–Whitehead Equation 285
- 6.5 Remarks on the Stability of Linearly Implicit Schemes 288

7 Advanced Topic III: Further Remarks 293
- 7.1 Solving System of Nonlinear Equations 293
  - 7.1.1 Use of Numerical Newton Method Libraries 294
  - 7.1.2 Variants of Newton Method 295
  - 7.1.3 Spectral Residual Methods 296
  - 7.1.4 Implementation as a Predictor–Corrector Method ... 298
- 7.2 Switch to Galerkin Framework 298
  - 7.2.1 Design of Galerkin Schemes 299
  - 7.2.2 Application Examples 309
- 7.3 Extension to Non-Rectangular Meshes on 2D Region 348

Appendix A Semi-Discrete Schemes in Space 353

Appendix B Proof of Proposition 3.4 357

Bibliography 359

Index 373

Preface

This book describes a numerical method, called the "discrete variational derivative method," which is for designing numerical schemes for certain partial differential equations (PDEs, for short). The targets include, for example, (i) the Korteweg–de Vries equation:

$$\frac{\partial u}{\partial t} = \frac{\partial}{\partial x}\left(\frac{1}{2}u^2 + \frac{\partial^2 u}{\partial x^2}\right)$$

which describes shallow water waves, (ii) the nonlinear Schrödinger equation:

$$\mathrm{i}\frac{\partial u}{\partial t} = -\frac{\partial^2 u}{\partial x^2} - \gamma |u|^{p-1}u, \qquad \gamma \in \mathbb{R},\ p = 3, 4, \ldots$$

for modeling optical waves, (iii) the Cahn–Hilliard equation:

$$\frac{\partial u}{\partial t} = \frac{\partial^2}{\partial x^2}\left(pu + ru^3 + q\frac{\partial^2 u}{\partial x^2}\right), \qquad p < 0,\ q < 0,\ r > 0$$

which is a model of certain phase separation phenomena, and (iv) the Newell–Whitehead equation:

$$\frac{\partial u}{\partial t}(t, x, y) = \mu u - |u|^2 u + \left(\frac{\partial}{\partial x} - \frac{\mathrm{i}}{2k_c}\frac{\partial^2}{\partial y^2}\right)^2 u, \qquad \mu, k_c \in \mathbb{R}$$

which simulates two-dimensional Bénard convection flow. Reflecting these physical backgrounds, the PDEs have one striking feature in common; associated with the PDEs there are scalar functions, often referred to as "energies," that strictly remain constant or monotonically decrease as time evolves. In fact, under appropriate conditions (i) the Korteweg–de Vries equation has the energy conservation property:

$$\frac{\mathrm{d}}{\mathrm{d}t}\int\left(\frac{1}{6}u^3 - \frac{1}{2}(u_x)^2\right)\mathrm{d}x = 0,$$

and (ii) the nonlinear Schrödinger equation has the property:

$$\frac{\mathrm{d}}{\mathrm{d}t}\int\left(-\frac{1}{2}|u_x|^2 + \frac{1}{p}|u|^p\right)\mathrm{d}x = 0.$$

Similarly, (iii) the Cahn–Hilliard equation has the energy dissipation property:

$$\frac{\mathrm{d}}{\mathrm{d}t}\int\left(\frac{p}{2}u^2 + \frac{r}{4}u^4 - \frac{q}{2}(u_x)^2\right)\mathrm{d}x \leq 0,$$

and (iv) the Newell–Whitehead equation has the property:

$$\frac{\mathrm{d}}{\mathrm{d}t} \iint \left(\frac{\mu}{2}|u|^2 - \frac{1}{4}|u|^4 + \left| u_x - \frac{\mathrm{i}}{2k_c} u_{yy} \right|^2 \right) \mathrm{d}x\mathrm{d}y \leq 0.$$

In this book those PDEs are said to be "conservative" or "dissipative" PDEs, respectively. (Note that this definition of "dissipative" is slightly different from the definition in dynamical systems theory, where dissipative property is defined with absorbing sets.)

In the numerical computation of such conservative or dissipative PDEs, it is often preferable to employ some special numerical schemes that retain the conservation/dissipation properties in a discrete sense; they are called "conservative" or "dissipative" schemes throughout this book. The reason for this preference is that, from the numerical point of view, the properties often lead us to stabler computation; and for practitioners such as physicists and engineers the motivation is that the properties themselves may be quite important since they reflect important physical aspects of the modeled phenomena. Thus, since around the 1970s, much effort has been devoted to the development of conservative and dissipative schemes for various PDEs. In the early phase of these researches, studies had been carried out for each individual PDE; it was only during and after 1990s that more unified approaches that can be applied to a certain large class of PDEs had been found. The main topic of this book, the discrete variational derivative method, is one of such newer developments.

Here we have to mention the case of *ordinary* differential equations (ODEs), for which the history of research in the above context dates back to several decades ago, and consequently the corresponding literature is far richer than that of PDEs. For ODEs several unified approaches have been established, not only for conservative and dissipative ODEs, but for many classes of ODEs with various geometric structures. They include, for example, the symplectic method for Hamiltonian systems, the Lie group method for constrained mechanical systems, methods that preserve first-integrals, and methods for ODEs on manifolds, among many others. Nowadays the methods are regarded to form a big group called "structure-preserving methods for ODEs," or "geometric numerical integration methods," and more and more efforts are being devoted to this area at an ever-increasing rate. An excellent textbook for both beginners and experts is also available, which surveys the history and the whole picture of structure-preserving methods for ODEs [83].

Compared to this maturity, the research in the PDE context seems to be still at its beginning stage. Few classes for which structure-preserving integration is possible have been identified so far, and accordingly, "structure-preserving method for PDEs" is not a popular expression yet. There is no question, however, about the increasing importance of PDEs themselves, both in mathematical and practical senses. We thus strongly believe that in the next decade structure-preserving methods for PDEs will draw more and more interest, especially as the methods for ODEs come close to maximum maturity. In accordance with this belief, this book is written as the first one that is entirely dedicated to a structure-preserving method for PDEs.

This book is intended for both experts and non-experts. For both readers an introductory Chapter 1 is prepared, where all central ideas and essential examples are summarized. We believe that just glancing at this chapter will suffice to enable the reader to understand the essence of the discrete variational derivative method. The subsequent chapters, 2 to 4, are devoted to full description of the method: in Chapter 2 the PDEs which the method covers are classified; in Chapter 3 the procedure of the method is described in detail; in Chapter 4 the application examples are shown. Practitioners may, after reading Chapter 1, jump to Chapter 4 and see how the method is applied to typical problems. Chapters 5 to 7, including appendices, are for especially interested readers; there some advanced topics and technical details are summarized, which are too complicated to be included in the main sections.

We hope to thank all those who have helped this project. In particular, Kazuo Murota and Masaaki Sugihara for encouraging us to write this book, and continuously giving the authors many valuable comments. Masatake Mori, for guiding the authors to the rich world of numerical analysis. Our sincere thanks also go to Tetsuya Ishiwata, Toshiyuki Koto, Taketomo Mitsui, Yoshihisa Morita, Masaharu Nagayama, Shinji Odanaka, Takayoshi Ogawa, Masami Okada, Hisashi Okamoto, Norikazu Saito, Takashi Sakajo, and Takashi Suzuki, for valuable information related to the contents of this book. We also thank our colleagues Chris Budd, Jialin Hong, Takanori Ide, Brynjulf Owren, Reinout Quispel, Takaharu Yaguchi, among others, for fruitful discussions and valuable suggestions. We are also grateful to some of our students for drawing graphs, in particular Masayuki Hayashi, Satoshi Koide, Yohei Kubo, Yuto Miyatake, Yuki Sawada, Yuuki Sekino, Ken Takeya, Genta Tanaka, Eitaro Torii, Kenta Ueda, and Norio Yamaguchi. Leong LiMing, our editor at CRC Press, was so patient about our delayed manuscript, and very helpful during the whole project period. Finally, we acknowledge that this book was partially supported by the Global COE "The Research and Training Center for New Development in Mathematics."

We hope that this book be a help for all readers facing their problems and looking for "good" numerical solvers.

Osaka and Tokyo, December 2010

The Authors

Chapter 1

Introduction and Summary

The key ideas of the discrete variational derivative method are summarized with some illustrative examples. This chapter is a self-contained summary of this book. After reading this introductory chapter, readers are suggested to proceed to one of the subsequent chapters according to their points of interest.

1.1 An Introductory Example: Spinodal Decomposition

Let us have a look at an illustrative example, the "spinodal decomposition." This is a chemi-physical phenomenon which occurs when two liquids with different specific gravities are mixed. For example, when we put some oil and water in a glass and shake it well, the two ingredients first intermingle with each other, and then they are gradually separated. Figure 1.1[1] is a schematic view of that process, where, for example, the ingredient A is water and B is oil. Figure 1.2 shows an experimental result with polymer mixtures.

Mathematically, the phenomenon is modeled by the Cahn–Hilliard equation:

$$\frac{\partial u}{\partial t} = \frac{\partial^2}{\partial x^2}\left(pu + ru^3 + q\frac{\partial^2 u}{\partial x^2}\right), \ x \in (0, L), \ t > 0, \ p < 0, \ q < 0, \ r > 0. \quad (1.1)$$

The solution $u(x,t)$ describes the ratio of one component (oil, for example) to the other (water). Here we limit ourselves to the one-dimensional case, for simplicity of argument. We impose the boundary conditions below on the problem:

$$\frac{\partial u}{\partial x} = \frac{\partial^3 u}{\partial x^3} = 0, \quad x = 0, L. \quad (1.2)$$

---

[1] Reprinted figure with permission from H. Tanaka and T. Nishi, Direct determination of the probability distribution function of concentration in polymer mixtures undergoing phase separation, *Phys. Rev. Lett.*, 59, 692-695(1987). Copyright (1987) by the American Physical Society.

2   *Discrete Variational Derivative Method*

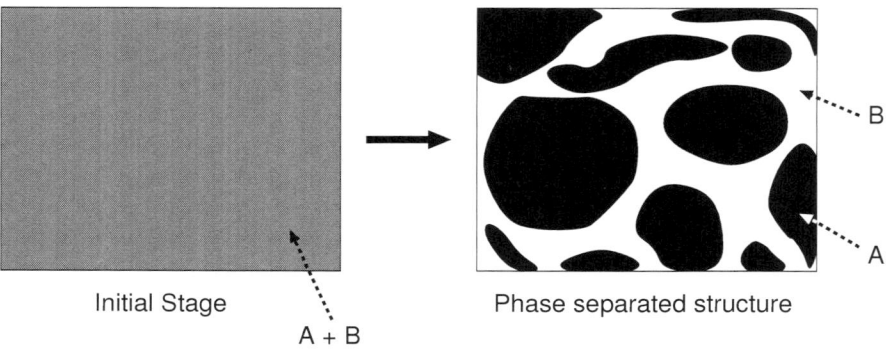

FIGURE 1.1: Schematic view of the spinodal decomposition.

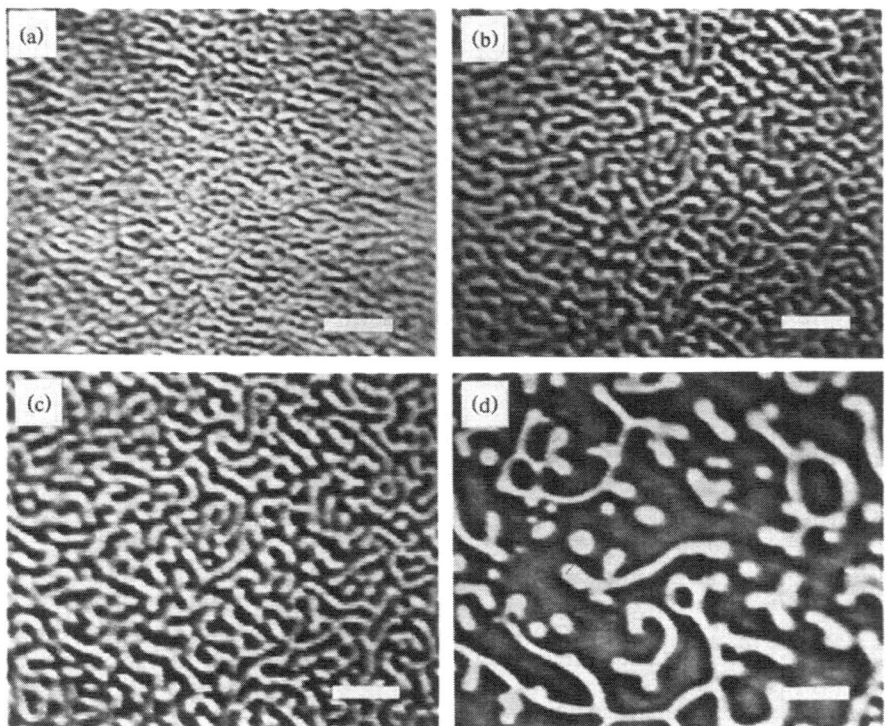

FIGURE 1.2: Temporal change of phase-separated structure of mixtures of polystyrene and polyvinyl methyl ether [163]. Bar corresponds to 20 μm. (a)–(d) structures at 480, 720, 1200, and 2400 s after quench, respectively.

It is not easy to integrate the Cahn–Hilliard equation numerically [70]. The right hand side of (1.1) includes a diffusion term pu_{xx} whose coefficient is *negative* (recall $p < 0$). This means that in the numerical integration we have to solve a diffusion equation in the *negative* time direction, which is obviously numerically unstable. In order to illustrate this, let us try an explicit Euler scheme as the simplest example. Let the spatial discretization width be $\Delta x = L/N$, where N is the number of the spatial grid points, and denote the time mesh width by $\Delta t > 0$. We denote the approximate solution by $U_k{}^{(m)} \simeq u(k\Delta x, m\Delta t)$ $(k = 0, 1, \ldots, N,\ m = 0, 1, 2, \ldots)$. We also write $\boldsymbol{U}^{(m)} = \left(U_0^{(m)}, \ldots, U_N^{(m)}\right)^\top$. Then the Euler scheme reads as follows.

Scheme 1.1 (Standard Euler scheme for Cahn–Hilliard equation)
Given an initial data $\boldsymbol{U}^{(0)}$, the approximate solutions $\boldsymbol{U}^{(m)}$ are calculated by, for $m = 0, 1, 2, \ldots$,

$$\frac{U_k{}^{(m+1)} - U_k{}^{(m)}}{\Delta t} = \delta_k^{\langle 2 \rangle} \left(p U_k{}^{(m)} + r \left(U_k{}^{(m)}\right)^3 + q \delta_k^{\langle 2 \rangle} U_k{}^{(m)} \right), \quad k = 0, \ldots, N, \tag{1.3}$$

with the discrete boundary condition corresponding to (1.2):

$$\delta_k^{\langle 1 \rangle} U_k{}^{(m)} = \delta_k^{\langle 3 \rangle} U_k{}^{(m)} = 0, \quad k = 0, N. \tag{1.4}$$

The symbols $\delta_k^{\langle p \rangle}$ ($p = 1, 2, 3$) mean the standard second-order central difference operators for $\partial^p / \partial x^p$, which are explicitly written as

$$\delta_k^{\langle 1 \rangle} f_k = \frac{f_{k+1} - f_{k-1}}{2\Delta x}, \tag{1.5}$$

$$\delta_k^{\langle 2 \rangle} f_k = \frac{f_{k+1} - 2f_k + f_{k-1}}{(\Delta x)^2}, \tag{1.6}$$

$$\delta_k^{\langle 3 \rangle} f_k = \frac{f_{k+2} - 2f_{k+1} + 2f_{k-1} - f_{k-2}}{2(\Delta x)^3}. \tag{1.7}$$

Figure 1.3 shows the result obtained by the scheme. In the example, the parameters are $p = -1.0$, $q = -0.001$, $r = 1.0$, and $L = 1$, $N = 50$ (thus $\Delta x = 1/50$). Two time mesh sizes: $\Delta t = 1/1200$ and $1/12000$ are tested. In both graphs, the staggered line lying around $u = 0$ line is the initial pattern:

$$u_0(x) = 0.1 \sin(2\pi x) + 0.01 \cos(4\pi x) + 0.06 \sin(4\pi x) + 0.02 \cos(10\pi x). \tag{1.8}$$

The numerical solution with $\Delta t = 1/1200$ (top graph) rapidly blows up, exhibiting strong oscillation in only four or five steps. This hardly improves even when we refine the time mesh; the numerical solution with $\Delta t = 1/12000$ (bottom graph) also blows up in only six or seven steps.

Facing this difficulty, we have two options: one is to use some reliable ODE solver which allows adaptive integration, after suitably discretizing the space

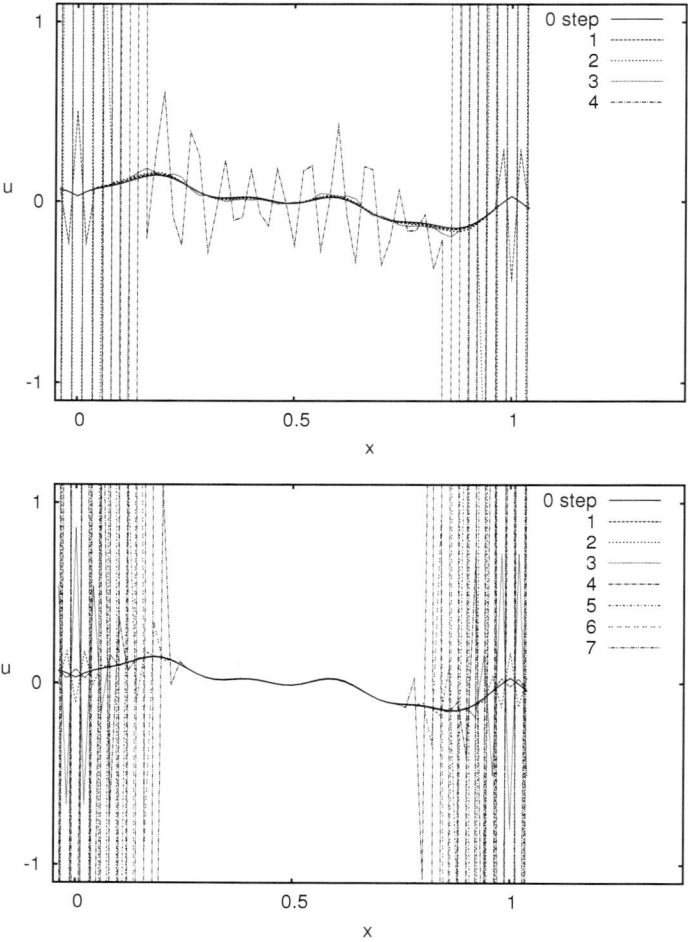

FIGURE 1.3: Numerical solutions of the Cahn–Hilliard equation by the explicit Euler scheme: (top) $\Delta t = 1/1200$; (bottom) $\Delta t = 1/12000$.

variable. This might work, though it may need considerable computation time because the package should be forced to choose very small time mesh size. The other option—which *is* the basic concept throughout this book—is to use some special scheme designed for stable integration of the equation.

To seek such a special scheme, let us cast a spotlight on a quantity, the "free energy" or "local energy" of the problem:

$$G(u, u_x) = \frac{1}{2}pu^2 + \frac{1}{4}ru^4 - \frac{1}{2}q(u_x)^2. \tag{1.9}$$

We call its spatial integration:

$$J(u) = \int_0^L G(u, u_x) \mathrm{d}x \tag{1.10}$$

the "global energy." Note that J is a functional of u, but at the same time it can be regarded as a function of t. The equation (1.1) can then be written as

$$\frac{\partial u}{\partial t} = \frac{\partial^2}{\partial x^2}\left(\frac{\delta G}{\delta u}\right), \tag{1.11}$$

where $\delta G/\delta u$ is the (first) variational derivative of $G(u, u_x)$ obtained from the following variation calculation.

$$\int_0^L (G(u + \delta u, u_x + \delta u_x) - G(u, u_x))\mathrm{d}x$$

$$= \int_0^L \left(\frac{\partial G}{\partial u}\delta u + \frac{\partial G}{\partial u_x}\delta u_x\right)\mathrm{d}x + O(\delta u^2)$$

$$= \int_0^L \left(\frac{\partial G}{\partial u} - \frac{\partial}{\partial x}\frac{\partial G}{\partial u_x}\right)\delta u\, \mathrm{d}x + \left[\frac{\partial G}{\partial u_x}\delta u\right]_0^L + O(\delta u^2)$$

$$= \int_0^L \frac{\delta G}{\delta u}\delta u\, \mathrm{d}x + \left[\frac{\partial G}{\partial u_x}\delta u\right]_0^L + O(\delta u^2). \tag{1.12}$$

The last equality defines $\delta G/\delta u$. The form (1.11) states that the evolution of the solution is roughly a "gradient-flow"; it evolves in such a direction that the global energy is decreased:

$$\frac{\mathrm{d}}{\mathrm{d}t}J(u) = \int_0^L \frac{\delta G}{\delta u}\frac{\partial u}{\partial t}\mathrm{d}x + \left[\frac{\partial G}{\partial u_x}\frac{\partial u}{\partial t}\right]_0^L$$

$$= -\int_0^L \left(\frac{\partial}{\partial x}\frac{\delta G}{\delta u}\right)^2 \mathrm{d}x + \left[\left(\frac{\delta G}{\delta u}\right)\frac{\partial}{\partial x}\left(\frac{\delta G}{\delta u}\right)\right]_0^L$$

$$\leq 0. \tag{1.13}$$

Note that $(\partial/\partial x)u = 0$ and $(\partial/\partial x)^3 u = 0$ mean $\partial G/\partial u_x = 0$ and $(\partial/\partial x)\delta G/\delta u = 0$, and thus the boundary terms vanish thanks to the boundary condition (1.2).

From the dissipation property, the next important proposition immediately follows.

PROPOSITION 1.1 L^∞**-boundedness of solution**
As to the solution of (1.1) under the boundary condition (1.2), we have this a priori estimate:
$$\|u\|_\infty < \infty, \quad t > 0, \tag{1.14}$$
where $\|\cdot\|_p$ $(p = 1, 2, \ldots, \infty)$ *is the standard* L^p *norm.*

PROOF Recalling $p, q < 0, r > 0$, we have the trivial identity:
$$\frac{p}{2}u^2 + \frac{r}{4}u^4 \geq -pu^2 - \frac{9p^2}{4r}.$$

Then by the energy dissipation property (1.13) we know for any $t > 0$,
$$J(u(x,0)) \geq J(u(x,t))$$
$$= \int_0^L \left\{ \frac{1}{2}pu^2 + \frac{1}{4}ru^4 - \frac{1}{2}q(u_x)^2 \right\} \mathrm{d}x$$
$$\geq \int_0^L \left\{ -pu^2 - \frac{9p^2}{4r} - \frac{1}{2}q(u_x)^2 \right\} \mathrm{d}x$$
$$= -p\|u\|_2^2 - \frac{9p^2 L}{4r} - \frac{q}{2}\|u_x\|_2^2. \tag{1.15}$$

Thus we have
$$J(u(x,0)) + \frac{9p^2 L}{4r} \geq -p\|u\|_2^2 - \frac{q}{2}\|u_x\|_2^2.$$

Again recalling $p, q < 0, r > 0$, we see that
$$\|u\|_2, \|u_x\|_2 < \infty. \tag{1.16}$$

Then with the aid of the Sobolev type inequality (see, for example, John [94]):
$$\|u\|_\infty^2 \leq c\left(\|u\|_2^2 + \|u_x\|_2^2\right),$$

which holds for every function $u(\cdot, t) \in H^1(0, L)$, we obtain $\|u\|_\infty < \infty$. □

In other words,

[Key observation 1]
The dissipation property prevents the solution's blow-up.

This observation encourages us to seek a scheme which retains the dissipation property, because it may also prevent the blow-up of the approximate solution. We here present such a scheme. (At the time being, we do not discuss how it is constructed. It will be covered in Chapter 4.)

Scheme 1.2 (Dissipative scheme for Cahn–Hilliard equation) *Given an initial data $\boldsymbol{U}^{(0)}$, the approximate solutions $\boldsymbol{U}^{(m)}$ are calculated by, for $m = 0, 1, 2, \ldots,$*

$$\frac{U_k^{(m+1)} - U_k^{(m)}}{\Delta t} = \delta_k^{\langle 2 \rangle} \left\{ p \left(\frac{U_k^{(m+1)} + U_k^{(m)}}{2} \right) + q \delta_k^{\langle 2 \rangle} \left(\frac{U_k^{(m+1)} + U_k^{(m)}}{2} \right) \right.$$
$$\left. + r \left(\frac{(U_k^{(m+1)})^3 + (U_k^{(m+1)})^2 U_k^{(m)} + U_k^{(m+1)} (U_k^{(m)})^2 + (U_k^{(m)})^3}{4} \right) \right\},$$
$$k = 0, \ldots, N, \qquad (1.17)$$

with the discrete boundary condition (1.4).

Scheme 1.2 has the desired discrete dissipation property.

PROPOSITION 1.2 Dissipation property of Scheme 1.2
Let us define a "discrete local energy" $G_{\mathrm{d}} : \mathbb{R}^{N+1} \to \mathbb{R}^{N+1}$ by

$$G_{\mathrm{d},k}(\boldsymbol{U}^{(m)}) \stackrel{\mathrm{d}}{=} \frac{p}{2}(U_k^{(m)})^2 + \frac{r}{4}(U_k^{(m)})^4 - \frac{q}{2} \left(\frac{\left(\delta_k^+ U_k^{(m)}\right)^2 + \left(\delta_k^- U_k^{(m)}\right)^2}{2} \right), \tag{1.18}$$

where $G_{\mathrm{d},k}(\boldsymbol{U}^{(m)})$ denotes the k-th element (the detail of this expression will be explained soon in Section 1.3). We also define the discrete global energy accordingly by

$$J_{\mathrm{d}}(\boldsymbol{U}^{(m)}) \stackrel{\mathrm{d}}{=} \sum_{k=0}^{N} {}'' G_{\mathrm{d},k}(\boldsymbol{U}^{(m)}) \Delta x, \tag{1.19}$$

where

$$\sum_{k=0}^{N} {}'' f_k \stackrel{\mathrm{d}}{=} \frac{1}{2} f_0 + f_1 + \cdots + f_{N-1} + \frac{1}{2} f_N \tag{1.20}$$

is the trapezoidal rule. Then the solution by Scheme 1.2 satisfies the following inequality.

$$J_{\mathrm{d}}(\boldsymbol{U}^{(m+1)}) \leq J_{\mathrm{d}}(\boldsymbol{U}^{(m)}), \quad m = 0, 1, 2, \ldots. \tag{1.21}$$

REMARK 1.1 Throughout this book, we basically adopt the trapezoidal rule as our main summation rule. Other rules, for example, the rectangle rule, can be also adopted. For example, when the periodic boundary condition is applied, the trapezoidal rule naturally coincides with the rectangle rule, and the latter is more convenient. Another example is the case where the use of the rectangle rule substantially simplifies the treatment of discrete boundary condition. This will be illustrated in Section 3.2.3.2. □

The proof of the proposition is left to Chapter 3 (generic theory) or Chapter 4 (the specific Cahn–Hilliard case). With this dissipation property, and a *discrete* Sobolev type inequality, we can prove that numerical solution *never* blows up (we leave the detail of this discussion to Chapter 4.)

For the moment we only show some numerical results. Figure 1.4 shows the result by Scheme 1.2 with a coarse time mesh $\Delta t = 1/1000$ (other parameters are the same as in the explicit Euler case). The calculation proceeds quite stably, and a physically correct pattern (a phase separation) is obtained. Figure 1.5 shows the evolution of the energy. The discrete energy is properly dissipated. For comparison, we present in Figure 1.6 the result obtained by the explicit scheme. There the energy is not dissipative at all; it even blows up. This fact agrees with the failure of the numerical computation.

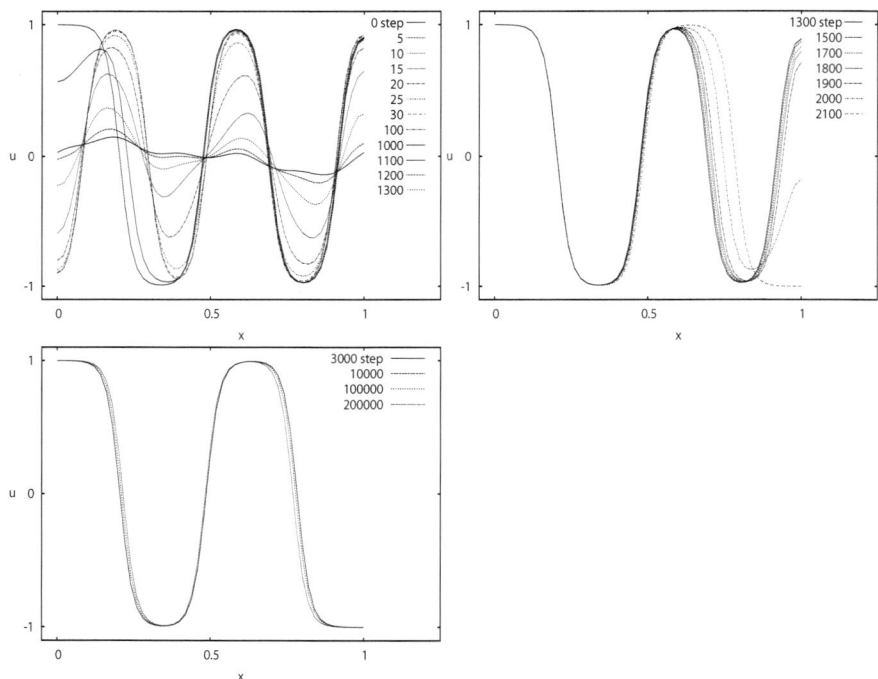

FIGURE 1.4: Numerical solutions of the Cahn–Hilliard equation by Scheme 1.2 ($\Delta t = 1/1000$): (top-left) steps 0 to 1300 (top-right) 1300 to 2100 (bottom) 3000 to 200,000.

The Cahn–Hilliard example clearly shows the superiority of the specialized scheme. The scheme preserves a discrete counterpart of the energy dissipation

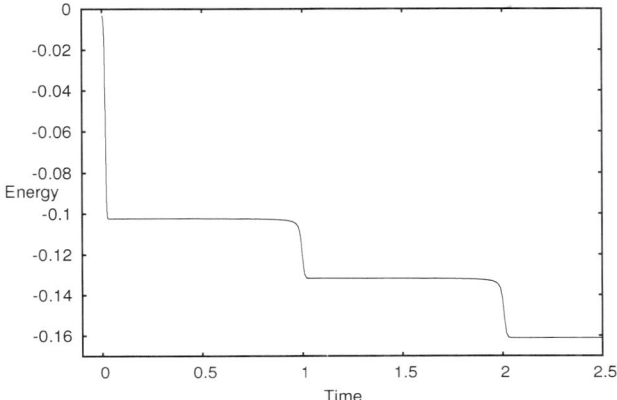

FIGURE 1.5:   The evolution of the discrete energy in Scheme 1.2.

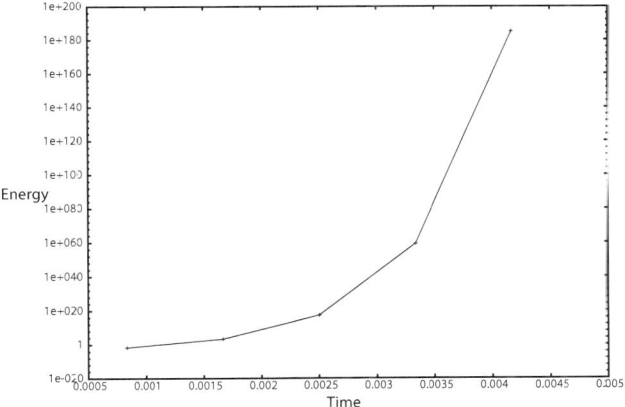

FIGURE 1.6:   The evolution of the discrete energy in Scheme 1.1.

property, and the property is quite crucial for better numerical integration. The same thing often happens also in many conservative problems (i.e., problems with conservation laws). Next we will see how dissipative or conservative schemes, such as Scheme 1.2, can be constructed.

1.2 History

In this section, we briefly mention the related studies on the main subject of this book.

First attempts on dissipative/conservative schemes, or more generally on structure-preserving algorithms, focused on ordinary differential equations such as Hamiltonian systems. For example, in the beginning of the 1970's Greenspan [77] considered strictly conservative discretization of some mechanical systems. The method was then extended to general mechanical systems by Gonzalez [74] and McLachlan–Quispel–Robidoux [126, 127] decades later. A strong alternative to these works is the so-called symplectic method, which is a specialized numerical method for Hamiltonian systems. Though symplectic schemes are not strictly conservative, they are nearly conservative, and provide us very effective ways to integrate Hamiltonian systems. For the symplectic method, see Hairer–Lubich–Wanner [83], Sanz-Serna–Calvo [151] and Leimkuhler–Reich [104]. Related interesting studies on nearly conservative numerical schemes include: Faou–Hairer–Pham [52] and Hairer [81].

After these successes on Hamiltonian ODEs, many other classes of ODEs that have some intrinsic geometric structure have been identified, and structure-preserving algorithms for these ODEs have been extensively studied. These activities for ODEs are now also referred to as the "geometric numerical integration of ODEs," and form a big trend in numerical analysis. Interested readers may refer to Hairer–Lubich–Wanner [83] and Budd–Piggott [23].

In the PDE context, a number of studies on dissipative/conservative schemes have been carried out on individual dissipative or conservative PDEs, since around the 1970's. Below are quite limited examples. Strauss–Vazquez [155] presented a conservative finite difference scheme for the nonlinear Klein–Gordon equation. Hughes–Caughey–Liu [89] presented a conservative finite element scheme for the nonlinear elastodynamics problem. Delfour–Fortin–Payre [35] presented a conservative finite difference scheme for the nonlinear Schrödinger equation, then Akrivis–Dougalis–Karakashian [8] presented a finite element version of the scheme and proved the convergence of the finite element scheme. Sanz-Serna [150] considered the nonlinear Schrödinger equation as well. Taha–Ablowitz [159, 160] presented conservative finite difference schemes for the nonlinear Schrödinger equation and the Korteweg–de Vries equation. Du–Nicolaides [39] presented a dissipative finite element scheme for the Cahn–Hilliard equation. Around the same time, in a completely different context from above, studies on soliton PDEs such as the KdV equation were done to find finite difference schemes that preserved discrete bilinear form or Wronskian form, corresponding to the original equations; see, for example, Hirota [85, 86]. They can be also regarded as structure-preserving methods.

Then during the 1990s, more general approaches that cover not only several individual PDEs but also a wide class of PDEs have been independently introduced by several groups. The discrete variational derivative method— the main subject of the present book—is one of such methods, proposed by Furihata–Mori [63, 64, 69, 65] around 1996 for PDEs with variational structure. The method has then been extended in various ways mainly by a Japanese group including Furihata, Matsuo, Ide, and Yaguchi [66, 67, 68, 90, 91, 116, 119, 120, 121, 122, 165, 166, 167], and succeeded in proving its effectiveness in various applications. At the same time, Gonzalez [75] proposed a conservative method for some general class of PDEs describing finite-deformation elastodynamics. There, the key is a special technique in time discretization devised for ODEs by Gonzalez [74]. Another excellent set of studies were given by McLachlan [129] and McLachlan–Robidoux [128], where a general method for designing conservative schemes for conservative PDEs based on their techniques on ODEs [126, 127] (and the related basic studies Quispel–Turner [145] and Quispel–Capel [144]) was developed (see also the recent related results: McLaren–Quispel [130], Quispel–McLaren [146], Celledoni et al. [26]). Jimenez [92] has also studied a systematic approach to obtain discrete conservation laws for certain finite difference schemes.

Aside from strictly conservative or dissipative methods, several interesting approaches for structure-preserving integration of PDEs have emerged as of the writing of the present book. For a very comprehensive review including these topics, see Budd–Piggot [23]. For Hamiltonian PDEs, a unique approach was proposed by Marsden–Patrick–Shkoller [112] (see also Marsden–West [113] for a good review), and it has been intensively studied by their group. Their method is based on the discretization of the variational principle. Its name "variational integrator" is quite close to the discrete variational derivative method, but these methods are quite different. For Hamiltonian PDEs, there is another interesting emerging method, the "multi-symplectic method," developed by Bridges–Reich [22]. In the method, Hamiltonian PDEs are transformed into a special "multi-symplectic form," and then integrated in such a way that the multi-symplecticity is conserved. This method can be regarded as a generalization of the symplectic method for ODEs (see also McLachlan [124]). For the recent literature in this context, see, for example, [27, 87, 88] and the references therein.

Finally we would like to note that in this short summary we could by no means cover all of the related studies. We recommend that interested readers refer to several key reviews, such as Hairer–Lubich–Wanner [83], Budd–Piggott [23], Leimkuhler–Reich [104], and Lubich [110], and consult their references as well.

1.3 Derivation of Dissipative or Conservative Schemes

In this section we demonstrate how numerical schemes that retain dissipation or conservation properties are constructed. To avoid exhaustive discussion involving cumbersome symbols, we here limit ourselves to some typical PDEs, possibly ignoring some details. More precise description will be found in Chapter 3 (generic theory) and Chapter 4 (application examples). We consider the following four cases: first-order real-valued PDEs, first-order complex-valued PDEs, systems of first-order PDEs, and second-order PDEs.

1.3.1 Procedure for First-Order Real-Valued PDEs

Suppose that $u(x,t)$ is a real-valued function, and the local energy function is given as a real-valued function $G(u, u_x)$. We define the associated global energy by

$$J(u) \stackrel{\mathrm{d}}{\equiv} \int_0^L G(u, u_x)\mathrm{d}x. \qquad (1.22)$$

Let us consider a real-valued PDE:

$$\frac{\partial u}{\partial t} = -\frac{\delta G}{\delta u}, \quad x \in (0, L), \ t > 0. \qquad (1.23)$$

The equation (1.23) is dissipative in the sense that

$$\frac{\mathrm{d}}{\mathrm{d}t}J(u) = \int_0^L \frac{\delta G}{\delta u}\frac{\partial u}{\partial t}\mathrm{d}x + \left[\frac{\partial G}{\partial u_x}\frac{\partial u}{\partial t}\right]_0^L = -\int_0^L \left(\frac{\delta G}{\delta u}\right)^2 \mathrm{d}x \leq 0, \qquad (1.24)$$

if boundary conditions are set so that the boundary term $[\,\cdot\,]_0^L$ vanishes. In fact it does, for example, under the Dirichlet boundary condition $u(0,t) = u(L,t) = 0$. Throughout this introductory chapter, we basically neglect boundary terms for simplicity.

Let us construct a dissipative scheme, i.e., a scheme that keeps a discrete version of the dissipation property, for the equation. Our strategy is based on the following important observation:

> [Key observation 2]
> *The dissipation property* (1.24) *immediately follows from the variational form* (1.23).

In fact, in the proof of the dissipation property (1.24), the concrete form of the energy G, and accordingly the concrete form of the PDE, are not relevant. The variational form itself *is* the key in the dissipation property. This observation leads us to a strategy summarized in Figure 1.7. The left half of the diagram summarizes the continuous PDE case, which reads (starting from the top) [step 1] Define an energy G.

[step 2] Take its variation to obtain the variational derivative $\delta G/\delta u$.

[step 3] Define a PDE with the variational derivative. Then, as a consequence (the up-pointing arrow), the energy dissipation property follows.

Our idea here is to simulate this round trip structure, from the energy $G(u)$ to its dissipation property via variational derivative, in a discrete setting. In this way the method is "structure-preserving." The right half of the diagram reads

[step 1_d] Define a *discrete* energy (as an approximation of the continuous energy G).

[step 2_d] Take its *discrete* variation to obtain the *discrete* variational derivative.

[step 3_d] Define a scheme with the *discrete* variational derivative. Then, the *discrete* dissipation property should follow (again, denoted by the up-pointing arrow).

As opposed to this structure-preserving strategy, the usual way of constructing a scheme is to directly discretize the concrete form of the PDE (the bottom right-pointing arrow from PDE to finite difference scheme). In such an way, however, the beautiful round trip structure is highly likely to be destroyed, and thus generally the desired dissipation property is lost.

Let us actually follow the strategy to construct a dissipative scheme for the equation (1.23). To illustrate how the calculation goes exactly, we pick the linear diffusion equation:
$$\frac{\partial u}{\partial t} = \frac{\partial^2 u}{\partial x^2} \tag{1.25}$$
as a concrete example, which is of the form (1.23) with $G(u, u_x) = (u_x)^2/2$.

[step 1_d] *Defining a discrete energy*

By simply replacing u in $G(u, u_x)$ with $U_k^{(m)}$, and u_x with some finite difference, we obtain a discrete energy $G_d(\boldsymbol{U}^{(m)})$. The subscript "d", standing for "discrete", is added to distinguish this quantity from the continuous energy G. The discrete energy G_d is a real-valued $(N+1)$-dimensional vector function of $\boldsymbol{U}^{(m)}$; we denote its each elements by $G_{d,k}$ ($k = 0, \ldots, N$). (See the example below.)

Note that there are several possibilities in approximating u_x, since there are many difference operators representing the same differentiation. For example, u_x^2 can be

$$(\delta_k^{\langle 1 \rangle} U_k^{(m)})^2, \ (\delta_k^+ U_k^{(m)})^2, \ (\delta_k^- U_k^{(m)})^2, \text{ or } \frac{(\delta_k^+ U_k^{(m)})^2 + (\delta_k^- U_k^{(m)})^2}{2}, \tag{1.26}$$

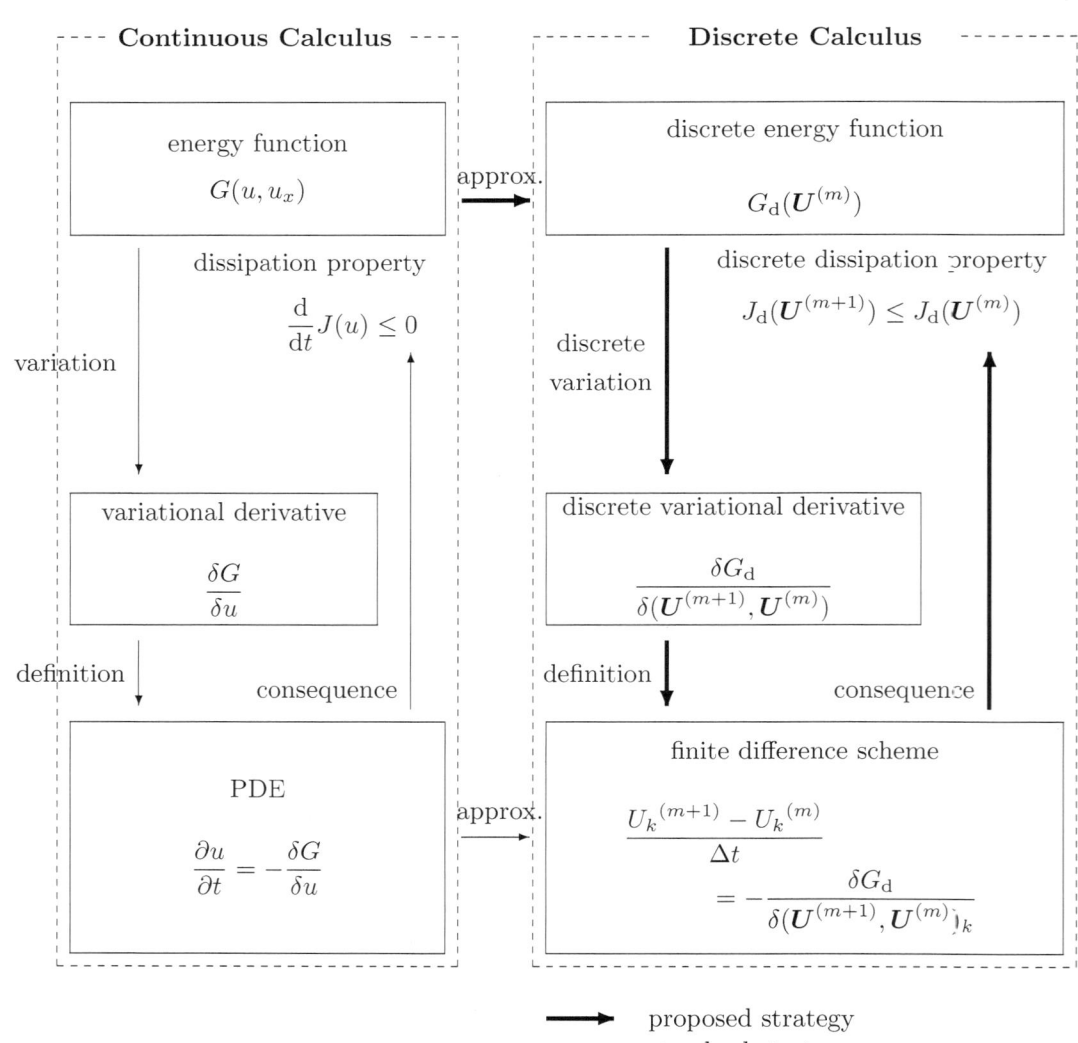

FIGURE 1.7: Standard strategy versus proposed strategy.

where
$$\delta_k^{\langle 1\rangle} f_k = \frac{f_{k+1} - f_{k-1}}{2\Delta x}, \quad \delta_k^+ f_k = \frac{f_{k+1} - f_k}{\Delta x}, \quad \delta_k^- f_k = \frac{f_k - f_{k-1}}{\Delta x} \qquad (1.27)$$

are the standard difference operators for $\partial/\partial x$. We can choose *any* of them. Regardless of the choice, we will obtain a dissipative scheme. We must note, however, that *a different choice leads to a different scheme* (see Remark 1.2).

In the concrete example of (1.25), let us choose a symmetric approximation:
$$(u_x)^2 \simeq \frac{(\delta_k^+ U_k^{(m)})^2 + (\delta_k^- U_k^{(m)})^2}{2}. \qquad (1.28)$$

Then the discrete local energy becomes
$$G_{\mathrm{d},k}(\boldsymbol{U}^{(m)}) = \frac{1}{2}\left(\frac{(\delta_k^+ U_k^{(m)})^2 + (\delta_k^- U_k^{(m)})^2}{2}\right), \qquad (1.29)$$

and the corresponding discrete global energy is
$$J_{\mathrm{d}}(\boldsymbol{U}^{(m)}) = \sum_{k=0}^{N}{}'' G_{\mathrm{d},k}(\boldsymbol{U}^{(m)})\Delta x. \qquad (1.30)$$

[**step 2$_\mathrm{d}$**] *Taking the discrete variation*

Recall the continuous variation calculation (1.12), which is summarized as
$$\int_0^L \{G(u+\delta u, u_x + \delta u_x) - G(u, u_x)\}\,\mathrm{d}x$$
$$= \int_0^L \frac{\delta G}{\delta u}\delta u\,\mathrm{d}x + (\text{boundary term}) + O(\delta u^2). \qquad (1.31)$$

We hope to simulate this in a discrete setting. That is, we hope to find an identity:
$$\sum_{k=0}^{N}{}''\left(G_{\mathrm{d},k}(\boldsymbol{U}^{(m+1)}) - G_{\mathrm{d},k}(\boldsymbol{U}^{(m)})\right)\Delta x =$$
$$\sum_{k=0}^{N}{}'' \frac{\delta G_\mathrm{d}}{\delta(\boldsymbol{U}^{(m+1)}, \boldsymbol{U}^{(m)})_k}\left(U_k^{(m+1)} - U_k^{(m)}\right)\Delta x + (\text{boundary term}). \quad (1.32)$$

At this point, readers need not fully understand the discrete symbols; they will be introduced in Chapter 3. For now it is sufficient to just recognize the correspondences between the continuous and discrete symbols:
$$G_{\mathrm{d},k}(\boldsymbol{U}^{(m+1)}) - G_{\mathrm{d},k}(\boldsymbol{U}^{(m)}) \Leftrightarrow G(u+\delta u, u_x+\delta u_x) - G(u, u_x),$$
$$\frac{\delta G_\mathrm{d}}{\delta(\boldsymbol{U}^{(m+1)}, \boldsymbol{U}^{(m)})_k} \Leftrightarrow \frac{\delta G}{\delta u},$$
$$U_k^{(m+1)} - U_k^{(m)} \Leftrightarrow \delta u.$$

The abstract identity (1.32) demands that the difference of the energies,

$$G_{\mathrm{d},k}(\boldsymbol{U}^{(m+1)}) - G_{\mathrm{d},k}(\boldsymbol{U}^{(m)}),$$

should be decomposable into the discrete version of δu,

$$U_k^{(m+1)} - U_k^{(m)},$$

and a discrete quantity which corresponds to the variational derivative, called the "discrete variational derivative,"

$$\frac{\delta G_{\mathrm{d}}}{\delta(\boldsymbol{U}^{(m+1)}, \boldsymbol{U}^{(m)})_k}.$$

Later in Chapter 3, it is shown that in fact for any given G_{d} this decomposition is possible.

In the case of example (1.25), the identity (1.32) can be easily found as follows.

$$\sum_{k=0}^{N} {}'' \left(G_{\mathrm{d},k}(\boldsymbol{U}^{(m+1)}) - G_{\mathrm{d},k}(\boldsymbol{U}^{(m)}) \right) \Delta x$$

$$= \frac{1}{2}\sum_{k=0}^{N} {}'' \left(\frac{(\delta_k^+ U_k^{(m+1)})^2 - (\delta_k^+ U_k^{(m)})^2}{2} + \frac{(\delta_k^- U_k^{(m+1)})^2 - (\delta_k^- U_k^{(m)})^2}{2} \right) \Delta x$$

$$= \frac{1}{2}\sum_{k=0}^{N} {}'' \left\{ \delta_k^+ \left(\frac{U_k^{(m+1)} + U_k^{(m)}}{2} \right) \cdot \delta_k^+ (U_k^{(m+1)} - U_k^{(m)}) \right.$$

$$\left. + \delta_k^- \left(\frac{U_k^{(m+1)} + U_k^{(m)}}{2} \right) \cdot \delta_k^- (U_k^{(m+1)} - U_k^{(m)}) \right\} \Delta x$$

$$= -\sum_{k=0}^{N} {}'' \left\{ \delta_k^{\langle 2 \rangle} \left(\frac{U_k^{(m+1)} + U_k^{(m)}}{2} \right) \right\} (U_k^{(m+1)} - U_k^{(m)}) \Delta x$$

$$+ \text{(boundary term)}. \quad (1.33)$$

The symbol $\delta_k^{\langle 2 \rangle}$ is the standard central difference operator for $\partial^2/\partial x^2$ defined by

$$\delta_k^{\langle 2 \rangle} f_k = \frac{f_{k+1} - 2f_k + f_{k-1}}{(\Delta x)^2}.$$

In (1.33) a trivial identity $\delta_k^+ \delta_k^- = \delta_k^- \delta_k^+ = \delta_k^{\langle 2 \rangle}$ is used. The summation-by-parts formula:

$$\sum_{k=0}^{N} {}''(\delta_k^+ f_k) g_k \Delta x = -\sum_{k=0}^{N} {}'' f_k (\delta_k^- g_k) \Delta x + \text{(boundary term)} \quad (1.34)$$

is used as well, which is a discrete analogue of the integration-by-parts formula. The precise form of the boundary term is omitted here in order to avoid complications. From (1.33), we find the concrete form of the discrete variational derivative in the case of (1.25) as

$$\frac{\delta G_{\mathrm{d}}}{\delta(\boldsymbol{U}^{(m+1)}, \boldsymbol{U}^{(m)})_k} = -\delta_k^{\langle 2 \rangle} \left(\frac{U_k^{(m+1)} + U_k^{(m)}}{2} \right). \tag{1.35}$$

Note that this in fact approximates the continuous version: $\delta G/\delta u = -u_{xx}$. This supports the view that the above calculation is in fact *discrete* variation.

[**step 3$_\mathrm{d}$**] *Defining a scheme*

Once the discrete variational derivative is found for a given discrete energy G_d, a scheme is defined with it in an abstract manner, analogously to the continuous one (1.23).

Scheme 1.3 (Dissipative scheme for (1.23)) *With given initial data $\boldsymbol{U}^{(0)}$ and appropriate boundary conditions, we compute $\boldsymbol{U}^{(m)}$ by, for $m = 0, 1, 2, \ldots$,*

$$\frac{U_k^{(m+1)} - U_k^{(m)}}{\Delta t} = -\frac{\delta G_\mathrm{d}}{\delta(\boldsymbol{U}^{(m+1)}, \boldsymbol{U}^{(m)})_k}, \qquad k = 0, \ldots, N. \tag{1.36}$$

This scheme keeps the desired dissipation property as follows. Observe that the proof proceeds exactly analogously to the continuous case (1.24); in particular, the concrete forms of the discrete energy function G_d and the discrete variational derivative $\delta G_\mathrm{d}/\delta(\boldsymbol{U}^{(m+1)}, \boldsymbol{U}^{(m)})$ are not relevant. Only the discrete variational structure matters.

PROPOSITION 1.3 Dissipation property of Scheme 1.3
Scheme 1.3 is dissipative in the sense that

$$J_\mathrm{d}(\boldsymbol{U}^{(m+1)}) \leq J_\mathrm{d}(\boldsymbol{U}^{(m)}), \quad m = 0, 1, 2 \ldots. \tag{1.37}$$

PROOF By the identity (1.32), we obtain

$$\begin{aligned}
&J_\mathrm{d}(\boldsymbol{U}^{(m+1)}) - J_\mathrm{d}(\boldsymbol{U}^{(m)}) \\
&= \frac{1}{\Delta t} \sum_{k=0}^{N} {}'' \left(G_{\mathrm{d},k}(\boldsymbol{U}^{(m+1)}) - G_{\mathrm{d},k}(\boldsymbol{U}^{(m)}) \right) \Delta x \\
&= \sum_{k=0}^{N} {}'' \frac{\delta G_\mathrm{d}}{\delta(\boldsymbol{U}^{(m+1)}, \boldsymbol{U}^{(m)})_k} \left(\frac{U_k^{(m+1)} - U_k^{(m)}}{\Delta t} \right) \Delta x + (\text{boundary term}) \\
&= -\sum_{k=0}^{N} {}'' \left(\frac{\delta G_\mathrm{d}}{\delta(\boldsymbol{U}^{(m+1)}, \boldsymbol{U}^{(m)})_k} \right)^2 \Delta x \\
&\leq 0.
\end{aligned} \tag{1.38}$$

In the second equality the boundary terms are assumed to vanish due to some appropriate boundary conditions. In the last equality the scheme in variational form (1.36) is used. □

In the case of the linear diffusion equation (1.25), Scheme 1.3 reads

$$\frac{U_k^{(m+1)} - U_k^{(m)}}{\Delta t} = -\frac{\delta G_d}{\delta(\boldsymbol{U}^{(m+1)}, \boldsymbol{U}^{(m)})_k} = \delta_k^{\langle 2 \rangle}\left(\frac{U_k^{(m+1)} + U_k^{(m)}}{2}\right). \tag{1.39}$$

The concrete form of the discrete variational derivative has been obtained in (1.35). The dissipation property is guaranteed by Proposition 1.3, where the discrete energy function is given by (1.29).

Note that in this case the resulting scheme is just the standard Crank–Nicolson scheme. Although we can say the project has successfully completed in the sense that we obtained a stable scheme (the stability of this Crank–Nicolson scheme is widely known, while it is also possible to prove it directly by utilizing the discrete dissipation property), it is not so exciting in that the obtained scheme is a trivial one. In more generic nonlinear problems, however, resulting schemes are non-trivial, and that is exactly where the discrete variational derivative method is of considerable benefit.

REMARK 1.2 As noted [**step 1d**] (page 13), the definition of discrete energy function is not unique, and a different choice will generally leads us to a different scheme. For example, let us approximate

$$(u_x)^2 \simeq (\delta_k^{\langle 1 \rangle} U_k^{(m)})^2, \tag{1.40}$$

instead of (1.28); that is, we start from the discrete energy function

$$G_{d,k}(\boldsymbol{U}^{(m)}) = \frac{(\delta_k^{\langle 1 \rangle} U_k^{(m)})^2}{2}. \tag{1.41}$$

Then the associated discrete variational derivative will be

$$\frac{\delta G_d}{\delta(\boldsymbol{U}^{(m+1)}, \boldsymbol{U}^{(m)})_k} = -(\delta_k^{\langle 1 \rangle})^2\left(\frac{U_k^{(m+1)} + U_k^{(m)}}{2}\right), \tag{1.42}$$

which then leads us to a scheme:

$$\frac{U_k^{(m+1)} - U_k^{(m)}}{\Delta t} = -\frac{\delta G_d}{\delta(\boldsymbol{U}^{(m+1)}, \boldsymbol{U}^{(m)})_k} = (\delta_k^{\langle 1 \rangle})^2\left(\frac{U_k^{(m+1)} + U_k^{(m)}}{2}\right). \tag{1.43}$$

This is different from (1.39). Still, the scheme is "dissipative," in the sense that Proposition 1.3 holds for the $G_d(\boldsymbol{U}^{(m)})$ defined in (1.41).

As this example illustrates, there is generally a degree of freedom in choosing a discrete energy function, and this is left to each user. Once it is fixed, however, by following the procedure of the discrete variational derivative method, we automatically obtain a scheme that preserves the desired dissipation (or in the conservative case, conservation) property *with respect to the specified discrete energy function*. The performance of the resulting scheme, such as stability and accuracy, often heavily depends on the choice. This issue will be discussed in detail in Section 3.2.3. □

Above procedure can be easily extended to more general real-valued dissipative or conservative PDEs of the following types.

[real-valued dissipative PDEs]

$$\frac{\partial u}{\partial t} = -(-1)^{s+1}\left(\frac{\partial}{\partial x}\right)^{2s}\frac{\delta G}{\delta u}, \quad \frac{\mathrm{d}}{\mathrm{d}t}\int_0^L G(u, u_x)\mathrm{d}x \leq 0, \tag{1.44}$$

where $s = 0, 1, 2, \ldots$. The linear diffusion equation belongs to this class with $s = 0$. The Cahn–Hilliard equation in the previous section is another example, where $s = 1$.

[real-valued conservative PDEs]

$$\frac{\partial u}{\partial t} = \left(\frac{\partial}{\partial x}\right)^{2s+1}\frac{\delta G}{\delta u}, \quad \frac{\mathrm{d}}{\mathrm{d}t}\int_0^L G(u, u_x)\mathrm{d}x = 0, \tag{1.45}$$

where $s = 0, 1, 2, \ldots$. This class includes, for example, the Korteweg–de Vries equation.

More detailed description on these PDEs is given in Chapter 2, and the full procedures for them are described in Chapter 3. Concrete examples will be found in Chapter 4.

1.3.2 Procedure for First-Order Complex-Valued PDEs

Several complex-valued PDEs have variational structure, and the idea described above can be utilized. Suppose that $u(x,t)$ is a complex-valued function, and a real-valued function $G(u, u_x)$ is given as the local energy function. As before, the associated global energy is defined by

$$J(u) \stackrel{\mathrm{d}}{\equiv} \int_0^L G(u, u_x)\mathrm{d}x. \tag{1.46}$$

Let us consider first-order complex-valued PDEs of the form

$$\mathrm{i}\frac{\partial u}{\partial t} = -\frac{\delta G}{\delta \bar{u}}, \quad x \in (0, L), \ t > 0, \tag{1.47}$$

where i $= \sqrt{-1}$. The symbol $\delta G/\delta \overline{u}$ is the variational derivative of G with respect to $\overline{u}$, which is obtained as follows.

$$\int_0^L (G(u+\delta u, u_x + \delta u_x) - G(u, u_x))\mathrm{d}x$$

$$= \int_0^L \left\{ \left(\frac{\partial G}{\partial u}\delta u + \frac{\partial G}{\partial u_x}\delta u_x \right) + \left(\frac{\partial G}{\partial \overline{u}}\delta \overline{u} + \frac{\partial G}{\partial \overline{u}_x}\delta \overline{u}_x \right) \right\} \mathrm{d}x + O(|\delta u|^2)$$

$$= \int_0^L \left\{ \left(\frac{\partial G}{\partial u} - \frac{\partial}{\partial x}\frac{\partial G}{\partial u_x} \right) \delta u + \left(\frac{\partial G}{\partial \overline{u}} - \frac{\partial}{\partial x}\frac{\partial G}{\partial \overline{u}_x} \right) \delta \overline{u} \right\} \mathrm{d}x$$

$$+ \left[\frac{\partial G}{\partial u_x}\delta u + \frac{\partial G}{\partial \overline{u}_x}\delta \overline{u} \right]_0^L + O(|\delta u|^2)$$

$$= \int_0^L \left(\frac{\delta G}{\delta u}\delta u + \frac{\delta G}{\delta \overline{u}}\delta \overline{u} \right) \mathrm{d}x + \left[\frac{\partial G}{\partial u_x}\delta u + \frac{\partial G}{\partial \overline{u}_x}\delta \overline{u} \right]_0^L + O(|\delta u|^2). \qquad (1.48)$$

The quantities

$$\frac{\delta G}{\delta u} = \frac{\partial G}{\partial u} - \frac{\partial}{\partial x}\frac{\partial G}{\partial u_x}, \qquad \frac{\delta G}{\delta \overline{u}} = \frac{\partial G}{\partial \overline{u}} - \frac{\partial}{\partial x}\frac{\partial G}{\partial \overline{u}_x}$$

are complex variational derivatives. Note that they are in general complex conjugates of each other:

$$\overline{\frac{\delta G}{\delta u}} = \frac{\delta G}{\delta \overline{u}}.$$

(Strictly speaking, we need some assumption on G so that the conjugacy holds; this is left to the discussion in Chapter 3.) The PDE (1.47) is conservative in the sense that

$$\frac{\mathrm{d}}{\mathrm{d}t}J(u) = \frac{\mathrm{d}}{\mathrm{d}t}\int_0^L G(u, u_x)\mathrm{d}x$$

$$= \int_0^L \left(\frac{\delta G}{\delta u}\frac{\partial u}{\partial t} + \frac{\delta G}{\delta \overline{u}}\frac{\partial \overline{u}}{\partial t} \right) \mathrm{d}x + \text{(boundary terms)}$$

$$= \int_0^L \left(\mathrm{i}\left|\frac{\delta G}{\delta u}\right|^2 - \mathrm{i}\left|\frac{\delta G}{\delta u}\right|^2 \right) \mathrm{d}x$$

$$= 0, \qquad (1.49)$$

provided some appropriate boundary conditions exist. For example, a linear conservative PDE

$$\mathrm{i}\frac{\partial u}{\partial t} = \frac{\partial^2 u}{\partial x^2}, \qquad G(u, u_x) = |u_x|^2, \qquad (1.50)$$

belongs to this class.

Let us see how a conservative scheme is derived for (1.47). Again we follow the three steps.

[step 1$_{\text{d}}$] Defining a discrete energy

As in the real-valued case, we obtain a discrete energy $G_{\text{d}}(\boldsymbol{U}^{(m)})$ by simply replacing u in $G(u, u_x)$ with $U_k{}^{(m)}$, and u_x with some finite difference. For the case of the linear PDE (1.50), we define a discrete local energy by, for example,

$$G_{\text{d},k}(\boldsymbol{U}^{(m)}) = \frac{|\delta_k^+ U_k{}^{(m)}|^2 + |\delta_k^- U_k{}^{(m)}|^2}{2}. \tag{1.51}$$

This only differs from (1.29) in having $|\cdot|$ (absolute value) in place of $(\cdot)$. The associated discrete global energy is defined by

$$J_{\text{d}}(\boldsymbol{U}^{(m)}) \overset{\text{d}}{\equiv} \sum_{k=0}^{N}{}'' G_{\text{d},k}(\boldsymbol{U}^{(m)}) \Delta x. \tag{1.52}$$

[step 2$_{\text{d}}$] Taking the discrete variation

As in the real-valued case, we hope to simulate the variation calculation (1.48) which can be summarized as

$$\int_0^L \{G(u + \delta u, u_x + \delta u_x) - G(u, u_x)\}\, \mathrm{d}x =$$

$$\int_0^L \left(\frac{\delta G}{\delta u}\delta u + \frac{\delta G}{\delta \overline{u}}\delta \overline{u}\right) \mathrm{d}x + \text{(boundary terms)}, \tag{1.53}$$

in a discrete setting to find a discrete identity:

$$\sum_{k=0}^{N}{}'' \left(G_{\text{d},k}(\boldsymbol{U}^{(m+1)}) - G_{\text{d},k}(\boldsymbol{U}^{(m)})\right) \Delta x =$$

$$\sum_{k=0}^{N}{}'' \left\{\frac{\delta G_{\text{d}}}{\delta(\boldsymbol{U}^{(m+1)}, \boldsymbol{U}^{(m)})_k} \left(U_k{}^{(m+1)} - U_k{}^{(m)}\right) + \right.$$

$$\left. \frac{\delta G_{\text{d}}}{\delta(\overline{\boldsymbol{U}^{(m+1)}}, \overline{\boldsymbol{U}^{(m)}})_k} \left(\overline{U_k{}^{(m+1)} - U_k{}^{(m)}}\right)\right\} \Delta x$$

$$+ \text{(boundary terms)}. \tag{1.54}$$

In the above identity, there are new discrete symbols whose correspondences are

$$\frac{\delta G_{\text{d}}}{\delta(\boldsymbol{U}^{(m+1)}, \boldsymbol{U}^{(m)})_k} \Leftrightarrow \frac{\delta G}{\delta u}, \tag{1.55a}$$

$$\frac{\delta G_{\text{d}}}{\delta(\overline{\boldsymbol{U}^{(m+1)}}, \overline{\boldsymbol{U}^{(m)}})_k} \Leftrightarrow \frac{\delta G}{\delta \overline{u}}. \tag{1.55b}$$

They are called "complex discrete variational derivatives."

In the case of the linear PDE (1.50), we find an identity corresponding to (1.54) as follows.

$$\sum_{k=0}^{N}{}'' \left(G_{\mathrm{d},k}(\boldsymbol{U}^{(m+1)}) - G_{\mathrm{d},k}(\boldsymbol{U}^{(m)}) \right) \Delta x$$

$$= \sum_{k=0}^{N}{}'' \left(\frac{|\delta_k^+ U_k^{(m+1)}|^2 - |\delta_k^+ U_k^{(m)}|^2}{2} + \frac{|\delta_k^- U_k^{(m+1)}|^2 - |\delta_k^- U_k^{(m)}|^2}{2} \right) \Delta x$$

$$= \frac{1}{2} \sum_{k=0}^{N}{}'' \left\{ \delta_k^+ \left(\overline{\frac{U_k^{(m+1)} + U_k^{(m)}}{2}} \right) \cdot \delta_k^+ \left(U_k^{(m+1)} - U_k^{(m)} \right) + (\mathrm{c.c.}) \right.$$

$$\left. + \delta_k^- \left(\overline{\frac{U_k^{(m+1)} + U_k^{(m)}}{2}} \right) \cdot \delta_k^- \left(U_k^{(m+1)} - U_k^{(m)} \right) + (\mathrm{c.c.}) \right\} \Delta x$$

$$= -\sum_{k=0}^{N}{}'' \left[\left\{ \delta_k^{\langle 2 \rangle} \left(\overline{\frac{U_k^{(m+1)} + U_k^{(m)}}{2}} \right) \right\} \left(U_k^{(m+1)} - U_k^{(m)} \right) + (\mathrm{c.c.}) \right] \Delta x$$

$$+ \text{(boundary terms)}. \tag{1.56}$$

The expression "(c.c.)" denotes the complex conjugate of the preceding term(s). In the above calculation, a trivial identity

$$\frac{|a|^2 - |b|^2}{2} = \frac{1}{2} \left\{ \frac{\overline{a+b}}{2}(a-b) + (\mathrm{c.c.}) \right\},$$

and the summation-by-parts formula (1.34) are used. As a result we find the discrete versions of the complex variational derivatives:

$$\frac{\delta G_{\mathrm{d}}}{\delta(\boldsymbol{U}^{(m+1)}, \boldsymbol{U}^{(m)})_k} = -\delta_k^{\langle 2 \rangle} \left(\overline{\frac{U_k^{(m+1)} + U_k^{(m)}}{2}} \right), \tag{1.57a}$$

$$\frac{\delta G_{\mathrm{d}}}{\delta(\overline{\boldsymbol{U}^{(m+1)}}, \overline{\boldsymbol{U}^{(m)}})_k} = -\delta_k^{\langle 2 \rangle} \left(\frac{U_k^{(m+1)} + U_k^{(m)}}{2} \right). \tag{1.57b}$$

[**step 3$_\mathrm{d}$**] *Defining a scheme*

With the discrete variational derivative, we define an abstract scheme analogously to (1.47).

Scheme 1.4 (Conservative scheme for (1.47)) *With given initial data $\boldsymbol{U}^{(0)}$ and appropriate boundary conditions, we compute $\boldsymbol{U}^{(m)}$ by, for $m = 0, 1, 2, \ldots$,*

$$\mathrm{i} \left(\frac{U_k^{(m+1)} - U_k^{(m)}}{\Delta t} \right) = -\frac{\delta G_{\mathrm{d}}}{\delta(\overline{\boldsymbol{U}^{(m+1)}}, \overline{\boldsymbol{U}^{(m)}})_k}, \qquad k = 0, \ldots, N. \tag{1.58}$$

The scheme automatically becomes conservative as follows.

PROPOSITION 1.4 Conservation property of Scheme 1.4
Scheme 1.4 is conservative in the sense that

$$J_{\rm d}(\boldsymbol{U}^{(m)}) = J_{\rm d}(\boldsymbol{U}^{(0)}), \quad m = 1, 2, 3 \ldots. \tag{1.59}$$

PROOF By the identity (1.54), we obtain

$$J_{\rm d}(\boldsymbol{U}^{(m+1)}) - J_{\rm d}(\boldsymbol{U}^{(m)})$$
$$= \frac{1}{\Delta t}\sum_{k=0}^{N}{}''\left(G_{{\rm d},k}(\boldsymbol{U}^{(m+1)}) - G_{{\rm d},k}(\boldsymbol{U}^{(m)})\right)\Delta x$$
$$= \sum_{k=0}^{N}{}''\left\{\overline{\frac{\delta G_{\rm d}}{\delta(\boldsymbol{U}^{(m+1)},\boldsymbol{U}^{(m)})_k}}\left(\frac{U_k^{(m+1)} - U_k^{(m)}}{\Delta t}\right) + ({\rm c.c.})\right\}\Delta x$$
$$+ ({\rm boundary\ terms})$$
$$= \sum_{k=0}^{N}{}''\left\{{\rm i}\left|\frac{\delta G_{\rm d}}{\delta(\boldsymbol{U}^{(m+1)},\boldsymbol{U}^{(m)})_k}\right|^2 + ({\rm c.c.})\right\}\Delta x$$
$$= 0. \tag{1.60}$$

In the second equality the boundary terms are assumed to vanish with appropriate boundary conditions, and in the last equality (1.58) is used. □

In the case of the linear PDE (1.50), Scheme 1.4 reads

$${\rm i}\left(\frac{U_k^{(m+1)} - U_k^{(m)}}{\Delta t}\right) = -\overline{\frac{\delta G_{\rm d}}{\delta(\boldsymbol{U}^{(m+1)},\boldsymbol{U}^{(m)})_k}} = \delta_k^{\langle 2\rangle}\left(\frac{U_k^{(m+1)} + U_k^{(m)}}{2}\right),$$

where the concrete form of the complex discrete variational derivative has been obtained in (1.57b). Again this is just the standard Crank–Nicolson scheme. The conservation property is guaranteed by Proposition 1.4, where the discrete local energy is given by (1.51).

In the subsequent chapters, we will deal with the following complex-valued PDEs.

[complex-valued dissipative PDEs]

$$\frac{\partial u}{\partial t} = -\frac{\delta G}{\delta \bar{u}}, \quad \frac{\rm d}{{\rm d}t}\int_0^L G(u,u_x){\rm d}x \leq 0. \tag{1.61}$$

This includes, for example, the complex Ginzburg–Landau equation and the Newell–Whitehead equation.

[complex-valued conservative PDEs]

$$\mathrm{i}\frac{\partial u}{\partial t} = -\frac{\delta G}{\delta \overline{u}}, \quad \frac{\mathrm{d}}{\mathrm{d}t}\int_0^L G(u, u_x)\mathrm{d}x = 0. \tag{1.62}$$

The linear PDE (1.50) and the nonlinear Schrödinger equation belong to this class.

1.3.3 Procedure for Systems of First-Order PDEs

The idea described in the previous sections can be also applied to the systems of PDEs. The following is an example.

Let us consider the Zakharov equations [72],

$$\mathrm{i}\frac{\partial E}{\partial t} + \frac{\partial^2 E}{\partial x^2} = nE, \quad \frac{\partial^2 n}{\partial t^2} - \frac{\partial^2 n}{\partial x^2} = \frac{\partial^2}{\partial x^2}|E|^2, \quad x \in (0, L),\ t > 0, \tag{1.63}$$

where $E(x,t)$ is complex-valued, and $n(x,t)$ is real-valued. The equations can be written with variational derivatives as

$$\frac{\mathrm{d}}{\mathrm{d}t}\begin{pmatrix} E \\ \overline{E} \\ n \\ v \end{pmatrix} = \begin{pmatrix} 0 & -\mathrm{i} & 0 & 0 \\ \mathrm{i} & 0 & 0 & 0 \\ 0 & 0 & 0 & -1 \\ 0 & 0 & 1 & 0 \end{pmatrix}\begin{pmatrix} \delta G/\delta E \\ \delta G/\delta \overline{E} \\ \delta G/\delta n \\ \delta G/\delta v \end{pmatrix}, \tag{1.64}$$

where $v(x,t)$ is a real-valued intermediate variable such that $v_t = n + |E|^2$, and $G(E, n, v)$ is the energy function defined by

$$G(E, n, v) \stackrel{\mathrm{d}}{=} |E_x|^2 + n|E|^2 + \frac{1}{2}(n^2 + (v_x)^2). \tag{1.65}$$

The concrete forms of the variational derivatives are

$$\frac{\delta G}{\delta E} = -\overline{E}_{xx} + n\overline{E}, \tag{1.66a}$$

$$\frac{\delta G}{\delta \overline{E}} = \overline{\frac{\delta G}{\delta E}}, \tag{1.66b}$$

$$\frac{\delta G}{\delta n} = n + |E|^2, \tag{1.66c}$$

$$\frac{\delta G}{\delta v} = -v_{xx}. \tag{1.66d}$$

It is easy to see that

$$\begin{aligned}
&\frac{\mathrm{d}}{\mathrm{d}t}\int_0^L G(E,n,v)\mathrm{d}x\\
&= \int_0^L \left(\frac{\delta G}{\delta E}\frac{\partial E}{\partial t} + (\text{c.c.}) + \frac{\delta G}{\delta n}\frac{\partial n}{\partial t} + \frac{\delta G}{\delta v}\frac{\partial v}{\partial t}\right)\mathrm{d}x + (\text{boundary terms})\\
&= \int_0^L \left(-\mathrm{i}\left|\frac{\delta G}{\delta E}\right|^2 + (\text{c.c.}) - \frac{\delta G}{\delta n}\frac{\delta G}{\delta v} + \frac{\delta G}{\delta v}\frac{\delta G}{\delta n}\right)\mathrm{d}x\\
&= 0.
\end{aligned}$$

Thus the Zakharov equations conserve the energy

$$J(E,n,v) \stackrel{\mathrm{d}}{\equiv} \int_0^L G(E,n,v)\mathrm{d}x. \tag{1.67}$$

As in the single PDE cases, we can construct discrete versions of the above variational derivatives, $\delta G/\delta E$, $\delta G/\delta \overline{E}$, $\delta G/\delta n$, and $\delta G/\delta v$, by which a conservative scheme for the Zakharov equations can be defined. Let us denote numerical solutions by $E_k{}^{(m)}$, $n_k{}^{(m)}$, $v_k^{(m)}$. Then we follow the three steps again as follows.

[**step 1**$_\mathrm{d}$] *Defining a discrete energy*

We define the discrete local energy by

$$\begin{aligned}
G_{\mathrm{d},k}(\boldsymbol{E}^{(m)}, \boldsymbol{n}^{(m)}, \boldsymbol{v}^{(m)}) =& \\
&\frac{|\delta_k^+ E_k{}^{(m)}|^2 + |\delta_k^- E_k{}^{(m)}|^2}{2} + n_k{}^{(m)}|E_k{}^{(m)}|^2\\
&+ \frac{1}{2}\left(n_k{}^{(m)2} + \frac{(\delta_k^+ v_k^{(m)})^2 + (\delta_k^- v_k^{(m)})^2}{2}\right).
\end{aligned} \tag{1.68}$$

We define the discrete global energy accordingly by

$$J_\mathrm{d}(\boldsymbol{E}^{(m)}, \boldsymbol{n}^{(m)}, \boldsymbol{v}^{(m)}) \stackrel{\mathrm{d}}{\equiv} \sum_{k=0}^{N} {}''G_{\mathrm{d},k}(\boldsymbol{E}^{(m)}, \boldsymbol{n}^{(m)}, \boldsymbol{v}^{(m)})\Delta x. \tag{1.69}$$

[**step 2**$_\mathrm{d}$] *Taking the discrete variation*

Taking discrete variation, we have

$$\sum_{k=0}^{N}{}'' \left\{ G_{d,k}(\boldsymbol{E}^{(m+1)}, \boldsymbol{n}^{(m+1)}, \boldsymbol{v}^{(m+1)}) - G_{d,k}(\boldsymbol{E}^{(m)}, \boldsymbol{n}^{(m)}, \boldsymbol{v}^{(m)}) \right\} \Delta x$$

$$= \sum_{k=0}^{N}{}'' \left\{ \frac{\delta G_d}{\delta(\boldsymbol{E}^{(m+1)}, \boldsymbol{E}^{(m)})_k} (E_k^{(m+1)} - E_k^{(m)}) \right.$$

$$+ \frac{\delta G_d}{\delta(\overline{\boldsymbol{E}^{(m+1)}, \boldsymbol{E}^{(m)}})_k} \overline{(E_k^{(m+1)} - E_k^{(m)})}$$

$$+ \frac{\delta G_d}{\delta(\boldsymbol{n}^{(m+1)}, \boldsymbol{n}^{(m)})_k} (n_k^{(m+1)} - n_k^{(m)})$$

$$\left. + \frac{\delta G_d}{\delta(\boldsymbol{v}^{(m+1)}, \boldsymbol{v}^{(m)})_k} (v_k^{(m+1)} - v_k^{(m)}) \right\} \Delta x, \quad (1.70)$$

where

$$\frac{\delta G_d}{\delta(\boldsymbol{E}^{(m+1)}, \boldsymbol{E}^{(m)})_k} = -\delta_k^{\langle 2 \rangle} \left(\overline{\frac{E_k^{(m+1)} + E_k^{(m)}}{2}} \right)$$

$$+ \left(\frac{E_k^{(m)} + E_k^{(m)}}{2} \right) \left(\frac{n_k^{(m+1)} + n_k^{(m)}}{2} \right), \quad (1.71a)$$

$$\frac{\delta G_d}{\delta(\overline{\boldsymbol{E}^{(m+1)}, \boldsymbol{E}^{(m)}})_k} = \overline{\left(\frac{\delta G_d}{\delta(\boldsymbol{E}^{(m+1)}, \boldsymbol{E}^{(m)})_k} \right)}, \quad (1.71b)$$

$$\frac{\delta G_d}{\delta(\boldsymbol{n}^{(m+1)}, \boldsymbol{n}^{(m)})_k} = \frac{n_k^{(m+1)} + n_k^{(m)}}{2}$$

$$+ \frac{|E_k^{(m+1)}|^2 + |E_k^{(m)}|^2}{2}, \quad (1.71c)$$

$$\frac{\delta G_d}{\delta(\boldsymbol{v}^{(m+1)}, \boldsymbol{v}^{(m)})_k} = -\delta_k^{\langle 2 \rangle} \left(\frac{v_k^{(m+1)} + v_k^{(m)}}{2} \right). \quad (1.71d)$$

They are obviously discrete analogues of (1.66a) through (1.66d).

[**step 3$_d$**] *Defining a scheme*

With the discrete variational derivatives, we define a numerical scheme.

Scheme 1.5 (Conservative scheme for the Zakharov equations) *With given initial data $\boldsymbol{E}^{(0)}, \boldsymbol{n}^{(0)}, \boldsymbol{u}^{(0)}$ and appropriate boundary conditions, we* *compute numerical solutions by, for $m = 0, 1, 2, \ldots$,*

$$i\left(\frac{E_k^{(m+1)} - E_k^{(m)}}{\Delta t}\right) = \frac{\delta G_d}{\delta(\overline{\boldsymbol{E}^{(m+1)}, \boldsymbol{E}^{(m)}})_k}$$
$$= -\delta_k^{\langle 2 \rangle}\left(\frac{E_k^{(m+1)} + E_k^{(m)}}{2}\right) + \left(\frac{E_k^{(m)} + E_k^{(m)}}{2}\right)\left(\frac{n_k^{(m+1)} + n_k^{(m)}}{2}\right),$$
(1.72a)

$$\frac{n_k^{(m+1)} - n_k^{(m)}}{\Delta t} = -\frac{\delta G_d}{\delta(\boldsymbol{v}^{(m+1)}, \boldsymbol{v}^{(m)})_k} = \delta_k^{\langle 2 \rangle}\left(\frac{v_k^{(m+1)} + v_k^{(m)}}{2}\right), \quad (1.72b)$$

$$\frac{v_k^{(m+1)} - v_k^{(m)}}{\Delta t} = \frac{\delta G_d}{\delta(\boldsymbol{n}^{(m+1)}, \boldsymbol{n}^{(m)})_k}$$
$$= \frac{n_k^{(m+1)} + n_k^{(m)}}{2} + \frac{|E_k^{(m+1)}|^2 + |E_k^{(m)}|^2}{2}, \quad (1.72c)$$

where $k = 0, \ldots, N$.

Then the scheme automatically becomes conservative as follows. We omit the proof, which is again the discrete analogue of the continuous version.

PROPOSITION 1.5 Conservation property of Scheme 1.5
Scheme 1.5 is conservative in the sense that

$$J_d(\boldsymbol{E}^{(m)}, \boldsymbol{n}^{(m)}, \boldsymbol{v}^{(m)}) = J_d(\boldsymbol{E}^{(0)}, \boldsymbol{n}^{(0)}, \boldsymbol{v}^{(0)}), \quad m = 1, 2, 3 \ldots. \quad (1.73)$$

More general cases are described in Chapter 2, and the procedure is presented in Chapter 3.

1.3.4 Procedure for Second-Order PDEs

So far the first-order PDEs of the form $u_t = \cdots$ have been considered. Let us consider next the cases where u_{tt} is concerned; we call such PDEs *second-order*. Let $u(x,t)$ be a real-valued function, and suppose that a real-valued function $G(u, u_x)$ is given. We here consider the PDE of the form

$$\frac{\partial^2 u}{\partial t^2} = -\frac{\delta G}{\delta u}, \quad x \in (0, L), \ t > 0. \quad (1.74)$$

This PDE has a conservation property

$$\frac{d}{dt}\int_0^L \left\{\frac{(u_t)^2}{2} + G(u, u_x)\right\} dx = \int_0^L \left(u_{tt} + \frac{\delta G}{\delta u}\right) u_t \, dx + \text{(boundary term)}$$
$$= 0, \quad (1.75)$$

provided some appropriate boundary conditions are given. This class of PDEs includes, for example, the linear wave equation

$$\frac{\partial^2 u}{\partial t^2} = \frac{\partial^2 u}{\partial x^2}, \quad G(u, u_x) = \frac{(u_x)^2}{2}. \tag{1.76}$$

Obviously the procedure used so far cannot be applied as is, since the conservation property (1.75) now includes not only $G(u, u_x)$ but also u_t in its integrand. In order to handle these problems, we have two options.

- We can consider directly the variation process with respect to the global energy:

$$J(u) \stackrel{\mathrm{d}}{\equiv} \int_0^L \left\{ \frac{(u_t)^2}{2} + G(u, u_x) \right\} \mathrm{d}x. \tag{1.77}$$

- Otherwise we can introduce a new variable $v = u_t$ to rewrite the equation into a system of first-order PDEs. The advantage of this option is that in this way we can avoid the explicit appearance of u_t in the energy and thus can apply the known procedure.

We demonstrate these two options below in turn. Let us first consider the first option.

[**step 1$_\mathrm{d}$**] *Defining a discrete energy*

We define a discrete local energy G_d, again replacing u with $U_k^{(m+1)}$ and $U_k^{(m)}$, and u_x with some differences.

In the case of the linear wave equation, let us define a local energy as

$$G_{\mathrm{d},k}(\boldsymbol{U}^{(m+1)}, \boldsymbol{U}^{(m)})$$
$$\stackrel{\mathrm{d}}{\equiv} \frac{1}{4} \left\{ \left(\frac{(\delta_k^+ U_k^{(m+1)})^2 + (\delta_k^- U_k^{(m+1)})^2}{2} \right) \right.$$
$$\left. + \left(\frac{(\delta_k^+ U_k^{(m)})^2 + (\delta_k^- U_k^{(m)})^2}{2} \right) \right\}, \tag{1.78}$$

and accordingly a global energy as

$$J_\mathrm{d}(\boldsymbol{U}^{(m+1)}, \boldsymbol{U}^{(m)})$$
$$\stackrel{\mathrm{d}}{\equiv} \sum_{k=0}^{N}{}'' \left\{ \frac{1}{2} \left(\frac{U_k^{(m+1)} - U_k^{(m)}}{\Delta t} \right)^2 + G_{\mathrm{d},k}(\boldsymbol{U}^{(m+1)}, \boldsymbol{U}^{(m)}) \right\} \Delta x. \tag{1.79}$$

Note that now the discrete local energy function is defined with two numerical solutions; this is a trick in order to consider the direct variation of (1.77). Its detail is left to Chapter 3.

[step 2$_\mathrm{d}$] *Taking the discrete variation*

Let us first recall the continuous case. The variation becomes

$$\int_0^L \left\{ \frac{1}{2}(u_t + \delta u_t)^2 + G(u + \delta u, u_x + \delta u_x) - \frac{1}{2}(u_t)^2 - G(u, u_x) \right\} \mathrm{d}x$$
$$= \int_0^L \left(u_t \delta u_t + \frac{\delta G}{\delta u} \delta u \right) \mathrm{d}x + \text{(boundary term)} + O(\delta u^2), \quad (1.80)$$

which implies when $\delta u/\Delta t \to u_t$ (as $\Delta t \to 0$),

$$\frac{\mathrm{d}}{\mathrm{d}t} J(u) = \int_0^L \left(u_{tt} + \frac{\delta G}{\delta u} \right) u_t \mathrm{d}x + \text{(boundary term)}. \quad (1.81)$$

By copying this calculation, we see for the term $u_t^2/2$

$$\sum_{k=0}^{N} {}''\left\{ \frac{(\delta_m^+ U_k^{(m)})^2}{2} - \frac{(\delta_m^+ U_k^{(m-1)})^2}{2} \right\}$$
$$= \sum_{k=0}^{N} {}''\left\{ \delta_m^+ \left(\frac{U_k^{(m)} + U_k^{(m-1)}}{2} \right) \cdot \delta_m^+ \left(U_k^{(m)} - U_k^{(m-1)} \right) \right\} \Delta x$$
$$= \sum_{k=0}^{N} {}''\left\{ \delta_m^{\langle 2 \rangle} U_k^{(m)} \cdot \left(\frac{U_k^{(m+1)} - U_k^{(m-1)}}{2} \right) \right\} \Delta x, \quad (1.82)$$

and for G

$$\sum_{k=0}^{N} {}''\left\{ G_{\mathrm{d},k}(\boldsymbol{U}^{(m+1)}, \boldsymbol{U}^{(m)}) - G_{\mathrm{d},k}(\boldsymbol{U}^{(m)}, \boldsymbol{U}^{(m-1)}) \right\} \Delta x$$
$$= \sum_{k=0}^{N} {}''\left\{ \frac{\delta G_\mathrm{d}}{\delta(\boldsymbol{U}^{(m+1)}, \boldsymbol{U}^{(m)}, \boldsymbol{U}^{(m-1)})_k} \left(\frac{U_k^{(m+1)} - U_k^{(m-1)}}{2} \right) \right\} \Delta x$$
$$+ \text{(boundary term)}. \quad (1.83)$$

The operator δ_m^+ is the first order difference operator in time direction (see (1.27)). The detail of this calculation will be shown in Section 3.5.1. The last equality should be regarded to define the discrete variational derivative:

$$\frac{\delta G_\mathrm{d}}{\delta(\boldsymbol{U}^{(m+1)}, \boldsymbol{U}^{(m)}, \boldsymbol{U}^{(m-1)})_k},$$

as in (1.32). Since the discrete variational derivative refers three numerical solutions, we call it the "*three-points* discrete variational derivative," to distinguish it from the standard discrete variational derivatives which refer only two solutions. (We do not further get into the detail of the new concept here. See also Section 1.4.2.)

In the case of the linear wave equation, it becomes

$$\sum_{k=0}^{N}{}'' \left\{ G_{d,k}(\boldsymbol{U}^{(m+1)}, \boldsymbol{U}^{(m)}) - G_{d,k}(\boldsymbol{U}^{(m)}, \boldsymbol{U}^{(m-1)}) \right\} \Delta x$$

$$= \frac{1}{8} \sum_{k=0}^{N}{}'' \left\{ (\delta_k^+ U_k^{(m+1)})^2 - (\delta_k^+ U_k^{(m-1)})^2 \right.$$
$$\left. + (\delta_k^- U_k^{(m+1)})^2 - (\delta_k^- U_k^{(m-1)})^2 \right\} \Delta x$$

$$= \frac{1}{8} \sum_{k=0}^{N}{}'' \left\{ \delta_k^+ (U_k^{(m+1)} + U_k^{(m-1)}) \cdot \delta_k^+ (U_k^{(m+1)} - U_k^{(m-1)}) \right.$$
$$\left. + \delta_k^- (U_k^{(m+1)} + U_k^{(m-1)}) \cdot \delta_k^- (U_k^{(m+1)} - U_k^{(m-1)}) \right\} \Delta x$$

$$= -\frac{1}{4} \sum_{k=0}^{N}{}'' \left\{ \delta_k^{\langle 2 \rangle}(U_k^{(m+1)} + U_k^{(m-1)}) \cdot (U_k^{(m+1)} - U_k^{(m-1)}) \right\} \Delta x$$
$$+ (\text{boundary term})$$

$$= -\sum_{k=0}^{N}{}'' \left\{ \delta_k^{\langle 2 \rangle} \left(\frac{U_k^{(m+1)} + U_k^{(m-1)}}{2} \right) \cdot \left(\frac{U_k^{(m+1)} - U_k^{(m-1)}}{2} \right) \right\} \Delta x$$
$$+ (\text{boundary term}). \qquad (1.84)$$

Again we used the summation-by-parts formula (1.34). Thus we find that

$$\frac{\delta G_d}{\delta(\boldsymbol{U}^{(m+1)}, \boldsymbol{U}^{(m)}, \boldsymbol{U}^{(m-1)})_k} = -\delta_k^{\langle 2 \rangle} \left(\frac{U_k^{(m+1)} + U_k^{(m-1)}}{2} \right). \qquad (1.85)$$

[**step 3$_d$**] *Defining a scheme*

Corresponding to the continuous equation (1.74), we define a discrete scheme as follows.

Scheme 1.6 (Conservative scheme for (1.74)) *With given initial data* $\boldsymbol{U}^{(0)}$, $\boldsymbol{U}^{(1)}$ *and appropriate boundary conditions, we compute numerical solutions by, for* $m = 1, 2, \ldots$,

$$\frac{U_k^{(m+1)} - 2U_k^{(m)} + U_k^{(m-1)}}{(\Delta t)^2} = -\frac{\delta G_d}{\delta(\boldsymbol{U}^{(m+1)}, \boldsymbol{U}^{(m)}, \boldsymbol{U}^{(m-1)})_k}, \qquad (1.86)$$

where $k = 0, \ldots, N$.

Then the scheme automatically becomes conservative as follows. We omit the proof.

PROPOSITION 1.6 Conservation property of Scheme 1.6
Scheme 1.6 is conservative in the sense

$$J_{\mathrm{d}}(\boldsymbol{U}^{(m+1)}, \boldsymbol{U}^{(m)}) = J_{\mathrm{d}}(\boldsymbol{U}^{(1)}, \boldsymbol{U}^{(0)}), \quad m = 1, 2, 3 \ldots. \qquad (1.87)$$

In the case of the linear wave equation, Scheme 1.6 reads

$$\frac{U_k^{(m+1)} - 2U_k^{(m)} + U_k^{(m-1)}}{(\Delta t)^2} = \delta_k^{\langle 2 \rangle} \left(\frac{U_k^{(m+1)} + U_k^{(m-1)}}{2} \right), \qquad (1.88)$$

whose conservation property is guaranteed by Proposition 1.6 with the energy function (1.78).

Next, let us here adopt the second option. By introducing a new variable $v(x,t) = u_t$, we can rewrite the equation (1.74) into a system of first-order PDEs:

$$\frac{\partial u}{\partial t} = v, \qquad (1.89\mathrm{a})$$

$$\frac{\partial v}{\partial t} = -\frac{\delta G}{\delta u}. \qquad (1.89\mathrm{b})$$

If we introduce a "modified local energy"

$$\widetilde{G}(u, u_x, v) = \frac{v^2}{2} + G(u, u_x), \qquad (1.90)$$

we can rewrite the equations (1.89) as a system of first-order PDEs:

$$\frac{\partial u}{\partial t} = \frac{\partial \widetilde{G}}{\partial v}, \qquad (1.91\mathrm{a})$$

$$\frac{\partial v}{\partial t} = -\frac{\partial \widetilde{G}}{\partial u}. \qquad (1.91\mathrm{b})$$

Then let us define a global energy associated with (1.91) by

$$J(u, v) \stackrel{\mathrm{d}}{\equiv} \int_0^L \widetilde{G}(u, u_x, v) \mathrm{d}x. \qquad (1.92)$$

The conservation property (1.75) is then

$$\frac{\mathrm{d}}{\mathrm{d}t} J(u, v) = 0. \qquad (1.93)$$

In the case of the linear wave equation (1.76), the modified local energy becomes

$$\widetilde{G}(u, u_x, v) = \frac{v^2}{2} + \frac{(u_x)^2}{2}. \qquad (1.94)$$

Let us construct a conservative scheme to the equations (1.91).

[**step 1$_\mathrm{d}$**] *Defining a discrete energy*

We define a discrete modified local energy $\widetilde{G}_\mathrm{d}$, again replacing u, v with $U_k^{(m)}, V_k^{(m)}$, and u_x with some differences.

In the case of the linear wave equation, let us define a modified local energy as

$$\widetilde{G}_{\mathrm{d},k}(\boldsymbol{U}^{(m)}, \boldsymbol{V}^{(m)}) \stackrel{\mathrm{d}}{=} \frac{(V_k^{(m)})^2}{2} + \frac{1}{2}\left(\frac{(\delta_k^+ U_k^{(m)})^2 + (\delta_k^- U_k^{(m)})^2}{2}\right), \quad (1.95)$$

and accordingly a global energy as

$$J_\mathrm{d}(\boldsymbol{U}^{(m)}, \boldsymbol{V}^{(m)}) \stackrel{\mathrm{d}}{=} \sum_{k=0}^{N} {}''\widetilde{G}_{\mathrm{d},k}(\boldsymbol{U}^{(m)}, \boldsymbol{V}^{(m)})\Delta x. \quad (1.96)$$

[**step 2$_\mathrm{d}$**] *Taking the discrete variation*

We aim at the identity

$$\int_0^L \widetilde{G}(u+\delta u, u_x + \delta u_x, v + \delta v) - \widetilde{G}(u, u_x, v)\mathrm{d}x$$
$$= \int_0^L \left(\frac{\partial \widetilde{G}}{\partial u}\delta u + \frac{\partial \widetilde{G}}{\partial v}\delta v\right)\mathrm{d}x + (\text{boundary term}), \quad (1.97)$$

in a discrete setting. For the purpose, we consider the difference

$$\sum_{k=0}^{N}{}''\left\{\widetilde{G}_{\mathrm{d},k}(\boldsymbol{U}^{(m+1)}, \boldsymbol{V}^{(m+1)}) - \widetilde{G}_{\mathrm{d},k}(\boldsymbol{U}^{(m)}, \boldsymbol{V}^{(m)})\right\}\Delta x$$
$$= \sum_{k=0}^{N}{}''\left\{\frac{\delta \widetilde{G}_\mathrm{d}}{\delta(\boldsymbol{U}^{(m+1)}, \boldsymbol{U}^{(m)})_k}(U_k^{(m+1)} - U_k^{(m)})\right.$$
$$\left.+ \frac{\delta \widetilde{G}_\mathrm{d}}{\delta(\boldsymbol{V}^{(m+1)}, \boldsymbol{V}^{(m)})_k}(V_k^{(m+1)} - V_k^{(m)})\right\}\Delta x$$
$$+ (\text{boundary term}), \quad (1.98)$$

to find the discrete variational derivatives,

$$\frac{\delta \widetilde{G}_\mathrm{d}}{\delta(\boldsymbol{U}^{(m+1)}, \boldsymbol{U}^{(m)})_k} \quad \text{and} \quad \frac{\delta \widetilde{G}_\mathrm{d}}{\delta(\boldsymbol{V}^{(m+1)}, \boldsymbol{V}^{(m)})_k}.$$

In the case of the linear wave equation, it becomes

$$\sum_{k=0}^{N} {}'' \left\{ \widetilde{G}_{\mathrm{d},k}(\boldsymbol{U}^{(m+1)}, \boldsymbol{V}^{(m+1)}) - \widetilde{G}_{\mathrm{d},k}(\boldsymbol{U}^{(m)}, \boldsymbol{V}^{(m)}) \right\} \Delta x$$

$$= \sum_{k=0}^{N} {}'' \left\{ \frac{(V_k^{(m+1)})^2}{2} - \frac{(V_k^{(m)})^2}{2} \right.$$

$$+ \frac{1}{2} \left(\frac{(\delta_k^+ U_k^{(m+1)})^2 + (\delta_k^- U_k^{(m+1)})^2}{2} \right)$$

$$\left. - \frac{1}{2} \left(\frac{(\delta_k^+ U_k^{(m)})^2 + (\delta_k^- U_k^{(m)})^2}{2} \right) \right\} \Delta x$$

$$= \sum_{k=0}^{N} {}'' \left\{ \left(\frac{V_k^{(m+1)} + V_k^{(m)}}{2} \right) (V_k^{(m+1)} - V_k^{(m)}) \right.$$

$$\left. - \delta_k^{\langle 2 \rangle} \left(\frac{U_k^{(m+1)} + U_k^{(m)}}{2} \right) \cdot (U_k^{(m+1)} - U_k^{(m)}) \right\} \Delta x$$

$$+ \text{(boundary term).} \tag{1.99}$$

Again we used the summation-by-parts formula (1.34). Thus we find that

$$\frac{\delta \widetilde{G}_{\mathrm{d}}}{\delta(\boldsymbol{U}^{(m+1)}, \boldsymbol{U}^{(m)})_k} = -\delta_k^{\langle 2 \rangle} \left(\frac{U_k^{(m+1)} + U_k^{(m)}}{2} \right), \tag{1.100a}$$

$$\frac{\delta \widetilde{G}_{\mathrm{d}}}{\delta(\boldsymbol{V}^{(m+1)}, \boldsymbol{V}^{(m)})_k} = \frac{V_k^{(m+1)} + V_k^{(m)}}{2}. \tag{1.100b}$$

[**step 3$_\mathrm{d}$**] *Defining a scheme*

Corresponding to the continuous equations (1.91), we define a discrete scheme as follows.

Scheme 1.7 (Conservative scheme for (1.91)) *With given initial data $\boldsymbol{U}^{(0)}$, $\boldsymbol{V}^{(0)}$ and appropriate boundary conditions, we compute numerical solutions by, for $m = 0, 1, 2, \ldots$,*

$$\frac{U_k^{(m+1)} - U_k^{(m)}}{\Delta t} = \frac{\delta \widetilde{G}_{\mathrm{d}}}{\delta(\boldsymbol{V}^{(m+1)}, \boldsymbol{V}^{(m)})_k}, \tag{1.101a}$$

$$\frac{V_k^{(m+1)} - V_k^{(m)}}{\Delta t} = -\frac{\delta \widetilde{G}_{\mathrm{d}}}{\delta(\boldsymbol{U}^{(m+1)}, \boldsymbol{U}^{(m)})_k}, \tag{1.101b}$$

where $k = 0, \ldots, N$.

Then the scheme automatically becomes conservative as follows. We omit the proof.

PROPOSITION 1.7 Conservation property of Scheme 1.7
Scheme 1.7 is conservative in the sense

$$J_d(\boldsymbol{U}^{(m)}, \boldsymbol{V}^{(m)}) = J_d(\boldsymbol{U}^{(0)}, \boldsymbol{V}^{(0)}), \quad m = 1, 2, 3 \ldots. \quad (1.102)$$

In the case of the linear wave equation, Scheme 1.7 reads

$$\frac{U_k^{(m+1)} - U_k^{(m)}}{\Delta t} = \frac{\delta \widetilde{G}_d}{\delta(\boldsymbol{V}^{(m+1)}, \boldsymbol{V}^{(m)})_k} = \frac{V_k^{(m+1)} + V_k^{(m)}}{2}, \quad (1.103a)$$

$$\frac{V_k^{(m+1)} - V_k^{(m)}}{\Delta t} = -\frac{\delta \widetilde{G}_d}{\delta(\boldsymbol{U}^{(m+1)}, \boldsymbol{U}^{(m)})_k}$$

$$= \delta_k^{\langle 2 \rangle} \left(\frac{U_k^{(m+1)} + U_k^{(m)}}{2} \right), \quad (1.103b)$$

whose conservation property is guaranteed by Proposition 1.7 with the energy function (1.95).

Other examples of the PDEs (1.74) will be described in Chapter 2. More rigorous procedures of constructing conservative schemes will be given in Chapter 3.

1.4 Advanced Topics

So far we have glanced through the basics of the discrete variational derivative method and its various examples. In this section we briefly comment on more advanced topics that will be covered in the second part of this book. The topics are on the designs of higher-order schemes (Section 1.4.1, whose detail will be given in Chapter 5), linearly implicit schemes (Section 1.4.2, Chapter 6), and other remarks (Section 1.4.3, Chapter 7).

1.4.1 Design of Higher-Order Schemes

Dissipative or conservative schemes designed with the standard discrete variational method are usually second-order both spatially and temporally; that is,

$$|U_k^{(m)} - u(k\Delta x, m\Delta t)| = O(\Delta x^2, \Delta t^2). \quad (1.104)$$

If more accuracy is demanded, we can increase the order under appropriate conditions. Below we try to give the readers a picture of these high-order

1.4.1.1 Design of Spatially High-Order Schemes

In designing spatially high-order schemes, we make one big assumption: we assume the periodic boundary condition. This is to utilize the high-order difference operators, $\delta_k^{\langle 1 \rangle, 2p}$ (the superscript $2p$ denotes the spatial accuracy), first three of which are

$$\text{2nd: } \delta_k^{\langle 1 \rangle, 2} f_k = \frac{\frac{1}{2} f_{k+1} - \frac{1}{2} f_{k-1}}{\Delta x}, \tag{1.105a}$$

$$\text{4th: } \delta_k^{\langle 1 \rangle, 4} f_k = \frac{-\frac{1}{12} f_{k+2} + \frac{2}{3} f_{k+1} - \frac{2}{3} f_{k-1} + \frac{1}{12} f_{k-1}}{\Delta x}, \tag{1.105b}$$

$$\text{6th: } \delta_k^{\langle 1 \rangle, 6} f_k = \frac{\frac{1}{60} f_{k+3} - \frac{3}{20} f_{k+2} + \frac{3}{4} f_{k+1} - \frac{3}{4} f_{k-1} + \frac{3}{20} f_{k-1} - \frac{1}{60} f_{k-3}}{\Delta x}. \tag{1.105c}$$

Note that the subscript k of the operator denotes that it operates on the spatial subindex k. These operators are $2p$-th order approximations of $\partial/\partial x$, i.e.,

$$\delta_k^{\langle 1 \rangle, 2p} u(k\Delta x, t) = u_x(k\Delta x, t) + O(\Delta x^{2p}). \tag{1.106}$$

These operators include as a special case $p = \infty$ the so-called "spectral differentiation" operator:

$$\delta_k^{\langle 1 \rangle, \infty} = (F^{-1} \widetilde{D} F)_k, \tag{1.107}$$

where F is the matrix of the discrete Fourier transform, and $\widetilde{D}$ is, roughly speaking, a diagonal matrix with the elements $(\widetilde{D})_{jj} = \mathrm{i} j$ (see Chapter 5 for exact definition). The spectral differentiation operator is quite accurate in the sense that the discretization error decreases *exponentially* as $\Delta x \to 0$, as far as the target function is sufficiently smooth. Note that the more the order $2p$ is increased, the more the number of the referenced points becomes. In particular, the number of the points outside the original domain $0 \leq k \leq N$, which are referenced unexpectedly by such a "wide" difference operator, increases, and it becomes extremely difficult to eliminate them with the limited number of discrete boundary conditions. The periodic boundary condition is assumed for simplicity so that we can forget this issue.

Rewriting the procedure of the standard discrete variational derivative method with the high-order difference operator $\delta_k^{\langle 1 \rangle, 2p}$, we obtain a new procedure for a *spatially higher-order* discrete variational derivative method. Let us see this in the case of the first-order real-valued PDEs:

$$\frac{\partial u}{\partial t} = -\frac{\delta G}{\delta u}, \quad x \in (0, L), \ t > 0. \tag{1.23}$$

We pick the linear diffusion equation again:
$$\frac{\partial u}{\partial t} = \frac{\partial^2 u}{\partial x^2}, \tag{1.25}$$
as a concrete example, which is of the form (1.23) with $G(u, u_x) = (u_x)^2/2$.

[step 1_d] *Defining a discrete energy*

By simply replacing u in $G(u, u_x)$ with $U_k^{(m)}$, and u_x with $\delta_k^{\langle 1\rangle, 2p} U_k^{(m)}$, we obtain a discrete energy $G_d(\boldsymbol{U}^{(m)})$.

For the linear diffusion equation, we define a discrete local energy by
$$G_{d,k}(\boldsymbol{U}^{(m)}) = \frac{1}{2}(\delta_k^{\langle 1\rangle, 2p} U_k^{(m)})^2, \tag{1.108}$$
and the corresponding discrete global energy by
$$J_d(\boldsymbol{U}^{(m)}) = \sum_{k=0}^{N}{}'' G_{d,k}(\boldsymbol{U}^{(m)}) \Delta x. \tag{1.109}$$

[step 2_d] *Taking the discrete variation*

We find an identity
$$\sum_{k=0}^{N}{}'' \left(G_{d,k}(\boldsymbol{U}^{(m+1)}) - G_{d,k}(\boldsymbol{U}^{(m)}) \right) \Delta x =$$
$$\sum_{k=0}^{N}{}'' \frac{\delta G_d}{\delta(\boldsymbol{U}^{(m+1)}, \boldsymbol{U}^{(m)})_k} \left(U_k^{(m+1)} - U_k^{(m)} \right) \Delta x, \tag{1.110}$$
by calculating the difference in the left hand side. Note that this seems to be identical to the identity (1.32) in Section 1.3.1, but now the spatial difference operator in the discrete variational derivative is replaced with higher-order one, and the boundary terms are trivially canceled due to the discrete periodic boundary condition.

In the linear diffusion equation case, the above identity becomes
$$\sum_{k=0}^{N}{}'' \left(G_{d,k}(\boldsymbol{U}^{(m+1)}) - G_{d,k}(\boldsymbol{U}^{(m)}) \right) \Delta x$$
$$= \sum_{k=0}^{N}{}'' \left(\frac{(\delta_k^{\langle 1\rangle, 2p} U_k^{(m+1)})^2 - (\delta_k^{\langle 1\rangle, 2p} U_k^{(m)})^2}{2} \right) \Delta x$$
$$= \sum_{k=0}^{N}{}'' \left\{ \delta_k^{\langle 1\rangle, 2p}\left(\frac{U_k^{(m+1)} + U_k^{(m)}}{2}\right) \cdot \delta_k^{\langle 1\rangle, 2p}(U_k^{(m+1)} - U_k^{(m)}) \right\}$$
$$= -\sum_{k=0}^{N}{}'' \left\{ \left(\delta_k^{\langle 1\rangle, 2p}\right)^2 \left(\frac{U_k^{(m+1)} + U_k^{(m)}}{2}\right) \right\} (U_k^{(m+1)} - U_k^{(m)}) \Delta x.$$
$$\tag{1.111}$$

Here we used a summation-by-parts formula with respect to $\delta_k^{\langle 1 \rangle, 2p}$:

$$\sum_{k=0}^{N}{}''(\delta_k^{\langle 1 \rangle, 2p} f_k) g_k \Delta x = -\sum_{k=0}^{N}{}'' f_k (\delta_k^{\langle 1 \rangle, 2p} g_k) \Delta x, \qquad (1.112)$$

which holds under the discrete periodic boundary condition. Thus, we find a discrete variational derivative.

$$\frac{\delta G_{\mathrm{d}}}{\delta(\boldsymbol{U}^{(m+1)}, \boldsymbol{U}^{(m)})_k} = -\left(\delta_k^{\langle 1 \rangle, 2p}\right)^2 \left(\frac{U_k^{(m+1)} + U_k^{(m)}}{2}\right). \qquad (1.113)$$

[**step 3**$_\mathrm{d}$] *Defining a scheme*

With the discrete variational derivative, we define a scheme.

Scheme 1.8 (Spatially high-order dissipative scheme for (1.23)) *With given initial data $\boldsymbol{U}^{(0)}$ and appropriate boundary conditions, we compute $\boldsymbol{U}^{(m)}$ by, for $m = 0, 1, \ldots$,*

$$\frac{U_k^{(m+1)} - U_k^{(m)}}{\Delta t} = -\frac{\delta G_{\mathrm{d}}}{\delta(\boldsymbol{U}^{(m+1)}, \boldsymbol{U}^{(m)})_k}, \qquad k = 0, \ldots, N. \qquad (1.114)$$

Then the scheme automatically becomes dissipative as follows.

PROPOSITION 1.8 Dissipation property of Scheme 1.8
Scheme 1.8 is dissipative in the sense that

$$J_{\mathrm{d}}(\boldsymbol{U}^{(m+1)}) \leq J_{\mathrm{d}}(\boldsymbol{U}^{(m)}), \qquad m = 0, 1, 2 \ldots. \qquad (1.115)$$

PROOF It is straightforward by the identity (1.110). □

In the case of the linear diffusion equation, with the discrete derivative (1.113), Scheme 1.8 becomes

$$\frac{U_k^{(m+1)} - U_k^{(m)}}{\Delta t} = -\frac{\delta G_{\mathrm{d}}}{\delta(\boldsymbol{U}^{(m+1)}, \boldsymbol{U}^{(m)})_k} = (\delta_k^{\langle 1 \rangle, 2p})^2 \left(\frac{U_k^{(m+1)} + U_k^{(m)}}{2}\right). \qquad (1.116)$$

The dissipation property is guaranteed by Proposition 1.8, where $G_{\mathrm{c},k}(\boldsymbol{U}^{(m)})$ is given by (1.108).

It can be proved that Scheme 1.8 is really spatially high-order just by considering Taylor expansion (see Chapter 5). The above procedure can be easily extended to other types of PDEs.

1.4.1.2 Design of Temporally High-Order Schemes

Accuracy in time direction can be increased in two ways; by utilizing the "composition method," and by generalizing the concept of discrete variation to temporally high-order ones.

1.4.1.2.1 By the composition method

Suppose that we already have a second-order dissipative or conservative scheme, which is obtained by, for example, the standard discrete variational derivative method. Then, by the so-called "composition method," we can easily increase its temporal order to 4th, 6th, and so on.

The composition method was first introduced independently by Suzuki [157] and Yoshida [168], for high-order numerical integrators of ordinary differential equations, and then further developed by various authors. The method is based on the decomposition of exponential operator. In the easiest case, it reads:

Suppose we have a second-order temporally-symmetric integrator $\phi(\Delta t) : \boldsymbol{U}^{(m)} \mapsto \boldsymbol{U}^{(m+1)}$.
Then, a composed integrator : $\phi(c_1 \Delta t)\phi(c_2 \Delta t)\phi(c_1 \Delta t)$, where $c_1 \simeq 1.3512072$, $c_2 \simeq -1.7024143$, is a fourth-order integrator.

The composition method is quite simple, at least from the practical point of view; we just use the second-order scheme repeatedly with the specified time steps, and that automatically makes a fourth-order scheme. If we already have a program code for the second-order scheme, we can easily update it for the fourth-order scheme just by adding a few lines. In a similar manner, we can obtain higher-order schemes. For more complete description of the composition method, see Chapter 5.

It is obvious that when we apply the composition method to a conservative second-order scheme, the resulting fourth- and higher-order schemes automatically become conservative as well.

The composition method has, however, two disadvantages. One is that the computational cost considerably increases as we repeat the composition. For example, it is obvious that the fourth-order scheme obtained in the above way requires as three times much cost as the second-order one for the same single step of Δt. The other disadvantage is that basically the scheme cannot be applied to *dissipative* problems. In any composition method, there is at least one *negative* step, where the time mesh size must be chosen to a negative value. This negative step can destroy the overall dissipation property, even if the original second-order scheme is strictly dissipative.

1.4.1.2.2 By generalizing the discrete variation process

The second way of designing temporally high-order schemes is to generalize the concept of discrete variation so that it allows temporally high-order approximations.

The new procedure in this case, however, becomes far more complicated than any of the procedures mentioned so far. We here describe the outline briefly.

The basic idea is, as in the spatially high-order case, to introduce *temporally high-order* difference operators, say $\delta_m^{\langle 1 \rangle, q}$ (the superscript q denotes the temporal accuracy), and rewrite the whole procedure with them. Note that the subscript m denotes the time subindex m. As opposed to the spatially high-order case, this update requires a completely new discrete variation process as follows.

$$\delta_m^{\langle 1 \rangle, q} \sum_{k=0}^{N}{}'' G_{\mathrm{d},k}(\boldsymbol{U}^{(m)}) \Delta x$$
$$= \sum_{k=0}^{N}{}'' \left(\frac{\delta G_\mathrm{d}}{\delta(\boldsymbol{U}^{(m+l_2)}, \ldots, \boldsymbol{U}^{(m-l_1)})_k} \cdot \delta_m^{\langle 1 \rangle, q} U_k^{(m)} \right) \Delta x + \text{(boundary terms)}.$$
(1.117)

This is a generalization of the standard discrete variation, for example (1.32). Note that the difference $(U_k^{(m+1)} - U_k^{(m)})$ in (1.32), which implicitly means that the resulting scheme is second-order, is now replaced with higher-order difference operator $\delta_m^{\langle 1 \rangle, q}$. The symbol

$$\frac{\delta G_\mathrm{d}}{\delta(\boldsymbol{U}^{(m+l_2)}, \ldots, \boldsymbol{U}^{(m-l_1)})_k}$$

is a *temporally high-order discrete variational derivative*, which is also a generalization of the standard discrete variational derivative

$$\frac{\delta G_\mathrm{d}}{\delta(\boldsymbol{U}^{(m+1)}, \boldsymbol{U}^{(m)})_k}.$$

To answer the important question whether it is always possible to find a temporally high-order discrete variational derivative which satisfies the discrete variation identity (1.117) is not an easy thing. The answer is, however, fortunately *yes*; it will be discussed in Chapter 5.

If we admit that we can find such a temporally high-order discrete variational derivative, we can define a temporally high-order dissipative or conservative scheme with it. For example, for the first-order real-valued dissipative PDE (1.23), a temporally high-order dissipative scheme is given as follows.

Scheme 1.9 (Temporally high-order dissipative scheme for (1.23))
Suppose we are given initial data $\boldsymbol{U}^{(0)}, \boldsymbol{U}^{(1)}, \ldots, \boldsymbol{U}^{(l-2)}$, where l is the number of points that the temporally high-order difference operator $\delta_m^{\langle 1 \rangle, q}$ refers, and appropriate boundary conditions. Then we compute $\boldsymbol{U}^{(l-1)}, \boldsymbol{U}^{(l)}, \ldots$ by

$$\delta_m^{\langle 1 \rangle, q} U_k^{(m)} = -\frac{\delta G_\mathrm{d}}{\delta(\boldsymbol{U}^{(m+l_2)}, \ldots, \boldsymbol{U}^{(m-l_1)})_k}, \qquad k = 0, \ldots, N. \qquad (1.118)$$

PROPOSITION 1.9 Dissipation property of Scheme 1.9
Scheme 1.9 is dissipative in the sense that

$$\delta_m^{\langle 1 \rangle, q} J_{\mathrm{d}}(\boldsymbol{U}^{(m)}) \leq 0. \tag{1.119}$$

We can combine spatially and temporally high-order techniques to obtain dissipative or conservative schemes which are high-order both spatially and temporally. The detail will be given in Chapter 5.

Finally, we would like to briefly compare the two approaches of designing temporally high-order schemes: one by the composition method, and one by the method presented here. As noted in the preceding paragraph, the composition method has two drawbacks: relatively high computational cost and incapability of handling dissipative problems. These are not problems in the second approach; the computational cost will not increase practically even if the order is increased, and it can handle dissipative cases as well. On the other hand, the second approach has its own difficulty that its procedure is quite complicated and not easy to implement on computers, while the implementation of the composition method is always easy.

REMARK 1.3   The high-order technique described here was first introduced for *ordinary* differential equations in Matsuo [114]. There he considered dissipative or conservative ordinary differential equations and their high-order discretizations inheriting the dissipation or conservation property. To understand the basic idea of the temporally high-order technique, it is better to first consult this paper, since the situation is much simpler in ODE cases. □

1.4.2   Design of Linearly Implicit Schemes

Various conservative or dissipative schemes we have seen so far were basically *nonlinear*; that is, they form systems of nonlinear equations at each time step, and some iterative solver such as the Newton method is indispensable in time evolution process. This can be quite time-consuming when the problem is large. In this subsection we are going to see that by a simple trick we can avoid this difficulty; we can make *linearly implicit* schemes while keeping the desired conservation or dissipation property in some sense. This is a short summary of Chapter 6.

Before starting, we would like to clarify the difference between "linear vs. nonlinear" and "explicit vs. implicit," to make our point clear. We call a scheme *linear* if the scheme is linear with respect to the unknown variable, i.e., the approximate solution at the next time step; if not, we call it *nonlinear*. We call a scheme *implicit* if we need to solve some equation to know the value of the unknown variable; if not, we call it *explicit*. As opposed to the definition of linearity/nonlinearity, which is simple, the definition of implicitness/explicitness requires further consideration.

Implicitness can come from two different sources: (i) from nonlinearity, possibly inherited from the original PDE; and (ii) from any couplings between equations with respect to the spatial subindex k, which are often caused by spatial difference operators, such as $\delta_k^{\langle 2 \rangle}$. These different types of implicitness can occur independently. We must always make sure what type(s) of implicitness we are facing. We summarize the four possible types of numerical schemes in Table 1.1.

TABLE 1.1: Classification of numerical schemes

Name	Nonlinearity	Implicitness	Type of Imp.
Explicit	linear	explicit	—
Linearly implicit	linear	implicit	(ii)
Nonlinear (uncoupled)	nonlinear	implicit	(i)
Nonlinear (coupled)	nonlinear	implicit	(i)+(ii)

Let us cast a glance at the four cases in turn, taking the nonlinear PDE:

$$\frac{\partial u}{\partial t} = \frac{\partial^2 u}{\partial x^2} - u^3 \tag{1.120}$$

as an example.

- *Explicit* schemes are the schemes where we can directly compute the unknown variable without solving any equations. They are quite fast, but often unstable. For example, a scheme for the above PDE,

$$\frac{U_k^{(m+1)} - U_k^{(m)}}{\Delta t} = \delta_k^{\langle 2 \rangle} U_k^{(m)} - (U_k^{(m)})^3,$$

is explicit.

- *Linearly implicit* schemes are the schemes where we need to solve a system of linear equations in each time step. They can have both advantages of explicit and nonlinear schemes; they are usually cheaper than nonlinear (coupled) schemes, and often more stable than explicit schemes. For example, a scheme for the above PDE,

$$\frac{U_k^{(m+1)} - U_k^{(m)}}{\Delta t} = \delta_k^{\langle 2 \rangle} \left(\frac{U_k^{(m+1)} + U_k^{(m)}}{2} \right) - (U_k^{(m)})^3,$$

is linearly implicit. Note there is a term $\delta_k^{\langle 2 \rangle} U_k^{(m+1)}$, which makes the scheme implicit.

- *Nonlinear (uncoupled)* schemes are the schemes where we have to solve nonlinear equations in each time step, but each equation in the system can be solved *independently*. For example, a scheme for the above PDE,

$$\frac{U_k^{(m+1)} - U_k^{(m)}}{\Delta t} = \delta_k^{\langle 2 \rangle} U_k^{(m)} - (U_k^{(m+1)})^3,$$

is nonlinear but uncoupled. Observe that the spatial difference operator $\delta_k^{\langle 2 \rangle}$ which can cause the coupling of equations does not operate on $U_k^{(m+1)}$, thus the equations are not coupled with each other. These types of schemes are often as cheap as linearly implicit schemes, but rarely used because they are not much more stable than explicit schemes.

- *Nonlinear (coupled)* schemes are the schemes where we have to solve a system of nonlinear equations in each time step. Generally this type of numerical scheme is the most expensive but most stable. When we simply say "a nonlinear scheme," we usually mean this type of scheme (it is also often called as a "fully implicit scheme"). A scheme for the above PDE,

$$\frac{U_k^{(m+1)} - U_k^{(m)}}{\Delta t} = \delta_k^{\langle 2 \rangle} \left(\frac{U_k^{(m+1)} + U_k^{(m)}}{2} \right) - \left(\frac{U_k^{(m+1)} + U_k^{(m)}}{2} \right)^3, \tag{1.121}$$

is nonlinear and coupled. Note that the difference operator $\delta_k^{\langle 2 \rangle}$ causes couplings of the equations.

When we apply the standard discrete variational derivative method to a *linear* PDE, we obtain a *linearly implicit* scheme. The implicitness is caused by the spatial difference operators, inherited from the original PDE. In the case of a *nonlinear* PDE, we will always obtain *nonlinear (coupled)* scheme. The nonlinearity and the spatial couplings of equations are both inherited from the original PDE.

As we focus on dissipative or conservative schemes, being nonlinear and coupled is not necessarily bad news, because dissipative or conservative schemes are usually quite stable and thus time mesh size can be large to reduce the overall computational cost. It is also possible to design *linearly implicit* schemes for *nonlinear* problems, which are still dissipative or conservative in some sense. We can also design explicit, or nonlinear but decoupled schemes, but from the view point of stability, we mainly focus on linearly implicit ones throughout this book. In what follows, we briefly explain how linearly implicit dissipative or conservative schemes can be designed for nonlinear problems.

Let us consider the nonlinear PDE:

$$\frac{\partial u}{\partial t} = \frac{\partial^2 u}{\partial x^2} - u^3, \qquad x \in (0, L),\ t > 0, \tag{1.120}$$

which belongs to the class of dissipative PDEs (see Section 1.3.1):

$$\frac{\partial u}{\partial t} = -\frac{\delta G}{\delta u}, \qquad (1.23)$$

with

$$G(u, u_x) = \frac{(u_x)^2}{2} + \frac{u^4}{4}. \qquad (1.122)$$

Let us consider first the nonlinear dissipative scheme obtained by the procedure described in Section 1.3.1. To this end, let us define a discrete local energy by

$$G_{\mathrm{d},k}(\boldsymbol{U}^{(m)}) \stackrel{\mathrm{d}}{\equiv} \frac{1}{2}\left(\frac{(\delta_k^+ U_k^{(m)})^2 + (\delta_k^- U_k^{(m)})^2}{2}\right) + \frac{(U_k^{(m)})^4}{4}. \qquad (1.123)$$

Then we obtain the nonlinear coupled scheme:

$$\frac{U_k^{(m+1)} - U_k^{(m)}}{\Delta t} = \delta_k^{\langle 2 \rangle}\left(\frac{U_k^{(m+1)} + U_k^{(m)}}{2}\right)$$
$$- \left(\frac{(U_k^{(m+1)})^2 + (U_k^{(m)})^2}{2}\right)\left(\frac{U_k^{(m+1)} - U_k^{(m)}}{2}\right), \qquad (1.124)$$

which is dissipative in the sense that

$$\sum_{k=0}^{N} {}''G_{\mathrm{d},k}(\boldsymbol{U}^{(m+1)})\Delta x \leq \sum_{k=0}^{N} {}''G_{\mathrm{d},k}(\boldsymbol{U}^{(m)})\Delta x, \quad m = 0, 1, 2, \ldots. \qquad (1.125)$$

Now observe carefully how nonlinearity is "passed down" from the energy function $G(u, u_x)$ to the resulting PDE, and at the same time, the discrete energy function G_{d} to the scheme. In the continuous case, the equation (1.23) is defined with the variational derivative of the energy G. Because variational derivative is a kind of derivative with respect to u, the nonlinearity in the resulting PDE, u^3, is *one order lower* than the nonlinear term in the energy, $u^4/4$. In the nonlinear scheme case, the highest nonlinearity in the resulting scheme, $(U_k^{(m+1)})^3/4$, is also one order lower than the nonlinear term in the discrete energy, $(U_k^{(m)})^4/4$. This is natural since discrete variation process is just a discrete version of variation. Thus we reach an important observation:

> [Key observation]
> *The order of nonlinearity in the resulting scheme is always one order lower than in the discrete energy.*

This observation, in turn, suggests the next strategy for designing linearly implicit schemes.

[Idea]
To design a linearly implicit scheme, the nonlinearity in the discrete energy should be quadratic *at most.*

In the case of the nonlinear PDE (1.120), the quartic nonlinearity, $(U_k^{(m)})^4/4$, in the discrete local energy (1.123) should be reduced to quadratic. For the purpose, let us define a slightly different discrete local energy as follows.

$$G_{d,k}(\boldsymbol{U}^{(m+1)}, \boldsymbol{U}^{(m)}) \stackrel{d}{=} \frac{1}{4}\left(\frac{(\delta_k^+ U_k^{(m+1)})^2 + (\delta_k^- U_k^{(m+1)})^2}{2}\right)$$
$$+ \frac{1}{4}\left(\frac{(\delta_k^+ U_k^{(m)})^2 + (\delta_k^- U_k^{(m)})^2}{2}\right)$$
$$+ \frac{(U_k^{(m+1)})^2 (U_k^{(m)})^2}{4}. \qquad (1.126)$$

Note that this discrete energy is defined with two consecutive approximate solutions $U_k^{(m+1)}$ and $U_k^{(m)}$, where previously it was defined with a single solution $U_k^{(m)}$. With this trick the quartic nonlinear term is now *quadratic* with respect to the newest value $U_k^{(m+1)}$ (i.e. the term $(U_k^{(m+1)})^2 (U_k^{(m)})^2/4$). We define associated discrete global energy by

$$J_d(\boldsymbol{U}^{(m+1)}, \boldsymbol{U}^{(m)}) \stackrel{d}{=} \sum_{k=0}^{N}{}'' G_{d,k}(\boldsymbol{U}^{(m+1)}, \boldsymbol{U}^{(m)}) \Delta x, \qquad (1.127)$$

which accordingly refers two approximate solutions.

Let us consider discrete variation of $G_d(\boldsymbol{U}^{(m+1)}, \boldsymbol{U}^{(m)})$. We consider the difference

$$\sum_{k=0}^{N}{}'' \left(G_{d,k}(\boldsymbol{U}^{(m+1)}, \boldsymbol{U}^{(m)}) - G_{d,k}(\boldsymbol{U}^{(m)}, \boldsymbol{U}^{(m-1)})\right) \Delta x$$
$$= \sum_{k=0}^{N}{}'' \left\{\frac{1}{4}\left(\frac{(\delta_k^+ U_k^{(m+1)})^2 - (\delta_k^+ U_k^{(m-1)})^2}{2}\right)\right.$$
$$+ \frac{1}{4}\left(\frac{(\delta_k^- U_k^{(m+1)})^2 - (\delta_k^- U_k^{(m-1)})^2}{2}\right)$$
$$\left. + \frac{(U_k^{(m)})^2}{4}\left((U_k^{(m+1)})^2 - (U_k^{(m-1)})^2\right)\right\} \Delta x \qquad (1.128)$$

$$= \sum_{k=0}^{N}{}'' \left\{ \frac{1}{2}\delta_k^+ \left(\frac{U_k^{(m+1)} + U_k^{(m-1)}}{2} \right) \cdot \delta_k^+ \left(\frac{U_k^{(m+1)} - U_k^{(m-1)}}{2} \right) \right.$$

$$+ \frac{1}{2}\delta_k^- \left(\frac{U_k^{(m+1)} + U_k^{(m-1)}}{2} \right) \cdot \delta_k^- \left(\frac{U_k^{(m+1)} - U_k^{(m-1)}}{2} \right)$$

$$\left. + (U_k^{(m)})^2 \left(\frac{U_k^{(m+1)} + U_k^{(m-1)}}{2} \right) \left(\frac{U_k^{(m+1)} - U_k^{(m-1)}}{2} \right) \right\} \Delta x$$

$$= \sum_{k=0}^{N}{}'' \left\{ -\delta_k^{\langle 2 \rangle} \left(\frac{U_k^{(m+1)} + U_k^{(m-1)}}{2} \right) + (U_k^{(m)})^2 \left(\frac{U_k^{(m+1)} + U_k^{(m-1)}}{2} \right) \right\}$$

$$\times \left(\frac{U_k^{(m+1)} - U_k^{(m-1)}}{2} \right) \Delta x + \text{(boundary terms)}. \qquad (1.129)$$

In the last equality we used the summation-by-parts formula (1.34). Assuming some boundary condition with which the boundary terms vanish, we can summarize the above calculation as

$$\sum_{k=0}^{N}{}'' \left(G_{\mathrm{d},k}(\boldsymbol{U}^{(m+1)}, \boldsymbol{U}^{(m)}) - G_{\mathrm{d},k}(\boldsymbol{U}^{(m)}, \boldsymbol{U}^{(m-1)}) \right) \Delta x =$$

$$\sum_{k=0}^{N}{}'' \frac{\delta G_{\mathrm{d}}}{\delta (\boldsymbol{U}^{(m+1)}, \boldsymbol{U}^{(m)}, \boldsymbol{U}^{(m-1)})_k} \left(\frac{U_k^{(m+1)} - U_k^{(m-1)}}{2} \right) \Delta x, \quad (1.130)$$

where

$$\frac{\delta G_{\mathrm{d}}}{\delta (\boldsymbol{U}^{(m+1)}, \boldsymbol{U}^{(m)}, \boldsymbol{U}^{(m-1)})_k} \overset{\mathrm{d}}{=}$$

$$- \delta_k^{\langle 2 \rangle} \left(\frac{U_k^{(m+1)} + U_k^{(m-1)}}{2} \right) + (U_k^{(m)})^2 \left(\frac{U_k^{(m+1)} + U_k^{(m-1)}}{2} \right)$$

$$(1.131)$$

is a discrete variational derivative.

With the three-points discrete variational derivative (observe that it depends on three numerical solutions) we define a scheme as follows.

Scheme 1.10 (Dissipative scheme for (1.120)) *Suppose that the initial data $\boldsymbol{U}^{(0)}$ and the starting value $\boldsymbol{U}^{(1)}$ are given, and appropriate boundary* conditions are set. Then we compute numerical solutions by, for $m = 1, 2, \ldots$,

$$\frac{U_k^{(m+1)} - U_k^{(m-1)}}{2\Delta t} = -\frac{\delta G_{\mathrm{d}}}{\delta(\boldsymbol{U}^{(m+1)}, \boldsymbol{U}^{(m)}, \boldsymbol{U}^{(m-1)})_k}$$
$$= \delta_k^{\langle 2 \rangle} \left(\frac{U_k^{(m+1)} + U_k^{(m-1)}}{2} \right)$$
$$- (U_k^{(m)})^2 \left(\frac{U_k^{(m+1)} + U_k^{(m-1)}}{2} \right),$$
$$k = 0, \ldots, N. \quad (1.132)$$

Then the scheme automatically becomes dissipative as follows.

PROPOSITION 1.10 Dissipation property of Scheme 1.10
Scheme 1.10 is dissipative in the sense that

$$J_{\mathrm{d}}(\boldsymbol{U}^{(m+1)}, \boldsymbol{U}^{(m)}) \leq J_{\mathrm{d}}(\boldsymbol{U}^{(m)}, \boldsymbol{U}^{(m-1)}), \quad m = 1, 2, 3 \ldots. \quad (1.133)$$

PROOF Straightforward from the discrete variation equation (1.130). ☐

Observe that Scheme 1.10 is *linear* with respect to the unknown variable $U_k^{(m+1)}$, as expected.

The above idea can be further extended to more general cases where higher order nonlinearity involves. There, the concept of "three-points discrete variational derivative" will be further generalized to "*multiple-points* discrete variational derivative." This topic is fully discussed in Chapter 6.

REMARK 1.4 By slightly modifying the definition of discrete local energy, we can obtain various schemes that preserve the desired dissipation property in some ways. For example, if we start with a discrete local energy:

$$G_{\mathrm{d},k}(\boldsymbol{U}^{(m+1)}, \boldsymbol{U}^{(m)}) \stackrel{\mathrm{d}}{=} \frac{1}{2} \left(\frac{(\delta_k^+ U_k^{(m+1)})(\delta_k^+ U_k^{(m)}) + (\delta_k^- U_k^{(m+1)})(\delta_k^- U_k^{(m)})}{2} \right)$$
$$+ \frac{(U_k^{(m+1)})^4 + (U_k^{(m)})^4}{8}, \quad (1.134)$$

we obtain a *nonlinear uncoupled* scheme

$$\frac{U_k^{(m+1)} - U_k^{(m-1)}}{2\Delta t} = -\frac{\delta G_{\mathrm{d}}}{\delta(\boldsymbol{U}^{(m+1)}, \boldsymbol{U}^{(m)}, \boldsymbol{U}^{(m-1)})_k}$$
$$= \delta_k^{\langle 2 \rangle} U_k^{(m)} - \left(\frac{(U_k^{(m+1)})^2 + (U_k^{(m-1)})^2}{2} \right) \left(\frac{U_k^{(m+1)} + U_k^{(m-1)}}{2} \right),$$
$$(1.135)$$

which is still dissipative in the sense that

$$J_{\mathrm{d}}(\boldsymbol{U}^{(m+1)}, \boldsymbol{U}^{(m)}) \leq J_{\mathrm{d}}(\boldsymbol{U}^{(m)}, \boldsymbol{U}^{(m-1)}), \quad m = 1, 2, 3 \ldots.$$

Note that J_{d} is defined with the local energy (1.134). □

REMARK 1.5 Similarly to Remark 1.4, if we start with a discrete local energy:

$$G_{\mathrm{d},k}(\boldsymbol{U}^{(m+1)}, \boldsymbol{U}^{(m)}) \stackrel{\mathrm{d}}{=} \frac{1}{2}\left(\frac{(\delta_k^+ U_k{}^{(m+1)})(\delta_k^+ U_k{}^{(m)}) + (\delta_k^- U_k{}^{(m+1)})(\delta_k^- U_k{}^{(m)})}{2}\right)$$
$$+ \frac{(U_k{}^{(m+1)})^2 (U_k{}^{(m)})^2}{4}, \quad (1.136)$$

then we obtain an explicit scheme

$$\frac{U_k{}^{(m+1)} - U_k{}^{(m-1)}}{2\Delta t} = -\frac{\delta G_{\mathrm{d}}}{\delta(\boldsymbol{U}^{(m+1)}, \boldsymbol{U}^{(m)}, \boldsymbol{U}^{(m-1)})_k}$$
$$= \delta_k^{\langle 2 \rangle} U_k{}^{(m)} - (U_k{}^{(m)})^2 \left(\frac{U_k{}^{(m+1)} + U_k{}^{(m-1)}}{2}\right), \quad (1.137)$$

which is still dissipative in the sense that

$$J_{\mathrm{d}}(\boldsymbol{U}^{(m+1)}, \boldsymbol{U}^{(m)}) \leq J_{\mathrm{d}}(\boldsymbol{U}^{(m)}, \boldsymbol{U}^{(m-1)}), \quad m = 1, 2, 3 \ldots.$$

□

REMARK 1.6 As Remark 1.4 and Remark 1.5 show, we can design various types of dissipative or conservative schemes by considering different forms of discrete local energy. But we must note that in the linearly implicit, or explicit or nonlinear unncoupled schemes, the dissipation or conservation properties do not necessarily guarantee the stability of approximate solution. We will consider this issue in Chapter 6. □

1.4.3 Further Remarks

The last chapter of this book, Chapter 7 is devoted to further remarks on the discrete variational derivative method. There, the following topics are covered.

1.4.3.1 Solving Nonlinear Equations

As mentioned earlier, the DVDM schemes usually inherit nonlinearity from nonlinear PDEs. One way of avoiding the heavy computational cost—the linearization technique—has been introduced in Section 1.4.2.

A simpler solution is to employ modern, efficient solvers for nonlinear equations. In Section 7.1, some information on this issue is summarized.

1.4.3.2 Galerkin Version of the DVDM

So far we have basically limited ourselves to spatially one-dimensional problems, and this is also our basic focus throughout this book.

It is, however, quite straightforward to see that the method can be extended to two- or three-dimensional problems, as far as rectangular spatial domains and their uniform rectangular meshes are sufficient for required analysis. An example will be shown in Section 4.1.1, where the Cahn–Hilliard problem is also considered on a rectangular domain.

Another natural way of extending the method is to switch to the Galerkin (or more specifically, finite element) framework in place of finite difference. A brief explanation on this extension will be given in Section 7.2. Several examples including the Ginzburg–Landau equation on two-dimensional domain describing superconductivity are shown.

1.4.3.3 Extension to Non-Uniform Meshes

In this last subsection, Section 7.3, we will briefly mention another extension for two-dimensional problems: the extension to non-uniform meshes. More specifically, here we consider Voronoi meshes and special finite differences on them such that a certain "summation-by-parts" formula holds.

Quite recently (as of writing this book) many related studies have been carried out on such special finite differences; they are called (depending on the context) "compatible spatial discretization" or "mimetic discretization." Interested readers may refer to the references mentioned in Section 7.3.

Chapter 2

Target Partial Differential Equations

In this chapter we summarize the target partial differential equations (PDEs) of the discrete variational derivative method. In short, they are the variational PDEs that are defined with variational derivatives of some "energy" functions. The PDEs fall in four categories: first-order (i.e. with u_t only) real-valued PDEs, first-order complex-valued PDEs, second-order (i.e. with u_{tt}) real-valued PDEs, and finally systems of the first three types of PDEs.

2.1 Variational Derivatives

Let us commence by briefly reviewing the concept of variational derivatives. Let $u(x,t)$ be a function of $(x,t) \in [0,L] \times [0,\infty)$, and $G(u, u_x)$ be a real-valued function of u, u_x. We often call G "local energy." We then define a "global energy" by

$$J(u) = \int_0^L G(u, u_x) \mathrm{d}x. \qquad (2.1)$$

Below we consider three cases: when u is a real-valued scalar function, a complex-valued scalar function, and vector of real- and complex-valued functions.

Suppose u is a real-valued scalar function. The variation of the global energy $J(u)$ is defined with the Gâteaux derivative:

$$\begin{aligned}
\delta J(u;\eta) &= \lim_{\varepsilon \to 0} \int_0^L \frac{G(u+\varepsilon\eta, u_x+\varepsilon\eta_x) - G(u,u_x)}{\varepsilon} \mathrm{d}x \\
&= \int_0^L \left\{ \frac{\partial G}{\partial u}\eta + \frac{\partial G}{\partial u_x}\eta_x \right\} \mathrm{d}x \\
&= \int_0^L \left(\frac{\partial G}{\partial u} - \frac{\partial}{\partial x}\frac{\partial G}{\partial u_x} \right) \eta \, \mathrm{d}x + \left[\frac{\partial G}{\partial u_x}\eta \right]_0^L,
\end{aligned} \qquad (2.2)$$

where $\eta : [0,\infty) \times [0,L] \to \mathbb{R}$ is a smooth function, and $\varepsilon \in \mathbb{R}$. Then the variational derivative of $G(u, u_x)$ is defined by

$$\frac{\delta G}{\delta u} \stackrel{\mathrm{d}}{\equiv} \frac{\partial G}{\partial u} - \frac{\partial}{\partial x}\frac{\partial G}{\partial u_x}. \qquad (2.3)$$

Note that this coincides with the expression obtained by the intuitive calculation (1.12) in Chapter 1. Also note that if we take $\eta = u_t$, then $\delta J(u; u_t) = (\mathrm{d}/\mathrm{d}t)J(u)$, and the above calculation becomes

$$\delta J(u; u_t) = \frac{\mathrm{d}}{\mathrm{d}t}J(u) = \int_0^L \left(\frac{\partial G}{\partial u} - \frac{\partial}{\partial x}\frac{\partial G}{\partial u_x}\right) u_t \mathrm{d}x + \left[\frac{\partial G}{\partial u_x}u_t\right]_0^L, \qquad (2.4)$$

which has been already seen in (1.13).

When u is a complex-valued scalar function, we also need to consider the variation with respect to the complex conjugate of u.

$$\delta J(u; \eta) = \lim_{\varepsilon \to 0}\int_0^L \frac{G(u + \varepsilon\eta, u_x + \varepsilon\eta_x) - G(u, u_x)}{\varepsilon}\mathrm{d}x$$

$$= \int_0^L \left\{\frac{\partial G}{\partial u}\eta + \frac{\partial G}{\partial \overline{u}}\overline{\eta} + \frac{\partial G}{\partial u_x}\eta_x + \frac{\partial G}{\partial \overline{u_x}}\overline{\eta_x}\right\}\mathrm{d}x$$

$$= \int_0^L \left\{\left(\frac{\partial G}{\partial u} - \frac{\partial}{\partial x}\frac{\partial G}{\partial u_x}\right)\eta + \left(\frac{\partial G}{\partial \overline{u}} - \frac{\partial}{\partial x}\frac{\partial G}{\partial \overline{u_x}}\right)\overline{\eta}\right\}\mathrm{d}x$$

$$+ \left[\frac{\partial G}{\partial u_x}\eta + \frac{\partial G}{\partial \overline{u_x}}\overline{\eta}\right]_0^L, \qquad (2.5)$$

where $\eta : [0, \infty) \times [0, L] \to \mathbb{C}$ is a smooth function, and $\overline{u}$ is the complex conjugate of u. Partial derivatives with respect to complex-variables are defined formally by $\partial/\partial z = ((\partial/\partial(\mathrm{Re}z) - \mathrm{i}\partial/\partial(\mathrm{Im}z))/2$, $\partial/\partial \overline{z} = ((\partial/\partial(\mathrm{Re}z) + \mathrm{i}\partial/\partial(\mathrm{Im}z))/2$ (see, for example, [7, Section 1.2]). The "complex variational derivative of G" is then defined as follows.

$$\frac{\delta G}{\delta u} \stackrel{\mathrm{d}}{\equiv} \frac{\partial G}{\partial u} - \frac{\partial}{\partial x}\frac{\partial G}{\partial u_x}, \qquad (2.6\mathrm{a})$$

$$\frac{\delta G}{\delta \overline{u}} \stackrel{\mathrm{d}}{\equiv} \frac{\partial G}{\partial \overline{u}} - \frac{\partial}{\partial x}\frac{\partial G}{\partial \overline{u_x}}. \qquad (2.6\mathrm{b})$$

When $G(u, u_x)$ is real-valued, the variational derivatives are complex conjugates of each other, that is,

$$\overline{\frac{\delta G}{\delta u}} = \frac{\delta G}{\delta \overline{u}}. \qquad (2.7)$$

When u is a vector of real- and/or complex-valued functions, we consider variations with respect to all variables, including variations with respect to the complex conjugates of complex-valued elements. To write the calculation explicitly, let u be a $(N_\mathrm{r} + N_\mathrm{i})$-dimensional vector, the first N_r of which are real-valued and the rest, N_i, are complex-valued:

$$u = (u_1, \ldots, u_{N_\mathrm{r}}, u_{N_\mathrm{r}+1}, \ldots, u_{N_\mathrm{r}+N_\mathrm{i}})^\top. \qquad (2.8)$$

Then the variation of $J(u)$ is as follows.

$$\delta J(u;\eta) = \lim_{\varepsilon \to 0} \int_0^L \frac{G(u+\varepsilon\eta, u_x+\varepsilon\eta_x) - G(u, u_x)}{\varepsilon} \mathrm{d}x$$

$$= \sum_{i=1}^{N_\mathrm{r}} \int_0^L \left\{ \frac{\partial G}{\partial u_i}\eta_i + \frac{\partial G}{\partial u_{i,x}}\eta_{i,x} \right\} \mathrm{d}x$$

$$+ \sum_{i=N_\mathrm{r}+1}^{N_\mathrm{r}+N_\mathrm{i}} \int_0^L \left\{ \frac{\partial G}{\partial u_i}\eta_i + \frac{\partial G}{\partial \overline{u_i}}\overline{\eta_i} + \frac{\partial G}{\partial u_{i,x}}\eta_{i,x} + \frac{\partial G}{\partial \overline{u_{i,x}}}\overline{\eta_{i,x}} \right\} \mathrm{d}x$$

$$= \sum_{i=1}^{N_\mathrm{r}} \int_0^L \left\{ \left(\frac{\partial G}{\partial u_i} - \frac{\partial}{\partial x}\frac{\partial G}{\partial u_{i,x}} \right) \eta_i \right\} \mathrm{d}x$$

$$+ \sum_{N_\mathrm{r}+1}^{N_\mathrm{r}+N_\mathrm{i}} \int_0^L \left\{ \left(\frac{\partial G}{\partial u_i} - \frac{\partial}{\partial x}\frac{\partial G}{\partial u_{i,x}} \right) \eta_i + \left(\frac{\partial G}{\partial \overline{u_i}} - \frac{\partial}{\partial x}\frac{\partial G}{\partial \overline{u_{i,x}}} \right) \overline{\eta_i} \right\} \mathrm{d}x$$

$$+ \sum_{i=1}^{N_\mathrm{r}} \left[\frac{\partial G}{\partial u_{i,x}}\eta_i \right]_0^L + \sum_{i=N_\mathrm{r}+1}^{N_\mathrm{r}+N_\mathrm{i}} \left[\frac{\partial G}{\partial u_{i,x}}\eta_i + \frac{\partial G}{\partial \overline{u_{i,x}}}\overline{\eta_i} \right]_0^L. \quad (2.9)$$

In the above calculation, $u_{i,x}$ denotes $\frac{\partial}{\partial x}(u_i)$, and η is a $(N_\mathrm{r}+N_\mathrm{i})$-dimensional function whose first N_r elements are real and the remaining (N_i) elements are complex.

It is often more convenient to rewrite the calculation introducing the extended solution vector of length $N_\mathrm{ex} = N_\mathrm{r} + 2N_\mathrm{i}$:

$$u = (u_1, \ldots, u_{N_\mathrm{r}}, u_{N_\mathrm{r}+1}, u_{N_\mathrm{r}+2}, \ldots, u_{N_\mathrm{r}+2(N_\mathrm{i}-1)+1}, u_{N_\mathrm{r}+2(N_\mathrm{i}-1)+2})^\top, \quad (2.10)$$

where $u_{N_\mathrm{r}+2(i-1)+2} = \overline{u_{N_\mathrm{r}+2(i-1)+1}}$ ($i = 1, 2, \ldots, N_\mathrm{i}$). That is, we consider the $(N_\mathrm{r} + 2N_\mathrm{i})$-dimensional extended vector where the complex conjugates of complex variables are explicitly included as its elements. Employing this notation, we can greatly simplify (2.9) as

$$\delta J(u;\eta) = \int_0^L \left\{ \frac{\delta G}{\delta u} \cdot \eta \right\} \mathrm{d}x + \left[\frac{\partial G}{\partial u_x} \cdot \eta \right]_0^L, \quad (2.11)$$

where

$$\frac{\delta G}{\delta u_i} = \frac{\partial G}{\partial u_i} - \frac{\partial}{\partial x}\frac{\partial G}{\partial u_{i,x}} \quad (i = 1, \ldots, N_\mathrm{ex}), \quad (2.12\mathrm{a})$$

$$\frac{\delta G}{\delta u} = \left(\frac{\delta G}{\delta u_1}, \ldots, \frac{\delta G}{\delta u_{N_\mathrm{ex}}} \right), \quad (2.12\mathrm{b})$$

$$\frac{\partial G}{\partial u} = \left(\frac{\partial G}{\partial u_1}, \ldots, \frac{\partial G}{\partial u_{N_\mathrm{ex}}} \right), \quad (2.12\mathrm{c})$$

$$\frac{\partial G}{\partial u_x} = \left(\frac{\partial G}{\partial u_{1,x}}, \ldots, \frac{\partial G}{\partial u_{N_\mathrm{ex},x}} \right), \quad (2.12\mathrm{d})$$

and $\eta = (\eta_1, \ldots, \eta_N)$ whose first N_r elements are real, and the rest are complex numbers which satisfy $\eta_{N_\mathrm{r}+2(i-1)+2} = \overline{\eta_{N_\mathrm{r}+2(i-1)+1}}$ ($i = 1, \ldots, N_\mathrm{i}$). The dot "·" denotes the standard (real) vector inner product. In particular, when $\eta = u_t$, (2.11) reduces to

$$\delta J(u; u_t) = \frac{\mathrm{d}}{\mathrm{d}t} J(u) = \int_0^L \left(\frac{\delta G}{\delta u} \cdot u_t\right) \mathrm{d}x + \left[\frac{\partial G}{\partial u_x} \cdot u_t\right]_0^L, \qquad (2.13)$$

which is a generalization of (2.4).

REMARK 2.1 We can consider the variation of $J(u)$ in more general cases where G also includes u_{xx} or more higher order derivatives of u, by repeatedly using an integration-by-parts formula. We do not explicitly consider such cases throughout this book for brevity, but like to emphasize that it is easy to extend the whole method for the general cases.

Similarly, as noted in Section 1.4.3, we also do not provide full descriptions for spatially two- or three-dimensional cases. Some two-dimensional examples will appear in Chapter 4, and treatment by the Galerkin framework will be covered in Chapter 7. □

2.2 First-Order Real-Valued PDEs

The PDEs of the form $u_t = \cdots$ where u is a scalar real-valued function belong to this category. Here we consider two subclasses. The first subclass of "dissipative" PDEs is of the following form.

Target PDEs 1 (Real-valued, single, dissipative PDEs)

$$\frac{\partial u}{\partial t} = (-1)^{s+1} \left(\frac{\partial}{\partial x}\right)^{2s} \frac{\delta G}{\delta u}, \quad x \in (0, L),\ t > 0, \qquad (2.14)$$

where $s = 0, 1, 2, \ldots$.

The PDEs 1 are "dissipative," in the sense that they satisfy the following inequality.

$$\frac{\mathrm{d}}{\mathrm{d}t} \int_0^L G(u, u_x)\,\mathrm{d}x$$
$$= \int_0^L \frac{\delta G}{\delta u}\frac{\partial u}{\partial t}\,\mathrm{d}x + \left[\frac{\partial G}{\partial u_x}\frac{\partial u}{\partial t}\right]_0^L$$
$$= \int_0^L \frac{\delta G}{\delta u}\cdot(-1)^{s+1}\left(\frac{\partial}{\partial x}\right)^{2s}\frac{\delta G}{\delta u}\,\mathrm{d}x$$
$$= -\int_0^L \left\{\left(\frac{\partial}{\partial x}\right)^s \frac{\delta G}{\delta u}\right\}^2 \mathrm{d}x + \left[\sum_{l=1}^{s}(-1)^{s+l}F^{\langle l-1\rangle}F^{\langle 2s-l\rangle}\right]_0^L$$
$$\le 0. \tag{2.15}$$

This holds true when the boundary conditions imposed on the PDEs satisfy the following two conditions:

$$\left[\frac{\partial G}{\partial u_x}\frac{\partial u}{\partial t}\right]_0^L = 0, \quad t > 0, \tag{2.16}$$

and

$$\left[\sum_{l=1}^{s}(-1)^{s+l}F^{\langle l-1\rangle}F^{\langle 2s-l\rangle}\right]_0^L = 0, \quad t > 0, \tag{2.17}$$

where $F^{\langle l\rangle} = \left(\frac{\partial}{\partial x}\right)^l \frac{\delta G}{\delta u}$. The condition (2.16) is relatively easy to be satisfied; for example, the Dirichlet boundary condition, the zero Neumann boundary condition, and the periodic boundary condition satisfy the condition. Whether the condition (2.17) is also satisfied or not under these boundary conditions depends on the concrete form of $G(u, u_x)$. The inequality (2.15) implies that the "global energy" decreases monotonically as time evolves; this property is called "dissipation" in this book.

Below are the examples of the PDEs in this category.

1. **Linear diffusion equation**: With $s = 0$ and $G(u, u_x) = \dfrac{(u_x)^2}{2}$,

$$\frac{\partial u}{\partial t} = \frac{\partial^2 u}{\partial x^2}.$$

2. **Allen–Cahn equation** [9]: With $s = 0$ and $G(u, u_x) = \dfrac{p}{2}u^2 + \dfrac{r}{4}u^4 - \dfrac{q}{2}(u_x)^2$,

$$\frac{\partial u}{\partial t} = pu + ru^3 + q\frac{\partial^2 u}{\partial x^2} \quad (p > 0, q > 0, r < 0).$$

This equation describes a microscopic diffusion theory for the motion of curved antiphase boundary. The theory is incorporated into a model for antiphase **domain coarsening**. The function $u(x,t)$ is an order profile and not conserved.

3. **Cahn–Hilliard equation** [25]:
With $s = 1$ and $G(u, u_x) = \dfrac{p}{2}u^2 + \dfrac{r}{4}u^4 - \dfrac{q}{2}(u_x)^2$,

$$\frac{\partial u}{\partial t} = \frac{\partial^2}{\partial x^2}\left(pu + ru^3 + q\frac{\partial^2 u}{\partial x^2}\right) \qquad (p<0, q<0, r>0).$$

This equation is a model equation of the spinodal decomposition that describes a conserved **domain decomposition** phenomenon. We have already picked this equation as an example in Chapter 1. We will demonstrate it again in Chapter 4.

4. **Prominence temperature equation** [10, p.7–8]:
With $s = 1$ and $G(u, u_x) = \dfrac{2}{9}u^{\frac{9}{2}}$,

$$\frac{\partial u}{\partial t} = \frac{\partial^2}{\partial x^2}\left(u^{\frac{7}{2}}\right).$$

This equation describes the variation of kinetic temperature in the prominence by means of a heat conduction equation. Since the conductivity depends on the temperature this equation becomes nonlinear.

REMARK 2.2 The Swift–Hohenberg equation [158]:

$$\frac{\partial u}{\partial t} = 2u - cu^2 - u^3 - 2\frac{\partial^2 u}{\partial x^2} - \frac{\partial^4 u}{\partial x^4}, \tag{2.18}$$

is an example of real-valued dissipative PDE with $s = 0$ and

$$G(u, u_x, u_{xx}) = -u^2 + \frac{c}{3}u^3 + \frac{1}{4}u^4 - (u_x)^2 + \frac{1}{2}(u_{xx})^2. \tag{2.19}$$

The parameter $c \in \mathbb{R}$ is a constant. This equation describes the effects of thermal and hydrodynamic fluctuations on the **convective instability**. As declared in Remark 2.1, we basically do not consider this equation, since it includes u_{xx} in G.

Another example of this kind is the extended Fisher–Kolmogorov equation [34]:

$$\frac{\partial u}{\partial t} = -\left(pu + ru^3 + q\frac{\partial^2 u}{\partial x^2} + \gamma\frac{\partial^4 u}{\partial x^4}\right), \tag{2.20}$$

where $p < 0, q < 0, r > 0$, and $\gamma > 0$, is a special case of $s = 0$ with

$$G(u, u_x, u_{xx}) = \frac{p}{2}u^2 + \frac{r}{4}u^4 - \frac{q}{2}(u_x)^2 + \frac{\gamma}{2}(u_{xx})^2. \tag{2.21}$$

This equation is a model of a dynamical transition in the propagation of fronts into an unstable state of a bistable system in biophysics and chemical waves. See also Section 4.1.3. □

REMARK 2.3 The Keller–Segel equation [95]:

$$\frac{\partial u}{\partial t} = \frac{\partial}{\partial x}\left(u_x - u\frac{\partial}{\partial x}\left\{a - \left(\frac{\partial}{\partial x}\right)^2\right\}^{-1}u\right), \qquad (2.22)$$

does not belong to the dissipative class discussed above, but it is dissipative in the sense that

$$\frac{\mathrm{d}}{\mathrm{d}t}\int_0^L G(u)\mathrm{d}x \leq 0, \quad \text{where} \quad G(u) = u\log u - u - \frac{1}{2}u\left\{a - \left(\frac{\partial}{\partial x}\right)^2\right\}^{-1}u. \qquad (2.23)$$

Due to the inverse of the Helmholtz operator in G, it formally does not belong to the target PDEs (1). This equation describes a mathematical model of the chemotactic interaction of small beings, for instance amoebae, used for their aggregation. The aggregation is viewed as a breakdown of stability caused by intrinsic changes. This equation will be mentioned later in Section 4.7.1. □

REMARK 2.4 By replacing t with $-t$, we obtain the PDEs

$$\frac{\partial u}{\partial t} = -(-1)^{s+1}\left(\frac{\partial}{\partial x}\right)^{2s}\frac{\delta G}{\delta u}, \quad x \in (0, L),\ t > 0, \qquad (2.24)$$

whose energy is now *increasing*:

$$\frac{\mathrm{d}}{\mathrm{d}t}\int_0^L G(u, u_x)\mathrm{d}x \geq 0. \qquad (2.25)$$

An interesting example in this category is the semi-linear parabolic PDE (often referred to as "Fujita-type"):

$$\frac{\partial u}{\partial t} = \frac{\partial^2 u}{\partial x^2} + u^p, \qquad (2.26)$$

where $p \geq 2$ is an integer. This equation is a special case of (2.24) with $s = 0$ and the local energy

$$G(u, u_x) = -\frac{(u_x)^2}{2} + \frac{u^{p+1}}{p+1}. \qquad (2.27)$$

This equation has drawn much interest for decades since its solution can "blow-up," i.e., some norm of u can be infinite in finite time. See also [90, 91]. □

The second subclass consists of the "conservative" PDEs defined as follows.

Target PDEs 2 (Real-valued, single, conservative PDEs)

$$\frac{\partial u}{\partial t} = \left(\frac{\partial}{\partial x}\right)^{2s+1}\frac{\delta G}{\delta u}, \quad x \in (0,L), \ t > 0, \tag{2.28}$$

where $s = 0, 1, 2, \ldots$.

It is assumed that the boundary condition imposed on the PDEs satisfies the following two conditions

$$\left[\frac{\partial G}{\partial u_x}\frac{\partial u}{\partial t}\right]_0^L = 0, \quad t > 0, \tag{2.16}$$

and

$$\left[(-1)^s \frac{1}{2} F^{\langle s \rangle} F^{\langle s \rangle}\right]_0^L + \left[\sum_{l=1}^{s}(-1)^{l-1} F^{\langle l-1 \rangle} F^{\langle 2s+1-l \rangle}\right]_0^L = 0, \quad t > 0. \tag{2.29}$$

Note that the first condition is the same as in the dissipative case. For the definition of $F^{\langle s \rangle}$, see (2.17). Under these conditions, the following identity holds:

$$\begin{aligned}
&\frac{\mathrm{d}}{\mathrm{d}t}\int_0^L G(u, u_x)\,\mathrm{d}x \\
&= \int_0^L \frac{\delta G}{\delta u}\frac{\partial u}{\partial t}\,\mathrm{d}x + \left[\frac{\partial G}{\partial u_x}\frac{\partial u}{\partial t}\right]_0^L \\
&= \int_0^L \frac{\delta G}{\delta u}\cdot\left(\frac{\partial}{\partial x}\right)^{2s+1}\frac{\delta G}{\delta u}\,\mathrm{d}x \\
&= \left[(-1)^s\frac{1}{2}F^{\langle s \rangle}F^{\langle s \rangle}\right]_0^L + \left[\sum_{l=1}^{s}(-1)^{l-1}F^{\langle l-1 \rangle}F^{\langle 2s+1-l \rangle}\right]_0^L \\
&= 0.
\end{aligned} \tag{2.30}$$

This identity implies that the global energy is conserved. Throughout this book, this is called "conservation property."

The examples belonging to this class include the following:

1. **Linear convection equation**: With $s = 0$ and $G(u, u_x) = \dfrac{u^2}{2}$,

$$\frac{\partial u}{\partial t} = \frac{\partial u}{\partial x}.$$

2. **Korteweg–de Vries equation** [98]:
 With $s = 0$ and $G(u, u_x) = \dfrac{u^3}{6} - \dfrac{(u_x)^2}{2}$,

$$\frac{\partial u}{\partial t} = \frac{\partial}{\partial x}\left(\frac{u^2}{2} + \frac{\partial^2 u}{\partial x^2}\right).$$

This celebrated equation describes nonlinear long waves propagating on shallow water surfaces. It has interesting solutions called "solitons," which behave as if they are independent particles. This is a typical example of nonlinear integrable PDEs, whose exact solutions can be found in analytic ways.

3. **Zakharov–Kuznetsov equation** [172]:
   With $s = 0$ and $G(u, \nabla u) = -\dfrac{u^3}{6} + \dfrac{(u_x)^2}{2} + \dfrac{(u_y)^2}{2}$,
   $$\frac{\partial u}{\partial t} = \frac{\partial}{\partial x}\left(-\frac{u^2}{2} - \frac{\partial^2 u}{\partial x^2} - \frac{\partial^2 u}{\partial y^2}\right). \tag{2.31}$$

   This equation also models waves on shallow water surfaces, for example, the KdV equation, but the spatial domain is now two-dimensional. The solitary waves in this equation do not behave as complete particles after collisions, and thus they are called "quasi-solitons". (In Chapter 4, several examples are shown.)

REMARK 2.5 Recently, an interesting class of equations which is not immediately covered by the conservative class above has emerged and been intensively studied:

$$\frac{\partial u}{\partial t} - \frac{\partial^3 u}{\partial x^2 \partial t} + \kappa \frac{\partial u}{\partial x} + 3u\frac{\partial u}{\partial x} = \gamma\left(2\frac{\partial u}{\partial x}\frac{\partial^2 u}{\partial x^2} + u\frac{\partial^3 u}{\partial x^3}\right), \tag{2.32}$$

where $\kappa, \gamma \in \mathbb{R}$. The equation with $\kappa \geq 0, \gamma = 1$ was first discovered by Fuchssteiner–Fokas [62] in the context of completely integrable systems and then rediscovered by Camassa–Holm [24] with physical derivation in the context of shallow water waves. After their discovery, Dai [33] re-rediscovered the equation with the parameters ranged in $\kappa = 0, \gamma \in \mathbb{R}$, in the study of finite-length and small-amplitude waves in cylindrical compressible hyper-elastic rods. Furthermore, when $\kappa = 1, \gamma = 0$, the equation is greatly simplified, and coincides with the well-known "BBM" (Benjamin–Bona–Mahony) equation [14] or a regularized long wave equation [138].

The equation (2.32) can be written with a discrete variational derivative as follows.
$$\left(1 - \frac{\partial^2}{\partial x^2}\right)\frac{\partial u}{\partial t} = \frac{\partial}{\partial x}\left(\frac{\delta G}{\delta u}\right), \tag{2.33}$$

where
$$G(u, u_x) = -\frac{\kappa u^2 + u^3 + \gamma u (u_x)^2}{2}. \tag{2.34}$$

Since now we have the additional differential operator $(1 - \partial^2/\partial x^2)$ in front of u_t, this does not belong to any classes of target PDEs described above, and accordingly the standard procedure of the DVDM shown in Chapter 3 would not apply as is. It is still possible, however, to construct conservative schemes; in fact, there are two possible solutions. The first solution is to appropriately modify the procedure of the DVDM so that it can also apply to the equation (2.33). Since this involves a lot of further technical details, in this book we do not formally cover this in the standard DVDM procedure in Chapter 3. A concrete example in the Camassa–Holm equation case will be presented in Chapter 4. The second approach is to introduce an inverse operator $\mathcal{K} = (1 - \partial^2/\partial x^2)^{-1}$, and rewrite the equation to

$$\frac{\partial u}{\partial t} = \mathcal{K}\frac{\partial}{\partial x}\left(\frac{\delta G}{\delta u}\right), \qquad (2.35)$$

which looks much more similar to the target PDEs 2. A crucial difference is that the inverse operator is now *nonlocal*. Recall that it is well known that the inverse operator is well-defined under some conditions, in particular the periodic boundary condition (see, for example, [19, VIII, example 8]). In Chapter 7, we construct several conservative Galerkin schemes, based on the representation (2.35). □

2.3 First-Order Complex-Valued PDEs

The PDEs of the form $u_t = \cdots$ where u is a scalar complex-valued function fall in this category. Here we consider two subclasses. The first class is the "dissipative" PDEs defined as follows.

Target PDEs 3 (Complex-valued, single, dissipative PDEs)

$$\frac{\partial u}{\partial t} = -\frac{\delta G}{\delta \overline{u}}, \quad x \in (0, L), t > 0. \qquad (2.36)$$

□

In this type of PDE, the global energy decreases as time evolves:

$$\begin{aligned}
\frac{\mathrm{d}}{\mathrm{d}t}J(u) &= \int_0^L \frac{\partial}{\partial t}G(u, u_x)\mathrm{d}x \\
&= \int_0^L \left\{\frac{\partial G}{\partial u}\frac{\partial u}{\partial t} + \frac{\partial G}{\partial \overline{u}}\frac{\partial \overline{u}}{\partial t} + \frac{\partial G}{\partial u_x}\frac{\partial u_x}{\partial t} + \frac{\partial G}{\partial \overline{u_x}}\frac{\partial \overline{u_x}}{\partial t}\right\}\mathrm{d}x \\
&= \int_0^L \left\{\frac{\delta G}{\delta u}\frac{\partial u}{\partial t} + \frac{\delta G}{\delta \overline{u}}\frac{\partial \overline{u}}{\partial t}\right\}\mathrm{d}x + \left[\frac{\partial G}{\partial u_x}\frac{\partial u}{\partial t} + \frac{\partial G}{\partial \overline{u_x}}\frac{\partial \overline{u}}{\partial t}\right]_0^L \\
&= -2\int_0^L \left|\frac{\delta G}{\delta u}\right|^2 \mathrm{d}x \\
&\leq 0, \qquad\qquad\qquad\qquad\qquad\qquad\qquad\qquad\qquad\qquad (2.37)
\end{aligned}$$

if the associated boundary conditions satisfy

$$\left[\frac{\partial G}{\partial u_x}\frac{\partial u}{\partial t} + \frac{\partial G}{\partial \overline{u}_x}\frac{\partial \overline{u}}{\partial t}\right]_0^L = 0. \tag{2.38}$$

The periodic boundary condition and the Dirichlet boundary condition satisfy (2.38).

Examples in this category are:

1. **(A variant of) Ginzburg–Landau equation** [106]:
   With $G(u, u_x) = \frac{p}{2}|u_x|^2 - \frac{q}{4}|u|^4 - \frac{r}{2}|u|^2$,

$$\frac{\partial u}{\partial t} = p\frac{\partial^2 u}{\partial x^2} + q|u|^2 u + ru \qquad (p > 0, q < 0, r \in \mathbb{R}).$$

   This equation is a special one-dimensional case of the famous Ginzburg–Landau theory of **superconductivity**. Originally the equation included the effect of magnetic field, but in the above simplified equation, the effect is ignored. (See also Remark 2.6.)

2. **Newell–Whitehead equation** [134]:
   With $G(u, u_x) = -\mu\frac{|u|^2}{2} + \frac{|u|^4}{4} + \left|u_x - \frac{\mathrm{i}}{2k_c}u_{yy}\right|^2$,

$$\frac{\partial u}{\partial t} = \mu u - |u|^2 u + \left(\frac{\partial}{\partial x} - \frac{\mathrm{i}}{2k_c}\frac{\partial^2}{\partial y^2}\right)^2 u \qquad (\mu, k_c \in \mathbb{R}).$$

   This two-dimensional equation describes **Bénard convection flow** in systems close to the threshold of instability. It shows various patterns such as roll, zigzag and so on, and is an interesting model from the view point of **pattern formation**. (In Chapter 4, several examples are shown.)

The second subclass of this category is the "conservative" PDE defined as follows.

Target PDEs 4 (Complex-valued, single, conservative PDEs)

$$\mathrm{i}\frac{\partial u}{\partial t} = -\frac{\delta G}{\delta \overline{u}}, \quad x \in (0, L),\ t > 0. \tag{2.39}$$

□

This equation is called "conservative" because it conserves the global energy J, as can be easily seen as

$$\frac{\mathrm{d}}{\mathrm{d}t} J(u) = \int_0^L \left\{ \frac{\delta G}{\delta u} \frac{\partial u}{\partial t} + \frac{\delta G}{\delta \overline{u}} \frac{\partial \overline{u}}{\partial t} \right\} \mathrm{d}x + \left[\frac{\partial G}{\partial u_x} \frac{\partial u}{\partial t} + \frac{\partial G}{\partial \overline{u}_x} \frac{\partial \overline{u}}{\partial t} \right]_0^L$$

$$= \int_0^L \left(\mathrm{i} \left| \frac{\delta G}{\delta u} \right|^2 - \mathrm{i} \left| \frac{\delta G}{\delta u} \right|^2 \right) \mathrm{d}x$$

$$= 0, \qquad (2.40)$$

under the condition (2.38).

Examples in this category include:

1. **Nonlinear Schrödinger equation** (see, for example, [18, 156]):
   With $G(u, u_x) = -|u_x|^2 + \dfrac{2}{p+1}|u|^{p+1}$,

   $$\mathrm{i}\frac{\partial u}{\partial t} = -\frac{\partial^2 u}{\partial x^2} - |u|^{p-1} u \qquad (p = 2, 3, \ldots).$$

   This is one of the most fundamental **nonlinear wave** equations, which arises in various physical contexts such as **optical fibers, plasma** and **ideal fluid**. It also has been attracting mathematical interest since it has rich mathematical structures; for example, it is **completely integrable** when $p = 3$, it has **Hamiltonian structure**, and the solution can **"blow-up"** (i.e. certain norms of the solution can be indefinite in finite time).

2. **Gross–Pitaevskii equation** [79, 140]:
   With $G(u, u_x) = -|u_x|^2 - \dfrac{1}{2}(1 - |u|^2)^2$,

   $$\mathrm{i}\frac{\partial u}{\partial t} = -\frac{\partial^2 u}{\partial x^2} - (|u|^2 - 1)u. \qquad (2.41)$$

   This equation describes the evolution of the order profile of quantum systems with weak interaction between particles. This can be regarded as an extension of the above nonlinear Schrödinger equation. This equation in particular describes the **Bose–Einstein condensation** (BEC) phenomenon, and it has been shown that simulation results well agree with experimental results. (In Chapter 4, several examples are shown.)

2.4 Systems of First-Order PDEs

So far only single PDEs have been considered. In some cases, however, the concept can be generalized to systems of PDEs. An example is the Zakharov equations [72, 171],

$$i\frac{\partial E}{\partial t} + \frac{\partial^2 E}{\partial x^2} = nE, \quad \frac{\partial^2 n}{\partial t^2} - \frac{\partial^2 n}{\partial x^2} = \frac{\partial^2}{\partial x^2}|E|^2, \quad (2.42)$$

where $E(x,t)$ is a complex-valued, and $n(x,t)$ is a real-valued function. They can be written in variational form:

$$\frac{d}{dt}\begin{pmatrix} E \\ \overline{E} \\ n \\ v \end{pmatrix} = \begin{pmatrix} 0 & -i & 0 & 0 \\ i & 0 & 0 & 0 \\ 0 & 0 & 0 & -1 \\ 0 & 0 & 1 & 0 \end{pmatrix} \begin{pmatrix} \delta G/\delta E \\ \delta G/\delta \overline{E} \\ \delta G/\delta n \\ \delta G/\delta v \end{pmatrix}, \quad (2.43)$$

where $v(x,t)$ is a real-valued intermediate variable such that $v_x = n + |E|^2$, and $G(E, n, v_x)$ is an energy function defined by

$$G(E, n, v_x) = |E_x|^2 + n|E|^2 + \frac{1}{2}(n^2 + (v_x)^2). \quad (2.44)$$

By easy calculation the system is shown to be conservative (again under appropriate boundary conditions):

$$\frac{d}{dt}\int_0^L G(E, n, v_x)\,dx = 0. \quad (2.45)$$

The discrete variational derivative method can handle this kind of PDEs.

Our aim in this section is to define general classes of such systems. To this end, we employ the notation of extended solution vector introduced in Section 2.1, and suppose that a real-valued energy function $G(u, u_x)$ and accordingly the global energy $J(u) = \int_0^L G(u, u_x)\,dx$ are defined.

Then the systems of PDEs we consider here are of the following form:

$$\begin{cases} \dfrac{\partial u}{\partial t} = A\dfrac{\delta G}{\delta u}, & x \in (0, L),\, t > 0, \\ Bu = 0, & x = 0, L,\, t > 0, \\ u(0, x) = u_0(x), & x \in (0, L), \end{cases} \quad (2.46)$$

where A is an $N_{\mathrm{ex}} \times N_{\mathrm{ex}}$ matrix ($N_{\mathrm{ex}} = N_{\mathrm{r}} + 2N_{\mathrm{i}}$), whose elements are either

- constants, or
- $c\partial_x{}^s$ ($s \in \{1, 2, \ldots\}$), where $\partial_x \stackrel{d}{\equiv} \partial/\partial x$ and $c \in \mathbb{R}$.

In the second equation of (2.46), $Bu = 0$ denotes a specific boundary condition. The above systems of PDEs become conservative or dissipative, when A and B are of certain special forms. Notice that when

$$\left[\left(\frac{\partial G}{\partial u_x}\right) \cdot \frac{\partial u}{\partial t}\right]_0^L = 0, \quad (2.47)$$

the identity (2.13) reads

$$\frac{\mathrm{d}}{\mathrm{d}t} J(u) = \int_0^L \left(\frac{\delta G}{\delta u} \cdot u_t \right) \mathrm{d}x. \qquad (2.48)$$

It is easy to understand that, the system is conservative when A is skew-symmetric, and dissipative when negative-semidefinite, provided appropriate boundary conditions. In what follows, we further clarify such examples.

We first consider conservative cases.

Target PDEs 5 (Conservative systems) *The following systems of PDEs are conservative under the specified assumptions.*

Type C1 *The real-valued systems (2.46) where*

$$N_{\mathrm{ex}} = N_{\mathrm{r}} = 1, N_{\mathrm{i}} = 0, \quad A = \partial_x^{2s+1}, \quad s = 0, 1, 2, \ldots.$$

We assume that the imposed boundary condition satisfies the condition (2.47) and the condition:

$$\frac{1}{2}(-1)^s \left[\left(\partial_x^s \frac{\delta G}{\delta u} \right)^2 \right]_0^L + \sum_{j=1}^s (-1)^{j-1} \left[\partial_x^{j-1} \left(\frac{\delta G}{\delta u} \right) \partial_x^{2s+1-j} \left(\frac{\delta G}{\delta u} \right) \right]_0^L = 0.$$

This is nothing but the real-valued single conservative PDEs 2.

Type C2 *The systems (2.46) where*

$$N_{\mathrm{ex}} = 2, \quad A = \begin{pmatrix} 0 & 1 \\ -1 & 0 \end{pmatrix}.$$

The variables u_1, u_2 are either

- *both real-valued (i.e., $N_{\mathrm{r}} = 2$), or*
- *both complex-valued (i.e. $N_{\mathrm{r}} = 0, N_{\mathrm{i}} = 1$), and $u_2 = \overline{u_1}$ (i.e., a pair of complex conjugate variables). Multiply matrix A with $\mathrm{i} = \sqrt{-1}$, we find this class of systems equates to the complex-valued single PDEs 4.*

We also assume that the imposed boundary condition satisfies the condition (2.47). These systems are often referred to as "Hamiltonian PDEs."

Type C3 *The real-valued systems (2.46), where*

$$N_{\mathrm{ex}} = N_{\mathrm{r}} = 2, \quad A = D_x^{2s+1}, \quad s = 0, 1, 2, \ldots.$$

We assume that the imposed boundary condition satisfies the condition (2.47) and the condition:

$$\frac{1}{2}(-1)^s \left[\left\{ D_x{}^s \frac{\partial G}{\partial u} \right\}^\top S D_x{}^s \frac{\partial G}{\partial u} \right] +$$

$$\sum_{j=1}^{s}(-1)^{j-1}\left[\left\{D_x{}^{j-1}\left(\frac{\delta G}{\delta u}\right)\right\}^\top SD_x{}^{2s+1-j}\left(\frac{\delta G}{\delta u}\right)\right]_0^L = 0. \quad (2.49)$$

where

$$D_x \stackrel{\mathrm{d}}{\equiv} \begin{pmatrix} 0 & \partial_x \\ \partial_x & 0 \end{pmatrix}, \quad S \stackrel{\mathrm{d}}{\equiv} \begin{pmatrix} 0 & 1 \\ 1 & 0 \end{pmatrix}.$$

Type C4 *The systems (2.46) that are combinations of the above C1–C3 systems (see the examples below).*

It is easy to see that the above PDEs are in fact conservative in the sense that the identity

$$\frac{\mathrm{d}}{\mathrm{d}t} J(u) = 0 \quad (2.50)$$

holds. For Type C1 systems, the identity (2.50) has been already proved in Section 2.2. In Type C2 systems, the obvious skew-symmetry of A immediately yields the conservation property. For Type C3 systems, we make use of the following identity, which holds for any sufficiently smooth functions $v(x) = (v_1(x), v_2(x))$ and $w(x) = (w_1(x), w_2(x))$:

$$\int_0^L v(x)^\top D_x w(x) \mathrm{d}x = -\int_0^L \{D_x v(x)\}^\top w(x) \mathrm{d}x + \left[v(x)^\top S w(x)\right]_0^L. \quad (2.51)$$

By repeatedly applying this identity, we can easily prove that the identity (2.50) holds for Type C3 systems under the assumption (2.49).

We list several examples below. The Zakharov equations (2.43) are a conservative system of PDEs of Type C4; they combine two Type C2 systems (one is a real-valued Hamiltonian PDE, and the other is a complex-valued Hamiltonian PDE).

The so-called "good" Boussinesq equation (e.g. [111]):

$$\frac{\partial^2 u}{\partial t^2} = \frac{\partial^2}{\partial x^2}\left(u + u^2 - \frac{\partial^2 u}{\partial x^2}\right) \quad (2.52)$$

can be regarded as another example of the conservative systems. If we introduce an intermediate variable v such that $v_x = u_t$, and define the energy function by $G = u^2/2 + u^3/3 + (u_x)^2/2 + v^2/2$, the equation can be rewritten as

$$\frac{\mathrm{d}}{\mathrm{d}t}\begin{pmatrix} u \\ v \end{pmatrix} = \begin{pmatrix} 0 & \partial_x \\ \partial_x & 0 \end{pmatrix}\begin{pmatrix} \delta G/\delta u \\ \delta G/\delta v \end{pmatrix}.$$

Hence it can be regarded as an example of Type C3 with $s = 0$.

The Boussinesq–Schrödinger equation (see [17]):

$$\frac{\mathrm{d}}{\mathrm{d}t}\begin{pmatrix} E \\ \overline{E} \\ n \\ u \end{pmatrix} = \begin{pmatrix} 0 & -\mathrm{i} & 0 & 0 \\ \mathrm{i} & 0 & 0 & 0 \\ 0 & 0 & 0 & \partial_x \\ 0 & 0 & \partial_x & 0 \end{pmatrix} \begin{pmatrix} \delta G/\delta E \\ \delta G/\delta \overline{E} \\ \delta G/\delta n \\ \delta G/\delta u \end{pmatrix},$$

where

$$G(E, E_x, n, u, n_x) = |E_x|^2 + n|E|^2 + \frac{1}{2}n^2 + \frac{1}{3}n^3 + \frac{1}{2}(n_x)^2 + \frac{1}{2}u^2,$$

is an example of Type C4; a combination of Type C2 and Type C3 systems.

Other examples include the coupled Klein–Gordon–Schrödinger equation (see, e.g., [12]), the Boussinesq–Schrödinger equation (see [17]), and the short- and long-wave interaction equation (see, e.g., [169]).

Next we consider dissipative cases.

Target PDEs 6 (Dissipative systems) *The following systems of PDEs are dissipative under the specified conditions.*

Type D1 *The real-valued systems* (2.46), *where*

$$N_{\mathrm{ex}} = N_{\mathrm{r}} = 1, N_{\mathrm{i}} = 0, \quad A = (-1)^{s+1}\partial_x^{2s}, \quad s = 0, 1, 2, \ldots.$$

We assume that the imposed boundary condition which satisfies the condition (2.47) *and the condition:*

$$\sum_{j=1}^{s}(-1)^{s+j}\left[\partial_x^{j-1}\left(\frac{\delta G}{\delta u}\right)\partial_x^{2s-j}\left(\frac{\delta G}{\delta u}\right)\right]_0^L = 0.$$

This is just the real-valued single PDEs 1.

Type D2 *The two-variable complex-valued systems* (2.46), *where* $u_2 = \overline{u_1}$, *and*

$$N_{\mathrm{ex}} = 2, N_{\mathrm{r}} = 0, N_{\mathrm{i}} = 1, \quad A = \begin{pmatrix} 0 & -1 \\ -1 & 0 \end{pmatrix}.$$

We assume that the imposed boundary condition satisfies the condition (2.47). *This is nothing but the complex-valued dissipative PDEs 3.*

Type D3 *The systems* (2.46) *which are combinations of the above Type C1–C3 conservative systems and Type D1 and D2 dissipative systems.*

□

It is easy to see that the Type D1–D3 systems are dissipative in the sense that the inequality

$$\frac{\mathrm{d}}{\mathrm{d}t} J(u(t)) \leq 0 \tag{2.53}$$

holds.

For example, the Eguchi–Oki–Matsumura equation [45]:

$$\frac{\partial u}{\partial t} = \frac{\partial^2}{\partial x^2}\left(\frac{\delta G}{\delta u}\right), \tag{2.54a}$$

$$\frac{\partial v}{\partial t} = -\frac{\delta G}{\delta v}, \tag{2.54b}$$

where $a > 0$, $b \in \mathbb{R}$, and

$$G(u, u_x, v, v_x) = \frac{\varepsilon^2}{2}(u_x)^2 + \frac{1}{2}(v_x)^2 + \frac{a}{2}u^2 + \frac{1}{4}v^4 - \frac{b}{2}v^2 + \frac{1}{2}u^2 v^2,$$

is a dissipative system which is a combination of two Type D1 systems (one with $s = 1$, and the other with $s = 0$). This equation describes the dynamics of order and composition profiles in phase separation processes, such as spinodal decomposition, on alloy or polymer mixture materials. This can be regarded as an extension of the Cahn–Hilliard equation, and has the same difficulty mentioned in Chapter 1. That is, it includes a *negative* diffusion term (note the $u^2/2$ term in G and the equation (2.54a)). For the studies on this equation using the discrete variational derivative method, see, for example, [80].

REMARK 2.6 The full Ginzburg–Landau equations with magnetic effect (see, e.g., [40]) form a dissipative system when appropriate gauge is chosen. Since they inevitably require vector calculus, which is not easy to mimic in finite-difference regime, we do not consider the finite-difference discretization of the equations in this book. A better approach is the Galerkin framework, which will be described in Section 7.2. □

2.5 Second-Order PDEs

The PDEs which involve u_{tt} belong to this category. Let $u(x,t)$ be a real-valued function of x, t and $G(u, u_x)$ be a real-valued function of u, u_x. Then the PDEs in this category can be written as follows.

Target PDEs 7 (Second-order conservative PDEs)

$$\frac{\partial^2 u}{\partial t^2} = -\frac{\delta G}{\delta u}, \quad x \in (0, L), \ t > 0. \tag{2.55}$$

□

These PDEs have associated conservation property as follows.

$$\frac{\mathrm{d}}{\mathrm{d}t} \int_0^L \left\{ \frac{1}{2}(u_t)^2 + G(u, u_x) \right\} \mathrm{d}x = \int_0^L \left(u_{tt} + \frac{\delta G}{\delta u} \right) u_t \mathrm{d}x + \left[\frac{\partial G}{\partial u_x} \frac{\partial u}{\partial t} \right]_0^L$$
$$= 0, \qquad (2.56)$$

which holds when the imposed boundary conditions satisfy the condition

$$\left[\frac{\partial G}{\partial u_x} \frac{\partial u}{\partial t} \right]_0^L = 0. \qquad (2.57)$$

Below are the examples belonging to this class.

1. **Linear wave equation**: With $G(u, u_x) = \frac{1}{2}(u_x)^2$,

$$\frac{\partial^2 u}{\partial t^2} = \frac{\partial^2 u}{\partial x^2}.$$

2. **Fermi–Pasta–Ulam equation I** [57]:
   With $G(u, u_x) = \frac{1}{2}(u_x)^2 + \frac{\varepsilon}{6}(u_x)^3$,

$$\frac{\partial^2 u}{\partial t^2} = \frac{\partial^2 u}{\partial x^2} \left(1 + \varepsilon \frac{\partial u}{\partial x} \right).$$

   This and the following equations model vibrating strings with nonlinear connectivity effects. These equations have interesting history in terms of numerical analysis. When the discoverers, Fermi, Pasta and Ulam, first considered this equation, they expected that these systems would asymptotically become **ergodic**, i.e. almost random, due to their nonlinearity. The discoverers then carried out numerical simulations to find, to their surprise, the systems' behaviors were **quasi-periodic**. Their report was an important milestone both in nonlinear science and numerical analysis; for nonlinear science, it demonstrated how nonlinearity can produce rich, unexpected dynamics. For numerical analysis, it proved numerical simulation was (and was going to be) an indispensable tool for other areas of science.

3. **Fermi–Pasta–Ulam equation II** [57]:
   With $G(u, u_x) = \frac{1}{2}(u_x)^2 + \frac{\varepsilon}{12}(u_x)^4$,

$$\frac{\partial^2 u}{\partial t^2} = \frac{\partial^2 u}{\partial x^2} \left(1 + \varepsilon \left(\frac{\partial u}{\partial x} \right)^2 \right).$$

4. **String vibration equation** [30]:
   With $G(u, u_x) = -\sqrt{1 + (u_x)^2}$,

   $$\frac{\partial^2 u}{\partial t^2} = \frac{\partial}{\partial x}\left(\frac{u_x}{\sqrt{1+(u_x)^2}}\right).$$

   This is one of the oldest nonlinear string equations. In this model, the changes in amplitude and in tension are assumed to be relatively large. Carrier [30] derived this equation via perturbation theory.

5. **Nonlinear Klein–Gordon equation** [59, 60, 76, 96]:
   With $G(u, u_x) = \frac{1}{2}(u_x)^2 + \phi(u)$,

   $$\frac{\partial^2 u}{\partial t^2} = \frac{\partial^2 u}{\partial x^2} - \phi'(u) \qquad (\phi : \text{given function}). \tag{2.58}$$

   This celebrated equation was first introduced to describe motion of relativistic quantum fields, and is also referred to as the relativistic Schrödinger equation. In this family, the so-called "**sine-Gordon equation**," in which $\phi'(u) = \sin(u)$, is the most famous example. Recently they are used in various contexts such as condensed matter physics and nonlinear optics. (In Chapter 4, several examples are shown.)

6. **Shimoji–Kawai equation** [154]:
   With $G(u, u_x) = \frac{1}{12}(u_x)^4$,

   $$\frac{\partial^2 u}{\partial t^2} = \left(\frac{\partial u}{\partial x}\right)^2 \frac{\partial^2 u}{\partial x^2}. \tag{2.59}$$

   This equation was discovered in Shimoji–Kawai [154], where they proved that beyond some time the solutions can be multi-valued. (In Chapter 4, several examples are shown.)

7. **Ebihara equation** [44]:
   With $G(u, u_x) = \frac{1}{2}x^\alpha (u_x)^2 - \frac{x^{-\gamma}}{2p+2}u^{2p+2}$,

   $$\frac{\partial^2 u}{\partial t^2} = \frac{\partial}{\partial x}\left(x^\alpha u_x\right) - x^{-\gamma} u^{2p+1} \quad (\alpha, \gamma \geq 0, p \in \mathbb{N}, \alpha + 2\gamma < 2p+2).$$

   This equation is a mathematical model of the wave propagation on the materials in which density depends on the distance from the origin. Ebihara showed that the existence of spherically symmetric global solutions to this equation.

REMARK 2.7 The PDEs 7 can be rewritten in systems of first-order PDEs by introducing new variable $v = u_t$.

$$\frac{\partial u}{\partial t} = v = \frac{\partial \widetilde{G}}{\partial v}, \tag{2.60a}$$

$$\frac{\partial v}{\partial t} = -\frac{\partial \widetilde{G}}{\partial u}, \tag{2.60b}$$

where $\widetilde{G} = v^2/2 + G(u, u_x)$ is a modified local energy. The conservation law

$$\frac{\mathrm{d}}{\mathrm{d}t} \int_0^L \left\{ \frac{1}{2}(u_t)^2 + G(u, u_x) \right\} \mathrm{d}x = 0, \tag{2.61}$$

is rewritten accordingly as

$$\frac{\mathrm{d}}{\mathrm{d}t} \int_0^L \widetilde{G}(u, u_x, v) \mathrm{d}x = 0. \tag{2.62}$$

□

REMARK 2.8 The equation proposed in Feng [55] (see also Feng–Doi–Kawahara [56]):

$$\frac{\partial^2 u}{\partial t^2} - \gamma \frac{\partial^4 u}{\partial t^2 \partial x^2} = -\Phi'(u), \tag{2.63}$$

where $\gamma > 0$ and $\Phi(u)$ is an arbitrary function, does not belong to the conservative class above, but is conservative in the sense that

$$\frac{\mathrm{d}}{\mathrm{d}t} \int_0^L \left(\frac{1}{2}(u_t)^2 + \frac{\gamma}{2}(u_{tx})^2 + \Phi(u) \right) \mathrm{d}x = 0. \tag{2.64}$$

This models chains of particles connected by nearest-neighborhood interactions with nonlinear non-harmonic potentials. This model was introduced to investigate discrete breathers or intrinsic localized modes in one-dimensional non-harmonic lattice. Feng *et al.* [56] found some exact solutions that form spatial stationary breathers.

□

Chapter 3

Discrete Variational Derivative Method

In this chapter we give the full description of the discrete variational derivative method. We first define discrete symbols and formulas required to describe the method. Then we present concrete and rigorous procedures of the method for each of the target PDEs introduced in Chapter 2 in turn. Readers can refer to Chapter 4 as needs arise, where many concrete examples that are useful for understanding the method are given. In the end of this chapter, a brief explanation on discrete functional analysis is given, which will be used in the subsequent chapters.

3.1 Discrete Symbols and Formulas

The discrete symbols and formulas frequently used in this book are defined. Let $\{f_k\}_{k \in \mathbb{Z}}$ be a sequence and $\Delta x > 0$ be the spatial mesh size. Below are the standard operators.

[Shift operators]

$$s^{\langle 0 \rangle} \stackrel{\mathrm{d}}{\equiv} 1, \tag{3.1a}$$

$$s_k^+ f_k \stackrel{\mathrm{d}}{\equiv} f_{k+1}, \tag{3.1b}$$

$$s_k^- f_k \stackrel{\mathrm{d}}{\equiv} f_{k-1}, \tag{3.1c}$$

$$s_k^{\langle 1 \rangle} f_k \stackrel{\mathrm{d}}{\equiv} \frac{f_{k+1} + f_{k-1}}{2}. \tag{3.1d}$$

[Difference operators]

$$\delta_k^{\langle 0 \rangle} \stackrel{\mathrm{d}}{\equiv} 1, \tag{3.2a}$$

$$\delta_k^+ f_k \stackrel{\mathrm{d}}{\equiv} \frac{f_{k+1} - f_k}{\Delta x}, \tag{3.2b}$$

$$\delta_k^- f_k \stackrel{\mathrm{d}}{\equiv} \frac{f_k - f_{k-1}}{\Delta x}, \tag{3.2c}$$

$$\delta_k^{\langle 1\rangle} f_k \stackrel{\mathrm{d}}{=} \frac{f_{k+1} - f_{k-1}}{2\Delta x}, \tag{3.2d}$$

$$\delta_k^{\langle 2\rangle} f_k \stackrel{\mathrm{d}}{=} \frac{f_{k+1} - 2f_k + f_{k-1}}{(\Delta x)^2}, \tag{3.2e}$$

$$\delta_k^{\langle 2s+1\rangle} \stackrel{\mathrm{d}}{=} \delta_k^{\langle 1\rangle} \delta_k^{\langle 2s\rangle}, \quad s = 1, 2, \ldots, \tag{3.2f}$$

$$\delta_k^{\langle 2s+2\rangle} \stackrel{\mathrm{d}}{=} \delta_k^{\langle 2\rangle} \delta_k^{\langle 2s\rangle}, \quad s = 1, 2, \ldots. \tag{3.2g}$$

[Averaging operators]

$$\mu_k^{\langle 0\rangle} \stackrel{\mathrm{d}}{=} 1, \tag{3.3a}$$

$$\mu_k^+ f_k \stackrel{\mathrm{d}}{=} \frac{f_k + f_{k+1}}{2}, \tag{3.3b}$$

$$\mu_k^- f_k \stackrel{\mathrm{d}}{=} \frac{f_k + f_{k-1}}{2}, \tag{3.3c}$$

$$\mu_k^{\langle 1\rangle} f_k \stackrel{\mathrm{d}}{=} \frac{f_{k+1} + 2f_k + f_{k-1}}{4}. \tag{3.3d}$$

Regarding the difference operators, it is easy to see that the identities

$$\delta_k^+ (\delta_k^- f_k) = \delta_k^- (\delta_k^+ f_k) = \delta_k^{\langle 2\rangle} f_k, \tag{3.4}$$

$$\left(\frac{\delta_k^+ + \delta_k^-}{2}\right) f_k = \delta_k^{\langle 1\rangle} f_k \tag{3.5}$$

hold. Sometimes we use the following difference operator as well.

$$\delta_k^{\langle 2+\rangle} f_k \stackrel{\mathrm{d}}{=} \frac{f_{k+2} - f_{k+1} - f_k + f_{k-1}}{2(\Delta x)^2}. \tag{3.6}$$

They will be quite frequently used in the subsequent sections.

As a discretization of integral, the following summation rule is used:

$$\sum_{k=0}^{N}{}'' f_k \Delta x \stackrel{\mathrm{d}}{=} \left(\frac{1}{2} f_0 + \sum_{k=1}^{N} f_k + \frac{1}{2} f_N\right) \Delta x. \tag{3.7}$$

This rule is called the *trapezoidal rule*. It is also possible to adopt other summation rules; actually, in some cases other choice is advantageous. See Section 3.2.3.2 for an example.

As to the summation rule, the following identity holds.

PROPOSITION 3.1 Summation of differences

$$\sum_{k=0}^{N}{}'' \delta_k^{\langle s\rangle} f_k \Delta x = \begin{cases} \left[\delta_k^{\langle s-1\rangle} f_k\right]_0^N & \text{if } s \text{ is even,} \\ \left[\mu_k^{\langle 1\rangle} \delta_k^{\langle s-1\rangle} f_k\right]_0^N & \text{if } s \text{ is odd,} \end{cases} \tag{3.8}$$

where
$$[f_k]_0^N \stackrel{\mathrm{d}}{=} f_N - f_0. \tag{3.9}$$

PROOF This proof is based on the general summation-by-parts formula shown in the next proposition. When s is even, using (3.12a) with $f_k = 1$ or $g_k = 1$, we see

$$\sum_{k=0}^{N}{}'' \delta_k^{\langle s \rangle} f_k \Delta x$$

$$= \sum_{k=0}^{N}{}'' \left(\frac{\delta_k^+ \delta_k^- + \delta_k^- \delta_k^+}{2} \right) \delta_k^{\langle s-2 \rangle} f_k \Delta x$$

$$= \sum_{k=0}^{N}{}'' \left\{ \delta_k^+ \left(\frac{1}{2} \delta_k^- \delta_k^{\langle s-2 \rangle} f_k \right) + \delta_k^- \left(\frac{1}{2} \delta_k^+ \delta_k^{\langle s-2 \rangle} f_k \right) \right\} \Delta x$$

$$= \left[\frac{s_k^+(\frac{1}{2} \delta_k^- \delta_k^{\langle s-2 \rangle} f_k) + (\frac{1}{2} \delta_k^- \delta_k^{\langle s-2 \rangle} f_k)}{2} + \frac{s_k^-(\frac{1}{2} \delta_k^+ \delta_k^{\langle s-2 \rangle} f_k) + (\frac{1}{2} \delta_k^+ \delta_k^{\langle s-2 \rangle} f_k)}{2} \right]_0^N$$

$$= \left[\frac{(\frac{1}{2} \delta_k^+ \delta_k^{\langle s-2 \rangle} f_k) + (\frac{1}{2} \delta_k^- \delta_k^{\langle s-2 \rangle} f_k)}{2} + \frac{(\frac{1}{2} \delta_k^- \delta_k^{\langle s-2 \rangle} f_k) + (\frac{1}{2} \delta_k^+ \delta_k^{\langle s-2 \rangle} f_k)}{2} \right]_0^N$$

$$= \left[\delta_k^{\langle 1 \rangle} \delta_k^{\langle s-2 \rangle} f_k \right]_0^N. \tag{3.10}$$

Similarly, when s is odd, we see

$$\sum_{k=0}^{N}{}'' \delta_k^{\langle s \rangle} f_k \Delta x$$

$$= \sum_{k=0}^{N}{}'' \left(\frac{\delta_k^+ + \delta_k^-}{2} \right) \delta_k^{\langle s-1 \rangle} f_k \Delta x$$

$$= \sum_{k=0}^{N}{}'' \left\{ \delta_k^+ \left(\frac{1}{2} \delta_k^{\langle s-1 \rangle} f_k \right) + \delta_k^- \left(\frac{1}{2} \delta_k^{\langle s-1 \rangle} f_k \right) \right\} \Delta x$$

$$= \left[\frac{s_k^+(\frac{1}{2} \delta_k^{\langle s-1 \rangle} f_k) + (\frac{1}{2} \delta_k^{\langle s-1 \rangle} f_k)}{2} + \frac{s_k^-(\frac{1}{2} \delta_k^{\langle s-1 \rangle} f_k) + (\frac{1}{2} \delta_k^{\langle s-1 \rangle} f_k)}{2} \right]_0^N$$

$$= \left[\left(\frac{s_k^+ + 2 + s_k^-}{4} \right) \delta_k^{\langle s-1 \rangle} f_k \right]_0^N. \tag{3.11}$$

This completes the proof. □

The above identity corresponds to the continuous integration

$$\int_0^L \left(\frac{\partial}{\partial x}\right)^s f(x)\,\mathrm{d}x = \left[\left(\frac{\partial}{\partial x}\right)^{s-1} f(x)\right]_0^L.$$

Next we introduce the so-called summation-by-parts formulas, which correspond to the integration-by-parts formula. These formulas hold in various forms depending on the choice of difference operators. Below we show several examples which are frequently used in this book. Let us introduce another sequence $\{g_k\}_{k\in\mathbb{Z}}$, in addition to $\{f_k\}$. The following first-order summation-by-parts formulas hold.

PROPOSITION 3.2 First-order summation-by-parts formulas

$$\sum_{k=0}^{N}{}'' f_k \left(\delta_k^+ g_k\right) \Delta x + \sum_{k=0}^{N}{}'' \left(\delta_k^- f_k\right) g_k \Delta x = \left[\frac{f_k(s_k^+ g_k) + (s_k^- f_k)g_k}{2}\right]_0^N. \quad (3.12a)$$

$$\sum_{k=0}^{N}{}'' f_k \left(\delta_k^{\langle 1\rangle} g_k\right) \Delta x + \sum_{k=0}^{N}{}'' \left(\delta_k^{\langle 1\rangle} f_k\right) g_k \Delta x = \left[\frac{f_k(s_k^{\langle 1\rangle} g_k) + (s_k^{\langle 1\rangle} f_k)g_k}{2}\right]_0^N. \quad (3.12b)$$

PROOF For (3.12a),

$$\sum_{k=0}^{N}{}'' f_k \left(\delta_k^+ g_k\right) \Delta x + \sum_{k=0}^{N}{}'' \left(\delta_k^- f_k\right) g_k \Delta x$$

$$= \sum_{k=0}^{N}{}'' \left\{ f_k(s_k^+ g_k - g_k) + (f_k - s_k^- f_k)g_k \right\}$$

$$= \sum_{k=0}^{N}{}'' \left\{ f_k(s_k^+ g_k) - (s_k^- f_k)g_k \right\}$$

$$= \sum_{k=0}^{N}{}'' f_k(s_k^+ g_k) - \sum_{k=-1}^{N-1}{}'' f_k(s_k^+ g_k)$$

$$= \left[\frac{f_k(s_k^+ g_k) + (s_k^- f_k)g_k}{2}\right]_0^N. \quad (3.13)$$

Exchanging f_k and g_k in (3.12a), and adding it on (3.12a) itself, we obtain (3.12b). □

Based on the first-order summation-by-parts formulas, second-order summation-by-parts formulas can be obtained.

PROPOSITION 3.3 Second-order summation-by-parts formulas

$$\sum_{k=0}^{N}{}'' \left(\frac{(\delta_k^+ f_k)(\delta_k^+ g_k) + (\delta_k^- f_k)(\delta_k^- g_k)}{2} \right) \Delta x + \sum_{k=0}^{N}{}'' \left(\delta_k^{\langle 2 \rangle} f_k \right) g_k \Delta x$$
$$= \left[\frac{(\delta_k^+ f_k)(\mu_k^+ g_k) + (\delta_k^- f_k)(\mu_k^- g_k)}{2} \right]_0^N.$$
(3.14a)

In particular, when $f_k = g_k$,

$$\sum_{k=0}^{N}{}'' \left(\frac{(\delta_k^+ f_k)^2 + (\delta_k^- f_k)^2}{2} \right) \Delta x + \sum_{k=0}^{N}{}'' \left(\delta_k^{\langle 2 \rangle} f_k \right) f_k \Delta x = \left[(\delta_k^{\langle 1 \rangle} f_k)(s_k^{\langle 1 \rangle} f_k) \right]_0^N.$$
(3.14b)

PROOF Substituting $\delta_k^+ f_k$ into f_k in (3.12a), we have

$$\sum_{k=0}^{N}{}'' \left\{ (\delta_k^+ f_k)(\delta_k^+ g_k) + (\delta_k^{\langle 2 \rangle} f_k) g_k \right\} \Delta x = \left[\frac{(\delta_k^+ f_k)(s_k^+ g_k) + (\delta_k^- f_k) g_k}{2} \right]_0^N.$$
(3.15)

Similarly, reversing f_k and g_k and substituting $\delta_k^- f_k$ into f_k in (3.12a), we see

$$\sum_{k=0}^{N}{}'' \left\{ g_k (\delta_k^{\langle 2 \rangle} f_k) + (\delta_k^- g_k)(\delta_k^- f_k) \right\} \Delta x = \left[\frac{g_k (\delta_k^+ f_k) + (s_k^- g_k)(\delta_k^- f_k)}{2} \right]_0^N.$$
(3.16)

From these two equations we immediately obtain the claim. □

Although in most practical problems the first- and second-order summation-by-parts formulas are adequate, it is also possible to construct further higher-order formulas by repeatedly using (3.12b).

PROPOSITION 3.4 Higher-order summation-by-parts formula
Suppose $s \in \{1, 2, 3, \ldots\}$. When s is even,

$$\sum_{k=0}^{N}{}'' f_k \delta_k^{\langle s \rangle} f_k \Delta x = (-1)^{s/2} \sum_{k=0}^{N}{}'' F_k^{(s,s/2)} \Delta x$$
$$+ \left[-\sum_{\substack{1 \le l \le s/2 \\ l: \text{even}}} \frac{2 f_k^{\langle l-1 \rangle} f_k^{\langle s-l \rangle} + \left(\delta_k^+ f_k^{\langle l-2 \rangle} \right) \left(s_k^+ f_k^{\langle s-l \rangle} \right) + \left(\delta_k^- f_k^{\langle l-2 \rangle} \right) \left(s_k^- f_k^{\langle s-l \rangle} \right)}{4} \right.$$

$$+\sum_{\substack{1\leq l\leq s/2 \\ l:\text{odd}}} \frac{2f_k^{\langle l-1\rangle}f_k^{\langle s-l\rangle} + \left(s_k^+ f_k^{\langle l-1\rangle}\right)\left(\delta_k^+ f_k^{\langle s-l-1\rangle}\right) + \left(s_k^- f_k^{\langle l-1\rangle}\right)\left(\delta_k^- f_k^{\langle s-l-1\rangle}\right)}{4}\Bigg]_0^N.$$

(3.17)

Otherwise (when s is odd),

$$\sum_{k=0}^{N}{}'' f_k \delta_k^{\langle s\rangle} f_k \Delta x =$$

$$\Bigg[-\sum_{\substack{1\leq l\leq (s-1)/2 \\ l:\text{even}}} \frac{\left(\delta_k^+ f_k^{\langle l-2\rangle}\right)\left(\delta_k^+ f_k^{\langle s-l-1\rangle}\right) + \left(\delta_k^- f_k^{\langle l-2\rangle}\right)\left(\delta_k^- f_k^{\langle s-l-1\rangle}\right)}{2}$$

$$+\sum_{\substack{1\leq l\leq (s-1)/2 \\ l:\text{odd}}} \frac{f_k^{\langle l-1\rangle}\left(s_k^{\langle 1\rangle} f_k^{\langle s-l\rangle}\right) + \left(s_k^{\langle 1\rangle} f_k^{\langle l-1\rangle}\right) f_k^{\langle s-l\rangle}}{2}$$

$$+\frac{1}{2}(-1)^{(s-1)/2} F_k^{(s,(s-1)/2)}\Bigg]_0^N. \qquad (3.18)$$

The symbols are defined by

$$f_k^{\langle l\rangle} \stackrel{\mathrm{d}}{\equiv} \delta_k^{\langle l\rangle} f_k, \qquad (3.19)$$

$$F_k^{(l,l')} \stackrel{\mathrm{d}}{\equiv} \begin{cases} f_k^{\langle l'\rangle} s_k^{\langle l \bmod 2\rangle} f_k^{\langle l'\rangle}, & \text{if } l' \text{ is even,} \\ \dfrac{1}{2}\left\{\left(\delta_k^+ f_k^{\langle l'-1\rangle}\right)^2 + \left(\delta_k^- f_k^{\langle l'-1\rangle}\right)^2\right\}, & \text{if } l' \text{ is odd,} \end{cases} \qquad (3.20)$$

for $l, l' \in \{0, 1, 2, \ldots\}$.

PROOF This is left to Appendix B. □

3.2 Procedure for First-Order Real-Valued PDEs

In this section the complete procedure of the discrete variational derivative method for the first-order real-valued PDEs 1 and PDEs 2 in Section 2.2 is presented. First, the concept of "*discrete variational derivative*," which corresponds to the continuous variational derivative, is introduced for real-valued and scalar function u. Then dissipative and conservative schemes are defined with the discrete variational derivative analogously to the original (continuous) equation.

3.2.1 Discrete Variational Derivative: Real-Valued Case

Numerical solutions are denoted as

$$U_k^{(m)} \simeq u(k\Delta x, m\Delta t), \qquad k = 0, 1, \ldots, N, \ m = 0, 1, 2, \ldots, \tag{3.21}$$

where N is the number of spatial mesh points (i.e. $\Delta x = L/N$), and Δt is the time-mesh size. They are also written in vector as $\boldsymbol{U}^{(m)} = (U_0^{(m)}, \ldots, U_N^{(m)})^\top$. The superscript (m) is omitted where no confusion occurs.

Here an assumption is set: the energy function $G(u, u_x)$ is assumed to be of the form:

$$G(u, u_x) = \sum_{l=1}^{\widetilde{M}} f_l(u) g_l(u_x), \quad \widetilde{M} \in \mathbb{N}. \tag{3.22}$$

That is, the energy function is assumed to be a combination of $\widetilde{M}$ terms, each of which can be split into functions of u and u_x; for example, for $G(u, u_x) = u^3/6 - u_x^2/2$ (the KdV equation), we can take $f_1(u) = u^3/6$, $g_1(u_x) = 1$, $f_2(u) = 1$, $g_2(u_x) = u_x^2/2$. Practically this assumption is not restrictive at all. In fact, all the examples described in Chapter 2 meet the assumption. Although it is also possible to generalize the theory to allow more general cases, it is generally much more advantageous to stand by the above "special" cases (see Remark 3.4).

For such a $G(u, u_x)$, suppose that we can define a discrete analogue of G as follows.

$$G_{\mathrm{d},k}(\boldsymbol{U}^{(m)}) = \sum_{l=1}^{M} f_l(U_k^{(m)}) g_l^+(\delta_k^+ U_k^{(m)}) g_l^-(\delta_k^- U_k^{(m)}), \quad k = 0, \ldots, N. \tag{3.23}$$

The discrete quantity $G_{\mathrm{d},k}$ is called the *discrete energy (function)*. The subscript "d" denotes that it is a discrete quantity, and "k" the spatial index. Since generally the first derivative u_x is approximated by some combination of $\delta_k^+ U_k^{(m)}$ and $\delta_k^- U_k^{(m)}$, the term $g_l(u_x)$ in (3.22) now consists of two terms

g_l^+, g_l^-. Recall, for example, (1.26) and (1.28). Since now there is a possibility that one derivative term can be approximated by *two* finite-differences, for example,

$$u_x{}^2 \simeq \frac{(\delta_k^+ U_k)^2 + (\delta_k^- U_k)^2}{2},$$

we suppose the discrete energy function is a combination of M ($\geq \widetilde{M}$) terms. Note also that the central difference operator $\delta_k^{\langle 1 \rangle}$ is covered by the expression (3.23) as well, since it can be written as $\delta_k^{\langle 1 \rangle} = (\delta_k^+ + \delta_k^-)/2$. The discrete energy function is a real-valued scalar function of $\boldsymbol{U}^{(m)}$ which approximates $G(u, u_x)$ at $x = k\Delta x$, $t = m\Delta t$. We also write $G_\mathrm{d}(\boldsymbol{U}^{(m)})$ as a vector function[1]. In order that G_d is an appropriate approximation of $G(u, u_x)$ in the form of (3.22), g_l^+ and g_l^- should be carefully chosen, such that, for example,

$$g_l^+(\delta_k^+ U_k{}^{(m)}) g_l^-(\delta_k^- U_k{}^{(m)}) \simeq g_l(u_x|_{\substack{x=k\Delta x \\ t=m\Delta t}}). \tag{3.24}$$

Note that at boundaries $k = 0$ and N, $G_{\mathrm{d},k}(\boldsymbol{U}^{(m)})$ possibly refers to undefined values $U_{-1}^{(m)}$ and $U_{N+1}^{(m)}$ (recall that in (3.21) approximate solutions are defined only on $k = 0, \ldots, N$). We assume that these values are resolved with the known values $U_0^{(m)}, \ldots, U_N^{(m)}$, in accordance with the imposed discrete boundary condition (see, for example, Example 3.1). Based on the discrete local energy, an associated discrete global energy is defined by

$$J_\mathrm{d}(\boldsymbol{U}^{(m)}) \stackrel{\mathrm{d}}{=} \sum_{k=0}^{N} {}''G_{\mathrm{d},k}(\boldsymbol{U}^{(m)})\Delta x. \tag{3.25}$$

Notice that this corresponds to the definition of the continuous global energy (2.1). Again, the subscript "d" indicates it is a discrete quantity.

Next we consider a discrete version of the variation (or Gâteaux differentiation) process (2.2) to get a *discrete* variational derivative. For the purpose, let us consider the following difference:

$$\sum_{k=0}^{N} {}''\{G_{\mathrm{d},k}(\boldsymbol{U}) - G_{\mathrm{d},k}(\boldsymbol{V})\}\Delta x,$$

where $\boldsymbol{U}, \boldsymbol{V} \in \mathbb{R}^{N+1}$. Notice also that this directly corresponds to the elemental variation calculation (1.12) in Chapter 1:

$$\int_0^L \{G(u + \delta u, u_x + \delta u_x) - G(u, u_x)\}\mathrm{d}x.$$

---

[1] To summarize, $G_\mathrm{d} : \mathbb{R}^{N+1} \to \mathbb{R}^{N+1}$, and $G_{\mathrm{d},k} : \mathbb{R}^{N+1} \to \mathbb{R}$. Note that $G_{\mathrm{d},k}(\boldsymbol{U}^{(m)})$ is an element of the vector $G_\mathrm{d}(\boldsymbol{U}^{(m)})$, i.e., $G_{\mathrm{d},k}(\boldsymbol{U}^{(m)}) \stackrel{\mathrm{d}}{=} \{G_\mathrm{d}(\boldsymbol{U}^{(m)})\}_k$. Although the notation might be a bit confusing, we employ it for the sake of saving space.

Discrete Variational Derivative Method

In the present setting, we know that the discrete local energy is of the form (3.23), and for such an energy function, the difference can be explicitly calculated by factorization as follows.

$$\sum_{k=0}^{N}{}''\{G_{d,k}(\boldsymbol{U}) - G_{d,k}(\boldsymbol{V})\}\Delta x$$

$$= \sum_{k=0}^{N}{}'' \left[\sum_{l=1}^{M} \left(\frac{f_l(U_k) + f_l(V_k)}{2} \right) \times \right.$$

$$\left\{ \left(\frac{g_l^+(\delta_k^+ U_k) + g_l^+(\delta_k^+ V_k)}{2} \right) \left(\frac{g_l^-(\delta_k^- U_k) - g_l^-(\delta_k^- V_k)}{\delta_k^-(U_k - V_k)} \right) \delta_k^-(U_k - V_k) \right.$$

$$\left. + \left(\frac{g_l^-(\delta_k^- U_k) + g_l^-(\delta_k^- V_k)}{2} \right) \left(\frac{g_l^+(\delta_k^+ U_k) - g_l^+(\delta_k^+ V_k)}{\delta_k^+(U_k - V_k)} \right) \delta_k^+(U_k - V_k) \right\}$$

$$+ \sum_{l=1}^{M} \left(\frac{f_l(U_k) - f_l(V_k)}{U_k - V_k} \right) \left(\frac{g_l^+(\delta_k^+ U_k) g_l^-(\delta_k^- U_k) + g_l^+(\delta_k^+ V_k) g_l^-(\delta_k^- V_k)}{2} \right)$$

$$\left. \times (U_k - V_k) \right] \Delta x. \quad (3.26)$$

In the above factorization, a trivial identity:

$$ab - cd = \left(\frac{a+c}{2} \right)(b-d) + (a-c)\left(\frac{b+d}{2} \right)$$

which holds for any constants a, b, c, d, is repeatedly used.

In order to simplify this expression, let us introduce new symbols:

$$\frac{\partial G_d}{\partial (\boldsymbol{U}, \boldsymbol{V})_k} \stackrel{\mathrm{d}}{\equiv} \sum_{l=1}^{M} \left(\frac{f_l(U_k) - f_l(V_k)}{U_k - V_k} \right)$$

$$\times \left(\frac{g_l^+(\delta_k^+ U_k) g_l^-(\delta_k^- U_k) + g_l^+(\delta_k^+ V_k) g_l^-(\delta_k^- V_k)}{2} \right), \quad (3.27a)$$

$$\frac{\partial G_d}{\partial \delta^-(\boldsymbol{U}, \boldsymbol{V})_k} \stackrel{\mathrm{d}}{\equiv} \sum_{l=1}^{M} \left(\frac{f_l(U_k) + f_l(V_k)}{2} \right) \left(\frac{g_l^+(\delta_k^+ U_k) + g_l^+(\delta_k^+ V_k)}{2} \right)$$

$$\times \left(\frac{g_l^-(\delta_k^- U_k) - g_l^-(\delta_k^- V_k)}{\delta_k^-(U_k - V_k)} \right), \quad (3.27b)$$

$$\frac{\partial G_d}{\partial \delta^+(\boldsymbol{U}, \boldsymbol{V})_k} \stackrel{\mathrm{d}}{\equiv} \sum_{i=1}^{M} \left(\frac{f_l(U_k) + f_l(V_k)}{2} \right) \left(\frac{g_l^-(\delta_k^- U_k) + g_l^-(\delta_k^- V_k)}{2} \right)$$

$$\times \left(\frac{g_l^+(\delta_k^+ U_k) - g_l^+(\delta_k^+ V_k)}{\delta_k^+(U_k - V_k)} \right). \quad (3.27c)$$

The first one is a discrete approximation of $\partial G/\partial u$. The other two approximate $\partial G/\partial u_x$, based on the difference operators δ_k^+ and δ_k^-, respectively. With these symbols, the difference (3.26) can be simply written as

$$\sum_{k=0}^{N}{}'' \{G_{\mathrm{d},k}(\boldsymbol{U}) - G_{\mathrm{d},k}(\boldsymbol{V})\} \Delta x = \sum_{k=0}^{N}{}'' \left[\frac{\partial G_{\mathrm{d}}}{\partial (\boldsymbol{U},\boldsymbol{V})_k}(U_k - V_k) + \frac{\partial G_{\mathrm{d}}}{\partial \delta^+(\boldsymbol{U},\boldsymbol{V})_k} \left(\delta_k^+(U_k - V_k) \right) + \frac{\partial G_{\mathrm{d}}}{\partial \delta^-(\boldsymbol{U},\boldsymbol{V})_k} \left(\delta_k^-(U_k - V_k) \right) \right] \Delta x. \quad (3.28)$$

This corresponds to the second line of (1.12). Applying the summation-by-parts formula (3.12a), we have

$$\sum_{k=0}^{N}{}'' \{G_{\mathrm{d},k}(\boldsymbol{U}) - G_{\mathrm{d},k}(\boldsymbol{V})\} \Delta x =$$

$$\sum_{k=0}^{N}{}'' \left[\left\{ \frac{\partial G_{\mathrm{d}}}{\partial (\boldsymbol{U},\boldsymbol{V})_k} - \delta_k^- \left(\frac{\partial G_{\mathrm{d}}}{\partial \delta^+(\boldsymbol{U},\boldsymbol{V})_k} \right) - \delta_k^+ \left(\frac{\partial G_{\mathrm{d}}}{\partial \delta^-(\boldsymbol{U},\boldsymbol{V})_k} \right) \right\} (U_k - V_k) \right] \Delta x$$

$$+ \frac{1}{2} \left[\frac{\partial G_{\mathrm{d}}}{\partial \delta^+(\boldsymbol{U},\boldsymbol{V})_k}(s_k^+(U_k - V_k)) + \left\{ s_k^- \left(\frac{\partial G_{\mathrm{d}}}{\partial \delta^+(\boldsymbol{U},\boldsymbol{V})_k} \right) \right\} (U_k - V_k) \right.$$

$$+ \frac{\partial G_{\mathrm{d}}}{\partial \delta^-(\boldsymbol{U},\boldsymbol{V})_k}(s_k^-(U_k - V_k)) + \left\{ s_k^+ \left(\frac{\partial G_{\mathrm{d}}}{\partial \delta^-(\boldsymbol{U},\boldsymbol{V})_k} \right) \right\} (U_k - V_k) \left. \right]_0^N, \quad (3.29)$$

which corresponds to the third line of (1.12). If we further introduce new symbols:

$$\frac{\delta G_{\mathrm{d}}}{\delta (\boldsymbol{U},\boldsymbol{V})_k} \stackrel{\mathrm{d}}{\equiv} \frac{\partial G_{\mathrm{d}}}{\partial (\boldsymbol{U},\boldsymbol{V})_k} - \delta_k^- \left(\frac{\partial G_{\mathrm{d}}}{\partial \delta^+(\boldsymbol{U},\boldsymbol{V})_k} \right) - \delta_k^+ \left(\frac{\partial G_{\mathrm{d}}}{\partial \delta^-(\boldsymbol{U},\boldsymbol{V})_k} \right), \quad (3.30)$$

$$B_{\mathrm{r},1}(\boldsymbol{U},\boldsymbol{V}) \stackrel{\mathrm{d}}{\equiv} \frac{1}{2} \left[\frac{\partial G_{\mathrm{d}}}{\partial \delta^+(\boldsymbol{U},\boldsymbol{V})_k}(s_k^+(U_k - V_k)) \right.$$

$$+ \left\{ s_k^- \left(\frac{\partial G_{\mathrm{d}}}{\partial \delta^+(\boldsymbol{U},\boldsymbol{V})_k} \right) \right\} (U_k - V_k)$$

$$+ \frac{\partial G_{\mathrm{d}}}{\partial \delta^-(\boldsymbol{U},\boldsymbol{V})_k}(s_k^-(U_k - V_k))$$

$$+ \left\{ s_k^+ \left(\frac{\partial G_{\mathrm{d}}}{\partial \delta^-(\boldsymbol{U},\boldsymbol{V})_k} \right) \right\} (U_k - V_k) \left. \right]_0^N, \quad (3.31)$$

we finally obtain the following expression:

$$\sum_{k=0}^{N}{}'' \{G_{\mathrm{d},k}(\boldsymbol{U}) - G_{\mathrm{d},k}(\boldsymbol{V})\} \Delta x =$$
$$\sum_{k=0}^{N}{}'' \left[\left(\frac{\delta G_{\mathrm{d}}}{\delta(\boldsymbol{U},\boldsymbol{V})_k}\right)(U_k - V_k)\right]\Delta x + B_{\mathrm{r},1}(\boldsymbol{U},\boldsymbol{V}). \qquad (3.32)$$

This corresponds to the last line of (2.2). The symbol $\delta G_{\mathrm{d}}/\delta(\boldsymbol{U},\boldsymbol{V})$ approximates $\delta G/\delta u$, and hence is called the *discrete variational derivative* of G_{d}. This and the discrete variation identity (3.32) play a central role in the discrete variational derivative method. The symbol $B_{\mathrm{r},1}(\boldsymbol{U},\boldsymbol{V})$ denotes the boundary values, corresponding to the last term in (2.2). "B" is for "Boundary," "r" denotes "real-valued case," and "1" is the index for distinguishing it from other boundary values that will appear later.

REMARK 3.1 In the above calculations and definitions, we find quantities in the form $(f(a) - f(b))/(a - b)$, such as $(f_l(U_k) - f_l(V_k))/(U_k - V_k)$ in (3.26). Such quantities should be understood as $f'(a)$ when $a = b$. This remark applies to all the similar expressions in this book. □

REMARK 3.2 In more general cases where G involves $u_{xx}, u_{xxx}, \ldots$, or where the spatial dimension is more than one, the discrete variational derivative of G can be defined in a similar manner. □

REMARK 3.3 For a given discrete energy function $G_{\mathrm{d},k}$, the discrete variational derivative can be automatically calculated by (3.30), with associated definitions (3.27a), (3.27b) and (3.27c). In practical situations, however, it is often more convenient to directly consider the factorization of the difference of the given discrete energies, the LHS of (3.32), as will be shown in, for example, Example 3.1 and Remark 4.5. The above formal expressions are mainly for mathematical completeness. □

REMARK 3.4 The most crucial property demanded for "discrete variational derivative" is that it has an associated discrete variation identity such as (3.32). The discrete variational derivative shown above indeed keeps it, but we would like to mention here that it is *not* the only possibility.

Let us suppose the given PDE is already discretized in space and we have a semi-discrete scheme, as will be discussed in Appendix A, and we have a concrete form of semi-discrete variational derivative

$$\frac{\delta G_{\mathrm{d}}}{\delta(\boldsymbol{U})_k} \simeq \left.\frac{\delta G}{\delta u}\right|_{x=k\Delta x},$$

which satisfies the semi-discrete variation equality (A.10). (For the notation, see Appendix A.) Then the quantity

$$\frac{\widehat{\delta G_\mathrm{d}}}{\delta(\boldsymbol{U},\boldsymbol{V})_k} = \frac{\delta G_\mathrm{d}}{\delta(\boldsymbol{U})_k} + \frac{G_{\mathrm{d},k}(\boldsymbol{U}) - G_{\mathrm{d},k}(\boldsymbol{V}) - \sum_{k=0}^{N}{}'' \frac{\delta G_\mathrm{d}}{\delta(\boldsymbol{U})_k}(U_k - V_k)\Delta x}{\|\boldsymbol{U}-\boldsymbol{V}\|_2^2}(U_k - V_k), \quad (3.33)$$

also satisfies the full-discrete variation equality

$$\sum_{k=0}^{N}{}'' \left(G_{\mathrm{d},k}(\boldsymbol{U}) - G_{\mathrm{d},k}(\boldsymbol{V})\right)\Delta x = \sum_{k=0}^{N}{}'' \frac{\widehat{\delta G_\mathrm{d}}}{\delta(\boldsymbol{U},\boldsymbol{V})_k}(U_k - V_k)\Delta x,$$

provided that some appropriate boundary condition is imposed so that the boundary terms vanish. We can rewrite the whole procedure of the discrete variational derivative method utilizing the above definition.

The definition is in one point superior to the one defined in (3.30); if we employ this definition, we can drop the assumption (3.22), which was required to factorize the difference of energies. In fact, (3.33) only refers to $G_{\mathrm{d},k}$ itself and the semi-discrete $\delta G_\mathrm{d}/\delta \boldsymbol{U}$, and does not need any factorization. Thus it can also handle, for example, $G(u, u_x) = \sin(uu_x)$. If such energy functions are important, the above definition deserves consideration.

Usually, however, we do not adopt the definition from the following two reasons. First, as stated earlier, we barely find such PDEs in practice that unfactorizable terms like $\sin(uu_x)$ are inevitable. Second and more importantly, the definition (3.33) is always nonlinear with respect to U_k, V_k, even when the original variational derivative $\delta G/\delta u$ is linear, and at the same time, it causes global couplings of variables $U_1, \ldots, U_N$ (and $V_1, \ldots, V_N$) even in separable cases. Due to these drawbacks, we essentially do not utilize the definition. □

3.2.2 Design of Schemes

Now we are in a position to define dissipative or conservative finite difference schemes.

Let us first define a dissipative scheme for the dissipative PDEs 1 (Section 2.2, page 52) as follows.

Scheme 3.1 (Scheme for the PDEs 1) *Let $U_k^{(0)} = u(k\Delta x, 0)$ be initial data. Then, a dissipative scheme for the PDE 1 is given by, for $m = 0, 1, 2, \ldots$,*

$$\frac{U_k^{(m+1)} - U_k^{(m)}}{\Delta t} = (-1)^{s+1}\delta_k^{\langle 2s \rangle}\frac{\delta G_\mathrm{d}}{\delta(\boldsymbol{U}^{(m+1)}, \boldsymbol{U}^{(m)})_k}, \quad k = 0, \ldots, N. \quad (3.34)$$

In order to state the dissipation property of the scheme, it is inevitable to introduce the following discrete quantity:

$$B_{\mathrm{r},2}^{\langle 2s \rangle}(\boldsymbol{U}^{(m+1)}, \boldsymbol{U}^{(m)}) =$$

$$\left[-\sum_{\substack{1 \le l \le s \\ l:\text{even}}} \frac{2\varphi_k^{\langle l-1 \rangle} \varphi_k^{\langle 2s-l \rangle} + \left(s_k^+ \varphi_k^{\langle l-2 \rangle}\right)\left(s_k^+ \varphi_k^{\langle 2s-l \rangle}\right) + \left(s_k^- \varphi_k^{\langle l-2 \rangle}\right)\left(s_k^- \varphi_k^{\langle 2s-l \rangle}\right)}{4} \right.$$

$$\left. + \sum_{\substack{1 \le l \le s \\ l:\text{odd}}} \frac{2\varphi_k^{\langle l-1 \rangle} \varphi_k^{\langle 2s-l \rangle} + \left(s_k^+ \varphi_k^{\langle l-1 \rangle}\right)\left(\delta_k^+ \varphi_k^{\langle 2s-l-1 \rangle}\right) + \left(s_k^- \varphi_k^{\langle l-1 \rangle}\right)\left(\delta_k^- \varphi_k^{\langle 2s-l-1 \rangle}\right)}{4} \right]_0^N,$$

(3.35a)

where

$$\varphi_k^{\langle l \rangle} \stackrel{\mathrm{d}}{\equiv} \delta_k^{\langle l \rangle} \frac{\delta G_{\mathrm{d}}}{\delta (\boldsymbol{U}^{(m+1)}, \boldsymbol{U}^{(m)})_k}, \qquad (3.35\text{b})$$

for $l \in \{0, 1, 2, \ldots\}$. The quantity $B_{\mathrm{r},2}^{\langle 2s \rangle}(\boldsymbol{U}^{(m+1)}, \boldsymbol{U}^{(m)})$ corresponds to the boundary terms appearing in (2.17) and (2.29) (see the subsequent theorems below). The subindex "2" and superscript "$\langle 2s \rangle$" are for distinguishing it from (3.31).

THEOREM 3.1 Discrete dissipation property of Scheme 3.1

Assume that a discrete boundary condition satisfying the following two conditions is imposed on Scheme 3.1:

(i) $B_{\mathrm{r},1}(\boldsymbol{U}^{(m+1)}, \boldsymbol{U}^{(m)}) = 0 \quad (m = 0, 1, 2, \ldots)$, *and*

(ii) $B_{\mathrm{r},2}^{\langle 2s \rangle}(\boldsymbol{U}^{(m+1)}, \boldsymbol{U}^{(m)}) = 0 \quad (m = 0, 1, 2, \ldots)$ *when* $s = 1, 2, \ldots$.

Then the scheme is dissipative in the sense that the inequality

$$J_{\mathrm{d}}(\boldsymbol{U}^{(m+1)}) \le J_{\mathrm{d}}(\boldsymbol{U}^{(m)}), \quad m = 0, 1, 2, \ldots \qquad (3.36)$$

holds.

PROOF Substituting $\boldsymbol{U}^{(m+1)}$ into $\boldsymbol{U}$ and $\boldsymbol{U}^{(m)}$ into $\boldsymbol{V}$ in (3.32), we have

$$\frac{1}{\Delta t} \sum_{k=0}^{N} {}''\{G_{\mathrm{d},k}(\boldsymbol{U}^{(m+1)}) - G_{\mathrm{d},k}(\boldsymbol{U}^{(m)})\} \Delta x$$

$$= \sum_{k=0}^{N}{}''\left[\left(\frac{\delta G_{\mathrm{d}}}{\delta(\boldsymbol{U}^{(m+1)},\boldsymbol{U}^{(m)})_k}\right)\left(\frac{U_k^{(m+1)} - U_k^{(m)}}{\Delta t}\right)\right]\Delta x$$
$$+ B_{\mathrm{r},1}(\boldsymbol{U}^{(m+1)},\boldsymbol{U}^{(m)})$$
$$= \sum_{k=0}^{N}{}''\left[\left(\frac{\delta G_{\mathrm{d}}}{\delta(\boldsymbol{U}^{(m+1)},\boldsymbol{U}^{(m)})_k}\right) \cdot (-1)^{s+1}\delta_k^{\langle 2s\rangle}\frac{\delta G_{\mathrm{d}}}{\delta(\boldsymbol{U}^{(m+1)},\boldsymbol{U}^{(m)})_k}\right]\Delta x$$
$$= -\sum_{k=0}^{N}{}''\Psi_k^{(s)}\Delta x + (-1)^{s+1}B_{\mathrm{r},2}^{\langle 2s\rangle}(\boldsymbol{U}^{(m+1)},\boldsymbol{U}^{(m)})$$
$$\leq 0, \tag{3.37}$$

where

$$\Psi_k^{(s)} \stackrel{\mathrm{d}}{=} \begin{cases} \left(\varphi_k^{\langle s\rangle}\right)^2 & \text{if } s : \text{even}, \\ \dfrac{1}{2}\left\{\left(\delta_k^+\varphi_k^{\langle s-1\rangle}\right)^2 + \left(\delta_k^-\varphi_k^{\langle s-1\rangle}\right)^2\right\} & \text{if } s : \text{odd}, \end{cases} \tag{3.38}$$

In the second equality, the condition (i) is used. In the last inequality, the summation-by-parts formula in Proposition 3.4 and the condition (ii) are used.
□

We here present an easy example to help readers' understanding. More practical examples will be found in Chapter 4.

Example 3.1 Linear diffusion equation
Let us consider the linear diffusion equation. We have already seen the example in Section 1.3.1, but this time let us try a more rigorous approach explicitly clarifying the boundary term. To this end, let us set the Neumann boundary condition as follows.

$$\begin{cases} u_t = u_{xx}, & x \in (0, L), \ t > 0, \\ u_x = 0, & x = 0, L, \ t > 0. \end{cases} \tag{3.39}$$

This is an example of the dissipative PDE 1, where $s = 0$, $G(u, u_x) = (u_x)^2/2$. Let us define a discrete energy function by

$$G_{\mathrm{d},k}(\boldsymbol{U}) \stackrel{\mathrm{d}}{=} \frac{\left(\delta_k^+ U_k\right)^2 + \left(\delta_k^- U_k\right)^2}{4}, \tag{3.40}$$

which means that we take in (3.23) $M = 2$ and

$$f_1 = f_2 = 1,$$

$$g_1^+(\delta_k^+ U_k) = \frac{(\delta_k^+ U_k)^2}{4}, \quad g_1^- = 1, \quad g_2^+ = 1, \quad g_2^-(\delta_k^- U_k) = \frac{(\delta_k^- U_k)^2}{4}.$$

Putting them into (3.30) (and the associated expressions (3.27a)–(3.27c)), we obtain

$$\frac{\delta G_{\mathrm{d}}}{\delta(\boldsymbol{U},\boldsymbol{V})_k} = -\delta_k^{\langle 2 \rangle}\left(\frac{U_k + V_k}{2}\right). \tag{3.41}$$

This agrees with (1.35) if we put $U_k = U_k^{(m)}$ and $V_k = U_k^{(m+1)}$. Recall that the formal expressions (3.30) and (3.27a)–(3.27c) are derived by factorizing the difference of G_{d}'s in (3.26), and it is natural that they agree with the direct result (1.33). We here would like to stress again that, as commented in Remark 3.3, we usually do not refer to the formal expressions; they are mainly for the completeness of the theory, and in practice we always consider the factorization for each given energy function.

Now Scheme 3.1 reads: for $m = 0, 1, 2, \ldots$,

$$\frac{U_k^{(m+1)} - U_k^{(m)}}{\Delta t} = -\frac{\delta G_{\mathrm{d}}}{\delta(\boldsymbol{U}^{(m+1)}, \boldsymbol{U}^{(m)})_k} = \delta_k^{\langle 2 \rangle}\left(\frac{U_k^{(m+1)} + U_k^{(m)}}{2}\right), \tag{3.42}$$

for $k = 0, \ldots, N$, which is known as standard Crank–Nicolson scheme [147]. Let us check if it is really dissipative under the discrete Neumann boundary condition:

$$\delta_k^{\langle 1 \rangle} U_k^{(m)} = 0, \quad k = 0, N, \quad m = 0, 1, 2, \ldots. \tag{3.43}$$

According to Theorem 3.1, it is sufficient to check if $B_{\mathrm{r},1}(\boldsymbol{U}^{(m+1)}, \boldsymbol{U}^{(m)}) = 0$. By easy calculation we see

$$\begin{aligned}
&B_{\mathrm{r},1}(\boldsymbol{U}^{(m+1)}, \boldsymbol{U}^{(m)}) \\
&= \frac{1}{2}\Bigg[\frac{\delta_k^+(U_k^{(m+1)} + U_k^{(m)})}{4}(s_k^+(U_k^{(m+1)} - U_k^{(m)})) \\
&\quad + \left\{s_k^-\left(\frac{\delta_k^+(U_k^{(m+1)} + U_k^{(m)})}{4}\right)\right\}(U_k^{(m+1)} - U_k^{(m)}) \\
&\quad + \frac{\delta_k^-(U_k^{(m+1)} + U_k^{(m)})}{4}(s_k^-(U_k^{(m+1)} - U_k^{(m)})) \\
&\quad + \left\{s_k^+\left(\frac{\delta_k^-(U_k^{(m+1)} + U_k^{(m)})}{4}\right)\right\}(U_k^{(m+1)} - U_k^{(m)})\Bigg]_0^N \\
&= \frac{1}{2}\Bigg[\frac{\delta_k^+(U_k^{(m+1)} + U_k^{(m)})}{2} \cdot \mu_k^+(U_k^{(m+1)} - U_k^{(m)}) \\
&\quad + \frac{\delta_k^-(U_k^{(m+1)} + U_k^{(m)})}{2} \cdot \mu_k^-(U_k^{(m+1)} - U_k^{(m)})\Bigg]_0^N.
\end{aligned}$$

The last term vanishes due to the discrete Neumann boundary condition, which implies

$$\delta_k^+(U_k{}^{(m+1)} + U_k{}^{(m)}) = -\delta_k^-(U_k{}^{(m+1)} + U_k{}^{(m)}),$$
$$\mu_k^+(U_k{}^{(m+1)} - U_k{}^{(m)}) = \mu_k^-(U_k{}^{(m+1)} - U_k{}^{(m)})$$

at $k = 0, N$.

Note that in (3.42), undefined exterior numerical solutions U_{-1} and U_{N+1} are used. We regard that they are eliminated by the boundary condition (3.43). This is a concrete example of the notice given right after (3.24). □

Similarly, the conservative scheme for the conservative PDEs 2 is given as follows.

Scheme 3.2 (Scheme for the PDEs 2) *Let $U_k^{(0)} = u(k\Delta x, 0)$ be initial values. Then, a conservative scheme for the PDE 2 is given by, for $m = 0, 1, 2, \ldots$,*

$$\frac{U_k{}^{(m+1)} - U_k{}^{(m)}}{\Delta t} = \delta_k^{\langle 2s+1 \rangle} \frac{\delta G_d}{\delta(\boldsymbol{U}^{(m+1)}, \boldsymbol{U}^{(m)})_k}, \quad k = 0, \ldots, N. \tag{3.44}$$

In order to state the conservation property of the scheme, it is also inevitable to introduce the following discrete quantity:

$$B_{\mathrm{r},2}^{\langle 2s+1 \rangle}(\boldsymbol{U}^{(m+1)}, \boldsymbol{U}^{(m)}) =$$
$$\left[-\sum_{\substack{1 \leq l \leq s \\ l:\text{even}}} \frac{\left(\delta_k^+ \varphi_k^{\langle l-2 \rangle}\right)\left(\delta_k^+ \varphi_k^{\langle 2s-l \rangle}\right) + \left(\delta_k^- \varphi_k^{\langle l-2 \rangle}\right)\left(\delta_k^- \varphi_k^{\langle 2s-l \rangle}\right)}{2} \right.$$
$$+ \sum_{\substack{1 \leq l \leq s \\ l:\text{odd}}} \frac{\varphi_k^{\langle l-1 \rangle}\left(s_k^{\langle 1 \rangle} \varphi_k^{\langle 2s-l+1 \rangle}\right) + \left(s_k^{\langle 1 \rangle} \varphi_k^{\langle l-1 \rangle}\right) \varphi_k^{\langle 2s-l+1 \rangle}}{2}$$
$$\left. + \frac{1}{2}(-1)^s \widetilde{\Psi}_k^{(s)} \right]_0^N, \tag{3.45a}$$

where $\varphi_k^{\langle l \rangle}$ is defined in (3.35b) and

$$\widetilde{\Psi}_k^{(s)} \stackrel{\mathrm{d}}{\equiv} \begin{cases} \varphi_k^{\langle s \rangle}\left(s_k^{\langle 1 \rangle} \varphi_k^{\langle s \rangle}\right), & \text{if } s : \text{even}, \\ \dfrac{1}{2}\left\{\left(\delta_k^+ \varphi_k^{\langle s-1 \rangle}\right)^2 + \left(\delta_k^- \varphi_k^{\langle s-1 \rangle}\right)^2\right\}, & \text{if } s : \text{odd}. \end{cases} \tag{3.45b}$$

THEOREM 3.2 Discrete conservation property of Scheme 3.2

Assume that a discrete boundary condition satisfying the following two conditions is imposed on Scheme 3.2:

(i) $B_{\mathrm{r},1}(\boldsymbol{U}^{(m+1)}, \boldsymbol{U}^{(m)}) = 0 \quad (m = 0, 1, 2, \ldots)$, and

(ii) $B_{\mathrm{r},2}^{\langle 2s+1 \rangle}(\boldsymbol{U}^{(m+1)}, \boldsymbol{U}^{(m)}) = 0 \quad (m = 0, 1, 2, \ldots)$.

Then the scheme is conservative in the sense that the inequality

$$J_{\mathrm{d}}(\boldsymbol{U}^{(m)}) = J_{\mathrm{d}}(\boldsymbol{U}^{(0)}), \quad m = 1, 2, \ldots, \tag{3.46}$$

holds.

PROOF Substituting $\boldsymbol{U}^{(m+1)}$ into $\boldsymbol{U}$ and $\boldsymbol{U}^{(m)}$ into $\boldsymbol{V}$ in (3.32), we have

$$\frac{1}{\Delta t} \sum_{k=0}^{N} {}'' \left\{ G_{\mathrm{d},k}(\boldsymbol{U}^{(m+1)}) - G_{\mathrm{d},k}(\boldsymbol{U}^{(m)}) \right\} \Delta x$$

$$= \sum_{k=0}^{N} {}'' \left[\left(\frac{\delta G_{\mathrm{d}}}{\delta (\boldsymbol{U}^{(m+1)}, \boldsymbol{U}^{(m)})_k} \right) \left(\frac{U_k^{(m+1)} - U_k^{(m)}}{\Delta t} \right) \right] \Delta x$$

$$\quad + B_{\mathrm{r},1}(\boldsymbol{U}^{(m+1)}, \boldsymbol{U}^{(m)})$$

$$= \sum_{k=0}^{N} {}'' \left[\left(\frac{\delta G_{\mathrm{d}}}{\delta (\boldsymbol{U}^{(m+1)}, \boldsymbol{U}^{(m)})_k} \right) \cdot \delta_k^{\langle 2s+1 \rangle} \frac{\delta G_{\mathrm{d}}}{\delta (\boldsymbol{U}^{(m+1)}, \boldsymbol{U}^{(m)})_k} \right] \Delta x$$

$$= B_{\mathrm{r},2}^{\langle 2s+1 \rangle}(\boldsymbol{U}^{(m+1)}, \boldsymbol{U}^{(m)})$$

$$= 0. \tag{3.47}$$

In the second equality, the condition (i) is used. In the last inequality, the condition (ii) is used. □

Example 3.2 Convection equation

Let us demonstrate Scheme 3.2 with the convection equation:

$$u_t = u_x, \quad x \in (0, L), \ t > 0,$$

under the L-periodic boundary condition:

$$u(0, t) = u(L, t), \ u_x(0, t) = u_x(L, t), \quad t > 0.$$

This is an example of PDEs 2 with $s = 0$, $G(u, u_x) = u^2/2$. Let us define a discrete energy function by

$$G_{\mathrm{d},k}(\boldsymbol{U}) \stackrel{\mathrm{d}}{=} \frac{(U_k)^2}{2}, \tag{3.48}$$

which means that we take in (3.23) $M = 1$ and

$$f_1 = \frac{(U_k)^2}{2}, \ g_1^+ = g_1^- = 1.$$

Then by the formal expression (3.30) (and the associated expressions (3.27a)–(3.27c)), we obtain

$$\frac{\delta G_{\mathrm{d}}}{\delta(\boldsymbol{U},\boldsymbol{V})_k} = \frac{U_k + V_k}{2}. \tag{3.49}$$

This can be also obtained by a direct factorization,

$$\frac{(U_k)^2}{2} - \frac{(V_k)^2}{2} = \left(\frac{U_k + V_k}{2}\right)(U_k - V_k),$$

which is again far simpler than to utilize the formal expressions.

Now Scheme 3.2 reads: for $m = 0, 1, \ldots,$

$$\frac{U_k{}^{(m+1)} - U_k{}^{(m)}}{\Delta t} = \delta_k^{\langle 1 \rangle}\frac{\delta G_{\mathrm{d}}}{\delta(\boldsymbol{U}^{(m+1)},\boldsymbol{U}^{(m)})_k} = \delta_k^{\langle 1 \rangle}\left(\frac{U_k{}^{(m+1)} + U_k{}^{(m)}}{2}\right), \tag{3.50}$$

for $k = 0, \ldots, N$. Suppose that the discrete periodic boundary condition:

$$U_0^{(m)} = U_N^{(m)}, \quad U_{-1}^{(m)} = U_{N-1}^{(m)}, \quad U_1^{(m)} = U_{N+1}^{(m)}, \tag{3.51}$$

is imposed. Then it is trivial that the assumptions (i) and (ii) in Theorem 3.2 are satisfied. In fact, when $s = 0$ the expression (3.45b) reduces to

$$B_{\mathrm{r},2}^{\langle 1 \rangle}(\boldsymbol{U}^{(m+1)}, \boldsymbol{U}^{(m)}) = \left[\frac{1}{2}\frac{\delta G_{\mathrm{d}}}{\delta(\boldsymbol{U}^{(m+1)},\boldsymbol{U}^{(m)})_k} \cdot s_k^{\langle 1 \rangle}\frac{\delta G_{\mathrm{d}}}{\delta(\boldsymbol{U}^{(m+1)},\boldsymbol{U}^{(m)})_k}\right]_0^N,$$

and it obviously vanishes due to the periodicity. This boundary term can be also more directly and easily obtained by following the proof of Theorem 3.2 with $s = 0$. Then by (3.12b) we immediately see

$$\frac{1}{\Delta t}\sum_{k=0}^{N}{}''\{G_{\mathrm{d},k}(\boldsymbol{U}^{(m+1)}) - G_{\mathrm{d},k}(\boldsymbol{U}^{(m)})\}\Delta x$$

$$= \sum_{k=0}^{N}{}''\left[\left(\frac{\delta G_{\mathrm{d}}}{\delta(\boldsymbol{U}^{(m+1)},\boldsymbol{U}^{(m)})_k}\right) \cdot \delta_k^{\langle 1 \rangle}\frac{\delta G_{\mathrm{d}}}{\delta(\boldsymbol{U}^{(m+1)},\boldsymbol{U}^{(m)})_k}\right]\Delta x$$

$$= \left[\frac{1}{2}\frac{\delta G_{\mathrm{d}}}{\delta(\boldsymbol{U}^{(m+1)},\boldsymbol{U}^{(m)})_k} \cdot s_k^{\langle 1 \rangle}\frac{\delta G_{\mathrm{d}}}{\delta(\boldsymbol{U}^{(m+1)},\boldsymbol{U}^{(m)})_k}\right]_0^N. \tag{3.52}$$

As in the calculation of discrete variational derivative, in practice it is often much easier to derive boundary terms by directly following the proof with the summation-by-parts formulas.

Finally, it should be mentioned that the resulting scheme above is nothing but the standard Crank–Nicolson scheme, and it seems a trivial result. In nonlinear problems, however, resulting schemes are generally non-trivial. □

REMARK 3.5 Scheme 3.1 and 3.2 have one favorable feature that the dissipation or conservation property is kept even when the time-mesh size is changed during the time evolution process. We can easily observe this since the inequality (3.36) and the identity (3.46) involve only the numerical solutions at two consecutive time steps $m+1$ and m, and thus the dissipation or conservation property is a *local* property with respect to this single time step. By exploiting this welcome feature, we can reduce overall computational cost by utilizing some adaptive techniques to control time-mesh size while keeping the dissipation or conservation property. □

REMARK 3.6 When the PDEs 1 or PDEs 2 are nonlinear, the resulting schemes are also nonlinear, and some iterative solver such as the Newton method is inevitable. It does not necessarily imply that the schemes are too expensive (compared to standard schemes), since they generally allow larger time-mesh size Δt thanks to the dissipation or conservation property.

If one still hopes for a linear scheme, then it is also possible to consider linearly implicit versions of Scheme 3.1 and 3.2. Interested readers may refer to Chapter 5. □

3.2.3 User's Choices

In a nutshell, the procedures described above are used in the following way:

1. User defines a discrete energy G_d.

2. Then by the given procedure, the discrete variational derivative, (accordingly) the resulting scheme, and the necessary conditions on the discrete boundary conditions for dissipation/conservation are *automatically* derived.

3. Finally, user defines discrete boundary conditions so that the conditions above are satisfied (and of course consistent with the original continuous boundary conditions).

The second step is automatic and there is no freedom (except the possibility of other definitions of discrete variational derivative; see Remark 3.4). In this subsection, we would like to make further useful remarks on the freedoms in the first and third steps.

3.2.3.1 Choice of the Discrete Energy

As emphasized several times (see, for example, Remark 1.2), there are degrees of freedom in defining discrete energy function. It can be chosen arbitrarily as far as the consistency condition like (3.24) is satisfied. For example, in the linear diffusion equation case where $G(u, u_x) = u_x^2/2$, there are at least four possibilities as (1.26) suggests. Let us see how things would go differently with these choices.

[First choice]
$$G_{\mathrm{d},k} = \frac{(\delta_k^+ U_k^{(m)})^2 + (\delta_k^- U_k^{(m)})^2}{4}.$$

In this case,
$$M = 2,\ f_1 = 1,\ g_1^+ = \frac{(\delta_k^+ U_k^{(m)})^2}{4},\ g_1^- = 1,$$
$$f_2 = 1,\ g_2^+ = 1,\ g_2^- = \frac{(\delta_k^- U_k^{(m)})^2}{4},$$

and from which the discrete variational derivative is automatically calculated as
$$\frac{\delta G_\mathrm{d}}{\delta(\boldsymbol{U}^{(m+1)}, \boldsymbol{U}^{(m)})_k} = -\delta_k^{\langle 2 \rangle} \left(\frac{U_k^{(m+1)} + U_k^{(m)}}{2} \right).$$

Thus the resulting scheme is
$$\frac{U_k^{(m+1)} - U_k^{(m)}}{\Delta t} = -\frac{\delta G_\mathrm{d}}{\delta(\boldsymbol{U}^{(m+1)}, \boldsymbol{U}^{(m)})_k} = \delta_k^{\langle 2 \rangle} \left(\frac{U_k^{(m+1)} + U_k^{(m)}}{2} \right).$$

The condition (i) in Theorem 3.1 reads
$$B_{\mathrm{r},1}(\boldsymbol{U}^{(m+1)}, \boldsymbol{U}^{(m)}) = \frac{1}{2} \left[\frac{\delta_k^+ (U_k^{(m+1)} + U_k^{(m)})}{2} \cdot \mu_k^+ (U_k^{(m+1)} - U_k^{(m)}) \right.$$
$$\left. + \frac{\delta_k^- (U_k^{(m+1)} + U_k^{(m)})}{2} \cdot \mu_k^- (U_k^{(m+1)} - U_k^{(m)}) \right]_0^N$$
$$= 0. \tag{3.53}$$

This is the case in Example 3.1.

[Second choice]
$$G_{\mathrm{d},k} = \frac{(\delta_k^+ U_k^{(m)})^2}{2}.$$

In this case,
$$M = 1,\ f_1 = 1,\ g_1^+ = \frac{(\delta_k^+ U_k^{(m)})^2}{2},\ g_1^- = 1,$$

from which we obtain the *same* discrete variational derivative and accordingly the *same* scheme as in the first choice case. However, the condition (i) in

Theorem 3.1 is different:

$$
\begin{aligned}
B_{r,1}(\boldsymbol{U}^{(m+1)}, \boldsymbol{U}^{(m)}) &= \frac{1}{2}\left[\frac{\delta_k^+(U_k{}^{(m+1)} + U_k{}^{(m)})}{2}(s_k^+(U_k{}^{(m+1)} - U_k{}^{(m)}))\right.\\
&\quad \left.+ \left\{s_k^-\left(\frac{\delta_k^+(U_k{}^{(m+1)} + U_k{}^{(m)})}{2}\right)\right\}(U_k{}^{(m+1)} - U_k{}^{(m)})\right]_0^N\\
&= 0. \quad (3.54)
\end{aligned}
$$

[Third choice]
$$G_{\mathrm{d},k} = \frac{(\delta_k^- U_k{}^{(m)})^2}{2}.$$

In this case,

$$M = 1,\ f_1 = 1,\ g_1^+ = 1,\ g_1^- = \frac{(\delta_k^+ U_k{}^{(m)})^2}{2},$$

and from which we obtain the same discrete variational derivative and scheme as in the first and second choices, again. In this case, however, the condition (i) in Theorem 3.1 reads:

$$
\begin{aligned}
B_{r,1}(\boldsymbol{U}^{(m+1)}, \boldsymbol{U}^{(m)}) &= \left[\frac{\delta_k^-(U_k{}^{(m+1)} + U_k{}^{(m)})}{2}(s_k^-(U_k{}^{(m+1)} - U_k{}^{(m)}))\right.\\
&\quad \left.+ \left\{s_k^+\left(\frac{\delta_k^-(U_k{}^{(m+1)} + U_k{}^{(m)})}{2}\right)\right\}(U_k{}^{(m+1)} - U_k{}^{(m)})\right]_0^N\\
&= 0. \quad (3.55)
\end{aligned}
$$

[Fourth choice]
$$G_{\mathrm{d},k} = \frac{(\delta_k^{\langle 1\rangle} U_k{}^{(m)})^2}{2} = \frac{1}{2}\left(\frac{\delta_k^+ U_k{}^{(m)} + \delta_k^- U_k{}^{(m)}}{2}\right)^2.$$

In this case,

$$M = 3,\ f_1 = 1,\ g_1^+ = \frac{(\delta_k^+ U_k{}^{(m)})^2}{8},\ g_1^- = 1,$$

$$f_2 = 1,\ g_2^+ = \frac{\delta_k^+ U_k{}^{(m)}}{2},\ g_2^- = \frac{\delta_k^- U_k{}^{(m)}}{2},$$

$$f_3 = 1,\ g_3^+ = 1,\ g_3^- = \frac{(\delta_k^- U_k{}^{(m)})^2}{8}.$$

Then we have

$$\frac{\delta G_\mathrm{d}}{\delta(\boldsymbol{U}^{(m+1)}, \boldsymbol{U}^{(m)})_k} = -(\delta_k^{\langle 1 \rangle})^2 \left(\frac{U_k^{(m+1)} + U_k^{(m)}}{2} \right),$$

and accordingly

$$\frac{U_k^{(m+1)} - U_k^{(m)}}{\Delta t} = -\frac{\delta G_\mathrm{d}}{\delta(\boldsymbol{U}^{(m+1)}, \boldsymbol{U}^{(m)})_k} = (\delta_k^{\langle 1 \rangle})^2 \left(\frac{U_k^{(m+1)} + U_k^{(m)}}{2} \right).$$

The condition (i) in Theorem 3.1 reads

$$B_{\mathrm{r},1}(\boldsymbol{U}^{(m+1)}, \boldsymbol{U}^{(m)}) = \left[\frac{\delta_k^{\langle 1 \rangle}(U_k^{(m+1)} + U_k^{(m)})}{2} (s_k^{\langle 1 \rangle}(U_k^{(m+1)} - U_k^{(m)})) \right.$$

$$\left. + \left\{ s_k^{\langle 1 \rangle} \left(\frac{\delta_k^{\langle 1 \rangle}(U_k^{(m+1)} + U_k^{(m)})}{2} \right) \right\} (U_k^{(m+1)} - U_k^{(m)}) \right]_0^N$$

$$= 0. \tag{3.56}$$

Thus in the present problem, we see that from the four different energy functions we obtain the *two* different schemes, and the *four* different conditions for discrete dissipation property. Then a natural question arises: *which choice is the best? Or more generally, are there any general principles by which even the choice of discrete energy function can be automated?* In fact, this is one of the most frequently asked questions the authors have received regarding the method.

This is really a difficult question to answer, and here we would like to explain our attitude in the following way. Whether a choice of discrete energy function is good or not depends on several factors:

1. *How the form itself matters*:
   Sometimes the concrete form of the discrete energy function itself can have serious meaning in mathematical analysis. For example, suppose an energy function $G(u, u_x) = (u^2 + (u_x)^2)/2$, which is obviously the H^1-norm of the solution, is given (the Camassa–Holm equation [24] has this energy function; see Chapter 4). In this case the boundedness of the energy function implies by the Sobolev lemma ($\forall u \in H^1, \exists c > 0, \|u\|_\infty \leq c\|u\|_{H^1}$) the uniform boundedness of the solution. Then it is natural to expect that its discrete version would serve as discrete H^1-norm, and assure the uniform boundedness of *discrete* solutions. On this issue, interested readers may refer to Chapter 4, where in some examples mathematical analyses utilizing *discrete* functional analysis are given.

2. *Symmetry*:
   Keeping the energy function spatially symmetric is generally advantageous, since by doing so the resulting schemes always become spatially

symmetric as well, which in turn means that they are of second-order accuracy in space.

In the former easy example of *linear* diffusion equation, the resulting schemes are all spatially symmetric regardless of the discrete energy functions. In general *nonlinear* cases, however, it is not the case: for example, consider a conservative equation $u_t = (\delta G/\delta u)_x = 3(u_x)^2 u_{xx}$ with an energy function $G(u, u_x) = -(u_x)^4/4$. If we define a discrete energy by

$$G_{\mathrm{d},k} = -\frac{(\delta_k^+ U_k^{(m)})^4}{4},$$

then we would obtain a scheme:

$$\frac{U_k^{(m+1)} - U_k^{(m)}}{\Delta t} = \delta_k^{\langle 1 \rangle} \delta_k^- \left\{ \left(\frac{(\delta_k^+ U_k^{(m+1)})^2 + (\delta_k^+ U_k^{(m)})^2}{2} \right) \right.$$
$$\left. \times \left(\frac{\delta_k^+ U_k^{(m+1)} + \delta_k^+ U_k^{(m)}}{2} \right) \right\},$$

which is *not* symmetric (and thus it is only of first-order in space).

3. *Relation with discrete boundary conditions*:
   In the former example, we have seen that the choice of discrete energy function substantially affects the conditions needed for discrete dissipation property. In fact, with the four different discrete energy functions, we had the four different conditions (3.53)–(3.56). This is also the case in general problems, and from this point of view, it is considerably advantageous to choose such an energy function that *the conditions can be easily satisfied by natural choices of discrete boundary conditions*.

   For example, let us carefully investigate the four conditions (3.53)–(3.56). As demonstrated earlier, the first condition (3.53) can be satisfied by the standard discrete Neumann boundary condition (3.43). The second condition (3.54) cannot be satisfied by the same discrete Neumann boundary condition; in fact, due to an obvious identity $s_k^- \delta_k^+ = \delta_k^-$, the terms $\delta_k^+((U_k^{(m+1)} + U_k^{(m)})/2)$ and $\delta_k^-((U_k^{(m+1)} + U_k^{(m)})/2)$ can be handled similarly to the first case, but now the other terms $s_k^+(U_k^{(m+1)} - U_k^{(m)})$ and $(U_k^{(m+1)} - U_k^{(m)})$ are not equal under the Neumann boundary condition. Notice also that the condition (3.54) is not symmetric with respect to the spatial index k, which is clearly caused by the asymmetry of the discrete energy function $G_{\mathrm{d},k}$. The third condition (3.55) cannot be satisfied by the discrete Neumann boundary condition (3.43), either. The fourth condition (3.56) even refers outer stencils at $k = -2, N + 2$ because of the operators $s_k^{\langle 1 \rangle} \delta_k^{\langle 1 \rangle}$, and thus obviously cannot be fulfilled by (3.43), even though the condition itself is symmetric as opposed to the second and third conditions.

On the treatment of discrete boundary conditions, see also the next subsection.

3.2.3.2 Treatment of Boundaries

As demonstrated above, it is generally quite essential to choose appropriate discrete energy function so that conditions for discrete dissipation/conservation properties can be easily fulfilled by "natural" discrete boundary conditions. In some cases, however, it turns out to be difficult to choose appropriate (natural) set of discrete energy function and discrete boundary conditions, and it is more convenient to consider a slightly modified procedure.

For example, let us consider the linear diffusion equation (3.39) again, but this time under the Dirichlet boundary condition:

$$\begin{cases} u_t = u_{xx} & x \in (0, L), \ t > 0, \\ u(0,t) = a, \ u(L,t) = b, & t > 0, \ a, b \in \mathbb{R}. \end{cases} \quad (3.57)$$

This problem is still dissipative, since

$$\frac{\mathrm{d}}{\mathrm{d}t} \int_0^L G(u, u_x)\mathrm{d}x = -\int_0^L (u_{xx})^2 \mathrm{d}x + [u_x u_t]_0^L \leq 0. \quad (3.58)$$

Note that the boundary term vanishes since $u_t = 0$ at $x = 0$ and L. If we follow the proposed method to obtain a dissipative scheme, we reach the same condition (3.44) as in the case of the Neumann boundary condition. It seems, however, considerably difficult to find a set of discrete boundary conditions which is an approximation to the Dirichlet boundary conditions and also satisfies the condition (3.44). The terms $\mu_k^+(U_k^{(m+1)} - U_k^{(m)})$ and $\mu_k^-(U_k^{(m+1)} - U_k^{(m)})$ are the discrete versions of the term u_t, but if we assume both terms vanish by the discrete Dirichlet boundary condition, a difficulty arises that the system becomes overdetermined.

One way of circumventing this difficulty is to slightly modify the whole process by exploiting the following identity:

$$\sum_{k=0}^N (\delta_k^+ f_k)(\delta_k^+ g_k)\Delta x = -\sum_{k=1}^N (\delta_k^{\langle 2 \rangle} f_k) g_k \Delta x + (\delta_k^+ f_N)g_{N+1} - (\delta_k^+ f_{-1})g_0, \quad (3.59)$$

which is a variant of the summation-by-parts formulas (compare it with (3.14a)). Let us define the discrete energy function by

$$G_{\mathrm{d},k}(\boldsymbol{U}) \stackrel{\mathrm{d}}{=} -\frac{(\delta_k^+ U_k)^2}{2}, \quad (3.60)$$

and the discrete global energy by

$$\sum_{k=0}^N G_{\mathrm{d},k}(\boldsymbol{U})\Delta x. \quad (3.61)$$

Note that in the above expressions, the rectangle rule is used as the summation rule instead of the trapezoidal rule. Under these conditions, let us consider the discrete variation:

$$\sum_{k=0}^{N}\{G_{\mathrm{d},k}(\boldsymbol{U}) - G_{\mathrm{d},k}(\boldsymbol{V})\}\Delta x$$

$$= -\sum_{k=1}^{N}\left\{\delta_{k}^{\langle 2\rangle}\left(\frac{U_{k}+V_{k}}{2}\right)\right\}(U_{k}-V_{k})\Delta x$$

$$+ \left\{\delta_{k}^{+}\left(\frac{U_{N}+V_{N}}{2}\right)\right\}(U_{N+1}-V_{N+1}) - \left\{\delta_{k}^{+}\left(\frac{U_{-1}+V_{-1}}{2}\right)\right\}(U_{0}-V_{0}). \tag{3.62}$$

Then we obtain a scheme: for $m = 0, 1, 2, \ldots$

$$\begin{cases} \dfrac{U_{k}^{(m+1)} - U_{k}^{(m)}}{\Delta t} = \delta_{k}^{\langle 2\rangle}\left(\dfrac{U_{k}^{(m+1)} + U_{k}^{(m)}}{2}\right), & k = 1, \ldots, N-1, \\ U_{0}^{(m)} = a, \ U_{N+1}^{(m)} = b. \end{cases} \tag{3.63}$$

The scheme is dissipative in the sense that the inequality:

$$\sum_{k=0}^{N}\{G_{\mathrm{d},k}(\boldsymbol{U}^{(m+1)}) - G_{\mathrm{d},k}(\boldsymbol{U}^{(m)})\}\Delta x \leq 0 \tag{3.64}$$

holds, which can be easily verified in light of (3.62). To summarize, it is often convenient to consider *another summation rule* and *another summation-by-parts formula* so that a discrete boundary condition can be implemented in a natural way.

3.3 Procedure for First-Order Complex-Valued PDEs

In this section we consider the first-order complex-valued PDEs 3 and PDEs 4, described in Section 2.3. The main tool is the concept of the "complex discrete variational derivative," i.e., a rigorous discretization of the complex variational derivative.

3.3.1 Discrete Variational Derivative: Complex-Valued Case

As in the real-valued case, suppose that $G(u, u_x)$ is in the form:

$$G(u, u_x) = \sum_{l=1}^{\widetilde{M}} \widetilde{c}_l |p_l(u)|^{N_l^P} |q_l(u_x)|^{N_l^Q}, \tag{3.65}$$

94 *Discrete Variational Derivative Method*

where $\widetilde{M} \in \mathbb{N}$, $N_l^P, N_l^Q \in \{2,3,4,\ldots\}$, $\widetilde{c}_l \in \mathbb{R}$, and p_l, q_l are assumed to be complex-valued functions that satisfy $p_l(\overline{u}) = \overline{p_l(u)}, q_l(\overline{u}) = \overline{q_l(u)}$. Though it could be more general, we do this to keep discussion simple and to give the explicit forms of the complex discrete variational derivatives.

Suppose that we can define a discrete local energy by

$$G_{\mathrm{d},k}(\boldsymbol{U}^{(m)}) = \sum_{l=1}^{M} c_l |p_l(U_k^{(m)})|^{N_l^P} |q_l^+(\delta_k^+ U_k^{(m)})|^{N_l^+} |q_l^-(\delta_k^- U_k^{(m)})|^{N_l^-}, \tag{3.66}$$

where $N_l^Q = N_l^+ + N_l^-$, $c_l \in \mathbb{R}$, and q_l^+, q_l^- are functions approximating q_l. They must be appropriate approximations, for example,

$$q_l^+(\delta_k^+ U_k^{(m)}) q_l^-(\delta_k^- U_k^{(m)}) \simeq q_l(u_x|_{\substack{x=k\Delta x \\ t=m\Delta t}}). \tag{3.67}$$

Hereafter we abbreviate $|p_l(U_k^{(m)})|^{N_l^P}$ as $P_l(U_k^{(m)})$, $|q_l^+(\delta_k^+ U_k^{(m)})|^{N_l^+}$ as $Q_l^+(U_k^{(m)})$, and $|q_l^-(\delta_k^- U_k^{(m)})|^{N_l^-}$ as $Q_l^-(U_k^{(m)})$. We define an associated global energy by

$$J_{\mathrm{d}}(\boldsymbol{U}^{(m)}) \stackrel{\mathrm{d}}{\equiv} \sum_{k=0}^{N} {}'' G_{\mathrm{d},k}(\boldsymbol{U}^{(m)}) \Delta x. \tag{3.68}$$

To follow the continuous variation calculation, let us consider the difference of the discrete energies at the different points $\boldsymbol{U}$ and $\boldsymbol{V}$, as in the real-valued case.

$$\sum_{k=0}^{N} {}'' \{G_{\mathrm{d},k}(\boldsymbol{U}) - G_{\mathrm{d},k}(\boldsymbol{V})\} \Delta x$$

$$= \sum_{k=0}^{N} {}'' \left\{ \frac{\partial G_{\mathrm{d}}}{\partial (\boldsymbol{U},\boldsymbol{V})_k}(U_k - V_k) + \frac{\partial G_{\mathrm{d}}}{\partial (\overline{\boldsymbol{U}},\overline{\boldsymbol{V}})_k}(\overline{U_k} - \overline{V_k}) + \right.$$

$$+ \frac{\partial G_{\mathrm{d}}}{\partial \delta^+(\boldsymbol{U},\boldsymbol{V})_k}(\delta_k^+ U_k - \delta_k^+ V_k) + \frac{\partial G_{\mathrm{d}}}{\partial \delta^+(\overline{\boldsymbol{U}},\overline{\boldsymbol{V}})_k}(\overline{\delta_k^+ U_k} - \overline{\delta_k^+ V_k})$$

$$\left. + \frac{\partial G_{\mathrm{d}}}{\partial \delta^-(\boldsymbol{U},\boldsymbol{V})_k}(\delta_k^- U_k - \delta_k^- V_k) + \frac{\partial G_{\mathrm{d}}}{\partial \delta^-(\overline{\boldsymbol{U}},\overline{\boldsymbol{V}})_k}(\overline{\delta_k^- U_k} - \overline{\delta_k^- V_k}) \right\} \Delta x$$

$$= \sum_{k=0}^{N} {}'' \left[\left\{ \frac{\partial G_{\mathrm{d}}}{\partial (\boldsymbol{U},\boldsymbol{V})_k} - \delta_k^- \left(\frac{\partial G_{\mathrm{d}}}{\partial \delta^+(\boldsymbol{U},\boldsymbol{V})_k} \right) - \delta_k^+ \left(\frac{\partial G_{\mathrm{d}}}{\partial \delta^-(\boldsymbol{U},\boldsymbol{V})_k} \right) \right\} (U_k - V_k) \right.$$

$$\left. + \left\{ \frac{\partial G_{\mathrm{d}}}{\partial (\overline{\boldsymbol{U}},\overline{\boldsymbol{V}})_k} - \delta_k^- \left(\frac{\partial G_{\mathrm{d}}}{\partial \delta^+(\overline{\boldsymbol{U}},\overline{\boldsymbol{V}})_k} \right) - \delta_k^+ \left(\frac{\partial G_{\mathrm{d}}}{\partial \delta^-(\overline{\boldsymbol{U}},\overline{\boldsymbol{V}})_k} \right) \right\} (\overline{U_k} - \overline{V_k}) \right] \Delta x$$

$$+ B_{\mathrm{c}}(\boldsymbol{U},\boldsymbol{V}), \tag{3.69}$$

where

$$\frac{\partial G_{\mathrm{d}}}{\partial (\boldsymbol{U},\boldsymbol{V})_k} \stackrel{\mathrm{d}}{\equiv} \sum_{l=1}^{M} c_l \left(\frac{Q_l^+(U_k)Q_l^-(U_k) + Q_l^+(V_k)Q_l^-(V_k)}{2} \right)$$
$$\times \left(\frac{p_l(U_k) - p_l(V_k)}{U_k - V_k} \right) f\left(N_l^P; p_l(U_k), p_l(V_k)\right), \quad (3.70\mathrm{a})$$

$$\frac{\partial G_{\mathrm{d}}}{\partial \delta^+(\boldsymbol{U},\boldsymbol{V})_k} \stackrel{\mathrm{d}}{\equiv} \sum_{l=1}^{M} c_l \left(\frac{P_l(U_k) + P_l(V_k)}{2} \right) \left(\frac{Q_l^-(U_k) + Q_l^-(V_k)}{2} \right)$$
$$\times \left(\frac{q_l^+(\delta_k^+ U_k) - q_l^+(\delta_k^+ V_k)}{\delta_k^+ U_k - \delta_k^+ V_k} \right) f\left(N_l^+; q_l^+(\delta_k^+ U_k), q_l^+(\delta_k^+ V_k)\right),$$
$$(3.70\mathrm{b})$$

$$\frac{\partial G_{\mathrm{d}}}{\partial \delta^-(\boldsymbol{U},\boldsymbol{V})_k} \stackrel{\mathrm{d}}{\equiv} \sum_{l=1}^{M} c_l \left(\frac{P_l(U_k) + P_l(V_k)}{2} \right) \left(\frac{Q_l^+(U_k) + Q_l^+(V_k)}{2} \right)$$
$$\times \left(\frac{q_l^-(\delta_k^- U_k) - q_l^-(\delta_k^- V_k)}{\delta_k^- U_k - \delta_k^- V_k} \right) f\left(N_l^-; q_l^-(\delta_k^- U_k), q_l^-(\delta_k^- V_k)\right),$$
$$(3.70\mathrm{c})$$

$$B_{\mathrm{c}}(\boldsymbol{U},\boldsymbol{V}) = \frac{1}{2}\left[\frac{\partial G_{\mathrm{d}}}{\partial \delta^+(\boldsymbol{U},\boldsymbol{V})_k} \cdot s_k^+(U_k - V_k) + s_k^- \left(\frac{\partial G_{\mathrm{d}}}{\partial \delta^+(\boldsymbol{U},\boldsymbol{V})_k} \right) \cdot (U_k - V_k) \right.$$
$$+ \frac{\partial G_{\mathrm{d}}}{\partial \delta^+(\overline{\boldsymbol{U}},\overline{\boldsymbol{V}})_k} \cdot s_k^+(\overline{U_k - V_k}) + s_k^- \left(\frac{\partial G_{\mathrm{d}}}{\partial \delta^+(\overline{\boldsymbol{U}},\overline{\boldsymbol{V}})_k} \right) \cdot (\overline{U_k - V_k})$$
$$+ \frac{\partial G_{\mathrm{d}}}{\partial \delta^-(\boldsymbol{U},\boldsymbol{V})_k} \cdot s_k^-(U_k - V_k) + s_k^+ \left(\frac{\partial G_{\mathrm{d}}}{\partial \delta^-(\boldsymbol{U},\boldsymbol{V})_k} \right) \cdot (U_k - V_k)$$
$$\left. + \frac{\partial G_{\mathrm{d}}}{\partial \delta^-(\overline{\boldsymbol{U}},\overline{\boldsymbol{V}})_k} \cdot s_k^-(\overline{U_k - V_k}) + s_k^+ \left(\frac{\partial G_{\mathrm{d}}}{\partial \delta^-(\overline{\boldsymbol{U}},\overline{\boldsymbol{V}})_k} \right) \cdot (\overline{U_k - V_k}) \right]_0^N,$$
$$(3.70\mathrm{d})$$

and

$$f(n; z_1, z_2) \stackrel{\mathrm{d}}{\equiv} \begin{cases} \dfrac{z_1 + z_2}{2}(|z_1|^{n-2} + |z_1|^{n-4}|z_2|^2 + \cdots + |z_2|^{n-2}), & n : \text{even}, \\ \dfrac{z_1 + z_2}{2} \dfrac{|z_1|^{n-1} + |z_1|^{n-2}|z_2| + \cdots + |z_2|^{n-1}}{|z_1| + |z_2|}, & n : \text{odd}. \end{cases}$$
$$(3.71)$$

In the second equality of (3.69), a trivial equality $ab - cd = \frac{1}{2}(a-c)(b+d) + \frac{1}{2}(a+c)(b-d)$ is repeatedly used (note that in particular $|a|^2 - |b|^2 = \frac{1}{2}\overline{(a+c)}(b-d) + \frac{1}{2}(a-c)\overline{(b+d)}$). In the third equality, the summation-by-parts formula (3.12a) is applied. $\partial G_{\mathrm{d}}/\partial(\boldsymbol{U},\boldsymbol{V})$ corresponds to $\partial G/\partial u$, and both

$\partial G_\mathrm{d}/\partial \delta^+(\boldsymbol{U},\boldsymbol{V})_k$ and $\partial G_\mathrm{d}/\partial \delta^-(\boldsymbol{U},\boldsymbol{V})_k$ correspond to $\partial G/\partial u_x$. Obviously, the discrete variation calculation (3.69) corresponds to (2.5). The symbol B_c denotes the discrete version of the boundary term, where the subscript "c" is for "<u>c</u>omplex."

Now we define the complex discrete variational derivatives as follows.

$$\frac{\delta G_\mathrm{d}}{\delta(\boldsymbol{U},\boldsymbol{V})_k} \stackrel{\mathrm{d}}{\equiv} \frac{\partial G_\mathrm{d}}{\partial(\boldsymbol{U},\boldsymbol{V})_k} - \delta_k^-\left(\frac{\partial G_\mathrm{d}}{\partial \delta^+(\boldsymbol{U},\boldsymbol{V})_k}\right) - \delta_k^+\left(\frac{\partial G_\mathrm{d}}{\partial \delta^-(\boldsymbol{U},\boldsymbol{V})_k}\right), \tag{3.72a}$$

$$\frac{\delta G_\mathrm{d}}{\delta(\overline{\boldsymbol{U}},\overline{\boldsymbol{V}})_k} \stackrel{\mathrm{d}}{\equiv} \left(\frac{\partial G_\mathrm{d}}{\partial \delta^+(\overline{\boldsymbol{U}},\overline{\boldsymbol{V}})_k}\right) - \delta_k^-\left(\frac{\partial G_\mathrm{d}}{\partial \delta^+(\overline{\boldsymbol{U}},\overline{\boldsymbol{V}})_k}\right) - \delta_k^+\left(\frac{\partial G_\mathrm{d}}{\partial \delta^-(\overline{\boldsymbol{U}},\overline{\boldsymbol{V}})_k}\right). \tag{3.72b}$$

Note that the complex discrete variational derivatives are complex conjugates of each other (as in the continuous case), that is,

$$\overline{\frac{\delta G_\mathrm{d}}{\delta(\boldsymbol{U},\boldsymbol{V})_k}} = \frac{\delta G_\mathrm{d}}{\delta(\overline{\boldsymbol{U}},\overline{\boldsymbol{V}})_k}. \tag{3.73}$$

It should be also noted that when

$$B_\mathrm{c}(\boldsymbol{U},\boldsymbol{V}) = 0, \tag{3.74}$$

the discrete variation (3.69) becomes

$$\sum_{k=0}^{N}{}'' \{G_{\mathrm{d},k}(\boldsymbol{U}) - G_{\mathrm{d},k}(\boldsymbol{V})\}\Delta x$$

$$= \sum_{k=0}^{N}{}'' \left[\left\{\frac{\partial G_\mathrm{d}}{\partial(\boldsymbol{U},\boldsymbol{V})_k} - \delta_k^-\left(\frac{\partial G_\mathrm{d}}{\partial \delta^+(\boldsymbol{U},\boldsymbol{V})_k}\right) - \delta_k^+\left(\frac{\partial G_\mathrm{d}}{\partial \delta^-(\boldsymbol{U},\boldsymbol{V})_k}\right)\right\}(U_k - V_k)\right.$$
$$\left. + \left\{\left(\frac{\partial G_\mathrm{d}}{\partial(\overline{\boldsymbol{U}},\overline{\boldsymbol{V}})_k}\right)_k - \delta_k^-\left(\frac{\partial G_\mathrm{d}}{\partial \delta^+(\overline{\boldsymbol{U}},\overline{\boldsymbol{V}})_k}\right) - \delta_k^+\left(\frac{\partial G_\mathrm{d}}{\partial \delta^-(\overline{\boldsymbol{U}},\overline{\boldsymbol{V}})_k}\right)\right\}(\overline{U_k} - \overline{V_k})\right]\Delta x$$

$$= \sum_{k=0}^{N}{}'' \left[\frac{\delta G_\mathrm{d}}{\delta(\boldsymbol{U},\boldsymbol{V})_k}(U_k - V_k) + \frac{\delta G_\mathrm{d}}{\delta(\overline{\boldsymbol{U}},\overline{\boldsymbol{V}})_k}(\overline{U_k} - \overline{V_k})\right]\Delta x. \tag{3.75}$$

This will be frequently referred to in the subsequent procedures.

3.3.2 Design of Schemes

With the complex discrete variational derivatives, dissipative or conservative finite difference schemes for the target PDEs 3 and PDEs 4 are defined below. A dissipative finite difference scheme for the PDEs 3 is given as follows.

Discrete Variational Derivative Method

Scheme 3.3 (Scheme for the PDEs 3) *Let $U_k^{(0)} = u(k\Delta x, 0)$ be initial values. Then, a dissipative scheme for the PDE 3 is given by, for $m = 0, 1, 2, \ldots$,*

$$\frac{U_k^{(m+1)} - U_k^{(m)}}{\Delta t} = -\frac{\delta G_\mathrm{d}}{\delta(\boldsymbol{U}^{(m+1)}, \boldsymbol{U}^{(m)})_k}, \quad k = 0, \ldots, N. \qquad (3.76)$$

THEOREM 3.3 Discrete dissipation property of Scheme 3.3
Assume that a discrete boundary condition satisfying the following condition is imposed on Scheme 3.3:

$$B_\mathrm{c}(\boldsymbol{U}^{(m+1)}, \boldsymbol{U}^{(m)}) = 0, \quad m = 0, 1, 2, \ldots. \qquad (3.77)$$

Then the scheme is dissipative, in the sense that the inequality

$$J_\mathrm{d}(\boldsymbol{U}^{(m+1)}) \leq J_\mathrm{d}(\boldsymbol{U}^{(m)}), \quad m = 0, 1, 2, \ldots \qquad (3.78)$$

holds.

PROOF In light of (3.75),

$$\frac{1}{\Delta t} \sum_{k=0}^{N}{}'' \left\{ G_{\mathrm{d},k}(\boldsymbol{U}^{(m+1)}) - G_{\mathrm{d},k}(\boldsymbol{U}^{(m)}) \right\} \Delta x$$

$$= \sum_{k=0}^{N}{}'' \left\{ \frac{\delta G_\mathrm{d}}{\delta(\boldsymbol{U}^{(m+1)}, \boldsymbol{U}^{(m)})_k} \left(\frac{U_k^{(m+1)} - U_k^{(m)}}{\Delta t} \right) \right.$$

$$\left. + \overline{\frac{\delta G_\mathrm{d}}{\delta(\boldsymbol{U}^{(m+1)}, \boldsymbol{U}^{(m)})_k}} \left(\overline{\frac{U_k^{(m+1)} - U_k^{(m)}}{\Delta t}} \right) \right\} \Delta x$$

$$= -2 \sum_{k=0}^{N}{}'' \left| \frac{\delta G_\mathrm{d}}{\delta(\boldsymbol{U}^{(m+1)}, \boldsymbol{U}^{(m)})_k} \right|^2 \Delta x$$

$$\leq 0. \qquad (3.79)$$

□

Example 3.3
Let us consider this example:

$$u_t = u_{xx}, \quad x \in (0, L), \ t > 0,$$

under the L-periodic boundary condition:

$$u^{(j)}(0, t) = u^{(j)}(L, t), \quad j = 0, 1, 2.$$

The superscript "(j)" denotes the j-th derivative. This is an example of PDEs 3 with $G(u, u_x) = |u_x|^2$. This can be regarded as the linearized version of the complex Ginzburg–Landau equation (see Section 2.3). It is easy to see directly that

$$\frac{\mathrm{d}}{\mathrm{d}t} \int_0^L G(u, u_x) \mathrm{d}x = \int_0^L (-\overline{u_{xx}} u_t - u_{xx} \overline{u}_t) \mathrm{d}x = -2 \int_0^L |u_t|^2 \mathrm{d}x \leq 0.$$

In order to construct a dissipative scheme, let us define a discrete energy function by

$$G_{\mathrm{d},k}(\boldsymbol{U}) \stackrel{\mathrm{d}}{=} \frac{|\delta_k^+ U_k|^2 + |\delta_k^- U_k|^2}{2}, \quad (3.80)$$

which means that we take in (3.66) $M = 2$ and

$$p_l = 1, \quad N_l^P = 1, \quad (l = 1, 2)$$

$$c_1 = \frac{1}{2}, \quad q_1^+ = \delta_k^+ U_k, \quad N_1^+ = 2, \quad q_1^- = 1, \quad N_1^- = 1,$$

$$c_2 = \frac{1}{2}, \quad q_2^+ = 1, \quad N_2^+ = 1, \quad q_2^- = \delta_k^- U_k, \quad N_2^- = 2.$$

Then by the formal expression (3.72a) and (3.72b) (and the associated expressions (3.70a)–(3.70c)), we obtain

$$\frac{\delta G_\mathrm{d}}{\delta(\overline{\boldsymbol{U}}, \overline{\boldsymbol{V}})_k} = -\delta_k^{\langle 2 \rangle} \left(\frac{U_k + V_k}{2} \right). \quad (3.81)$$

This can be also obtained by a direct factorization:

$$\sum_{k=0}^{N}{}'' \frac{1}{2} \left(|\delta_k^+ U_k|^2 - |\delta_k^+ V_k|^2 \right) \Delta x$$

$$= \sum_{k=0}^{N}{}'' \frac{1}{2} \left\{ \delta_k^+ \left(\frac{\overline{U_k + V_k}}{2} \right) \cdot \delta_k^+ (U_k - V_k) + \delta_k^+ (\overline{U_k - V_k}) \cdot \delta_k^+ \left(\frac{U_k + V_k}{2} \right) \right\} \Delta x$$

$$= -\sum_{k=0}^{N}{}'' \frac{1}{2} \left\{ \delta_k^{\langle 2 \rangle} \left(\frac{\overline{U_k + V_k}}{2} \right) \cdot (U_k - V_k) + (\overline{U_k - V_k}) \cdot \delta_k^{\langle 2 \rangle} \left(\frac{U_k + V_k}{2} \right) \right\} \Delta x$$

$$+ \frac{1}{4} \left[\delta_k^+ \left(\frac{\overline{U_k + V_k}}{2} \right) \cdot s_k^+ (U_k - V_k) + s_k^- \delta_k^+ \left(\frac{\overline{U_k + V_k}}{2} \right) \cdot (U_k - V_k) \right.$$

$$\left. + \delta_k^+ \left(\frac{U_k + V_k}{2} \right) \cdot s_k^+ (\overline{U_k - V_k}) + s_k^- \delta_k^+ \left(\frac{U_k + V_k}{2} \right) \cdot (\overline{U_k - V_k}) \right]_0^N, \quad (3.82)$$

which is again far simpler than to utilize the formal expressions. The terms regarding $|\delta_k^- U_k|^2/2$ can be handled in a similar manner. Now Scheme 3.3

reads: for $m = 0, 1, \ldots$,

$$\frac{U_k^{(m+1)} - U_k^{(m)}}{\Delta t} = -\frac{\delta G_d}{\delta(\boldsymbol{U}^{(m+1)}, \boldsymbol{U}^{(m)})_k} = \delta_k^{\langle 2 \rangle} \left(\frac{U_k^{(m+1)} + U_k^{(m)}}{2} \right), \quad (3.83)$$

for $k = 0, \ldots, N$. Suppose that the discrete periodic boundary condition:

$$U_0^{(m)} = U_N^{(m)}, \quad U_{-1}^{(m)} = U_{N-1}^{(m)}, \quad U_1^{(m)} = U_{N+1}^{(m)}, \quad (3.84)$$

is imposed. The condition (3.77) in Theorem 3.3 reads (by the formal expression (3.70d))

$$B_c(\boldsymbol{U}^{(m+1)}, \boldsymbol{U}^{(m)}) =$$

$$\frac{1}{4} \left[\delta_k^+ \left(\overline{\frac{U_k^{(m+1)} + U_k^{(m)}}{2}} \right) \cdot s_k^+ (U_k^{(m+1)} - U_k^{(m)}) \right.$$

$$+ s_k^- \delta_k^- \left(\overline{\frac{U_k^{(m+1)} + U_k^{(m)}}{2}} \right) \cdot (U_k^{(m+1)} - U_k^{(m)})$$

$$+ \delta_k^+ \left(\frac{U_k^{(m+1)} + U_k^{(m)}}{2} \right) \cdot s_k^+ \overline{(U_k^{(m+1)} - U_k^{(m)})}$$

$$+ s_k^- \delta_k^- \left(\frac{U_k^{(m+1)} + U_k^{(m)}}{2} \right) \cdot \overline{(U_k^{(m+1)} - U_k^{(m)})}$$

$$\left. + \text{(similar terms regarding } |\delta_k^- U_k^{(m)}|^2) \right]_0^N. \quad (3.85)$$

This is obviously identical to the boundary term in (3.82). The boundary term vanishes due to the periodicity. □

A conservative scheme for the conservative PDEs 4 is defined as follows.

Scheme 3.4 (Scheme for the PDEs 4) Let $U_k^{(0)} = u(k\Delta x, 0)$ be initial values. Then, a conservative scheme for the PDE 4 is given by, for $m = 0, 1, 2, \ldots$,

$$\mathrm{i} \left(\frac{U_k^{(m+1)} - U_k^{(m)}}{\Delta t} \right) = -\frac{\delta G_d}{\delta(\boldsymbol{U}^{(m+1)}, \boldsymbol{U}^{(m)})_k}, \quad k = 0, \ldots, N. \quad (3.86)$$

□

THEOREM 3.4 Discrete conservation property of Scheme 3.4

Assume that a discrete boundary condition satisfying the following condition is imposed on Scheme 3.4:

$$B_c(\boldsymbol{U}^{(m+1)}, \boldsymbol{U}^{(m)}) = 0, \quad m = 0, 1, 2, \ldots, \quad (3.77)$$

for $m = 0, 1, 2, \ldots$. Then the scheme is conservative in the sense that the equality

$$J_{\mathrm{d}}(\boldsymbol{U}^{(m)}) = J_{\mathrm{d}}(\boldsymbol{U}^{(0)}), \quad m = 1, 2, \ldots \quad (3.87)$$

holds.

PROOF In light of (3.75),

$$\frac{1}{\Delta t} \sum_{k=0}^{N} {}'' \left\{ G_{\mathrm{d},k}(\boldsymbol{U}^{(m+1)}) - G_{\mathrm{d},k}(\boldsymbol{U}^{(m)}) \right\} \Delta x$$

$$= \sum_{k=0}^{N} {}'' \left\{ \frac{\delta G_{\mathrm{d}}}{\delta(\boldsymbol{U}^{(m+1)}, \boldsymbol{U}^{(m)})_k} \left(\frac{U_k^{(m+1)} - U_k^{(m)}}{\Delta t} \right) \right.$$

$$\left. + \frac{\delta G_{\mathrm{d}}}{\delta(\overline{\boldsymbol{U}^{(m+1)}, \boldsymbol{U}^{(m)}})_k} \overline{\left(\frac{U_k^{(m+1)} - U_k^{(m)}}{\Delta t} \right)} \right\} \Delta x$$

$$= \sum_{k=0}^{N} {}'' \left\{ \mathrm{i} \left| \frac{\delta G_{\mathrm{d}}}{\delta(\boldsymbol{U}^{(m+1)}, \boldsymbol{U}^{(m)})_k} \right|^2 - \mathrm{i} \left| \frac{\delta G_{\mathrm{d}}}{\delta(\boldsymbol{U}^{(m+1)}, \boldsymbol{U}^{(m)})_k} \right|^2 \right\} \Delta x$$

$$= 0. \quad (3.88)$$

□

Example 3.4
Let us demonstrate Scheme 3.4 with the linear wave equation:

$$\mathrm{i} u_t = -u_{xx}, \quad x \in (0, L), \ t > 0,$$

under the L-periodic boundary condition:

$$u^{(j)}(0, t) = u^{(j)}(L, t), \quad j = 0, 1, 2.$$

This is an example of PDEs 4 with $G(u, u_x) = -|u_x|^2$, and can be considered as a linearized version of the nonlinear Schrödinger equation (see Section 2.3). Let us define a discrete energy function by

$$G_{\mathrm{d},k}(\boldsymbol{U}) \stackrel{\mathrm{d}}{\equiv} \frac{|\delta_k^+ U_k|^2 + |\delta_k^- U_k|^2}{2}. \quad (3.89)$$

Then the story goes almost the same way as in the previous example to find the concrete form of Scheme 3.4: for $m = 0, 1, 2, \ldots$,

$$\mathrm{i} \left(\frac{U_k^{(m+1)} - U_k^{(m)}}{\Delta t} \right) = -\frac{\delta G_{\mathrm{d}}}{\delta(\boldsymbol{U}^{(m+1)}, \boldsymbol{U}^{(m)})_k} = -\delta_k^{\langle 2 \rangle} \left(\frac{U_k^{(m+1)} + U_k^{(m)}}{2} \right), \quad (3.90)$$

for $k = 0, \ldots, N$. This becomes conservative under the discrete periodic boundary conditions (which can be easily verified). □

3.4 Procedure for Systems of First-Order PDEs

In this section we consider the systems of first-order PDEs 5 and PDEs 6. In view of the extended vector expression (2.10) we denote the numerical solutions by

$$U_{j,k}^{(m)} \simeq u_j(k\Delta x, m\Delta t), \quad (j = 1, \ldots, N_{\text{ex}}, \ k = 0, \ldots, N, \ m = 0, 1, 2, \ldots). \tag{3.91}$$

The first subindex j corresponds to the j-th variable u_j, and the second one k denotes the spatial index. We also introduce a discrete version of the extended solution vector introduced in Section 2.1, which is a vector of length $N_{\text{ex}} \times (N+1)$:

$$\boldsymbol{U} = (U_{1,0}, \ldots, U_{1,N}, \ldots, U_{N_{\text{ex}},0}, \ldots, U_{N_{\text{ex}},N})^\top. \tag{3.92}$$

We assume that the discrete version of the energy function $G(u, u_x)$ is of the form

$$G_{\text{d},k}(\boldsymbol{U}) = \sum_{l=1}^{M} \prod_{j=1}^{N_{\text{ex}}} f_{l,j}(U_{j,k}) g_{l,j}^+(\delta_k^+ U_{j,k}) g_{l,j}^-(\delta_k^- U_{j,k}). \tag{3.93}$$

The discrete global energy is defined accordingly by

$$J_{\text{d}}(\boldsymbol{U}^{(m)}) \stackrel{\text{d}}{=} \sum_{k=0}^{N} {}'' G_{\text{d},k}(\boldsymbol{U}) \Delta x. \tag{3.94}$$

For the discrete energy function defined above, we hope to find an identity of the form:

$$\sum_{k=0}^{N} {}'' (G_{\text{d},k}(\boldsymbol{U}) - G_{\text{d},k}(\boldsymbol{V})) \Delta x$$

$$= \sum_{k=0}^{N} {}'' \left[\sum_{j=1}^{N_{\text{ex}}} \left\{ \frac{\partial G_{\text{d}}}{\partial (\boldsymbol{U}, \boldsymbol{V})} \bigg|_{j,k} (U_{j,k} - V_{j,k}) + \frac{\partial G_{\text{d}}}{\partial \delta^+ (\boldsymbol{U}, \boldsymbol{V})} \bigg|_{j,k} (\delta_k^+ U_{j,k} - \delta_k^+ V_{j,k}) \right. \right.$$

$$\left. \left. + \frac{\partial G_{\text{d}}}{\partial \delta^- (\boldsymbol{U}, \boldsymbol{V})} \bigg|_{j,k} (\delta_k^- U_{j,k} - \delta_k^- V_{j,k}) \right\} \right] \Delta x, \tag{3.95}$$

where $\partial G_{\text{d}}/\partial (\boldsymbol{U}, \boldsymbol{V})_{j,k}$ and $\partial G_{\text{d}}/\partial \delta^\pm (\boldsymbol{U}, \boldsymbol{V})_{j,k}$ are supposed to represent $\partial G/\partial u_j$ and $\partial G/\partial u_{j,x}$, respectively. If an identity in the form of (3.95) is realized, then by the summation-by-parts formula, we obtain a discrete variation identity:

$$\sum_{k=0}^{N}{}'' \left(G_{\mathrm{d},k}(\boldsymbol{U}) - G_{\mathrm{d},k}(\boldsymbol{V}) \right) \Delta x$$

$$= \sum_{k=0}^{N}{}'' \left[\sum_{j=1}^{N_{\mathrm{ex}}} \frac{\delta G_{\mathrm{d}}}{\delta(\boldsymbol{U},\boldsymbol{V})}\bigg|_{j,k} (U_{j,k} - V_{j,k}) \right] \Delta x + B_{\mathrm{sys}}(\boldsymbol{U},\boldsymbol{V}), \qquad (3.96)$$

where

$$\frac{\delta G_{\mathrm{d}}}{\delta(\boldsymbol{U},\boldsymbol{V})}\bigg|_{j,k} \stackrel{\mathrm{d}}{=} \frac{\partial G_{\mathrm{d}}}{\partial(\boldsymbol{U},\boldsymbol{V})}\bigg|_{j,k} - \delta_k^{-}\left(\frac{\partial G_{\mathrm{d}}}{\partial \delta^{+}(\boldsymbol{U},\boldsymbol{V})}\bigg|_{j,k} \right) - \delta_k^{+}\left(\frac{\partial G_{\mathrm{d}}}{\partial \delta^{-}(\boldsymbol{U},\boldsymbol{V})}\bigg|_{j,k} \right), \qquad (3.97)$$

and

$$B_{\mathrm{sys}}(\boldsymbol{U},\boldsymbol{V}) \stackrel{\mathrm{d}}{=}$$
$$\frac{1}{2}\sum_{j=1}^{N_{\mathrm{ex}}} \left[\frac{\partial G_{\mathrm{d}}}{\partial \delta^{+}(\boldsymbol{U},\boldsymbol{V})}\bigg|_{j,k} \{s_k^{+}(U_{j,k} - V_{j,k})\} + s_k^{-}\left(\frac{\partial G_{\mathrm{d}}}{\partial \delta^{+}(\boldsymbol{U},\boldsymbol{V})}\bigg|_{j,k} \right)(U_{j,k} - V_{j,k}) \right.$$
$$\left. + \frac{\partial G_{\mathrm{d}}}{\partial \delta^{-}(\boldsymbol{U},\boldsymbol{V})}\bigg|_{j,k} \{s_k^{-}(U_{j,k} - V_{j,k})\} + s_k^{+}\left(\frac{\partial G_{\mathrm{d}}}{\partial \delta^{-}(\boldsymbol{U},\boldsymbol{V})}\bigg|_{j,k} \right)(U_{j,k} - V_{j,k}) \right]_0^N. \qquad (3.98)$$

The symbol $\delta G_{\mathrm{d}}/\delta(\boldsymbol{U},\boldsymbol{V})_{j,k}$ is a discrete variational derivative in the multivariate case. The identity corresponds to the continuous case (2.9) (or more simply, (2.11).) The term B_{sys} corresponds to the boundary terms in (2.9), where "sys" is for "system".

A troublesome issue remains to be discussed: How can we fix an identity (3.95)? The trouble is *not* that it is hard to find one; the truth is that we find *too many*, and it seems impossible to determine a single choice as a universal template that adapts to wide range of problems.

The next lemma shows that at least one identity of the form (3.95) can be found.

LEMMA 3.1
Let J be a positive integer, $a_j : \mathbb{C} \to \mathbb{C}$ ($j = 1, \ldots, J$) be sufficiently smooth functions, and u_j, v_j ($j = 1, \ldots, J$) be their arguments. Then there exist functions F_j ($j = 1, \ldots, N_{\mathrm{ex}}$) that are polynomials of $a_j(u_j), a_j(v_j)$ ($i = 1, \ldots, J$, $i \neq j$) such that

(i) *for any u_j, v_j ($j = 1, \ldots, J$)*

$$\prod_{j=1}^{J} a_j(u_j) - \prod_{j=1}^{J} a_j(v_j) = \sum_{j=1}^{J} (a_j(u_j) - a_j(v_j)) F_j$$

holds;

(ii) when $u_i = v_i$ $(i = 1, \ldots, J,\ i \neq j)$,
$$F_j = \prod_{i \neq j} a_i(u_i).$$

Thus it follows that
$$\lim_{u_j \to v_j} \left(\frac{a_j(u_j) - a_j(v_j)}{u_j - v_j} \right) F_j = \frac{\partial}{\partial u_j} \prod_{i=1}^{J} a_i(u_i).$$

PROOF It can be explicitly constructed by repeatedly using the trivial identity $ab - cd = (a+c)(b-d)/2 + (a-c)(b+d)/2$. Let us abbreviate $a_j(u_j)$ as a_j and $a_j(v_j)$ as b_j to save space. If we choose to separate $(a_J - b_J)$ first, then

$$\prod_{j=1}^{J} a_j - \prod_{j=1}^{J} b_j = \left(\frac{\prod_{j=1}^{J-1} a_j + \prod_{j=1}^{J-1} b_j}{2} \right)(a_J - b_J) + \left(\prod_{j=1}^{J-1} a_j - \prod_{j=1}^{J-1} b_j \right)\left(\frac{a_J + b_J}{2} \right).$$

The same identity holds true with J replaced with $J-1$, by which the term $\prod_{j=1}^{J-1} a_j - \prod_{j=1}^{J-1} b_j$ can be further factorized. By repeating this process for $J-2, J-3, \ldots, 2$, we find that

$$F_j = \prod_{i=j+1}^{J} \left(\frac{a_i + b_i}{2} \right) \left(\frac{\prod_{i=1}^{j} a_i + \prod_{i=1}^{j} b_i}{2} \right) \qquad (j = 1, \ldots, J). \qquad (3.99)$$

□

Setting $J = 3N_{\text{ex}}$, and taking $f_{l,j}(U_{j,k}), g_{l,j}^{+}(\delta_k^{+} U_{j,k}), g_{l,j}^{-}(\delta_k^{-} U_{j,k})$ as $a_j(u_j)$ and $f_{l,j}(V_{j,k}), g_{l,j}^{+}(\delta_k^{+} V_{j,k}), g_{l,j}^{-}(\delta_k^{-} V_{j,k})$ as $a_j(v_j)$, we see by the above lemma that

$$\prod_{j=1}^{N_{\text{ex}}} f_{l,j}(U_{j,k}) g_{l,j}^{+}(\delta_k^{+} U_{j,k}) g_{l,j}^{-}(\delta_k^{-} U_{j,k}) - \prod_{j=1}^{N_{\text{ex}}} f_{l,j}(V_{j,k}) g_{l,j}^{+}(\delta_k^{+} V_{j,k}) g_{l,j}^{-}(\delta_k^{-} V_{j,k})$$

(cf. the energy (3.93)) can be factorized so that the terms $(U_{j,k} - V_{j,k})$, $\delta_k^{\pm}(U_{j,k} - V_{j,k})$ are separated. Then by summing them for $l = 1, \ldots, M$, we obtain the concrete forms of $\partial G_{\text{d}}/\partial (\boldsymbol{U}, \boldsymbol{V})_{j,k}$ and $\partial G_{\text{d}}/\partial \delta^{\pm}(\boldsymbol{U}, \boldsymbol{V})_{j,k}$. Moreover, they are in fact valid approximations to the continuous partial derivatives (due to (ii) of the above lemma).

The proof of the above lemma reveals that there are combinatorially many choices for F_j's. In fact, we can arbitrarily change the order of separation, such as $j = 1, 2, \ldots, J$ (instead of $j = J, J-1, \ldots, 1$ in the above proof), by which a different set of F_j's that are still valid approximations of the continuous partial derivatives are obtained. Furthermore, it is also possible to separate two or more terms at a time; for example,

$$\prod_{j=1}^{4} a_j - \prod_{j=1}^{4} b_j = \left(\frac{a_1 a_2 + b_1 b_2}{2}\right)(a_3 a_4 - b_3 b_4) + (a_1 a_2 - b_1 b_2)\left(\frac{a_3 a_4 + b_3 b_4}{2}\right). \tag{3.100}$$

If we allow these possibilities as well, the number of possible choices soon blows up as J (or equivalently, N_{ex}) increases. It also seems impossible (or useless) to give a single fixed choice such as (3.27a)–(3.27c) in the real-valued single PDE case, and (3.70a)–(3.70c) in the complex-valued single PDE case. The group of the systems of PDEs is too large to deal with by a single fixed pattern; the best choice should be judged on a case-by-case basis. In practical situations, this is not a hard task since appropriate factorizations are usually few, if not unique, for a given system of PDEs, in view of symmetry:

- *Symmetry as to $\delta_k^\pm$*: It is often preferable that in the resulting identity the symmetry as to $\delta_k^\pm$ is preserved (since it affects the spatial symmetry of the resulting schemes). To that end, the terms regarding $\delta_k^\pm U_{j,k}$ (and $\delta_k^\pm V_{j,k}$) should be grouped.

- *Symmetry as to conjugate variables*: If u_j and u_{j+1} are complex conjugate of each other, then they should be grouped.

- *Other physical symmetries*: If there are any symmetries in variables coming from physical derivation, it might be preferable to preserve them by appropriately grouping variables.

Observe that the factorizations in the single PDE cases (i.e. (3.27a)–(3.27c) and (3.70a)–(3.70c)), have been given with full respect for such symmetries. Note also that practically N_{ex} is often 3 or 4, at most (see, for example, the Zakharov case in Section 2.4).

REMARK 3.7 Lemma 3.1 can be generalized so that it allows the factorizations like (3.100); i.e., even when we utilize such factorizations, the resulting F_j's satisfy the conditions (i) and (ii). □

REMARK 3.8 If the symmetry as to u_j ($j = 1, \ldots, J$) is demanded, it is possible to recover the symmetry, for example, by averaging all the possible factorizations of the form (3.99). It is also possible to partially recover the symmetry for a subset of u_j's by appropriately averaging factorizations. But, by these averagings, the computational cost should increase accordingly. □

3.4.1 Design of Schemes

The general form of the proposed scheme for the target PDEs 5 and 6 is given as follows.

$$\frac{\boldsymbol{U}^{(m+1)} - \boldsymbol{U}^{(m)}}{\Delta t} = A_{\mathrm{d}} \left(\frac{\delta G_{\mathrm{d}}}{\delta(\boldsymbol{U}^{(m+1)}, \boldsymbol{U}^{(m)})} \right), \quad m = 0, 1, 2, \ldots. \quad (3.101)$$

The matrix A_{d} is defined automatically based on the original matrix A in PDEs 5 and 6. The subscript "d" stands for "discrete." The scheme becomes conservative or dissipative when the matrix A_{d} and the discrete boundary condition imposed on the scheme are of certain special forms. In what follows, we give the definitions of A_{d} and the necessary conditions for discrete dissipation or conservation, for each of Type C1–C4 and D1–D3 systems in turn.

Scheme 3.5 (Scheme for the PDEs 5) Let $U_{j,k}^{(0)} = u_j(k\Delta x, 0)$ be initial values. A scheme for PDEs 5 is given by (3.101) with the matrix A_{d} defined as follows:

For Type C1 A_{d} is an $(N+1) \times (N+1)$ matrix:

$$A_{\mathrm{d}} = \Delta_k^{\langle 2s+1 \rangle},$$

where $\Delta_k^{\langle s \rangle}$ is an $(N+1) \times (N+1)$ matrix which is defined as

$$\Delta_k^{\langle s \rangle} \stackrel{\mathrm{d}}{\equiv} \mathrm{diag}\left\{ \left(\delta_k^{\langle 1 \rangle} \right)^s \right\}.$$

For Type C2 A_{d} is a $(2N+2) \times (2N+2)$ matrix:

$$\begin{pmatrix} 0 & I \\ -I & 0 \end{pmatrix},$$

where I is the $(N+1) \times (N+1)$ identity matrix.

For Type C3 A_{d} is a $(2N+2) \times (2N+2)$ matrix:

$$D_k^{\langle 2s+1 \rangle},$$

where

$$D_k^{\langle s \rangle} \stackrel{\mathrm{d}}{\equiv} \begin{pmatrix} 0 & \Delta_k^{\langle s \rangle} \\ \Delta_k^{\langle s \rangle} & 0 \end{pmatrix}.$$

For Type C4 A_{d} is a block diagonal matrix where each block matrix is what is defined above for the corresponding Type C1–C3 systems.

□

THEOREM 3.5 Discrete conservation property of Scheme 3.5

Assume that discrete boundary conditions that satisfy the two conditions below are imposed. Then Scheme 3.5 is conservative in the sense that the equality:

$$J_\mathrm{d}(\boldsymbol{U}^{(m+1)}) = J_\mathrm{d}(\boldsymbol{U}^{(m)}), \quad m = 0, 1, 2, \ldots \qquad (3.102)$$

holds. The first condition is

$$B_\mathrm{sys}(\boldsymbol{U}^{(m+1)}, \boldsymbol{U}^{(m)}) = 0, \quad m = 0, 1, 2, \ldots. \qquad (3.103)$$

The second condition is given independently for each type as follows:

For Type C1

$$\frac{1}{2}(-1)^s \left[\left(\delta_k^{\langle 1 \rangle}\right)^s P_k \cdot s_k^{\langle 1 \rangle} \left(\delta_k^{\langle 1 \rangle}\right)^s P_k \right]_0^N$$
$$+ \frac{1}{2} \sum_{j=1}^{s} (-1)^{j-1} \left[\left(\delta_k^{\langle 1 \rangle}\right)^{j-1} P_k \cdot s_k^{\langle 1 \rangle} \left(\delta_k^{\langle 1 \rangle}\right)^{2s+1-j} P_k \right.$$
$$\left. + s_k^{\langle 1 \rangle} \left(\delta_k^{\langle 1 \rangle}\right)^{j-1} P_k \cdot \left(\delta_k^{\langle 1 \rangle}\right)^{2s+1-j} P_k \right]_0^N = 0, \qquad (3.104)$$

where

$$P_k \stackrel{\mathrm{d}}{\equiv} \frac{\delta G_\mathrm{d}}{\delta(\boldsymbol{U}^{(m+1)}, \boldsymbol{U}^{(m)})_k}.$$

(Because $M = 1$ in this case, we omit the subscript j here.).

For Type C2 *No additional conditions.*

For Type C3

$$\frac{1}{2}(-1)^s \left[\left(\widetilde{D}_k^s \boldsymbol{P}_k\right)^\top S^{\langle 1 \rangle} \left(\widetilde{D}_k^s \boldsymbol{P}_k\right) \right]_0^N$$
$$+ \frac{1}{2} \sum_{j=1}^{s} (-1)^{j-1} \left[\left(\widetilde{D}_k^{j-1} \boldsymbol{P}_k\right)^\top S^{\langle 1 \rangle} \left(\widetilde{D}_k^{2s+1-j} \boldsymbol{P}_k\right) \right.$$
$$\left. + S^{\langle 1 \rangle} \left(\widetilde{D}_k^{j-1} \boldsymbol{P}_k\right)^\top \left(\widetilde{D}_k^{2s+1-j} \boldsymbol{P}_k\right) \right]_0^N, \qquad (3.105\mathrm{a})$$

where

$$P_{j,k} \stackrel{\mathrm{d}}{\equiv} \frac{\delta G_\mathrm{d}}{\delta(\boldsymbol{U}^{(m+1)}, \boldsymbol{U}^{(m)})_{j,k}}, \quad j = 1, 2, \qquad (3.105\mathrm{b})$$

$$\boldsymbol{P}_k \stackrel{\mathrm{d}}{\equiv} (P_{1,k}, P_{2,k})^\top, \qquad (3.105\mathrm{c})$$

and

$$\widetilde{D}_k \overset{\mathrm{d}}{\equiv} \begin{pmatrix} 0 & \delta_k^{\langle 1 \rangle} \\ \delta_k^{\langle 1 \rangle} & 0 \end{pmatrix}, \quad S^{\langle 1 \rangle} \overset{\mathrm{d}}{\equiv} \begin{pmatrix} 0 & s_k^{\langle 1 \rangle} \\ s_k^{\langle 1 \rangle} & 0 \end{pmatrix} \quad (3.105\mathrm{d})$$

are 2×2 matrices.

For Type C4 We assume that for each block of A_d components, the corresponding conditions defined above are satisfied.

PROOF We consider Types C1 to C4 in order.

[Type C1]

By repeatedly using the summation-by-parts formula (3.12b), we have

$$\frac{1}{\Delta t} \sum_{k=0}^{N} {}'' G_{\mathrm{d},k}(\boldsymbol{U}^{(m+1)}) - G_{\mathrm{d},k}(\boldsymbol{U}^{(m)}) \Delta x$$

$$= \sum_{k=0}^{N} {}'' P_k \left(\frac{U_k^{(m+1)} - U_k^{(m)}}{\Delta t} \right) \Delta x + B_\mathrm{sys}(\boldsymbol{U}^{(m+1)}, \boldsymbol{U}^{(m)})$$

$$= \sum_{k=0}^{N} {}'' P_k \left(\delta_k^{\langle 1 \rangle} \right)^{2s+1} P_k \Delta x$$

$$= \frac{1}{2}(-1)^s \left[\left(\delta_k^{\langle 1 \rangle} \right)^s P_k \cdot s_k^{\langle 1 \rangle} \left(\delta_k^{\langle 1 \rangle} \right)^s P_k \right]_0^N$$

$$+ \frac{1}{2} \sum_{j=1}^{s} (-1)^{j-1} \left[\left(\delta_k^{\langle 1 \rangle} \right)^{j-1} P_k \cdot s_k^{\langle 1 \rangle} \left(\delta_k^{\langle 1 \rangle} \right)^{2s+1-j} P_k \right.$$

$$\left. + s_k^{\langle 1 \rangle} \left(\delta_k^{\langle 1 \rangle} \right)^{j-1} P_k \cdot \left(\delta_k^{\langle 1 \rangle} \right)^{2s+1-j} P_k \right]_0^N$$

$$= 0. \quad (3.106)$$

In the second equality, the assumption (3.103) is used. The last equality is from the assumption (3.104).

[Type C2]

We note the scheme is rewritten with $P_{j,k}$ as

$$\begin{cases} \dfrac{U_{1,k}^{(m+1)} - U_{1,k}^{(m)}}{\Delta t} = P_{2,k}, \\ \dfrac{U_{2,k}^{(m+1)} - U_{2,k}^{(m)}}{\Delta t} = -P_{1,k}. \end{cases} \quad (3.107)$$

Then, from (3.96),

$$\frac{1}{\Delta t}\sum_{k=0}^{N}{}''G_{\mathrm{d},k}(\boldsymbol{U}^{(m+1)}) - G_{\mathrm{d},k}(\boldsymbol{U}^{(m)})\Delta x$$
$$= \sum_{k=0}^{N}{}''\sum_{j=1}^{2}P_{j,k}\left(\frac{U_{j,k}^{(m+1)} - U_{j,k}^{(m)}}{\Delta t}\right)\Delta x + B_{\mathrm{sys}}(\boldsymbol{U}^{(m+1)}, \boldsymbol{U}^{(m)})$$
$$= \sum_{k=0}^{N}{}''P_{1,k}P_{2,k} - P_{2,k}P_{1,k}\Delta x$$
$$= 0. \qquad (3.108)$$

[Type C3]

Using the abbreviation $P_{j,k}$, the scheme is rewritten as

$$\begin{cases} \dfrac{U_{1,k}^{(m+1)} - U_{1,k}^{(m)}}{\Delta t} = \left(\delta_k^{\langle 1 \rangle}\right)^{2s+1} P_{2,k}, \\ \dfrac{U_{2,k}^{(m+1)} - U_{2,k}^{(m)}}{\Delta t} = \left(\delta_k^{\langle 1 \rangle}\right)^{2s+1} P_{1,k}. \end{cases} \qquad (3.109)$$

To prove the conservation property, we use the following summation-by-parts formula which corresponds to (2.51). For any sequences $\boldsymbol{F}_k = (F_{1,k}, F_{2,k})^\top$, $\boldsymbol{G}_k = (G_{1,k}, G_{2,k})^\top$,

$$\sum_{k=0}^{N}{}''\boldsymbol{F}_k^\top \widetilde{D}_k \boldsymbol{G}_k \Delta x = -\sum_{k=0}^{N}{}''\left(\widetilde{D}_k \boldsymbol{F}_k\right)^\top \boldsymbol{G}_k \Delta x$$
$$+ \frac{1}{2}\left[\boldsymbol{F}_k^\top S^{\langle 1 \rangle} \boldsymbol{G}_k + \left(S^{\langle 1 \rangle} \boldsymbol{F}_k\right)^\top \boldsymbol{G}_k\right]_0^N. \qquad (3.110)$$

Then, from (3.96),

$$\frac{1}{\Delta t}\sum_{k=0}^{N}{}''G_{\mathrm{d},k}(\boldsymbol{U}^{(m+1)}) - G_{\mathrm{d},k}(\boldsymbol{U}^{(m)})\Delta x$$
$$= \sum_{k=0}^{N}{}''\sum_{j=1}^{2}P_{j,k}\left(\frac{U_{j,k}^{(m+1)} - U_{1,k}^{(m)}}{\Delta t}\right)\Delta x + B_{\mathrm{sys}}(\boldsymbol{U}^{(m+1)}, \boldsymbol{U}^{(m)})$$
$$= \sum_{k=0}^{N}{}''\boldsymbol{P}_k^\top \left(\widetilde{D}_k\right)^{2s+1} \boldsymbol{P}_k \Delta x$$

$$= \frac{1}{2}(-1)^s \left[\left(\widetilde{D}_k^s \boldsymbol{P}_k\right)^\top S^{\langle 1 \rangle} \left(\widetilde{D}_k^s \boldsymbol{P}_k\right) \right]_0^N$$

$$+ \frac{1}{2} \sum_{j=1}^{s} (-1)^{j-1} \left[\left(\widetilde{D}_k^{j-1} \boldsymbol{P}_k\right)^\top S^{\langle 1 \rangle} \left(\widetilde{D}_k^{2s+1-j} \boldsymbol{P}_k\right) \right.$$

$$\left. + S^{\langle 1 \rangle} \left(\widetilde{D}_k^{j-1} \boldsymbol{P}_k\right)^\top \left(\widetilde{D}_k^{2s+1-j} \boldsymbol{P}_k\right) \right]_0^N$$

$$= 0. \qquad (3.111)$$

In the second equality, the assumption (3.103) is used. The last equality is from the assumption (3.105a).

[Type C4]

Trivial, because this is just a combination of the other types. □

Scheme 3.6 (Scheme for the PDEs 6) Let $U_{j,k}^{(0)} = u_j(k\Delta x, 0)$ be initial values. Then, a dissipative scheme for the PDE 5 of Type D1–D3 is given by (3.101), where the matrix A_d is given as follows.

Type D1 A_d is an $N \times N$ matrix:

$$(-1)^{s+1} \Delta_k^{\langle 2s \rangle}.$$

Type D2 A_d is a $2N \times 2N$ matrix:

$$\begin{pmatrix} 0 & -I \\ -I & 0 \end{pmatrix}.$$

Type D3 A_d is a block diagonal matrix where each block matrix is defined above for the corresponding Type C1–C3, D1, and D2 systems.

□

THEOREM 3.6 Discrete dissipation property of Scheme 3.1

Consider Scheme 3.5 for Types D1–D3 and assume that a discrete boundary condition is imposed which satisfies the two conditions stated below. Then the scheme is dissipative in the sense that the inequality:

$$J_\mathrm{d}(\boldsymbol{U}^{(m+1)}) \leq J_\mathrm{d}(\boldsymbol{U}^{(m)}), \quad m = 0, 1, 2, \ldots \qquad (3.112)$$

holds.

The first condition is

$$B_\mathrm{sys}(\boldsymbol{U}^{(m+1)}, \boldsymbol{U}^{(m)}) = 0 \quad (m = 0, 1, 2, \ldots). \qquad (3.113)$$

The second condition is as follows.

For Type D1

$$\frac{1}{2}\sum_{j=1}^{s}(-1)^{j-1}\left[\left(\delta_k^{\langle 1\rangle}\right)^{j-1}P_k \cdot s_k^{\langle 1\rangle}\left(\delta_k^{\langle 1\rangle}\right)^{2s-j}P_k\right.$$

$$\left.+s_k^{\langle 1\rangle}\left(\delta_k^{\langle 1\rangle}\right)^{j-1}P_k \cdot \left(\delta_k^{\langle 1\rangle}\right)^{2s-j}P_k\right]_0^N = 0. \qquad (3.114)$$

For Type D2 *We do not need any additional conditions.*

For Type D3 *We assume that for each block of $A_{\rm d}$ components, the corresponding conditions defined above are satisfied.*

PROOF We can easily prove the case of Type D1 just by repeatedly applying the summation-by-parts formula as in the conservative case. The other cases are trivial. □

3.5 Procedure for Second-Order PDEs

In this section we consider the second-order PDEs 7:

$$\frac{\partial^2 u}{\partial t^2} = -\frac{\delta G}{\delta u}, \quad x \in (0, L), \ t > 0, \qquad (2.55)$$

and their conservative schemes. The procedure essentially differs from those in the previous sections, since the PDEs themselves greatly differ; the conservation law is now

$$\frac{\mathrm{d}}{\mathrm{d}t}\int_0^L \left\{\frac{1}{2}(u_t)^2 + G(u, u_x)\right\}\mathrm{d}x = 0, \qquad (3.115)$$

which includes not only G but also u_t.

To design a conservative scheme for the PDEs 7, we have two different approaches. The first, direct way is to start from the conservation law (3.115) and consider its discrete variation. This is a completely new process because the local energy appearing in (3.115) includes not only u and u_x but also u_t. The second, somewhat tricky way is to rewrite (2.55) as a conservative system of PDEs and apply the procedure described in the preceding section. Basically these two ways lead us to different types of conservative schemes, though the latter can be considered to include the former one. In what follows we describe these two approaches.

3.5.1 First Approach: Direct Variation

In this approach we try to copy the continuous differentiation (2.56) directly in a discrete setting. Let $U_k{}^{(m)}$ be numerical solution. Now suppose that the local energy $G(u, u_x)$ be approximated by $G_{\mathrm{d},k}(\boldsymbol{U}^{(m)}, \boldsymbol{U}^{(m+1)})$. This is a completely new concept; so far the discrete local energy has been defined with only one approximate solution: $G_{\mathrm{d},k}(\boldsymbol{U}^{(m)})$. But now it is defined with *two* consecutive solutions. We then define a discrete global energy by

$$J_{\mathrm{d}}(\boldsymbol{U}^{(m+1)}, \boldsymbol{U}^{(m)}) \stackrel{\mathrm{d}}{=} \sum_{k=0}^{N}{}'' \left\{ \frac{1}{2}(\delta_m^+ U_k{}^{(m)})^2 + G_{\mathrm{d},k}(\boldsymbol{U}^{(m+1)}, \boldsymbol{U}^{(m)}) \right\} \Delta x. \quad (3.116)$$

The symbol δ_m^+ is the forward difference operator with respect to time index (m); i.e. $\delta_m^+ U_k{}^{(m)} = (U_k{}^{(m+1)} - U_k{}^{(m)})/\Delta t$. Note that the global energy J_{d} also refers to two solutions accordingly.

Let us consider the discrete version of (2.56) as follows.

$$J_{\mathrm{d}}(\boldsymbol{U}^{(m+1)}, \boldsymbol{U}^{(m)}) - J_{\mathrm{d}}(\boldsymbol{U}^{(m)}, \boldsymbol{U}^{(m-1)})$$
$$= \sum_{k=0}^{N}{}'' \left\{ \frac{(\delta_m^+ U_k{}^{(m)})^2}{2} - \frac{(\delta_m^+ U_k{}^{(m-1)})^2}{2} \right.$$
$$\left. + G_{\mathrm{d},k}(\boldsymbol{U}^{(m+1)}, \boldsymbol{U}^{(m)}) - G_{\mathrm{d},k}(\boldsymbol{U}^{(m)}, \boldsymbol{U}^{(m-1)}) \right\} \Delta x. \quad (3.117)$$

The first half of the right hand side of (3.117) is

$$\sum_{k=0}^{N}{}'' \left\{ \frac{(\delta_m^+ U_k{}^{(m)})^2}{2} - \frac{(\delta_m^+ U_k{}^{(m-1)})^2}{2} \right\}$$
$$= \sum_{k=0}^{N}{}'' \left\{ \delta_m^+ \left(\frac{U_k{}^{(m)} + U_k{}^{(m-1)}}{2} \right) \cdot \delta_m^+ \left(U_k{}^{(m)} - U_k{}^{(m-1)} \right) \right\} \Delta x$$
$$= \sum_{k=0}^{N}{}'' \left\{ \delta_m^{\langle 2 \rangle} U_k{}^{(m)} \cdot \delta_m^{\langle 1 \rangle} U_k{}^{(m)} \Delta t \right\} \Delta x. \quad (3.118)$$

In the last equality, note that

$$\delta_m^+ \left(\frac{U_k{}^{(m)} + U_k{}^{(m-1)}}{2} \right) = \frac{U_k{}^{(m+1)} - U_k{}^{(m-1)}}{2\Delta t} = \delta_m^{\langle 1 \rangle} U_k{}^{(m)},$$

and

$$\delta_m^+ \left(U_k{}^{(m)} - U_k{}^{(m-1)} \right) = \frac{U_k{}^{(m+1)} - 2U_k{}^{(m)} + U_k{}^{(m-1)}}{\Delta t} = \delta_m^{\langle 2 \rangle} U_k{}^{(m)} \Delta t.$$

To deal with the second half of the right hand side of (3.117), $G_{\mathrm{d},k}(\boldsymbol{U}^{(m+1)}, \boldsymbol{U}^{(m)}) - G_{\mathrm{d},k}(\boldsymbol{U}^{(m)}, \boldsymbol{U}^{(m-1)})$, we must extend the concept of discrete variational derivative to *three-points* discrete variational derivative. To this end, let us put the same assumption as before that the energy $G(u, u_x)$ is of the form

$$G(u, u_x) = \sum_{l=1}^{\widetilde{M}} f_l(u) g_l(u_x), \quad \widetilde{M} \in \mathbb{N}. \tag{3.22}$$

Let us define a discrete analogue of G with two solutions as

$$G_{\mathrm{d},k}(\boldsymbol{U}^{(m+1)}, \boldsymbol{U}^{(m)}) = \sum_{l=1}^{M} f_l(U_k^{(m+1)}, U_k^{(m)}) g_l^+(\delta_k^+ U_k^{(m+1)}, \delta_k^+ U_k^{(m)}) g_l^-(\delta_k^- U_k^{(m+1)}, \delta_k^- U_k^{(m)}). \tag{3.119}$$

For consistency, functions $g_l^\pm$ are supposed to satisfy, for example,

$$g_l^+(\delta_k^+ U_k^{(m+1)}, \delta_k^+ U_k^{(m)}) g_l^-(\delta_k^- U_k^{(m+1)}, \delta_k^- U_k^{(m)}) \simeq g_l(u_x)|_{x=k\Delta x, t=(m+1/2)\Delta t}. \tag{3.120}$$

Then after some calculation we obtain

$$\sum_{k=0}^{N}{}'' G_{\mathrm{d},k}(\boldsymbol{U}^{(m+1)}, \boldsymbol{U}^{(m)}) - G_{\mathrm{d},k}(\boldsymbol{U}^{(m)}, \boldsymbol{U}^{(m-1)}) \Delta x$$

$$= \sum_{k=0}^{N}{}'' \left\{ \frac{\partial G_{\mathrm{d}}}{\partial (\boldsymbol{U}^{(m+1)}, \boldsymbol{U}^{(m)}, \boldsymbol{U}^{(m-1)})_k} \left(\frac{U_k^{(m+1)} - U_k^{(m-1)}}{2} \right) \right.$$

$$+ \frac{\partial G_{\mathrm{d}}}{\partial \delta^+ (\boldsymbol{U}^{(m+1)}, \boldsymbol{U}^{(m)}, \boldsymbol{U}^{(m-1)})_k} \cdot \delta_k^+ \left(\frac{U_k^{(m+1)} - U_k^{(m-1)}}{2} \right)$$

$$\left. + \frac{\partial G_{\mathrm{d}}}{\partial \delta^- (\boldsymbol{U}^{(m+1)}, \boldsymbol{U}^{(m)}, \boldsymbol{U}^{(m-1)})_k} \cdot \delta_k^- \left(\frac{U_k^{(m+1)} - U_k^{(m-1)}}{2} \right) \right\} \Delta x$$

$$= \sum_{k=0}^{N}{}'' \left\{ \frac{\partial G_{\mathrm{d}}}{\partial (\boldsymbol{U}^{(m+1)}, \boldsymbol{U}^{(m)}, \boldsymbol{U}^{(m-1)})_k} \left(\delta_m^{\langle 1 \rangle} U_k^{(m)} \right) \right.$$

$$+ \frac{\partial G_{\mathrm{d}}}{\partial \delta^+ (\boldsymbol{U}^{(m+1)}, \boldsymbol{U}^{(m)}, \boldsymbol{U}^{(m-1)})_k} \cdot \delta_k^+ \left(\delta_m^{\langle 1 \rangle} U_k^{(m)} \right)$$

$$\left. + \frac{\partial G_{\mathrm{d}}}{\partial \delta^- (\boldsymbol{U}^{(m+1)}, \boldsymbol{U}^{(m)}, \boldsymbol{U}^{(m-1)})_k} \cdot \delta_k^- \left(\delta_m^{\langle 1 \rangle} U_k^{(m)} \right) \right\} \Delta t \, \Delta x, \tag{3.121}$$

where

$$\frac{\partial G_{\mathrm{d}}}{\partial (\boldsymbol{U}^{(m+1)}, \boldsymbol{U}^{(m)}, \boldsymbol{U}^{(m-1)})_k} =$$
$$\sum_{l=1}^{M} \frac{f_l^{(m,m+1)} - f_l^{(m,m-1)}}{\frac{1}{2}(U_k^{(m+1)} - U_k^{(m-1)})}$$
$$\times \left(\frac{g_l^{+,(m,m+1)} g_l^{-,(m,m+1)} + g_l^{+,(m,m-1)} g_l^{-,(m,m-1)}}{2} \right), \quad (3.122\mathrm{a})$$

$$\frac{\partial G_{\mathrm{d}}}{\partial \delta^{\pm}(\boldsymbol{U}^{(m+1)}, \boldsymbol{U}^{(m)}, \boldsymbol{U}^{(m-1)})_k} =$$
$$\sum_{l=1}^{M} \left(\frac{f_l^{(m,m+1)} + f_l^{(m,m-1)}}{2} \right) \left(\frac{g_l^{\mp,(m,m+1)} + g_l^{\mp,(m,m-1)}}{2} \right)$$
$$\times \left(\frac{g_l^{\pm,(m,m+1)} - g_l^{\pm,(m,m-1)}}{\frac{1}{2}\delta_k^{\pm}(U_k^{(m+1)} - U_k^{(m-1)})} \right). \quad (3.122\mathrm{b})$$

In the above notation we used the abbreviations

$$f_l^{(m,m+1)} \stackrel{\mathrm{d}}{\equiv} f_l(U_k^{(m+1)}, U_k^{(m)}),$$
$$g_l^{\pm,(m,m+1)} \stackrel{\mathrm{d}}{\equiv} g_l^{\pm}(\delta_k^{\pm} U_k^{(m+1)}, \delta_k^{\pm} U_k^{(m)}),$$

and so on, for the sake of space. Applying the summation-by-parts formula (3.12a), we obtain

$$\sum_{k=0}^{N} {}'' G_{\mathrm{d},k}(\boldsymbol{U}^{(m+1)}, \boldsymbol{U}^{(m)}) - G_{\mathrm{d},k}(\boldsymbol{U}^{(m)}, \boldsymbol{U}^{(m-1)}) \Delta x$$
$$= \sum_{k=0}^{N} {}'' \frac{\delta G_{\mathrm{d}}}{\delta(\boldsymbol{U}^{(m+1)}, \boldsymbol{U}^{(m)}, \boldsymbol{U}^{(m-1)})_k} \left(\frac{U_k^{(m+1)} - U_k^{(m-1)}}{2} \right)$$
$$+ B_{\mathrm{tt}}(\boldsymbol{U}^{(m+1)}, \boldsymbol{U}^{(m)}, \boldsymbol{U}^{(m-1)}), \quad (3.123)$$

where

$$\frac{\delta G_{\mathrm{d}}}{\delta(\boldsymbol{U}^{(m+1)}, \boldsymbol{U}^{(m)}, \boldsymbol{U}^{(m-1)})_k}$$
$$\stackrel{\mathrm{d}}{\equiv} \frac{\partial G_{\mathrm{d}}}{\partial(\boldsymbol{U}^{(m+1)}, \boldsymbol{U}^{(m)}, \boldsymbol{U}^{(m-1)})_k} - \delta_k^{-} \left(\frac{\partial G_{\mathrm{d}}}{\partial \delta^{+}(\boldsymbol{U}^{(m+1)}, \boldsymbol{U}^{(m)}, \boldsymbol{U}^{(m-1)})_k} \right)$$
$$- \delta_k^{+} \left(\frac{\partial G_{\mathrm{d}}}{\partial \delta^{-}(\boldsymbol{U}^{(m+1)}, \boldsymbol{U}^{(m)}, \boldsymbol{U}^{(m-1)})_k} \right), \quad (3.124)$$

$$B_{\text{tt}}(\boldsymbol{U}^{(m+1)}, \boldsymbol{U}^{(m)}, \boldsymbol{U}^{(m-1)}) \stackrel{\text{d}}{\equiv}$$
$$\frac{\Delta t}{2} \left[\frac{\partial G_{\text{d}}}{\partial (\boldsymbol{U}^{(m+1)}, \boldsymbol{U}^{(m)}, \boldsymbol{U}^{(m-1)})_k} \left\{ s_k^+ \left(\delta_m^{\langle 1 \rangle} U_k^{(m)} \right) \right\} \right.$$
$$+ \left\{ s_k^- \left(\frac{\partial G_{\text{d}}}{\partial (\boldsymbol{U}^{(m+1)}, \boldsymbol{U}^{(m)}, \boldsymbol{U}^{(m-1)})_k} \right) \right\} \left(\delta_m^{\langle 1 \rangle} U_k^{(m)} \right)$$
$$+ \frac{\partial G_{\text{d}}}{\partial \delta^-(\boldsymbol{U}^{(m+1)}, \boldsymbol{U}^{(m)}, \boldsymbol{U}^{(m-1)})_k} \left\{ s_k^- \left(\delta_m^{\langle 1 \rangle} U_k^{(m)} \right) \right\}$$
$$\left. + \left\{ s_k^+ \left(\frac{\partial G_{\text{d}}}{\partial \delta^-(\boldsymbol{U}^{(m+1)}, \boldsymbol{U}^{(m)}, \boldsymbol{U}^{(m-1)})_k} \right) \right\} \left(\delta_m^{\langle 1 \rangle} U_k^{(m)} \right) \right]_0^N. \tag{3.125}$$

We call the discrete quantity

$$\frac{\delta G_{\text{d}}}{\delta (\boldsymbol{U}^{(m+1)}, \boldsymbol{U}^{(m)}, \boldsymbol{U}^{(m-1)})_k},$$

the "three-points discrete variational derivative," since it refers three points $m-1$, m, $m+1$, in time. The identity (3.123) corresponds to the continuous variation (2.56). The term B_{tt} corresponds to the boundary term in (2.56), and the subscript "tt" is taken from the target PDE $u_{tt} = \cdots$.

Collecting (3.118) and (3.123), we can now summarize the difference (3.117) as

$$\frac{1}{\Delta t} \left(J_{\text{d}}(\boldsymbol{U}^{(m+1)}, \boldsymbol{U}^{(m)}) - J_{\text{d}}(\boldsymbol{U}^{(m)}, \boldsymbol{U}^{(m-1)}) \right)$$
$$= \sum_{k=0}^{N} {}'' \left\{ \delta_m^{\langle 2 \rangle} U_k^{(m)} + \frac{\delta G_{\text{d}}}{\delta (\boldsymbol{U}^{(m+1)}, \boldsymbol{U}^{(m)}, \boldsymbol{U}^{(m-1)})_k} \right\} \delta_m^{\langle 1 \rangle} U_k^{(m)} \Delta x$$
$$+ B_{\text{tt}}(\boldsymbol{U}^{(m+1)}, \boldsymbol{U}^{(m)}, \boldsymbol{U}^{(m-1)})/\Delta t. \tag{3.126}$$

This leads us to the next scheme.

Scheme 3.7 (Conservative scheme I for the PDEs 7) *Suppose the initial data $\boldsymbol{U}^{(0)}$ and the starting value $\boldsymbol{U}^{(1)}$ are given. Then, for $m = 1, 2, \ldots,$*

$$\delta_m^{\langle 2 \rangle} U_k^{(m)} = -\frac{\delta G_{\text{d}}}{\delta (\boldsymbol{U}^{(m+1)}, \boldsymbol{U}^{(m)}, \boldsymbol{U}^{(m-1)})_k}, \qquad k = 0, \ldots, N. \tag{3.127}$$

PROPOSITION 3.5 Conservation property of Scheme 3.7
Suppose that discrete boundary conditions are imposed so that

$$B_{\text{tt}}(\boldsymbol{U}^{(m+1)}, \boldsymbol{U}^{(m)}, \boldsymbol{U}^{(m-1)}) = 0, \qquad m = 1, 2, 3, \ldots.$$

Then Scheme 3.7 is conservative in the sense that

$$J_\mathrm{d}(\boldsymbol{U}^{(m+1)}, \boldsymbol{U}^{(m)}) = J_\mathrm{d}(\boldsymbol{U}^{(1)}, \boldsymbol{U}^{(0)}), \qquad m = 1, 2, \ldots \qquad (3.128)$$

holds.

PROOF   Clear from (3.126). $\square$

REMARK 3.9   Slightly modifying the procedure, we can also design an explicit scheme:

$$\frac{U_k^{(m+2)} - U_k^{(m+1)} - U_k^{(m)} + U_k^{(m-1)}}{2(\Delta t)^2} = -\frac{\delta G_\mathrm{d}}{\delta(\boldsymbol{U}^{(m+1)}, \boldsymbol{U}^{(m)})_k}, \qquad (3.129)$$

which is conservative in the sense that

$$\sum_{k=0}^{N}{}'' \left\{ \frac{(\delta_m^+ U_k^{(m)})(\delta_m^- U_k^{(m)})}{2} + G_{\mathrm{d},k}(\boldsymbol{U}^{(m)}) \right\} \Delta x$$

$$= \sum_{k=0}^{N}{}'' \left\{ \frac{(\delta_m^+ U_k^{(1)})(\delta_m^- U_k^{(1)})}{2} + G_{\mathrm{d},k}(\boldsymbol{U}^{(1)}) \right\} \Delta x \qquad (3.130)$$

holds for $m = 2, 3, \ldots$. The detail of this scheme is found in Furihata [67]. $\square$

3.5.2   Second Approach: System of PDEs

The PDE:

$$\frac{\partial^2 u}{\partial t^2} = -\frac{\delta G}{\delta u} \qquad (2.55)$$

can be rewritten into a system of PDEs by introducing a new variable $v = u_t$:

$$\frac{\partial u}{\partial t} = v = \frac{\partial \widetilde{G}}{\partial v}, \qquad (3.131\mathrm{a})$$

$$\frac{\partial v}{\partial t} = -\frac{\partial \widetilde{G}}{\partial u}, \qquad (3.131\mathrm{b})$$

where $\widetilde{G} = v^2/2 + G(u, u_x)$ is a modified local energy. The conservation law

$$\frac{\mathrm{d}}{\mathrm{d}t} \int_0^L \left\{ \frac{1}{2}(u_t)^2 + G(u, u_x) \right\} \mathrm{d}x = 0 \qquad (3.115)$$

is rewritten accordingly as

$$\frac{\mathrm{d}}{\mathrm{d}t} \int_0^L \widetilde{G}(u, u_x, v) \mathrm{d}x = 0. \qquad (3.132)$$

Since the matrix appearing in the right hand side of (3.131):

$$\begin{pmatrix} 0 & 1 \\ -1 & 0 \end{pmatrix} \tag{3.133}$$

is skew-symmetric, we can immediately apply the method described in the previous section to obtain a conservative scheme. Let $G_{\mathrm{d}}(\boldsymbol{U}^{(m)})$ be a discrete energy for $G(u, u_x)$, and

$$\widetilde{G}_{\mathrm{d},k}(\boldsymbol{U}^{(m)}, \boldsymbol{V}^{(m)}) = \frac{(V_k^{(m)})^2}{2} + G_{\mathrm{d},k}(\boldsymbol{U}^{(m)}), \tag{3.134}$$

be a discrete modified local energy.

Scheme 3.8 (Conservative scheme II for the PDEs 7) *For a given set of initial data* $\boldsymbol{U}^{(0)}, \boldsymbol{V}^{(0)}$, *we compute* $\boldsymbol{U}^{(m)}, \boldsymbol{U}^{(m)}$ $(k = 1, 2, \ldots)$ *by, for* $m = 1, 2, \ldots,$

$$\frac{U_k^{(m+1)} - U_k^{(m)}}{\Delta t} = \frac{\delta \widetilde{G}_{\mathrm{d}}}{\delta(\boldsymbol{V}^{(m+1)}, \boldsymbol{V}^{(m)})_k} = \frac{V_k^{(m+1)} + V_k^{(m)}}{2}, \tag{3.135a}$$

$$\frac{V_k^{(m+1)} - V_k^{(m)}}{\Delta t} = -\frac{\delta \widetilde{G}_{\mathrm{d}}}{\delta(\boldsymbol{U}^{(m+1)}, \boldsymbol{U}^{(m)})_k}, \tag{3.135b}$$

where $k = 0, \ldots, N$.

PROPOSITION 3.6 Conservation property of Scheme 3.8
Scheme 3.8 is conservative in the sense that

$$\sum_{k=0}^{N} {}''\widetilde{G}_{\mathrm{d},k}(\boldsymbol{U}^{(m)}, \boldsymbol{V}^{(m)})\Delta x = \sum_{k=0}^{N} {}''\widetilde{G}_{\mathrm{d},k}(\boldsymbol{U}^{(0)}, \boldsymbol{V}^{(0)}), \quad m = 1, 2, \ldots, \tag{3.136}$$

holds.

PROOF Straightforward and hence omitted. □

REMARK 3.10 By eliminating $V_k^{(m)}$ in (3.135a) by (3.135b), we obtain

$$\delta_m^{\langle 2 \rangle} U_k^{(m)} = -\frac{1}{2}\left(\frac{\delta \widetilde{G}_{\mathrm{d}}}{\delta(\boldsymbol{U}^{(m+1)}, \boldsymbol{U}^{(m)})_k} + \frac{\delta \widetilde{G}_{\mathrm{d}}}{\delta(\boldsymbol{U}^{(m)}, \boldsymbol{U}^{(m-1)})_k} \right).$$

This is slightly different from Scheme 3.7. □

REMARK 3.11 By considering a slightly modified procedure, we can obtain an explicit scheme as follows. Let us define a discrete modified energy

by
$$\widetilde{G}_{\mathrm{d},k}(\boldsymbol{U}^{(m+1)}, \boldsymbol{U}^{(m)}, \boldsymbol{V}^{(m+1)}, \boldsymbol{V}^{(m)}) = \frac{V_k^{(m+1)} V_k^{(m)}}{2} + G_{\mathrm{d},k}(\boldsymbol{U}^{(m+1)}, \boldsymbol{U}^{(m)}). \tag{3.137}$$

Then considering its discrete variation, we obtain a scheme

$$\frac{U_k^{(m+1)} - U_k^{(m-1)}}{2\Delta t} = V_k^{(m)}, \tag{3.138a}$$

$$\frac{V_k^{(m+1)} - V_k^{(m-1)}}{2\Delta t} = -\frac{\delta \widetilde{G}_{\mathrm{d}}}{\delta(\boldsymbol{U}^{(m+1)}, \boldsymbol{U}^{(m)}, \boldsymbol{U}^{(m-1)})_k}. \tag{3.138b}$$

This scheme is conservative in the sense that

$$\sum_{k=0}^{N} {}'' \widetilde{G}_{\mathrm{d},k}(\boldsymbol{U}^{(m+1)}, \boldsymbol{U}^{(m)}, \boldsymbol{V}^{(m+1)}, \boldsymbol{V}^{(m)}) \Delta x =$$
$$\sum_{k=0}^{N} {}'' \widetilde{G}_{\mathrm{d},k}(\boldsymbol{U}^{(1)}, \boldsymbol{U}^{(0)}, \boldsymbol{V}^{(1)}, \boldsymbol{V}^{(0)}) \Delta x, \tag{3.139}$$

holds for $m = 1, 2, \ldots$.

Another scheme can be obtained by replacing $V_k^{(m+1)} V_k^{(m)}/2$ in the discrete energy function by $\{(V_k^{(m+1)})^2 + (V_k^{(m)})^2\}/2$. We omit its detail. □

REMARK 3.12 If we introduce the so-called "staggered" grid, we can show that the resulting schemes obtained by the first approach can be interpreted as special cases of the resulting schemes by the second approach.

Let us start with the system of PDEs representation (3.131), but this time utilizing the staggered grid for discretizing the variable v; $V_k^{(m+\frac{1}{2})} \simeq v(k\Delta x, (m+\frac{1}{2})\Delta t)$. Then if we define a discrete modified energy by

$$\widetilde{G}_{\mathrm{d},k}(\boldsymbol{U}^{(m+1)}, \boldsymbol{U}^{(m)}, \boldsymbol{V}^{(m+\frac{1}{2})}) = \frac{(V_k^{(m+\frac{1}{2})})^2}{2} + G_{\mathrm{d},k}(\boldsymbol{U}^{(m+1)}, \boldsymbol{U}^{(m)}), \tag{3.140}$$

we have an implicit scheme:

$$\frac{U_k^{(m+1)} - U_k^{(m-1)}}{2\Delta t} = \frac{V_k^{(m+\frac{1}{2})} + V_k^{(m-\frac{1}{2})}}{2}, \tag{3.141a}$$

$$\frac{V_k^{(m+\frac{1}{2})} - V_k^{(m-\frac{1}{2})}}{\Delta t} = -\frac{\delta \widetilde{G}_{\mathrm{d}}}{\delta(\boldsymbol{U}^{(m+1)}, \boldsymbol{U}^{(m)}, \boldsymbol{U}^{(m-1)})_k}, \tag{3.141b}$$

through discrete variation calculation. This scheme is conservative in the sense that

$$\sum_{k=0}^{N} {}'' \widetilde{G}_{\mathrm{d},k}(\boldsymbol{U}^{(m+1)}, \boldsymbol{U}^{(m)}, \boldsymbol{V}^{(m+\frac{1}{2})}) \Delta x = \sum_{k=0}^{N} {}'' \widetilde{G}_{\mathrm{d},k}(\boldsymbol{U}^{(1)}, \boldsymbol{U}^{(0)}, \boldsymbol{V}^{(\frac{1}{2})}) \Delta x \tag{3.142}$$

holds for $m = 1, 2, \ldots$. Furthermore, if we assume the initial data satisfy the relation

$$\frac{U_k^{(1)} - U_k^{(0)}}{\Delta t} = V_k^{(\frac{1}{2})}, \qquad (3.143)$$

then (3.141a) reduces to

$$\frac{U_k^{(m+1)} - U_k^{(m)}}{\Delta t} = V_k^{(m+\frac{1}{2})}. \qquad (3.141a')$$

Subtracting the equations (3.141a') with $m+1$ and m, we obtain

$$\delta_m^{\langle 2 \rangle} U_k^{(m+1)} = \frac{V_k^{(m+\frac{1}{2})} - V_k^{(m-\frac{1}{2})}}{\Delta t} = -\frac{\delta \widetilde{G}_\mathrm{d}}{\delta (\boldsymbol{U}^{(m+1)}, \boldsymbol{U}^{(m)}, \boldsymbol{U}^{(m-1)})_k}, \qquad (3.144)$$

which is nothing but Scheme 3.7. □

REMARK 3.13 The explicit scheme (3.129) can be also derived by the system-of-PDEs approach, if we use the staggered grid in v. If we define a discrete modified energy by

$$\widetilde{G}_{\mathrm{d},k}(\boldsymbol{U}^{(m)}, \boldsymbol{V}^{(m+\frac{1}{2})}, \boldsymbol{V}^{(m-\frac{1}{2})}) = \frac{V_k^{(m+\frac{1}{2})} V_k^{(m-\frac{1}{2})}}{2} + G_{\mathrm{d},k}(\boldsymbol{U}^{(m)}), \qquad (3.145)$$

we have an explicit scheme:

$$\frac{U_k^{(m+1)} - U_k^{(m)}}{\Delta t} = V_k^{(m+\frac{1}{2})}, \qquad (3.146\mathrm{a})$$

$$\frac{V_k^{(m+\frac{3}{2})} - V_k^{(m-\frac{1}{2})}}{2\Delta t} = -\frac{\delta \widetilde{G}_\mathrm{d}}{\delta (\boldsymbol{U}^{(m+1)}, \boldsymbol{U}^{(m)})_k}, \qquad (3.146\mathrm{b})$$

through discrete variation calculation. Eliminating $V_k^{(m+\frac{1}{2})}$ we obtain

$$\frac{U_k^{(m+2)} - U_k^{(m+1)} - U_k^{(m)} + U_k^{(m-1)}}{2(\Delta t)^2} = -\frac{\delta \widetilde{G}_\mathrm{d}}{\delta (\boldsymbol{U}^{(m+1)}, \boldsymbol{U}^{(m)})_k}, \qquad (3.147)$$

which is the scheme (3.129). □

3.6 Preliminaries on Discrete Functional Analysis

In the subsequent chapters, we sometimes try the theoretical analyses of the constructed schemes. For those presentations, it is convenient to prepare some notation on the discrete version of functional analysis (which we call "*discrete* functional analysis").

As repeatedly declared, throughout this book we basically consider the one-dimensional case on $\Omega = [0, L]$, unless otherwise explicitly stated. We denote the standard Lebesgue space by $L^p(\Omega)$ ($p = 1, 2, \ldots, \infty$) with $\|\cdot\|_p$, its associated norm. For $L^2(\Omega)$, we denote the associated inner product by $(\,\cdot\,,\,\cdot\,)$. We also denote the Sobolev space by $H^s(\Omega)$ ($s = 1, 2, \ldots$) and its norm by $\|\cdot\|_{H^s}$. We define $L^p_{\mathrm{p}}(\Omega)$ and $H^s_{\mathrm{p}}(\Omega)$ as the sets of periodic functions in $L^p(\Omega)$ and $H^s(\Omega)$, respectively.

3.6.1 Discrete Function Spaces

Let us introduce the discrete counterparts of the above function spaces. We divide $\Omega = [0, L]$ into N meshes, i.e., $L = N\Delta x$, and denote numerical solutions by $U_k^{(m)} \simeq u(k\Delta x, m\Delta t)$ ($k = 0, \ldots, N$, $m = 0, 1, 2, \ldots$). We represent the mesh by Ω_N.

We define a finite-dimensional space $L^p(\Omega_N)$, which is a discrete version of $L^p(\Omega)$, by

$$L^p(\Omega_N) \stackrel{\mathrm{d}}{\equiv} \{ \boldsymbol{U} \mid \boldsymbol{U} \in \mathbb{C}^{N+1},\ \sum_{k=0}^{N}{}''|U_k|^p \Delta x < \infty \}. \tag{3.148}$$

Its associated norm is defined by

$$\|\boldsymbol{U}\|_p \stackrel{\mathrm{d}}{\equiv} \left(\sum_{k=0}^{N}{}''|U_k|^p \Delta x \right)^{1/p}. \tag{3.149}$$

We often omit the subscript 2 and write $\|\boldsymbol{U}\|$ when $p = 2$, and we also use the inner product:

$$(\boldsymbol{U}, \boldsymbol{V}) \stackrel{\mathrm{d}}{\equiv} \sum_{k=0}^{N}{}'' \overline{U_k} V_k \Delta x. \tag{3.150}$$

We also define the discrete sup space by

$$L^\infty(\Omega_N) \stackrel{\mathrm{d}}{\equiv} \{ \boldsymbol{U} \mid \boldsymbol{U} \in \mathbb{C}^{N+1},\ \max_{0 \le k \le N} |U_k| < \infty \} \tag{3.151}$$

and its associated norm is defined by

$$\|\boldsymbol{U}\|_\infty \stackrel{\mathrm{d}}{\equiv} \max_{0 \le k \le N} |U_k|. \tag{3.152}$$

A discrete version of $H^1(\Omega)$ is introduced as

$$H^1(\Omega_N) \stackrel{\mathrm{d}}{=} \left\{ \boldsymbol{U} \,\middle|\, \boldsymbol{U} \in \mathbb{C}^{N+1},\ \sum_{k=0}^{N}{}''|U_k|^2 \Delta x + \sum_{k=0}^{N-1}|\delta_k^+ U_k|^2 \Delta x < \infty \right\}. \quad (3.153)$$

Its norm is defined by

$$\|\boldsymbol{U}\|_{H^1} \stackrel{\mathrm{d}}{=} \left(\sum_{k=0}^{N}{}''|U_k|^2 \Delta x + \sum_{k=0}^{N-1}|\delta_k^+ U_k|^2 \Delta x \right)^{1/2}. \quad (3.154)$$

For convenience we often write this as

$$\|\boldsymbol{U}\|_{H^1} = \left(\|\boldsymbol{U}\|^2 + \|\boldsymbol{U}_x\|^2 \right)^{1/2} \quad (3.155)$$

where

$$\|\boldsymbol{U}_x\| \stackrel{\mathrm{d}}{=} \left(\sum_{k=0}^{N-1}|\delta_k^+ U_k|^2 \Delta x \right)^{1/2}. \quad (3.156)$$

When the periodic boundary condition is applied, we naturally assume $U_0 = U_N$, and the finite-dimensional space is substantially of dimension N, which we denote by $\mathbb{S}_N$:

$$\mathbb{S}_N \stackrel{\mathrm{d}}{=} \left\{ (V_k \in \mathbb{C})_{k \in \mathbb{Z}} \,\middle|\, V_k = V_{k \bmod N} \right\}. \quad (3.157)$$

Under the periodic boundary condition we define a discrete version of L_{p}^p by

$$L_{\mathrm{p}}^p(\Omega_N) \stackrel{\mathrm{d}}{=} \left\{ \boldsymbol{U} \,\middle|\, \boldsymbol{U} \in \mathbb{S}_N,\ \sum_{k=0}^{N-1}|U_k|^p \Delta x < \infty \right\} \quad (3.158)$$

and its associated norm is also defined by

$$\|\boldsymbol{U}\|_p \stackrel{\mathrm{d}}{=} \left(\sum_{k=0}^{N-1}|U_k|^p \Delta x \right)^{1/p}. \quad (3.159)$$

We note that this norm is equivalent to (3.149) under the periodic boundary condition $U_N = U_0$. We also introduce the inner product

$$(\boldsymbol{U}, \boldsymbol{V}) \stackrel{\mathrm{d}}{=} \sum_{k=0}^{N-1} \overline{U_k} V_k \Delta x \quad (3.160)$$

for $\boldsymbol{U}, \boldsymbol{V} \in \mathbb{S}_N$ and the discrete Sobolev space $H_{\mathrm{p}}^1(\Omega_N)$ as

$$H_{\mathrm{p}}^1(\Omega_N) \stackrel{\mathrm{d}}{=} \left\{ \boldsymbol{U} \,\middle|\, \boldsymbol{U} \in \mathbb{S}_N,\ \|\boldsymbol{U}\|^2 + \|\boldsymbol{U}_x\|^2 < \infty \right\} \quad (3.161)$$

where

$$\|\boldsymbol{U}_x\| \stackrel{\mathrm{d}}{=} \left\{ \left(\sum_{k=0}^{N-2} |\delta_k^+ U_k|^2 + \left| \frac{U_0 - U_{N-1}}{\Delta x} \right|^2 \right) \Delta x \right\}^{1/2}. \quad (3.162)$$

We also use the "$*$" product:

$$\boldsymbol{U} * \boldsymbol{V} \stackrel{\mathrm{d}}{=} \begin{cases} (U_0 V_0, U_1 V_1, \cdots, U_N V_N)^\mathrm{T} & \text{for } \boldsymbol{U}, \boldsymbol{V} \in \mathbb{C}^{N+1}, \\ (U_0 V_0, U_1 V_1, \cdots, U_{N-1} V_{N-1})^\mathrm{T} & \text{for } \boldsymbol{U}, \boldsymbol{V} \in \mathbb{S}_N. \end{cases} \quad (3.163)$$

It is straightforward to see that the product satisfies the following inequalities.

$$\|\boldsymbol{U} * \boldsymbol{V}\| \leq \begin{cases} \sqrt{\dfrac{2}{\Delta x}} \|\boldsymbol{U}\| \|\boldsymbol{V}\| & \text{for } \boldsymbol{U}, \boldsymbol{V} \in \mathbb{C}^{N+1}, \\ \dfrac{1}{\sqrt{\Delta x}} \|\boldsymbol{U}\| \|\boldsymbol{V}\| & \text{for } \boldsymbol{U}, \boldsymbol{V} \in \mathbb{S}_N. \end{cases} \quad (3.164)$$

3.6.2 Discrete Inequalities

3.6.2.1 Discrete Sobolev Lemma

LEMMA 3.2
With $L = N \Delta x$,

$$\|\boldsymbol{u}\|_\infty \leq \begin{cases} 2 \max\left(\dfrac{1}{\sqrt{L}}, \sqrt{\dfrac{L}{2}} \right) \|\boldsymbol{u}\|_{H^1} & \text{for } \boldsymbol{u} \in H^1(\Omega_N), \\ \sqrt{2} \max\left(\dfrac{1}{\sqrt{L}}, \sqrt{L} \right) \|\boldsymbol{u}\|_{H^1} & \text{for } \boldsymbol{u} \in H^1_\mathrm{p}(\Omega_N). \end{cases} \quad (3.165)$$

PROOF For simplicity we consider the real case: $\boldsymbol{u} \in \mathbb{R}^{N+1}$ (but the following proof can be easily extended to the complex case.) First we note

$$u_m - u_l = (u_m - u_{m-1}) + (u_{m-1} - u_{m-2}) + \cdots + (u_{l+1} - u_l)$$
$$= \Delta x \sum_{k=l}^{m-1} \delta_k^+ u_k, \quad (3.166)$$

which holds for $0 \leq l, m \leq N$. Squaring this and applying the Cauchy–Schwartz inequality, we obtain

$$(u_m - u_l)^2 \leq (\Delta x)^2 N \sum_{k=0}^{N-1} \left(\delta_k^+ u_k \right)^2. \quad (3.167)$$

Using $a^2/2 - b^2 \leq (a-b)^2$ gives

$$(u_m)^2 \leq 2(u_l)^2 + 2(\Delta x)^2 N \sum_{k=0}^{N-1} \left(\delta_k^+ u_k\right)^2. \tag{3.168}$$

From (3.168) we see

$$(u_m)^2 \leq 4(u_l)^2 + 2(\Delta x)^2 N \sum_{k=0}^{N-1} \left(\delta_k^+ u_k\right)^2. \tag{3.169}$$

Then, adding (3.168) and (3.169) we obtain

$$(N+1)(u_m)^2 \leq 4\sum_{k=0}^{N}{}''(u_k)^2 + 2(N+1)N(\Delta x)^2 \sum_{k=0}^{N-1} \left(\delta_k^+ u_k\right)^2. \tag{3.170}$$

Thus we have

$$\begin{aligned}
(u_m)^2 &\leq \frac{4}{N+1}\sum_{k=0}^{N}{}''(u_k)^2 + 2N(\Delta x)^2 \sum_{k=0}^{N-1} \left(\delta_k^+ u_k\right)^2 \\
&\leq \frac{4}{N}\sum_{k=0}^{N}{}''(u_k)^2 + 2N(\Delta x)^2 \sum_{k=0}^{N-1} \left(\delta_k^+ u_k\right)^2 \\
&= \frac{4\Delta x}{L}\sum_{k=0}^{N}{}''(u_k)^2 + 2L\Delta x \sum_{k=0}^{N-1} \left(\delta_k^+ u_k\right)^2 \\
&\leq 4\max\left(\frac{1}{L}, \frac{L}{2}\right)\left\{\sum_{k=0}^{N}{}''(u_k)^2\Delta x + \sum_{k=0}^{N-1}\left(\delta_k^+ u_k\right)^2 \Delta x\right\} \\
&= 4\max\left(\frac{1}{L}, \frac{L}{2}\right)\|\boldsymbol{u}\|_{H^1}^2.
\end{aligned} \tag{3.171}$$

The case $\boldsymbol{u} \in \mathbb{S}_N$ can be proved in a similar manner. □

In the section 8.6 of [93], a description can be found essentially equivalent to the above proof.

3.6.2.2 Discrete Poincaré–Wirtinger inequality

LEMMA 3.3
For any $\boldsymbol{u} \in \boldsymbol{R}^{N+1}$ and $0 \leq {}^\forall m \leq N$, the following inequality holds.

$$\frac{1}{L}\left(u_m - \frac{M}{L}\right)^2 \leq \|\boldsymbol{u}_x\|^2, \tag{3.172}$$

where
$$M \stackrel{\mathrm{d}}{=} \sum_{k=0}^{N} {}'' u_k \Delta x. \tag{3.173}$$

PROOF For any m such that $0 \le m \le N$ we have

$$u_m L - M = \sum_{k=0}^{N} {}'' (u_m - u_k) \Delta x = \sum_{k=0}^{N} {}'' \gamma_{k,m}(u) \Delta x, \tag{3.174}$$

where

$$\gamma_{k,m}(u) \stackrel{\mathrm{d}}{=} \begin{cases} \displaystyle\sum_{l=k}^{m-1} (\delta_l^+ u_l)\, \Delta x, & k \le m, \\ -\displaystyle\sum_{l=m}^{k-1} (\delta_l^+ u_l)\, \Delta x, & m < k. \end{cases} \tag{3.175}$$

This implies

$$\left| u_m - M/L \right| \le \sum_{k=0}^{N-1} \left| \delta_k^+ u_k \right| \Delta x, \tag{3.176}$$

since

$$\left| u_m L - M \right| \le \sum_{k=0}^{N} {}'' \left| \gamma_{k,m}(u) \right| \Delta x, \tag{3.177}$$

$$\left| \gamma_{k,m}(u) \right| \le \sum_{k=0}^{N-1} \left| \delta_k^+ u_k \right| \Delta x. \tag{3.178}$$

Finally, applying the Schwartz inequality to (3.176) we obtain the inequality (3.172). □

REMARK 3.14 The inequality (3.176) corresponds to the Poincaré–Wirtinger inequality [19, VII.1]. □

3.6.2.3 Discrete Gagliardo–Nirenberg Inequality

The next lemma is a discrete version of the Gagliardo–Nirenberg inequality under the zero Dirichlet boundary condition.

LEMMA 3.4 Discrete Gagliardo–Nirenberg inequality (I)
For any $\boldsymbol{V} \in \{\boldsymbol{U} \in H^1(\Omega_N) \,|\, U_0 = U_N = 0\}$,

$$\|\boldsymbol{V}\|_4^4 \le 2 \|\boldsymbol{V}_x\| \, \|\boldsymbol{V}\|^3. \tag{3.179}$$

PROOF First, we show that

$$\sup_k |V_k|^2 \leq 2\|\boldsymbol{V}_x\|\,\|\boldsymbol{V}\| \qquad (3.180)$$

holds. In fact, for any $n \leq N$,

$$|V_k|^2 = \sum_{l=1}^{n} \delta_k^- |V_l|^2 \Delta x$$

$$= \sum_{l=1}^{n} \left(|V_l|^2 - |V_{l-1}|^2\right)$$

$$= \sum_{l=1}^{n} \frac{1}{2}\left\{(V_l - V_{l-1})(\overline{V_l + V_{l-1}}) + (\overline{V_l - V_{l-1}})(V_l + V_{l-1})\right\}$$

$$= \sum_{l=1}^{n} \left(\delta_k^+ V_{l-1} \mu^+ \overline{V_{l-1}} + \delta_k^+ \overline{V_{l-1}} \mu^+ V_{l-1}\right)\Delta x$$

$$\leq 2\left|\sum_{l=1}^{n}\left(\delta_k^+ V_{l-1}\mu^+\overline{V_{l-1}}\right)\Delta x\right|$$

$$\leq 2\left(\sum_{l=1}^{n}|\delta_k^+ V_{l-1}|^2\Delta x\right)^{\frac{1}{2}}\left(\sum_{l=1}^{n}|\mu^+ V_{l-1}|^2\Delta x\right)^{\frac{1}{2}}$$

$$\leq 2\left(\sum_{l=1}^{N}|\delta_k^+ V_{l-1}|^2\Delta x\right)^{\frac{1}{2}}\left(\sum_{l=1}^{N}|\mu^+ V_{l-1}|^2\Delta x\right)^{\frac{1}{2}}$$

$$\leq 2\|\boldsymbol{V}_x\|\,\|\boldsymbol{V}\|.$$

Taking $\sup_k$ of both sides, we obtain (3.180). Therefore, we obtain

$$\sum_{k=1}^{N} |V_k|^4 \Delta x \leq \left(\sup_k |V_k|^2\right)\|\boldsymbol{V}\|^2$$

$$\leq 2\|\boldsymbol{V}_x\|\,\|\boldsymbol{V}\|^3.$$

□

The periodic boundary condition case can be proved based on the zero Dirichlet boundary condition case.

LEMMA 3.5 Discrete Gagliardo–Nirenberg inequality (II)
Let $N \geq 12$, and $b = 4\sqrt{2}\max(4/L, 1)$. Then,

$$\|\boldsymbol{V}\|_4^4 \leq b\|\boldsymbol{V}\|_{H^1}\|\boldsymbol{V}\|^3 \qquad (3.181)$$

for any $\boldsymbol{V} \in H_{\mathrm{p}}^1(\Omega_N)$.

PROOF Let us define a vector $\boldsymbol{\chi} = (\chi_0, \chi_1, \cdots, \chi_{N-1})^{\mathrm{T}}$ as follows (the condition $N \geq 12$ is required for this expression to be well-defined).

$$\chi_k = \begin{cases} 0, & 0 \leq k \leq [N/8], \\ \frac{4}{L}(x_k - x_{[N/8]}), & [N/8] + 1 \leq k \leq [N/4] - 1, \\ 1, & [N/4] \leq k \leq N - 1 - [N/4], \\ -\frac{4}{L}(x_k - x_{N-1-[N/8]}), & N - [N/4] \leq k \leq N - 2 - [N/8], \\ 0, & N - 1 - [N/8] \leq k \leq N - 1. \end{cases}$$

In the above definition, $[\cdot]$ marks denote Gauss's truncation symbols. Note that $0 \leq \chi_k \leq 1$, and $|\delta_k^+ \chi_k| \leq 4/L$.

With $\boldsymbol{\chi} * \boldsymbol{V}$ from Lemma 3.4,

$$\sum_{k=1}^{N-2} |\chi_k V_k|^4 \Delta x \leq 2 \left(\sum_{k=0}^{N-2} |\delta_k^+(\chi_k V_k)|^2 \Delta x \right)^{\frac{1}{2}} \left(\sum_{k=1}^{N-2} |\chi_k V_k|^2 \Delta x \right)^{\frac{3}{2}}. \quad (3.182)$$

In the above identity, (the second term at the right-hand side) $\leq \|\boldsymbol{V}\|^3$. The first term at the right-hand side is evaluated as

$$\left(\sum_{k=0}^{N-2} |\delta_k^+(\chi_k V_k)|^2 \Delta x \right)^{\frac{1}{2}} = \left(\sum_{k=0}^{N-2} |(\delta_k^+ \chi_k) V_{k+1} + \chi_k (\delta_k^+ V_k)|^2 \Delta x \right)^{\frac{1}{2}}$$

$$\leq \frac{4}{L} \left(\sum_{k=0}^{N-1} |V_{k+1}|^2 \Delta x \right)^{\frac{1}{2}} + \left(\sum_{k=0}^{N-1} |\chi_k (\delta_k^+ V_k)|^2 \Delta x \right)^{\frac{1}{2}}$$

$$\leq \frac{4}{L} \|\boldsymbol{V}\| + \|\boldsymbol{V}_x\|.$$

Thus (3.182) becomes

$$\|\boldsymbol{\chi} * \boldsymbol{V}\|_4^4 \leq 2 \left(\frac{4}{L} \|\boldsymbol{V}\| + \|\boldsymbol{V}_x\| \right) \|\boldsymbol{V}\|^3. \quad (3.183)$$

Next, let us consider a shifted vector $Z_k = (s_k^+)^{2[N/4]} V_k$ (note that now the periodic boundary condition is applied). For $\boldsymbol{Z}$, the same identity as above holds as follows.

$$\|\boldsymbol{\chi} * \boldsymbol{Z}\|_4^4 \leq 2 \left(\frac{4}{L} \|\boldsymbol{Z}\| + \|\boldsymbol{Z}_x\| \right) \|\boldsymbol{Z}\|^3$$

$$= 2 \left(\frac{4}{L} \|\boldsymbol{V}\| + \|\boldsymbol{V}_x\| \right) \|\boldsymbol{V}\|^3. \quad (3.184)$$

Note that, under the periodic boundary condition, the shift does not affect the value of the norm.

On the other hand, from the definition of χ, we have

$$\|V\|_4^4 = \sum_{k=[N/4]}^{N-1-[N/4]} |V_k|^4 \Delta x + \left(\sum_{k=0}^{[N/4]-1} |V_k|^4 \Delta x + \sum_{N-[N/4]}^{N-1} |V_k|^4 \Delta x \right)$$

$$= \sum_{k=[N/4]}^{N-1-[N/4]} |V_k|^4 \Delta x + \sum_{k=[N/4]}^{3[N/4]-1} |Z_k|^4 \Delta x$$

$$\leq \|\chi * V\|_4^4 + \|\chi * Z\|_4^4. \tag{3.185}$$

Thus, from (3.183), (3.184), and (3.185), we have

$$\|V\|_4^4 \leq 4 \left(\frac{4}{L} \|V\| + \|V_x\| \right) \|V\|^3$$

$$\leq 4\sqrt{2} \max(4/L, 1) \|V\|_{H^1} \|V\|^3$$

$$= b \|V\|_{H^1} \|V\|^3.$$

This completes the proof. □

3.6.3 Discrete Gronwall Lemma

The following is a discrete version of the Gronwall lemma.

LEMMA 3.6 Discrete Gronwall lemma [102]
Let $\omega^{(m)}$ and $\rho^{(m)}$ $(m = 0, 1, 2, \ldots)$ be non-negative sequences, and $\rho^{(m)}$ be a non-decreasing sequence. Then, if there exists $c > 0$ satisfying

$$\omega^{(m)} \leq \rho^{(m)} + c\Delta t \sum_{l=0}^{m-1} \omega^{(l)} \quad (m = 1, 2, 3, \ldots),$$

then for all $m = 1, 2, 3, \cdots$,

$$\omega^{(m)} \leq \rho^{(m)} e^{cm\Delta t}.$$

PROOF Let $\eta^{(m)} = \omega^{(m)} e^{-cm\Delta t}$ and $\eta^{(j)} = \max_{0 \leq l \leq m} \eta^{(l)}$. Then,

$$\omega^{(j)} \leq \rho^{(j)} + c\Delta t \sum_{l=0}^{j-1} \omega^{(l)}$$

$$\leq \rho^{(j)} + c\Delta t \eta^{(j)} \sum_{l=0}^{j-1} e^{cl\Delta t}$$

$$\leq \rho^{(m)} + c\eta^{(j)} \int_0^{j\Delta t} e^{cs} ds$$

$$= \rho^{(m)} + \eta^{(j)} \left(e^{cj\Delta t} - 1 \right)$$

holds. Thus, $\eta^{(j)} \leq \rho^{(m)}$. This immediately implies

$$\omega^{(m)} = \eta^{(m)} e^{cm\Delta t} \leq \max_{0 \leq l \leq m} \eta^{(l)} e^{cm\Delta t} \leq \rho^{(m)} e^{cm\Delta t}.$$

□

Chapter 4

Applications

In this chapter we present application examples of the discrete variational derivative method, in order to demonstrate how the method can be applied to actual problems. The examples are classified according to Chapter 2. In some examples, numerical examples and/or theoretical analyses are also given. In the last Section 4.7, we also give several examples for PDEs that are not directly covered by the classification in Chapter 2.

4.1 Target PDEs 1

In this section, examples for the target PDEs 1 (defined in Section 2.2; real-valued, single, dissipative PDEs) are shown. In the first example, in particular, the construction of the scheme (i.e., how we actually apply the discrete variational derivative method) is demonstrated with full detail. This example would help readers' understanding of the method.

4.1.1 Cahn–Hilliard Equation

4.1.1.1 Introduction to Problem

Let us consider the Cahn–Hilliard equation:

$$\frac{\partial u}{\partial t} = \frac{\partial^2}{\partial x^2}\left(pu + ru^3 + q\frac{\partial^2 u}{\partial x^2}\right), \qquad x \in (0, L),\ t > 0, \qquad (4.1)$$

under the boundary conditions:

$$\frac{\partial u}{\partial x} = 0,\ \ x = 0, L, \qquad (4.2)$$

$$\frac{\partial}{\partial x}\left(pu + ru^3 + q\frac{\partial^2 u}{\partial x^2}\right) = 0,\ \ x = 0, L. \qquad (4.3)$$

This is a dissipative PDE of the form 1 (Section 2.2), where

$$s = 1, \quad G(u, u_x) = \frac{1}{2}pu^2 + \frac{1}{4}ru^4 - \frac{1}{2}q\left(\frac{\partial u}{\partial x}\right)^2, \qquad (4.4)$$

which means that
$$\frac{\delta G}{\delta u} = pu + ru^3 + q\frac{\partial^2 u}{\partial x^2}. \quad (4.5)$$

Note that, with this expression, we can rewrite the second boundary condition as
$$\frac{\partial}{\partial x}\frac{\delta G}{\delta u} = 0, \quad x = 0, L. \quad (4.6)$$

Since now
$$\frac{\partial G}{\partial u_x} = -qu_x,$$

the condition (2.16) is satisfied. When $s = 1$, the second condition (2.17) for the dissipation reads as
$$\left[-\frac{\delta G}{\delta u} \cdot \frac{\partial}{\partial x}\frac{\delta G}{\delta u}\right]_0^L = 0,$$

which is assured by (4.6). Thus, the equation is in fact dissipative in view of (2.15). Further background of this equation can be found in Section 1.1.

4.1.1.2 Numerical Scheme

Let us construct a dissipative scheme for the Cahn–Hilliard equation, following the procedure in Chapter 3.

We first show the outline of the scheme construction. We commence by defining a discrete energy function by, for example,
$$G_{\mathrm{d},k}(\boldsymbol{U}) \stackrel{\mathrm{d}}{=} \frac{1}{2}p(U_k)^2 + \frac{1}{4}r(U_k)^4 - \frac{1}{2}q\left(\frac{(\delta_k^+ U_k)^2 + (\delta_k^- U_k)^2}{2}\right). \quad (4.7)$$

Then, by the discrete variation we obtain
$$\frac{\delta G_{\mathrm{d}}}{\delta(\boldsymbol{U}^{(m+1)}, \boldsymbol{U}^{(m)})_k}$$
$$= p\left(\frac{U_k^{(m+1)} + U_k^{(m)}}{2}\right)$$
$$+ r\left(\frac{(U_k^{(m+1)})^3 + (U_k^{(m+1)})^2 U_k^{(m)} + U_k^{(m+1)}(U_k^{(m)})^2 + (U_k^{(m)})^3}{4}\right)$$
$$+ q\delta_k^{\langle 2\rangle}\left(\frac{U_k^{(m+1)} + U_k^{(m)}}{2}\right), \quad (4.8)$$

and by Scheme 3.1, we have

$$\frac{U_k^{(m+1)} - U_k^{(m)}}{\Delta t} = \delta_k^{\langle 2 \rangle} \frac{\delta G_{\mathrm{d}}}{\delta(\boldsymbol{U}^{(m+1)}, \boldsymbol{U}^{(m)})_k}$$

$$= \delta_k^{\langle 2 \rangle} \left\{ p \left(\frac{U_k^{(m+1)} + U_k^{(m)}}{2} \right) \right.$$

$$+ r \left(\frac{(U_k^{(m+1)})^3 + (U_k^{(m+1)})^2 U_k^{(m)} + U_k^{(m+1)} (U_k^{(m)})^2 + (U_k^{(m)})^3}{4} \right)$$

$$\left. + q \delta_k^{\langle 2 \rangle} \left(\frac{U_k^{(m+1)} + U_k^{(m)}}{2} \right) \right\},$$

$$0 \le k \le N, \ m = 0, 1, 2, \ldots. \tag{4.9}$$

This has been already shown (without explanation) in Chapter 1 as Scheme 1.2.

The outline above can be realized in the following two ways. First, let us try the formal procedure of the discrete variational method, where (3.30) and the related definitions (3.27a)–(3.27c) play the central role. In this case, we first note that the discrete energy function (4.7) is of the form (3.23) with

$$M = 4, \ f_1 = \frac{1}{2} p(U_k)^2, g_1^+ = g_1^- = 1, \ f_2 = \frac{1}{4} r(U_k)^4, g_2^+ = g_2^- = 1,$$

$$f_3 = 1, g_3^+ = -\frac{1}{2} q \frac{(\delta_k^+ U_k)^2}{2}, g_3^- = 1,$$

$$f_4 = 1, g_4^+ = 1, g_4^- = -\frac{1}{2} q \frac{(\delta_k^- U_k)^2}{2}. \tag{4.10}$$

Note that although the original energy function consists of three terms (i.e., $\widetilde{M} = 3$, in the expression of (3.22)), the discrete version consists of four terms. This is caused by approximating the derivative term symmetrically by δ_k^+ and δ_k^-. In this way, generally $\widetilde{M} \le M$. Then, by substituting (4.10) into (3.27a)–(3.27c), we obtain

$$\frac{\partial G_{\mathrm{d}}}{\partial (\boldsymbol{U}, \boldsymbol{V})_k} = p \left(\frac{U_k + V_k}{2} \right) + r \left(\frac{(U_k)^3 + (U_k)^2 V_k + U_k (V_k)^2 + (V_k)^3}{4} \right), \tag{4.11}$$

$$\frac{\partial G_{\mathrm{d}}}{\partial \delta^+ (\boldsymbol{U}, \boldsymbol{V})_k} = \frac{\delta_k^+ U_k + \delta_k^+ V_k}{2}, \quad \frac{\partial G_{\mathrm{d}}}{\partial \delta^- (\boldsymbol{U}, \boldsymbol{V})_k} = \frac{\delta_k^- U_k + \delta_k^- V_k}{2}. \tag{4.12}$$

This, together with (3.30), allows us to reach the discrete variational derivative (4.8). Notice that although the expressions (3.27a)–(3.27c) involve the summations of $M = 4$ terms, many terms trivially vanish (see Remark 3.1). For example, in (3.27c), only the term $l = 3$ has a nonzero value, and other

terms vanish due to $g_1^+ = g_2^+ = g_4^+ = 1$, and
$$\frac{g_l^+(\delta_k^+ U_k) - g_l^+(\delta_k^+ V_k)}{\delta_k^+ U_k - \delta_k^+ V_k} = (g_l^+)'(\delta_k^+ U_k) = 0, \quad l = 1, 2, 4.$$

The first approach—the formal approach—has surely an advantage in that it goes automatic once a discrete energy function is given. No expertise is required there. However, the formal expressions such as (3.27a)–(3.27c) are considerably complicated for being generic, and quite often it is much easier to directly consider the discrete variation of the given discrete energy function, as repeatedly emphasized in Remark 3.3 and the subsequent easy examples there. This direct approach—the second approach here—goes as follows. For each of the three terms in the discrete energy function, we easily see

$$\sum_{k=0}^{N}{}''\left(\frac{1}{2}(U_k)^2 - \frac{1}{2}(V_k)^2\right)\Delta x = \sum_{k=0}^{N}{}''\left(\frac{U_k + V_k}{2}\right)(U_k - V_k)\Delta x, \quad (4.13\text{a})$$

$$\sum_{k=0}^{N}{}''\left(\frac{1}{4}(U_k)^4 - \frac{1}{4}(V_k)^4\right)\Delta x =$$
$$\sum_{k=0}^{N}{}''\left(\frac{(U_k)^3 + (U_k)^2 V_k + U_k(V_k)^2 + (V_k)^3}{4}\right)(U_k - V_k)\Delta x, \quad (4.13\text{b})$$

$$\sum_{k=0}^{N}{}''\left(\frac{1}{2}(\delta_k^\pm U_k)^2 - \frac{1}{2}(\delta_k^\pm V_k)^2\right)\Delta x$$
$$= \sum_{k=0}^{N}{}''\delta_k^\pm\left(\frac{U_k + V_k}{2}\right)\cdot\delta_k^\pm(U_k - V_k)\Delta x$$
$$= -\sum_{k=0}^{N}{}''\delta_k^{\langle 2\rangle}\left(\frac{U_k + V_k}{2}\right)(U_k - V_k)\Delta x$$
$$+ \frac{1}{2}\left[\delta_k^+\left(\frac{U_k + V_k}{2}\right)\cdot\mu_k^+(U_k - V_k) + \delta_k^-\left(\frac{U_k + V_k}{2}\right)\cdot\mu_k^-(U_k - V_k)\right]_0^N.$$
$$(4.13\text{c})$$

In the last equality, we use the second order summation-by-parts formula (3.14a). By collecting (4.13a)–(4.13c), and in view of (3.32), we readily obtain the expression (4.8).

Next, let us impose the following discrete boundary conditions:

$$\delta_k^{\langle 1\rangle} U_k^{(m)} = 0, \quad k = 0, N, \quad (4.14)$$

$$\delta_k^{\langle 1\rangle}\frac{\delta G_\text{d}}{\delta(\boldsymbol{U}^{(m+1)}, \boldsymbol{U}^{(m)})_k} = 0, \quad k = 0, N, \quad (4.15)$$

Applications

and check that the scheme is in fact dissipative under these conditions. By the statement of Theorem 3.1, it suffices to check that

$$B_{r,1}(\boldsymbol{U}^{(m+1)}, \boldsymbol{U}^{(m)}) = 0 \quad \text{and} \quad B_{r,2}^{\langle 2 \rangle}(\boldsymbol{U}^{(m+1)}, \boldsymbol{U}^{(m)}) = 0 \quad \text{for } m = 0, 1, 2, \ldots.$$

Again, the check can be done by the two different approaches—the formal approach, and the direct approach. Let us first demonstrate the first approach. By the definition (3.31), and the related definitions (3.27b)–(3.27c), the concrete form of the first condition becomes

$$\frac{1}{2}\left[\delta_k^+ \left(\frac{U_k^{(m+1)} + U_k^{(m)}}{2}\right) \cdot s_k^+ (U_k^{(m+1)} - U_k^{(m)}) \right.$$
$$+ s_k^- \delta_k^+ \left(\frac{U_k^{(m+1)} + U_k^{(m)}}{2}\right) \cdot (U_k^{(m+1)} - U_k^{(m)})$$
$$+ \delta_k^- \left(\frac{U_k^{(m+1)} + U_k^{(m)}}{2}\right) \cdot s_k^- (U_k^{(m+1)} - U_k^{(m)})$$
$$\left. + s_k^+ \delta_k^- \left(\frac{U_k^{(m+1)} + U_k^{(m)}}{2}\right) \cdot (U_k^{(m+1)} - U_k^{(m)}) \right]_0^N = 0. \quad (4.16)$$

It is an easy exercise on discrete operators to see that the above condition is equivalent to

$$\frac{1}{2}\left[\delta_k^+ \left(\frac{U_k^{(m+1)} + U_k^{(m)}}{2}\right) \cdot \mu_k^+ (U_k^{(m+1)} - U_k^{(m)}) \right.$$
$$\left. + \delta_k^- \left(\frac{U_k^{(m+1)} + U_k^{(m)}}{2}\right) \cdot \mu_k^- (U_k^{(m+1)} - U_k^{(m)}) \right]_0^N = 0. \quad (4.17)$$

Since the condition (4.14) implies for $k = 0, N$ that

$$\delta_k^+ U_k^{(m)} = -\delta_k^- U_k^{(m)}, \qquad \mu_k^+ U_k^{(m)} = \mu_k^- U_k^{(m)},$$

the identity (4.17) holds. For $B_{r,2}^{\langle 2 \rangle}(\boldsymbol{U}^{(m+1)}, \boldsymbol{U}^{(m)})$, by its definition in Scheme 3.1,

$$B_{r,2}^{\langle 2 \rangle}(\boldsymbol{U}^{(m+1)}, \boldsymbol{U}^{(m)}) = \frac{1}{4}\left[2\frac{\delta G_d}{\delta(\boldsymbol{U}^{(m+1)}, \boldsymbol{U}^{(m)})_k} \cdot \delta_k^{\langle 1 \rangle} \frac{\delta G_d}{\delta(\boldsymbol{U}^{(m+1)}, \boldsymbol{U}^{(m)})_k} \right.$$
$$+ s_k^+ \frac{\delta G_d}{\delta(\boldsymbol{U}^{(m+1)}, \boldsymbol{U}^{(m)})_k} \cdot \delta_k^+ \frac{\delta G_d}{\delta(\boldsymbol{U}^{(m+1)}, \boldsymbol{U}^{(m)})_k}$$
$$\left. + s_k^- \frac{\delta G_d}{\delta(\boldsymbol{U}^{(m+1)}, \boldsymbol{U}^{(m)})_k} \cdot \delta_k^- \frac{\delta G_d}{\delta(\boldsymbol{U}^{(m+1)}, \boldsymbol{U}^{(m)})_k} \right]_0^N$$
$$= \left[s_k^{\langle 1 \rangle} \frac{\delta G_d}{\delta(\boldsymbol{U}^{(m+1)}, \boldsymbol{U}^{(m)})_k} \cdot \delta_k^{\langle 1 \rangle} \frac{\delta G_d}{\delta(\boldsymbol{U}^{(m+1)}, \boldsymbol{U}^{(m)})_k} \right]_0^N$$
$$= 0.$$

This is guaranteed by the applied discrete boundary condition (4.15).

In the second, direct approach, we have already obtained in (4.13c) the left hand side of the identity (4.17). This completes the check for $B_{\mathrm{r},1}(\boldsymbol{U}^{(m+1)}, \boldsymbol{U}^{(m)})$. For $B_{\mathrm{r},2}^{\langle 2 \rangle}(\boldsymbol{U}^{(m+1)}, \boldsymbol{U}^{(m)})$, we note that by collecting (4.13a)–(4.13c), we already have (by neglecting $B_{\mathrm{r},1}(\boldsymbol{U}^{(m+1)}, \boldsymbol{U}^{(m)})$)

$$\frac{1}{\Delta t}\sum_{k=0}^{N}{}''\left(G_{\mathrm{d},k}(\boldsymbol{U}^{(m+1)}) - G_{\mathrm{d},k}(\boldsymbol{U}^{(m)})\right)\Delta x = \sum_{k=0}^{N}{}''\frac{\delta G_{\mathrm{d}}}{\delta(\boldsymbol{U}^{(m+1)}, \boldsymbol{U}^{(m)})_k}\frac{U_k^{(m+1)} - U_k^{(m)}}{\Delta t}\Delta x. \qquad (4.18)$$

Then by the scheme definition and the second-order summation-by-parts formula (3.14b), this equals

$$\sum_{k=0}^{N}{}''\frac{\delta G_{\mathrm{d}}}{\delta(\boldsymbol{U}^{(m+1)}, \boldsymbol{U}^{(m)})_k}\cdot\delta_k^{\langle 2 \rangle}\frac{\delta G_{\mathrm{d}}}{\delta(\boldsymbol{U}^{(m+1)}, \boldsymbol{U}^{(m)})_k}\Delta x$$
$$= -\frac{1}{2}\sum_{k=0}^{N}{}''\left\{\left(\delta_k^+\frac{\delta G_{\mathrm{d}}}{\delta(\boldsymbol{U}^{(m+1)}, \boldsymbol{U}^{(m)})_k}\right)^2 + \left(\delta_k^-\frac{\delta G_{\mathrm{d}}}{\delta(\boldsymbol{U}^{(m+1)}, \boldsymbol{U}^{(m)})_k}\right)^2\right\}\Delta x$$
$$+ \left[\delta_k^{\langle 1 \rangle}\frac{\delta G_{\mathrm{d}}}{\delta(\boldsymbol{U}^{(m+1)}, \boldsymbol{U}^{(m)})_k}\cdot s_k^{\langle 1 \rangle}\frac{\delta G_{\mathrm{d}}}{\delta(\boldsymbol{U}^{(m+1)}, \boldsymbol{U}^{(m)})_k}\right]_0^N. \qquad (4.19)$$

This coincides with the first case, and we hence complete the check. One can see that, as in the construction of the scheme, it is much simpler to directly check the dissipation property without the complicated formal expressions in Chapter 3. In this sense, *the formal procedure is for mathematical rigorousness, and not for practical use.*

Under the boundary condition (4.15), the scheme has the following additional conservation law.

THEOREM 4.1
The scheme (4.9) is "mass" conservative in the sense that

$$\sum_{k=0}^{N}{}''U_k^{(m)}\Delta x = \sum_{k=0}^{N}{}''U_k^{(0)}\Delta x, \qquad m = 0, 1, 2, \ldots, \qquad (4.20)$$

holds.

PROOF Easily shown as follows:

$$\frac{1}{\Delta t}\sum_{k=0}^{N}{}''\left(U_k^{(m)} - U_k^{(m-1)}\right)\Delta x$$
$$= \frac{1}{\Delta t}\sum_{k=0}^{N}{}''\delta_k^{\langle 2\rangle}\frac{\delta G_{\mathrm{d}}}{\delta(\boldsymbol{U}^{(m+1)},\boldsymbol{U}^{(m)})_k}\Delta x$$
$$= \frac{1}{\Delta t}\left[\delta_k^{\langle 1\rangle}\frac{\delta G_{\mathrm{d}}}{\delta(\boldsymbol{U}^{(m+1)},\boldsymbol{U}^{(m)})_k}\right]_0^N = 0, \qquad (4.21)$$

with the summation of difference (3.8). $\square$

This discrete law corresponds to

$$\frac{\mathrm{d}}{\mathrm{d}t}\int_0^L u\,\mathrm{d}x = 0,$$

in continuous context.

REMARK 4.1 The Cahn–Hilliard equation will be also mentioned in Chapter 6 (a linearly implicit scheme for the Cahn–Hilliard equation) and Chapter 7 (a scheme on non-rectangular mesh). $\square$

4.1.1.3 Numerical Examples

We have already seen the result in Figure 1.4, where the scheme (4.9) was run with a coarse time mesh $\Delta t = 1/1000$. Other parameters were taken to $p = -1.0$, $q = -0.001$, $r = 1.0$, and $L = 1$, $N = 50$ (thus $\Delta x = 1/50$). The discrete energy dissipation property of the scheme is confirmed in Figure 1.5.

4.1.1.4 Analysis of Scheme

In this subsection, we give a theoretical analysis of the scheme. The content of this subsection is based on [66]. For the notation of discrete functional analysis, readers should refer to Section 3.6.

First, let us prove the stability. We commence by the following a priori estimate of the numerical solutions.

PROPOSITION 4.1 Solutions' bounds with discrete Sobolev norm

The solutions, $\boldsymbol{U}^{(m)}$, satisfy the following a priori estimate:

$$\|\boldsymbol{U}^{(m)}\|_{H^1}^2 \leq \frac{1}{\min(-p, -\frac{1}{2}q)}\left\{\sum_{k=0}^{N}{}''G_{d,k}\left(\boldsymbol{U}^{(0)}\right)\Delta x + \frac{9}{4}\frac{p^2}{r}L\right\}, \qquad (4.22)$$

where $\|\bullet\|_{H^1}$ is the discrete first-order Sobolev–Hilbert norm which is defined in Section 3.6. We note them here again for readers' convenience.

$$\|\boldsymbol{f}\|_{H^1}^2 = \|\boldsymbol{f}\|^2 + \|\boldsymbol{f}_x\|^2 \tag{4.23}$$

where

$$\|\boldsymbol{f}\| = \left(\sum_{k=0}^{N}{}''|f_k|^2 \Delta x\right)^{1/2} \tag{4.24}$$

and

$$\|\boldsymbol{f}_x\| = \left(\sum_{k=0}^{N-1}|\delta_k^+ f_k|^2 \Delta x\right)^{1/2}. \tag{4.25}$$

PROOF Thanks to the dissipation property, we see

$$\sum_{k=0}^{N}{}'' G_{\mathrm{d},k}\left(\boldsymbol{U}^{(0)}\right) \Delta x$$

$$\geq \sum_{k=0}^{N}{}'' G_{\mathrm{d},k}\left(\boldsymbol{U}^{(m)}\right) \Delta x$$

$$\geq \sum_{k=0}^{N}{}'' \left\{-p(U_k^{(m)})^2 - \frac{9}{4}\frac{p^2}{r} - \frac{1}{2}q\frac{(\delta_k^+ U_k^{(m)})^2 + (\delta_k^- U_k^{(m)})^2}{2}\right\} \Delta x$$

$$\left(\text{ since } \frac{1}{2}pX^2 + \frac{1}{4}rX^4 \geq -pX^2 - \frac{9}{4}\frac{p^2}{r}\right)$$

$$\geq \min(-p, -\frac{1}{2}q)\sum_{k=0}^{N}{}'' \left\{(U_k^{(m)})^2 + \frac{(\delta_k^+ U_k^{(m)})^2 + (\delta_k^- U_k^{(m)})^2}{2}\right\} \Delta x$$

$$- \frac{9}{4}\frac{p^2}{r}L$$

$$= \min(-p, -\frac{1}{2}q)\|\boldsymbol{U}^{(m)}\|_{H^1}^2 - \frac{9}{4}\frac{p^2}{r}L, \tag{4.26}$$

where we have used the boundary condition (4.14) in the last equality. □

Recall the discussion in the continuous case (page 6). The proposition above means that a similar estimate holds for the discrete solutions as well, where the discrete Sobolev norm of the solutions is bounded by a constant depending only on the initial energy value.

With this and the discrete Sobolev lemma in Section 3.6.2.1, we obtain the following stability result.

THEOREM 4.2
The numerical solutions $\boldsymbol{U}^{(m)}$ by the scheme (4.9) satisfy for all $m \geq 0$,

$$\left\|\boldsymbol{U}^{(m)}\right\|_\infty \leq 2 \left[\frac{\max(1/L, L/2)}{\min(-p, -q/2)} \left\{\sum_{k=0}^{N} {}''G_{\mathrm{d},k}\left(\boldsymbol{U}^{(0)}\right)\Delta x + \frac{9}{4}\frac{p^2}{r}L\right\}\right]^{1/2}. \tag{4.27}$$

This theorem means that the scheme is numerically stable for any time step m (except for possible instabilities caused by rounding errors). This fact eloquently demonstrates that preserving the discrete energy dissipation *is* in fact advantageous. This is a typical illustrative example for our basic philosophy: "**structure-preserving provides superiority**" in computation.

The above estimate depends only on the initial data, and we can have more precise evaluation if the data is sufficiently smooth.

COROLLARY 4.1
If $U_k^{(0)} = u^{(0)}(k\Delta x)$ for a function $u^{(0)}(x) \in C^3[0, L]$, then it holds that

$$\max_{0\leq k\leq N}\left|U_k^{(m)}\right| \leq 2\sqrt{\frac{\max(1/L, L/2)}{\min(-p, -q/2)}\left\{\int_0^L G(u^{(0)})dx + C_0 L^2 + \frac{9}{4}\frac{p^2}{r}L\right\}}, \tag{4.28}$$

where

$$C_0 = \frac{1}{8}\int_0^L \left|\frac{\partial^2 G(u^{(0)})}{\partial x^2}\right|dx + \frac{-q}{2}L\left(\frac{1}{4}A_2^2 + \frac{1}{3}A_1 A_3 + \frac{L^2}{576}A_3^2\right), \tag{4.29}$$

$$A_{m'} = \max_{x\in[0,L]}\left|\frac{\partial^{m'}}{\partial x^{m'}}u^{(0)}\right|, \quad 1 \leq m' \leq 3. \tag{4.30}$$

Since the proof is an easy exercise by the Euler–Maclaurin summation formula, we skip it here.

Next we show the unique existence of the numerical solutions of the scheme. To prove that we use the contraction mapping theorem. Let us define a mapping $\mathcal{T}_{\boldsymbol{U}^{(m)}}: \mathbb{R}^{N+1} \to \mathbb{R}^{N+1}$ in terms of the following equation:

$$\left(1 - \frac{q\Delta t}{2}\delta_k^{\langle 4\rangle}\right)\{\mathcal{T}_{\boldsymbol{U}^{(m)}}\boldsymbol{V}\}_k = U_k^{(m)} + \frac{\Delta t}{2}\delta_k^{\langle 2\rangle}\left\{pV_k + r\{\mathcal{Q}_{\boldsymbol{U}^{(m)}}\boldsymbol{V}\}_k\right\}, \tag{4.31}$$

where the mapping $\mathcal{Q}_{\boldsymbol{U}^{(m)}}: \mathbb{R}^{N+1} \to \mathbb{R}^{N+1}$ is defined as

$$\{\mathcal{Q}_{\boldsymbol{U}^{(m)}}\boldsymbol{V}\}_k \stackrel{\mathrm{d}}{=} (V_k)^3 + V_k\left(V_k - U_k^{(m)}\right)^2. \tag{4.32}$$

We here promise that the operators in the above equation are defined under the boundary conditions (4.14) and (4.15), i.e.,

$$V_{-1} = V_1, \quad V_{N+1} = V_{N-1}, \tag{4.33}$$
$$V_{-2} = V_2, \quad V_{N+2} = V_{N-2}. \tag{4.34}$$

If the mapping $\mathcal{T}_{\boldsymbol{U}^{(m)}}$ has a fixed-point $\boldsymbol{V}^*$, then $2\boldsymbol{V}^* - \boldsymbol{U}^{(m)}$ is the solution $\boldsymbol{U}^{(m+1)}$ of the scheme (4.9). The following proposition implies that the mapping $\mathcal{T}_{\boldsymbol{U}^{(m)}}$ is well-defined for any $\boldsymbol{U}^{(m)}$.

PROPOSITION 4.2
The operator $\left(1 - \dfrac{q\Delta t}{2}\delta_k^{\langle 4 \rangle}\right)$ is nonsingular.

PROOF The $(N+1) \times (N+1)$ matrix expression of $\left(1 - \dfrac{q\Delta t}{2}\delta_k^{\langle 4 \rangle}\right)$ is $\left(I - \dfrac{q\Delta t}{2}D_2^2\right)$, where I is the identity matrix of order $N+1$ and D_2 is the expression matrix of the operator $\delta_k^{\langle 2 \rangle}$, which is defined by the following equality:

$$D_2 \stackrel{\mathrm{d}}{=} \frac{1}{(\Delta x)^2}\begin{pmatrix} -2 & 2 & & & 0 \\ 1 & -2 & 1 & & \\ & \ddots & \ddots & \ddots & \\ & & 1 & -2 & 1 \\ 0 & & & 2 & -2 \end{pmatrix} \tag{4.35}$$

under the boundary condition (4.14). Eigenvalues of D_2 are

$$\lambda_k \stackrel{\mathrm{d}}{=} \frac{2}{(\Delta x)^2}\left\{\cos(\frac{k}{N}\pi) - 1\right\}, \qquad k = 0, 1, \cdots, N \tag{4.36}$$

and accordingly the eigenvalues of $\left(I - \dfrac{q\Delta t}{2}D_2^2\right)$ are $1 - \dfrac{q\Delta t}{2}(\lambda_k)^2$, $k = 0, 1, \cdots, N$. The positiveness of the eigenvalues implies the nonsingularity of $\left(I - \dfrac{q\Delta t}{2}D_2^2\right)$. □

Now we have the following theorem, which states the unique existence of the numerical solution.

THEOREM 4.3
If

$$\Delta t < \min\left(\frac{-q(\Delta x)^2}{2\left(-p\Delta x + 82rM^2\right)^2}, \frac{-2q(\Delta x)^2}{(-p\Delta x + 226rM^2)^2}\right), \tag{4.37}$$

then the mapping $\mathcal{T}_{\boldsymbol{U}^{(m)}}$ has a unique fixed-point in the closed ball K, where

$$M \stackrel{\mathrm{d}}{\equiv} \|\boldsymbol{U}^{(m)}\|_2, \tag{4.38}$$

$$K \stackrel{\mathrm{d}}{\equiv} \left\{\boldsymbol{v} \in \mathbb{R}^{N+1} \Big| \ \|\boldsymbol{v}\|_2 \leq 4M\right\}. \tag{4.39}$$

REMARK 4.2 For the solution $U_k^{(m)}$ of the scheme (4.9), M is bounded as

$$M \leq \sqrt{\frac{1}{\min(-p,-q/2)}\left\{\sum_{k=0}^{N}{''}G_{d,k}\left(\boldsymbol{U}^{(0)}\right)\Delta x + \frac{9}{4}\frac{p^2}{r}L\right\}} \tag{4.40}$$

from the Theorem 4.2. □

REMARK 4.3 Since

$$M \sim \|u(m\Delta t, \cdot)\|_{L^2(0,L)}, \tag{4.41}$$

(4.37) implies that by taking $\Delta t = O(\Delta x^2)$ the unique solvability of the scheme is guaranteed. □

PROOF By the contraction mapping theorem it suffices to show that $\mathcal{T}_{\boldsymbol{U}^{(m)}}$ is a contraction mapping on K. We prove that $\mathcal{T}_{\boldsymbol{U}^{(m)}}$ is a mapping $K \to K$. We diagonalize the matrix D_2 as

$$D_2 = X\Lambda X^{-1}, \tag{4.42}$$

where X and Λ are matrices order $N+1$ as

$$X \stackrel{\mathrm{d}}{\equiv} \left(\cos\left(\frac{ij\pi}{N}\right)\right)_{i,j=0}^{N}, \tag{4.43}$$

$$\Lambda \stackrel{\mathrm{d}}{\equiv} \mathrm{diag}(\lambda_k), \tag{4.44}$$

with λ_k given by (4.36). Then the matrix expression of $\mathcal{T}_{\boldsymbol{U}^{(m)}}$ is given by

$$\mathcal{T}_{\boldsymbol{U}^{(m)}}\boldsymbol{V} = X\left(I - \frac{q\Delta t}{2}\Lambda^2\right)^{-1}X^{-1}\boldsymbol{U}^{(n)}$$
$$+ \frac{\Delta t}{2}X\left(I - \frac{q\Delta t}{2}\Lambda^2\right)^{-1}\Lambda X^{-1}\left\{p\boldsymbol{V} + r\mathcal{Q}_{\boldsymbol{U}^{(m)}}\boldsymbol{V}\right\}. \tag{4.45}$$

Hence

$$\|\mathcal{T}_{\boldsymbol{U}^{(m)}}\boldsymbol{V}\|_2$$
$$\leq \|X\|_2\|\left(I - \frac{q\Delta t}{2}\Lambda^2\right)^{-1}\|_2\|X^{-1}\|_2\|\boldsymbol{U}^{(m)}\|_2$$
$$+ \frac{\Delta t}{2}\|X\|_2\|\left(I - \frac{q\Delta t}{2}\Lambda^2\right)^{-1}\Lambda\|_2\|X^{-1}\|_2\left(-p\|\boldsymbol{V}\|_2 + r\|\mathcal{Q}_{\boldsymbol{U}^{(m)}}\boldsymbol{V}\|_2\right)$$
$$\leq 2 \max_{0\leq k\leq N}\left|\frac{1}{1 - \frac{q\Delta t}{2}\lambda_k^2}\right|\|\boldsymbol{U}^{(m)}\|_2$$
$$+ 2\frac{\Delta t}{2}\max_{0\leq k\leq N}\left|\frac{\lambda_k}{1 - \frac{q\Delta t}{2}\lambda_k^2}\right|\left(-p\|\boldsymbol{V}\|_2 + r\|\mathcal{Q}_{\boldsymbol{U}^{(m)}}\boldsymbol{V}\|_2\right)$$
$$\leq 2M\left\{1 + \sqrt{\frac{2\Delta t}{-q}}\left(-p + r\frac{82}{\Delta x}M^2\right)\right\}. \tag{4.46}$$

Here we have used the following estimates:

$$\|\mathrm{diag}(d_k)\|_2 = \max_k |d_k|, \tag{4.47}$$

$$\max_{0\leq k\leq N}\left|\frac{1}{1 - \frac{q\Delta t}{2}\lambda_k^2}\right| \leq 1, \tag{4.48}$$

$$\max_{0\leq k\leq N}\left|\frac{\lambda_k}{1 - \frac{q\Delta t}{2}\lambda_k^2}\right| \leq \frac{1}{\sqrt{-2q\Delta t}}, \tag{4.49}$$

$$\|\mathcal{Q}_{\boldsymbol{U}^{(m)}}\boldsymbol{V}\|_2 \leq \frac{328}{\Delta x}M^3, \tag{4.50}$$

$$\|X\|_2 \leq \sqrt{2N}, \tag{4.51}$$

$$\|X^{-1}\|_2 \leq \sqrt{\frac{2}{N}}, \tag{4.52}$$

that holds under the conditions $\|\boldsymbol{U}^{(m)}\|_2 = M$ and $\|\boldsymbol{V}\|_2 \leq 4M$. The evaluation of the nonlinear term (4.50) is obtained by (3.164). From (4.46) we see that $\mathcal{T}_{\boldsymbol{U}^{(m)}}$ is a mapping $K \to K$ if

$$\Delta t \leq \frac{-q(\Delta x)^2}{2\left(-p\Delta x + 82\,rM^2\right)^2}. \tag{4.53}$$

Next we prove that $\mathcal{T}_{\boldsymbol{U}^{(m)}}$ is contractive. Using (4.45) and the estimates above we can show

$$\|\mathcal{T}_{\boldsymbol{U}^{(m)}}\boldsymbol{V} - \mathcal{T}_{\boldsymbol{U}^{(m)}}\boldsymbol{V}'\|_2$$
$$\leq \sqrt{\frac{\Delta t}{-2q}}\{-p\|\boldsymbol{V}-\boldsymbol{V}'\|_2 + r\|\mathcal{Q}_{\boldsymbol{U}^{(m)}}\boldsymbol{V} - \mathcal{Q}_{\boldsymbol{U}^{(m)}}\boldsymbol{V}'\|_2\}$$
$$\leq \sqrt{\frac{\Delta t}{-2q}}\left(-p + \frac{226}{\Delta x}rM^2\right)\|\boldsymbol{V}-\boldsymbol{V}'\|_2, \tag{4.54}$$

because

$$\|\mathcal{Q}_{\boldsymbol{U}^{(m)}}\boldsymbol{V} - \mathcal{Q}_{\boldsymbol{U}^{(m)}}\boldsymbol{V}'\|_2 \leq \frac{226}{\Delta x}M^2\|\boldsymbol{V}-\boldsymbol{V}'\|_2. \tag{4.55}$$

Therefore $\mathcal{T}_{\boldsymbol{U}^{(m)}}$ is contractive if

$$\Delta t < \frac{-2q(\Delta x)^2}{(-p\Delta x + 226rM^2)^2}. \tag{4.56}$$

This completes the proof. □

Next, we evaluate the convergence of the scheme. Let us define the error by

$$e_k^{(m)} \stackrel{\mathrm{d}}{=} U_k^{(m)} - u(k\Delta x, m\Delta t), \quad k = -1, 0, 1, \cdots, N, N+1, \tag{4.57}$$

where $u(x,t)$ is the solution to the Cahn–Hilliard equation. We define an extension of u by

$$u(x,t) \stackrel{\mathrm{d}}{=} \begin{cases} u(x - 2lL, t) : 2lL \leq x \leq (2l+1)L, \\ u(2lL - x, t) : (2l-1)L < x < 2lL, \end{cases} \tag{4.58}$$

where $l \in \mathbf{Z}$ and

$$u(-\Delta x, t) \stackrel{\mathrm{d}}{=} u(\Delta x, t), \tag{4.59}$$
$$u((N+1)\Delta x, t) \stackrel{\mathrm{d}}{=} u((N-1)\Delta x, t). \tag{4.60}$$

The error is measured in terms of the discrete L^2-norm, $\|\boldsymbol{f}\|^2 = \sum_{k=0}^{N} {}''(f_k)^2 \Delta x$ for $\boldsymbol{f} = \{f_k\}_{k=-l}^{N+l} \in \mathbb{R}^{N+1+2l}$; $0 \leq l$. In what follows, we use the following special time-difference and -averaging operators (which are used only in this subsection):

$$\delta_m^{\langle 1 \rangle} f^{(m)} \stackrel{\mathrm{d}}{=} \frac{f^{(m+\frac{1}{2})} - f^{(m-\frac{1}{2})}}{\Delta t}, \tag{4.61a}$$

$$\delta_m^{\langle 2 \rangle} f^{(m)} \stackrel{\mathrm{d}}{=} \frac{f^{(m+\frac{1}{2})} - 2f^{(m)} + f^{(m-\frac{1}{2})}}{(\frac{1}{2}\Delta t)^2}, \tag{4.61b}$$

$$s_m^{\langle 1 \rangle} f^{(m)} \stackrel{\mathrm{d}}{=} \frac{f^{(m+\frac{1}{2})} + f^{(m-\frac{1}{2})}}{2}. \tag{4.61c}$$

We also promise that

$$\frac{\partial^2 u(-\Delta x, t)}{\partial x^2} \stackrel{\mathrm{d}}{\equiv} \frac{\partial^2 u(\Delta x, t)}{\partial x^2}, \tag{4.62}$$

$$\frac{\partial^2 u((N+1)\Delta x, t)}{\partial x^2} \stackrel{\mathrm{d}}{\equiv} \frac{\partial^2 u((N-1)\Delta x, t)}{\partial x^2}. \tag{4.63}$$

Before proceeding to the convergence estimates, we prepare several propositions.

PROPOSITION 4.3
The error $e^{(m)}$ satisfies

$$\frac{1}{\Delta t}\left\{\|e^{(m+1)}\|^2 - \|e^{(m)}\|^2\right\}$$
$$\leq \frac{1}{2}\left\{\|e^{(m+1)}\|^2 + \|e^{(m)}\|^2\right\}$$
$$-\frac{1}{q}\|\widetilde{\phi}(U^{(m+1)}; U^{(m)}) - \phi^{(m+\frac{1}{2})}\|^2$$
$$+\|\zeta_1^{(m+\frac{1}{2})}\|^2 + \|\zeta_2^{(m+\frac{1}{2})}\|^2, \tag{4.64}$$

where

$$\widetilde{\phi}(f_k; g_k) \stackrel{\mathrm{d}}{\equiv} p\left\{\frac{f_k + g_k}{2}\right\} + r\left\{\frac{(f_k)^3 + (f_k)^2 g_k + f_k(g_k)^2 + (g_k)^3}{4}\right\}, \tag{4.65}$$

$$\phi_k^{(m+\frac{1}{2})} \stackrel{\mathrm{d}}{\equiv} \left\{pu + ru^3\right\}\big|_{(x,t)=(k\Delta x, (m+\frac{1}{2})\Delta t)}, \tag{4.66}$$

$$\zeta_{1,k}^{(m+\frac{1}{2})} = \left\{\left(\frac{\partial}{\partial t} - \delta_m^{\langle 1 \rangle}\right) u - \left(\frac{\partial^2}{\partial x^2} - \delta_k^{\langle 2 \rangle}\right) \frac{\delta G}{\delta u}\right\}\bigg|_{(x,t)=(k\Delta x, (m+\frac{1}{2})\Delta t)}, \tag{4.67}$$

$$\zeta_{2,k}^{(m+\frac{1}{2})} = \sqrt{-q}\left\{\left(s_m^{\langle 1 \rangle}\delta_k^{\langle 2 \rangle} - \frac{\partial^2}{\partial x^2}\right) u\right\}\bigg|_{(x,t)=(k\Delta x, (m+\frac{1}{2})\Delta t)}, \tag{4.68}$$

for $k = 0, 1, \cdots, N$.

In (4.67), the term $\delta_m^{\langle 1 \rangle} u\big|_{(x,t)=(k\Delta x, (m+\frac{1}{2})\Delta t)}$ is defined as follows:

$$\delta_m^{\langle 1 \rangle} u\big|_{(x,t)=(k\Delta x, (m+\frac{1}{2})\Delta t)} \stackrel{\mathrm{d}}{\equiv} \delta_m^{\langle 1 \rangle} u(k\Delta x, (m+\frac{1}{2})\Delta t)$$
$$= \frac{u(k\Delta x, (m+1)\Delta t) - u(k\Delta x, m\Delta t)}{\Delta t}. \tag{4.69}$$

Other terms are defined similarly.

PROOF   Let us define

$$F_k^{(m+\frac{1}{2})} \stackrel{\mathrm{d}}{=} \frac{\delta G_\mathrm{d}}{\delta(\boldsymbol{U}^{(m+1)},\boldsymbol{U}^{(m)})_k} - \left.\frac{\delta G}{\delta u}\right|_{(x,t)=(k\Delta x,(m+\frac{1}{2})\Delta t)}, \qquad (4.70)$$

for $k = -1, 0, 1, \cdots, N, N+1$. From (4.1), (4.9), (4.57) and (4.70), we obtain

$$\frac{e_k^{(m+1)} - e_k^{(m)}}{\Delta t} = \delta_k^{\langle 2 \rangle} F_k^{(m+\frac{1}{2})} + \zeta_{1,k}^{(m+\frac{1}{2})}, \qquad (4.71)$$

for $k = 0, 1, \cdots, N$. From (4.5), (4.9) and (4.70) we obtain

$$F_k^{(m+\frac{1}{2})} = \widetilde{\phi}(U_k{}^{(m+1)}; U_k{}^{(m)}) - \phi_k^{(m+\frac{1}{2})} + q\delta_k^{\langle 2 \rangle}\frac{e_k^{(m+1)} + e_k^{(m)}}{2} - \sqrt{-q}\,\zeta_{2,k}^{(m+\frac{1}{2})}, \qquad (4.72)$$

for $k = 0, 1, \cdots, N$. From (4.71) and (4.72) we have

$$\frac{1}{2}\sum_{k=0}^{N}{}''\left\{\frac{(e_k^{(m+1)})^2 - (e_k^{(m)})^2}{\Delta t}\right\}\Delta x - \frac{1}{q}\sum_{k=0}^{N}{}''\left(F_k^{(m+\frac{1}{2})}\right)^2 \Delta x$$

$$= \sum_{k=0}^{N}{}''\left(\frac{e_k^{(m+1)} + e_k^{(m)}}{2}\times \mathrm{RHS}(4.71)\right)\Delta x - \frac{1}{q}\sum_{k=0}^{N}{}''\left(F_k^{(m+\frac{1}{2})}\times \mathrm{RHS}(4.72)\right)\Delta x$$

$$= \sum_{k=0}^{N}{}''\left\{\frac{e_k^{(m+1)} + e_k^{(m)}}{2}\delta_k^{\langle 2 \rangle} F_k^{(m+\frac{1}{2})} - F_k^{(m+\frac{1}{2})}\delta_k^{\langle 2 \rangle}\frac{e_k^{(m+1)} + e_k^{(m)}}{2}\right\}\Delta x$$

$$-\frac{1}{q}\sum_{k=0}^{N}{}''\left\{\left(\widetilde{\phi}(U_k{}^{(m+1)}; U_k{}^{(m)}) - \phi_k^{(m+\frac{1}{2})}\right) F_k^{(m+\frac{1}{2})}\right\}\Delta x$$

$$+\sum_{k=0}^{N}{}''\left\{\frac{e_k^{(m+1)} + e_k^{(m)}}{2}\zeta_{1,k}^{(m+\frac{1}{2})}\right\}\Delta x$$

$$-\frac{1}{q}\sum_{k=0}^{N}{}''\left\{F_k^{(m+\frac{1}{2})}(-\sqrt{-q})\zeta_{2,k}^{(m+\frac{1}{2})}\right\}\Delta x. \qquad (4.73)$$

Here the first term

$$\sum_{k=0}^{N}{}''\left\{\frac{e_k^{(m+1)} + e_k^{(m)}}{2}\delta_k^{\langle 2 \rangle} F_k^{(m+\frac{1}{2})} - F_k^{(m+\frac{1}{2})}\delta_k^{\langle 2 \rangle}\frac{e_k^{(m+1)} + e_k^{(m)}}{2}\right\}\Delta x$$

$$= \left[\frac{e_k^{(m+1)} + e_k^{(m)}}{2}\delta_k^{\langle 1 \rangle} F_k^{(m+\frac{1}{2})} - F_k^{(m+\frac{1}{2})}\delta_k^{\langle 1 \rangle}\frac{e_k^{(m+1)} + e_k^{(m)}}{2}\right]_{k=0}^{N} \qquad (4.74)$$

vanishes since

$$\left.\delta_k^{\langle 1 \rangle} F_k^{(m+\frac{1}{2})}\right|_{k=0} = \left.\delta_k^{\langle 1 \rangle} F_k^{(m+\frac{1}{2})}\right|_{k=N} = 0, \qquad (4.75)$$

$$\left.\delta_k^{\langle 1 \rangle} e_k^{(m)}\right|_{k=0} = \left.\delta_k^{\langle 1 \rangle} e_k^{(m)}\right|_{k=N} = 0, \qquad (4.76)$$

under the boundary condition (4.14) and definition (4.58), (4.59), (4.60), (4.62) and (4.63). The remaining terms can be bounded from above by the inequality $ab \leq \frac{1}{2}(a^2 + b^2)$. Hence we obtain the inequality (4.64).

$$\begin{aligned}
\text{RHS(4.73)} \leq &-\frac{1}{2q}\sum_{k=0}^{N}{}''\left\{\left(\widetilde{\phi}(U_k^{(m+1)};U_k^{(m)}) - \phi_k^{(m+\frac{1}{2})}\right)^2 + \left(F_k^{(m+\frac{1}{2})}\right)^2\right\}\Delta x \\
&+\frac{1}{2}\sum_{k=0}^{N}{}''\left\{\left(\frac{e_k^{(m+1)} + e_k^{(m)}}{2}\right)^2 + \left(\zeta_{1,k}^{(m+\frac{1}{2})}\right)^2\right\}\Delta x \\
&-\frac{1}{2q}\sum_{k=0}^{N}{}''\left\{\left(F_k^{(m+\frac{1}{2})}\right)^2 + \left(-\sqrt{-q}\zeta_{2,k}^{(m+\frac{1}{2})}\right)^2\right\}\Delta x \\
\leq &-\frac{1}{2q}\sum_{k=0}^{N}{}''\left\{\left(\widetilde{\phi}(U_k^{(m+1)};U_k^{(m)}) - \phi_k^{(m+\frac{1}{2})}\right)^2 + \left(F_k^{(m+\frac{1}{2})}\right)^2\right\}\Delta x \\
&+\frac{1}{2}\sum_{k=0}^{N}{}''\left\{\frac{1}{2}\left\{\left(e_k^{(m+1)}\right)^2 + \left(e_k^{(m)}\right)^2\right\} + \left(\zeta_{1,k}^{(m+\frac{1}{2})}\right)^2\right\}\Delta x \\
&-\frac{1}{2q}\sum_{k=0}^{N}{}''\left\{\left(F_k^{(m+\frac{1}{2})}\right)^2 - q\left(\zeta_{2,k}^{(m+\frac{1}{2})}\right)^2\right\}\Delta x \\
= &-\frac{1}{2q}\|\widetilde{\phi}(\boldsymbol{U}^{(m+1)};\boldsymbol{U}^{(m)}) - \phi^{(m+\frac{1}{2})}\|^2 - \frac{1}{q}\|\boldsymbol{F}^{(m+\frac{1}{2})}\|^2 \\
&+\frac{1}{4}\left(\|\boldsymbol{e}^{(m+1)}\|^2 + \|\boldsymbol{e}^{(m)}\|^2\right) \\
&+\frac{1}{2}\|\boldsymbol{\zeta}_1^{(m+\frac{1}{2})}\|^2 + \frac{1}{2}\|\boldsymbol{\zeta}_2^{(m+\frac{1}{2})}\|^2,
\end{aligned} \qquad (4.77)$$

since $(a+b)^2 \leq 2(a^2+b^2)$. $\square$

PROPOSITION 4.4

$$\begin{aligned}
\|\widetilde{\phi}(\boldsymbol{U}^{(m+1)};\boldsymbol{U}^{(m)}) &- \phi^{(m+\frac{1}{2})}\|^2 \\
&\leq \{-p + 3r(C_2)^2\}^2\left\{\|\boldsymbol{e}^{(m+1)}\|0,2^2 + \|\boldsymbol{e}^{(m)}\|^2\right\} \\
&\quad -q\left\{\|\boldsymbol{\zeta}_3^{(m+\frac{1}{2})}\|^2 + \|\boldsymbol{\zeta}_4^{(m+\frac{1}{2})}\|^2\right\},
\end{aligned} \qquad (4.78)$$

where

$$C_2 \stackrel{\mathrm{d}}{=} \max_{0\leq l\leq m+1}\left\{\max_{0\leq k\leq N}\left|U_k^{(l)}\right|, \sup_{x\in[0,L]}|u(x,l\Delta t)|\right\}, \qquad (4.79)$$

and $\|\bullet\|_4$ is a discrete L^4-norm which is defined as

$$\|f\|_4^4 = \sum_{k=0}^{N} {}''(f_k)^4 \Delta x, \qquad f = (f_k)_{k=-l}^{N+l} \in \mathbb{R}^{N+1+2l}; \ 0 \leq l, \qquad (4.80)$$

and

$$\zeta_{3,k}^{(m+\frac{1}{2})} \stackrel{d}{\equiv} \frac{r}{2\sqrt{-q}} C_2 \{u(k\Delta x, (m+1)\Delta t) - u(k\Delta x, m\Delta t)\}^2, \qquad (4.81)$$

$$\zeta_{4,k}^{(m+\frac{1}{2})} \stackrel{d}{\equiv} \frac{2}{\sqrt{-q}} \{-p + 3r(C_2)^2\} (s_m^{\langle 1 \rangle} - 1) u(k\Delta x, (m+\frac{1}{2})\Delta t), \qquad (4.82)$$

for $k = 0, 1, \cdots, N$.

REMARK 4.4 Note that C_2 is finite since the proposed scheme is numerically stable and the solution $u \in C^0([0, L])$. □

PROOF We denote $\widetilde{\phi} - \phi = \sum_{i=1}^{4} I_i$ where $I_i = (I_{i,k})_{k=0}^{N}$ with

$$I_{1,k} \stackrel{d}{\equiv} \widetilde{\phi}(U_k^{(m+1)}; U_k^{(m)}) - \widetilde{\phi}(u(k\Delta x, (m+1)\Delta t); U_k^{(m)}), \qquad (4.83)$$

$$I_{2,k} \stackrel{d}{\equiv} \widetilde{\phi}(u(k\Delta x, (m+1)\Delta t); U_k^{(m)}) \qquad (4.84)$$
$$\qquad - \widetilde{\phi}(u(k\Delta x, (m+1)\Delta t); u(k\Delta x, m\Delta t)),$$

$$I_{3,k} \stackrel{d}{\equiv} \widetilde{\phi}(u(k\Delta x, (m+1)\Delta t); u(k\Delta x, m\Delta t)) \qquad (4.85)$$
$$\qquad - \phi(\frac{u(k\Delta x, (m+1)\Delta t) + u(k\Delta x, m\Delta t)}{2}),$$

$$I_{4,k} \stackrel{d}{\equiv} \phi(\frac{u(k\Delta x, (m+1)\Delta t) + u(k\Delta x, (m+1)\Delta t)}{2}) \qquad (4.86)$$
$$\qquad - \phi(u(k\Delta x, (m+\frac{1}{2})\Delta t)).$$

The following estimates are easily obtained:

$$|I_{1,k}| \leq \frac{1}{2} \left(-p + 3r(C_2)^2\right) \left|e_k^{(m+1)}\right|, \qquad (4.87)$$

$$|I_{2,k}| \leq \frac{1}{2} \left(-p + 3r(C_2)^2\right) \left|e_k^{(m)}\right|, \qquad (4.88)$$

$$|I_{4,k}| \leq \left(-p + 3r(C_2)^2\right) (s_m^{\langle 1 \rangle} - 1) u(k\Delta x, (m+\frac{1}{2})\Delta t). \qquad (4.89)$$

The estimate for $(I_{3,k})_{k=0}^{N}$,

$$|I_{3,k}| \leq \frac{r}{4} C_2 \{u(k\Delta x, (m+1)\Delta t) - u(k\Delta x, m\Delta t)\}^2, \qquad (4.90)$$

is obtained by

$$\frac{u^3 + u^2v + uv^2 + v^3}{4} - \left(\frac{u+v}{2}\right)^3 = \frac{1}{8}(u+v)(u-v)^2. \qquad (4.91)$$

From these inequalities we obtain

$$\|I_1\|^2 \leq \frac{1}{4}\left(-p + 3r(C_2)^2\right)^2 \|e^{(m+1)}\|^2, \qquad (4.92)$$

$$\|I_2\|^2 \leq \frac{1}{4}\left(-p + 3r(C_2)^2\right)^2 \|e^{(m)}\|^2, \qquad (4.93)$$

$$\|I_3\|^2 \leq \frac{1}{16}r^2(C_2)^2 \|\{u(\cdot,(m+1)\Delta t) - u(\cdot,m\Delta t)\}^2\|^2, \qquad (4.94)$$

$$\|I_4\|^2 \leq \left(-p + 3r(C_2)^2\right)^2 \|(s_m^{\langle 1 \rangle} - 1)u(\cdot,(m+\frac{1}{2})\Delta t)\|^2. \qquad (4.95)$$

We obtain (4.78) by substituting (4.92)–(4.95) into

$$\|\widetilde{\phi}(\boldsymbol{U}^{(m+1)};\boldsymbol{U}^{(m)}) - \phi^{(m+\frac{1}{2})}\|^2 \leq 4\sum_{i=1}^{4}\|I_i\|^2. \qquad (4.96)$$

$\square$

PROPOSITION 4.5

$$\left\{1 - 2\Delta t\left(\frac{1}{2} + \frac{\{-p + 3r(C_2)^2\}^2}{-q}\right)\right\}\|e^{(m+1)}\|^2$$

$$\leq \|e^{(m)}\|^2 + \Delta t \sum_{i=1}^{4}\|\boldsymbol{\zeta}_i^{(m+\frac{1}{2})}\|^2. \qquad (4.97)$$

PROOF   From Proposition 4.3 and Proposition 4.4, we obtain

$$\frac{1}{\Delta t}\left\{\|e^{(m+1)}\|^2 - \|e^{(m)}\|^2\right\}$$
$$\leq \frac{1}{2}\left\{\|e^{(m+1)}\|^2 + \|e^{(m)}\|^2\right\}$$
$$-\frac{1}{q}\left[\{-p + 3r(C_2)^2\}^2\left\{\|e^{(m+1)}\|^2 + \|e^{(m)}\|^2\right\}\right.$$
$$\left. -q\left\{\|\boldsymbol{\zeta}_3^{(m+\frac{1}{2})}\|^2 + \|\boldsymbol{\zeta}_4^{(m+\frac{1}{2})}\|^2\right\}\right]$$
$$+\|\boldsymbol{\zeta}_1^{(m+\frac{1}{2})}\|^2 + \|\boldsymbol{\zeta}_2^{(m+\frac{1}{2})}\|^2 \qquad (4.98)$$

$$= \left(\frac{1}{2} + \frac{\{-p+3r(C_2)^2\}^2}{-q}\right)\left\{\|e^{(m+1)}\|^2 + \|e^{(m)}\|^2\right\}$$
$$+ \sum_{i=1}^{4}\|\zeta_i^{(m+\frac{1}{2})}\|^2. \tag{4.99}$$

Hence we obtain the following inequality

$$\frac{1}{\Delta t}\left\{\|e^{(m+1)}\|^2 - \|e^{(m)}\|^2\right\}$$
$$\leq \left(\frac{1}{2} + \frac{\{-p+3r(C_2)^2\}^2}{-q}\right)2\|e^{(m+1)}\|^2 + \sum_{m=1}^{4}\|\zeta_m^{(m+\frac{1}{2})}\|^2. \tag{4.100}$$

□

PROPOSITION 4.6
If
$$\Delta t < \frac{1}{2 + 4\frac{\{-p+3r(C_2)^2\}^2}{-q}}, \tag{4.101}$$

then
$$\|e^{(m)}\|^2 \leq \Delta t \sum_{l=1}^{m}(C_3)^l \sum_{i=1}^{4}\|\zeta_i^{(m+\frac{1}{2}-l)}\|^2, \tag{4.102}$$

where
$$C_3 \stackrel{\mathrm{d}}{=} 1 + \left(2 + 4\frac{\{-p+3r(C_2)^2\}^2}{-q}\right)\Delta t. \tag{4.103}$$

PROOF From Proposition 4.3 and Proposition 4.4 we obtain

$$\left\{1 - 2\Delta t\left(\frac{1}{2} + \frac{\{-p+3r(C_2)^2\}^2}{-q}\right)\right\}\|e^{(m+1)}\|^2$$
$$\leq \|e^{(m)}\|^2 + \Delta t\sum_{i=1}^{4}\|\zeta_i^{(m+\frac{1}{2})}\|^2. \tag{4.104}$$

If the inequality (4.101) is satisfied, from (4.104) we obtain

$$\|e^{(m)}\|^2 \leq C_3\left[\|e^{(m-1)}\|^2 + \Delta t\sum_{m=1}^{4}\|\zeta_m^{(m-\frac{1}{2})}\|^2\right]$$
$$\leq (C_3)^m\|e^{(0)}\|^2 + \Delta t\sum_{l=1}^{m}(C_3)^l\sum_{i=1}^{4}\|\zeta_i^{(m+\frac{1}{2}-l)}\|^2. \tag{4.105}$$

The term regarding $\boldsymbol{e}^{(0)}$ vanishes since the error of the initial data is zero. □

Now we are in a position to present the main convergence result.

THEOREM 4.4
Suppose that Δt is small enough to satisfy the condition (4.37) and (4.101). If (4.1) and (4.5) have a solution such that $u(x,t) \in C^6([0,L] \times [0,T])$, then the solution of the difference scheme (4.9) converges to the solution of (4.1) and (4.5) in the sense of discrete L^2-norm, and the convergence rate is $O((\Delta x)^2 + (\Delta t)^2)$.

PROOF From Proposition 4.6 we obtain

$$\|e^{(m)}\|^2 \leq \Delta t (C_3)^m \sum_{l=1}^{m} \sum_{i=1}^{4} \|\zeta_i^{(m+\frac{1}{2}-l)}\|^2. \qquad (4.106)$$

If (4.1) and (4.5) have a solution such that $u(x,t) \in C^6([0,L] \times [0,T])$, then

$$\zeta_{1,k}^{(m+\frac{1}{2})} = -\frac{(\Delta t)^2}{24} \frac{\partial^3 u}{\partial t^3}\bigg|_{\substack{x=k\Delta x \\ t=t_1}} + \frac{(\Delta x)^2}{12} \frac{\partial^4}{\partial x^4}\left(\frac{\delta G}{\delta u}\right)\bigg|_{\substack{x=x_1 \\ t=(m+\frac{1}{2})\Delta t}}, \qquad (4.107)$$

$$\zeta_{2,k}^{(m+\frac{1}{2})} = \sqrt{-q}\left(\frac{(\Delta x)^2}{12} \frac{\partial^4 u}{\partial x^4}\bigg|_{\substack{x=x_2 \\ t=(m+\frac{1}{2})\Delta t}} + \frac{(\Delta t)^2}{8} \frac{\partial^4 u}{\partial t^2 \partial x^2}\bigg|_{\substack{x=x_3 \\ t=t_2}}\right), \qquad (4.108)$$

$$\zeta_{3,k}^{(m+\frac{1}{2})} = \frac{r}{2\sqrt{-q}} C_2 (\Delta t)^2 \left(\frac{\partial u}{\partial t}\bigg|_{\substack{x=k\Delta x \\ t=t_3}}\right)^2, \qquad (4.109)$$

$$\zeta_{4,k}^{(m+\frac{1}{2})} = \frac{r}{\sqrt{-q}}\left\{-p + 3r(C_2)^2\right\}^2 \frac{(\Delta t)^2}{8} \frac{\partial^2 u}{\partial t^2}\bigg|_{\substack{x=k\Delta x \\ t=t_4}}, \qquad (4.110)$$

where $t_1, t_2, t_3, t_4 \in [m\Delta t, (m+1)\Delta t]$ and $x_1, x_2, x_3 \in [(k-1)\Delta x, (k+1)\Delta x]$. From these there is a constant C_4 such that

$$\sum_{i=1}^{4} \|\zeta_i^{(m+\frac{1}{2})}\|^2 \leq C_4 L \left(\Delta x^2 + \Delta t^2\right)^2. \qquad (4.111)$$

From (4.106) and (4.111) we obtain the following evaluation of the error,

$$\|e^{(m)}\|$$
$$\leq \sqrt{C_4 LT} \exp\left[\left(1 + \frac{2\{-p + 3r(C_2)^2\}^2}{-q}\right) T\right] \left(\Delta x^2 + \Delta t^2\right), \qquad (4.112)$$

where $T = m\Delta t$. □

4.1.1.5 Further Topic: Two-Dimensional Examples

The procedure in Chapter 3 is presented only for the one-dimensional cases. When the domain is rectangular, however, it is straightforward to extend the procedure to two- or three-dimensional cases, by simply applying the method for each direction.

For example, we can easily construct a dissipative scheme for the two-dimensional Cahn–Hilliard equation as follows.

Scheme 4.1 (Dissipative scheme for the 2D Cahn–Hilliard equation) *We here denote the numerical solution by $U_{k,l}^{(m)} \simeq u(k\Delta x, l\Delta y, m\Delta t)$ ($k = 0, 1, \ldots, N_x$, $l = 0, 1, \ldots, N_y$, $m = 0, 1, 2, \ldots$) and also write $\boldsymbol{U}^{(m)} = \left(U_{0,0}^{(m)}, \ldots, U_{N_x,N_y}^{(m)}\right)^\top$. Given an initial data $\boldsymbol{U}^{(0)}$, the approximate solutions $\boldsymbol{U}^{(1)}, \boldsymbol{U}^{(2)}, \ldots$ are calculated using the recurrence equation*

$$\frac{U_{k,l}^{(m+1)} - U_{k,l}^{(m)}}{\Delta t}$$
$$= \left(\delta_k^{\langle 2 \rangle} + \delta_l^{\langle 2 \rangle}\right) \left\{ p\left(\frac{U_{k,l}^{(m+1)} + U_{k,l}^{(m)}}{2}\right) + q\left(\delta_k^{\langle 2 \rangle} + \delta_l^{\langle 2 \rangle}\right)\left(\frac{U_{k,l}^{(m+1)} + U_{k,l}^{(m)}}{2}\right) \right.$$
$$\left. + r\left(\frac{(U_{k,l}^{(m+1)})^3 + (U_{k,l}^{(m+1)})^2 U_{k,l}^{(m)} + U_{k,l}^{(m+1)}(U_{k,l}^{(m)})^2 + (U_{k,l}^{(m)})^3}{4}\right) \right\}, \quad (4.113)$$

with discrete boundary conditions applied to two spatial directions corresponding to (1.4).

Scheme 4.1 keeps the desired discrete dissipation property, and the discrete mass conservation property (we omit their proof). We here show some numerical results in Figure 4.1.

4.1.2 Allen–Cahn Equation

4.1.2.1 Introduction to Problem

As noted in Section 2.2, the Allen–Cahn equation

$$\frac{\partial u}{\partial t} = pu + ru^3 + q\frac{\partial^2 u}{\partial x^2} \quad (4.114)$$

where $p > 0, q > 0$ and $r < 0$, is a special case of $s = 0$ of the target PDEs 1 with

$$G(u, u_x) = -\frac{p}{2}u^2 - \frac{r}{4}u^4 + \frac{q}{2}(u_x)^2. \quad (4.115)$$

This is also a mathematical model to some phase separation and domain coarsening phenomenon. This quite resembles the Cahn–Hilliard equation in

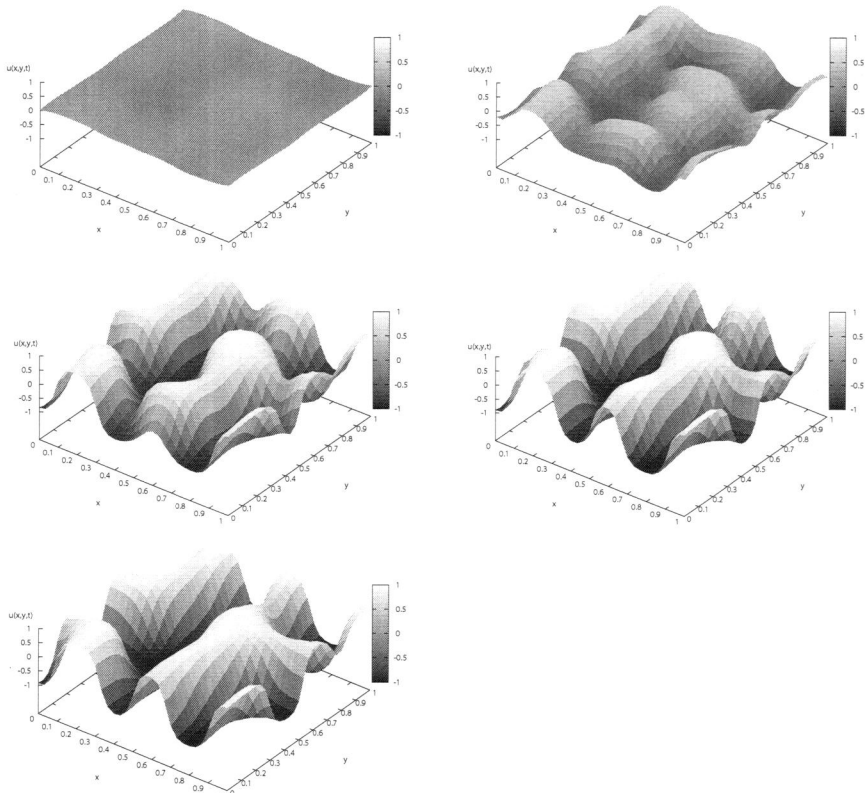

FIGURE 4.1: Numerical solutions of the Cahn–Hilliard equation using Scheme 4.1 on a 2D uniform rectangular mesh: $\Delta x = \Delta y = 1/30$ and $\Delta t = 3/2000$. Top left: profiles at time step $m = 0$, top right: at $m = 10$, middle left: at $m = 20$, middle right: at $m = 30$, bottom: at $m = 40$.

the previous subsection, but a difference is that the Allen–Cahn equation does not have a mass conservation law.

Here we apply the Neumann boundary condition:

$$u_x(0, t) = u_x(L, t), \qquad t > 0. \tag{4.116}$$

4.1.2.2 Numerical Scheme

To construct a numerical scheme, we take the discrete local energy function as

$$G_{\mathrm{d},k}(\boldsymbol{U}) \stackrel{\mathrm{d}}{\equiv} -\frac{p}{2}(U_k)^2 + \frac{r}{4}(U_k)^4 + \frac{q}{2}\left(\frac{(\delta_k^+ U_k)^2 + (\delta_k^- U_k)^2}{2}\right). \tag{4.117}$$

Following the same procedure in the previous subsection, we obtain a scheme:

$$\frac{U_k^{(m+1)} - U_k^{(m)}}{\Delta t} = -\frac{\delta G_{\mathrm{d}}}{\delta(\boldsymbol{U}^{(m+1)}, \boldsymbol{U}^{(m)})_k}, \tag{4.118}$$

where

$$\frac{\delta G_{\mathrm{d}}}{\delta(\boldsymbol{U}, \boldsymbol{V})_k} = -p\left(\frac{U_k + V_k}{2}\right) - r\left(\frac{(U_k)^3 + (U_k)^2 V_k + U_k(V_k)^2 + (V_k)^3}{4}\right)$$
$$- q \delta_k^{\langle 2 \rangle}\left(\frac{U_k + V_k}{2}\right). \tag{4.119}$$

Let us impose the discrete Neumann boundary condition:

$$\delta_k^{\langle 1 \rangle} U_0^{(m)} = \delta_k^{\langle 1 \rangle} U_N^{(m)}, \qquad m = 0, 1, 2, \ldots. \tag{4.120}$$

Then it is easy to confirm that the discrete global energy

$$J_{\mathrm{d}}(\boldsymbol{U}^{(m)}) \stackrel{\mathrm{d}}{\equiv} \sum_{k=0}^{N} {}''G_{\mathrm{d},k}(\boldsymbol{U}^{(m)}) \Delta x \tag{4.121}$$

is dissipated. In fact, we can follow exactly the same argument in the previous subsection for $B_{\mathrm{r},1}(\boldsymbol{U}^{(m+1)}, \boldsymbol{U}^{(m)}) = 0$ to find that Theorem 3.1 holds.

Let us here also try more advanced one. Note that the above scheme is *nonlinear* with respect to the unknown variable $\boldsymbol{U}^{(m+1)}$ (observe that the discrete variational derivative is quadratic with respect to U_k). This forces us to utilize some iterative solver in each time step. As a remedy for this computational difficulty, in Chapter 6 a linearization technique will be introduced (a brief introduction was also given in Section 1.4.2), by which we can design a *linearly implicit* (i.e. still implicit but linear with respect to the unknown variable) scheme, and accordingly save the computational effort to a considerable extent. In what follows, we demonstrate an example for the Allen–Cahn equation, without getting into the detail of the linearization technique.

For a linear scheme, we start by defining a discrete energy function such as

$$G_{d,k}(\boldsymbol{U},\boldsymbol{V}) \stackrel{\mathrm{d}}{=} -\frac{p}{2}U_k V_k - \frac{r}{4}(U_k)^2(V_k)^2$$
$$+\frac{q}{2}\frac{(\delta_k^+ U_k)^2 + (\delta_k^- U_k)^2 + (\delta_k^+ V_k)^2 + (\delta_k^- V_k)^2}{4}. \quad (4.122)$$

The "trick" here is to define a discrete energy function by *two* numerical solutions. If we carry out "discrete variation" for this multistep energy function, then we are able to derive the linearly implicit scheme:

$$\frac{U_k^{(m+1)} - U_k^{(m-1)}}{2\Delta t} = -\frac{\delta G_d}{\delta(\boldsymbol{U}^{(m+1)}, \boldsymbol{U}^{(m)}, \boldsymbol{U}^{(m-1)})_k}, \quad (4.123)$$

where

$$\frac{\delta G_d}{\delta(\boldsymbol{U},\boldsymbol{V},\boldsymbol{W})_k} = -pV_k - r(V_k)^2\left(\frac{U_k + W_k}{2}\right) - q\delta_k^{\langle 2 \rangle}\left(\frac{U_k + W_k}{2}\right). \quad (4.124)$$

Note that now the unknown variable $\boldsymbol{U}^{(m+1)}$ (which corresponds to $\boldsymbol{U}$ in the discrete variational derivative) appears only linearly in the scheme, and thus this is a *linear* scheme. In the next subsection, we will test this scheme in a two-dimensional setting.

4.1.2.3 Further Topic: Two-Dimensional Examples

Let us test the two-dimensional version of the above linearly implicit scheme.

Scheme 4.2 (Linearly implicit scheme for 2D Allen–Cahn equation)
Given an initial data $\boldsymbol{U}^{(0)}$, we propose the following scheme to obtain the approximate solutions $\boldsymbol{U}^{(1)}, \boldsymbol{U}^{(2)}, \ldots$.

$$\frac{U_{k,l}^{(m+1)} - U_{k,l}^{(m-1)}}{2\Delta t} = p\, U_{k,l}^{(m)} + r\left(U_{k,l}^{(m)}\right)^2 \left(\frac{U_{k,l}^{(m+1)} + U_{k,l}^{(m-1)}}{2}\right)$$
$$+ q\, \delta_{k,l}^{\langle 2 \rangle}\left(\frac{U_{k,l}^{(m+1)} + U_{k,l}^{(m-1)}}{2}\right) \quad (4.125)$$

under the following boundary conditions,

$$\left.\delta_k^{\langle 1 \rangle} U_{k,l}^{(m)}\right|_{k=0,M} = \left.\delta_l^{\langle 1 \rangle} U_{k,l}^{(m)}\right|_{l=0,N} = 0, \quad (4.126)$$

for $m = 1, 2, \ldots$.

Scheme 4.2 has the following discrete dissipation property. We omit the proof.

THEOREM 4.5
For the numerical solution $\boldsymbol{U}^{(m)}$ of the scheme, the following inequality holds.

$$J_{\mathrm{d}}(\boldsymbol{U}^{(m+1)}, \boldsymbol{U}^{(m)}) \leq J_{\mathrm{d}}(\boldsymbol{U}^{(m)}, \boldsymbol{U}^{(m-1)}), \qquad \text{for } m = 1, 2, \ldots. \tag{4.127}$$

Here we define

$$J_{\mathrm{d}}(\boldsymbol{U}, \boldsymbol{V}) \stackrel{\mathrm{d}}{\equiv} \sum_{k=0}^{N_x}{}'' \sum_{l=0}^{N_y}{}'' G_{\mathrm{d},k,l}(\boldsymbol{U}, \boldsymbol{V}) \Delta x \Delta y, \tag{4.128}$$

$$\begin{aligned}
G_{\mathrm{d},k,l}(\boldsymbol{U}, \boldsymbol{V}) &\stackrel{\mathrm{d}}{\equiv} \frac{p}{2} U_{k,l} V_{k,l} - \frac{r}{4}(U_{k,l})^2(V_{k,l})^2 \\
&+ \frac{q}{2} \frac{(\delta_k^+ U_{k,l})^2 + (\delta_k^- U_{k,l})^2 + (\delta_k^+ V_{k,l})^2 + (\delta_k^- V_{k,l})^2}{4} \\
&+ \frac{q}{2} \frac{(\delta_l^+ U_{k,l})^2 + (\delta_l^- U_{k,l})^2 + (\delta_l^+ V_{k,l})^2 + (\delta_l^- V_{k,l})^2}{4}.
\end{aligned} \tag{4.129}$$

We show a numerical example in Figure 4.2 with $\Omega = (0,4) \times (0,4)$, $p = 100.0$, $q = 1.0$, $r = -100.0$, $\Delta x = \Delta y = 0.08$ (i.e. $N_x = N_y = 50$) and $\Delta t = 10^{-4}$. The initial state is

$$u_0(x, y) = 0.5 \sin(\pi x) + 0.5 \sin(\pi y). \tag{4.130}$$

We observe that numerical phase separation occurs stably. (As we will see in Chapter 6, generally such linearization can cause numerical instability. In this case, however, the scheme happily runs without problems.)

4.1.3 Fisher–Kolmogorov Equation

4.1.3.1 Introduction to Problem

As noted in the Section 2.2, the extended Fisher–Kolmogorov equation

$$\frac{\partial u}{\partial t} = -\left(pu + ru^3 + q\frac{\partial^2 u}{\partial x^2} + \gamma \frac{\partial^4 u}{\partial x^4} \right), \tag{4.131}$$

where $p < 0, q < 0, r > 0$, and $\gamma > 0$, is a special case of the target PDEs 1 with $s = 0$ and

$$G(u, u_x, u_{xx}) = \frac{p}{2}u^2 + \frac{r}{4}u^4 - \frac{q}{2}(u_x)^2 + \frac{\gamma}{2}(u_{xx})^2. \tag{4.132}$$

The boundary conditions for this problem are

$$\frac{\partial u}{\partial x} = 0, \qquad \frac{\partial^3 u}{\partial x^3} = 0,$$

on boundary.

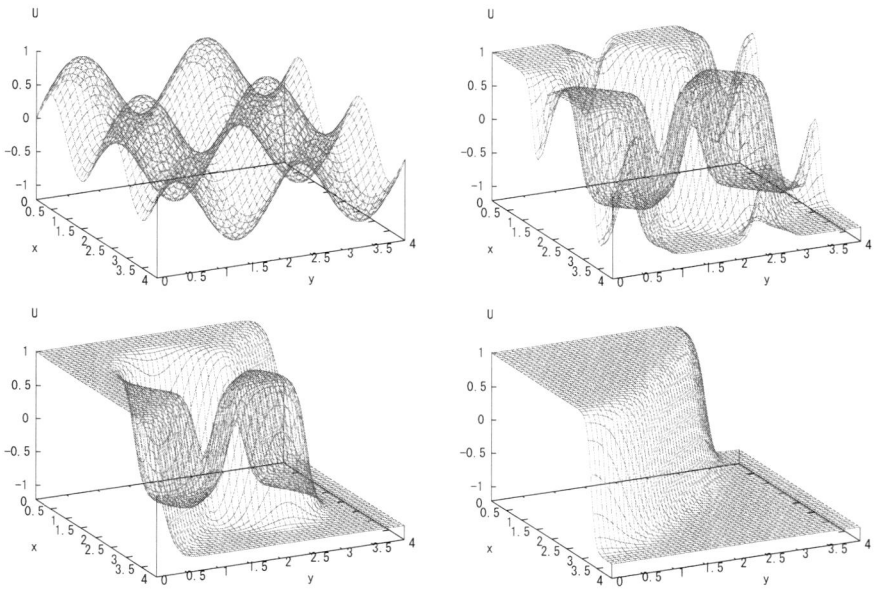

FIGURE 4.2: Numerical solutions of the Allen–Cahn equation by the linearly implicit scheme 4.2 on 2D rectangular region. The top left figure is the profile at time $t = 0$, the top right is $t = 0.08$, the bottom left is $t = 0.20$ and the bottom right is $t = 1.00$.

4.1.3.2 Numerical Scheme

To construct a numerical scheme, we take the discrete local energy function as

$$G_{\mathrm{d},k}(\boldsymbol{U}) \stackrel{\mathrm{d}}{=} \frac{p}{2}(U_k)^2 + \frac{r}{4}(U_k)^4 - \frac{q}{2}\frac{(\delta_k^+ U_k)^2 + (\delta_k^- U_k)^2}{2} + \frac{\gamma}{2}\left(\delta_k^{\langle 2 \rangle} U_k\right)^2. \quad (4.133)$$

The resulting scheme becomes

$$\frac{U_k{}^{(m+1)} - U_k{}^{(m)}}{\Delta t} = -\frac{\delta G_{\mathrm{d}}}{\delta(\boldsymbol{U}^{(m+1)}, \boldsymbol{U}^{(m)})_k}, \quad (4.134)$$

where

$$\left(\frac{\delta G_{\mathrm{d}}}{\delta(\boldsymbol{U},\boldsymbol{V})}\right)_k = p\left(\frac{U_k + V_k}{2}\right) + r\left(\frac{(U_k)^3 + (U_k)^2 V_k + U_k(V_k)^2 + (V_k)^3}{4}\right)$$

$$+ q\,\delta_k^{\langle 2 \rangle}\left(\frac{U_k + V_k}{2}\right) + \gamma\,\delta_k^{\langle 4 \rangle}\left(\frac{U_k + V_k}{2}\right). \quad (4.135)$$

We impose the following boundary conditions,

$$\delta_k^{\langle 1 \rangle} U_k{}^{(m)}\Big|_{k=0,N} = 0, \qquad \delta_k^{\langle 3 \rangle} U_k{}^{(m)}\Big|_{k=0,N} = 0. \quad (4.136)$$

These boundary conditions satisfy the conditions of the theorem 3.1 and this scheme is dissipative in the sense that the the discrete global energy

$$J_{\mathrm{d}}(\boldsymbol{U}^{(m)}) \stackrel{\mathrm{d}}{=} \sum_{k=0}^{N} {}''G_{\mathrm{d},k}(\boldsymbol{U}^{(m)})\,\Delta x \quad (4.137)$$

dissipates.

4.1.3.3 Numerical Examples

We show a numerical example in Figure 4.3 with $p = -100$, $q = -1$, $r = 100$, $\gamma = 0.01$, $\Delta x = 0.025$ and $\Delta t = 10^{-4}$. The initial state is taken to

$$u_0(x) = 0.1\sin(\pi x) - 0.1\cos(0.5\pi x). \quad (4.138)$$

For these numerical solutions we confirm that the global energy decreases monotonically as time evolves in Figure 4.4.

4.2 Target PDEs 2

In this section, examples for the target PDEs 2 (defined in Section 2.2; real-valued, single, conservative PDEs) are shown.

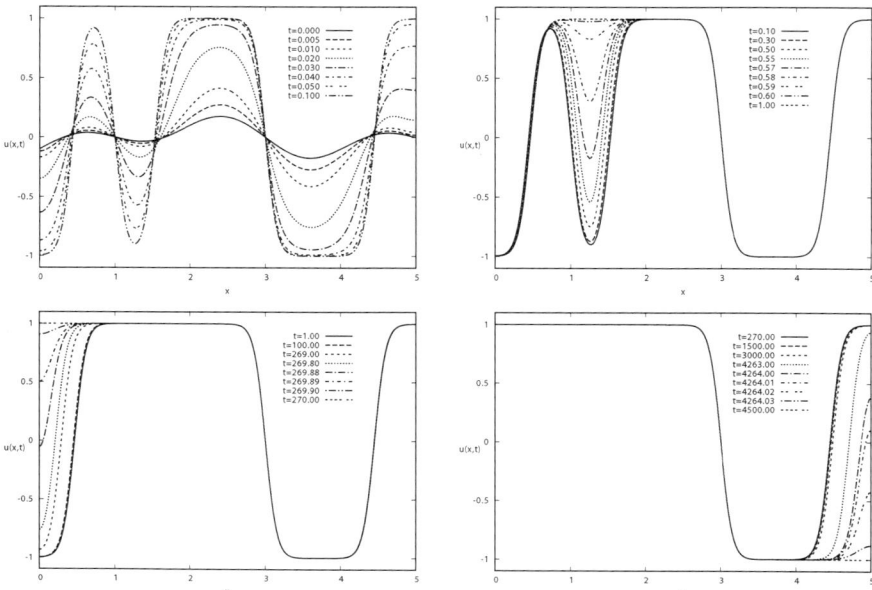

FIGURE 4.3: Numerical solutions for the extended Fisher–Kolmogorov equation. Top left: profiles at time $0 \leq t \leq 0.1$, top right: at $0.1 \leq t \leq 1.0$, bottom left: at $1.0 \leq t \leq 270.0$, and bottom right: at $270.0 \leq t \leq 4500.0$

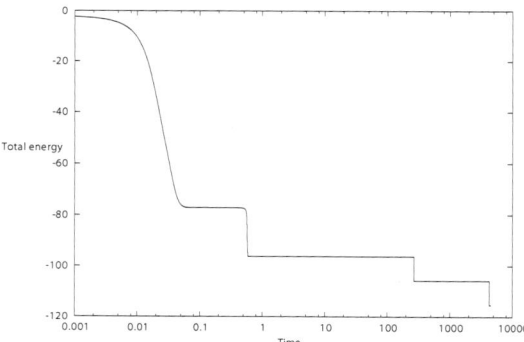

FIGURE 4.4: The evolution of the global energy. The time axis is in log-scale.

4.2.1 Korteweg–de Vries Equation

4.2.1.1 Introduction to Problem

Let us consider the famous Korteweg–de Vries equation (KdV):

$$\frac{\partial u}{\partial t} = \frac{\partial}{\partial x}\left(\frac{1}{2}u^2 + \frac{\partial^2 u}{\partial x^2}\right), \qquad x \in (0, L),\ t > 0. \tag{4.139}$$

There are a lot of studies about numerical integrators for this equation, including [71, 73, 78, 84, 108, 137, 142, 149, 160, 170]. This is an example of the conservative PDEs 2, where

$$s = 0, \quad G(u, u_x) = \frac{1}{6}u^3 - \frac{1}{2}\left(\frac{\partial u}{\partial x}\right)^2. \tag{4.140}$$

We impose the standard periodic boundary condition:

$$u^{(j)}(0, t) = u^{(j)}(L, t), \qquad j = 0, 1, 2,\ t > 0.$$

4.2.1.2 Numerical Scheme

We define a discrete energy function by

$$G_{\mathrm{d},k}(\boldsymbol{U}) \stackrel{\mathrm{d}}{=} \frac{1}{6}(U_k)^3 - \frac{1}{2}\left(\frac{(\delta_k^+ U_k)^2 + (\delta_k^- U_k)^2}{2}\right). \tag{4.141}$$

Then by (3.30),

$$\frac{\delta G_{\mathrm{d}}}{\delta(\boldsymbol{U}^{(m+1)}, \boldsymbol{U}^{(m)})_k} = \frac{1}{2}\left(\frac{(U_k^{(m+1)})^2 + U_k^{(m+1)}U_k^{(m)} + (U_k^{(m)})^2}{3}\right)$$
$$+ \delta_k^{\langle 2 \rangle}\left(\frac{U_k^{(m+1)} + U_k^{(m)}}{2}\right). \tag{4.142}$$

This concrete form can be calculated via (3.27a), (3.27b) and (3.27c). As emphasized in Section 4.1.1, it is also possible to calculate it by a direct discrete variation; see Remark 4.5 below for this. In either case, by Scheme 3.2 we have, for $m = 0, 1, 2, \ldots$,

$$\frac{U_k^{(m+1)} - U_k^{(m)}}{\Delta t} = \delta_k^{\langle 1 \rangle} \frac{\delta G_{\mathrm{d}}}{\delta(\boldsymbol{U}^{(m+1)}, \boldsymbol{U}^{(m)})_k}$$
$$= \delta_k^{\langle 1 \rangle}\left\{\frac{1}{2}\left(\frac{(U_k^{(m+1)})^2 + U_k^{(m+1)}U_k^{(m)} + (U_k^{(m)})^2}{3}\right)\right.$$
$$\left. + \delta_k^{\langle 2 \rangle}\left(\frac{U_k^{(m+1)} + U_k^{(m)}}{2}\right)\right\},$$
$$k = 0, \ldots, N - 1. \tag{4.143}$$

This scheme is conservative if the following conditions are satisfied:

$$B_{\mathrm{r},1}(\boldsymbol{U}^{(m+1)}, \boldsymbol{U}^{(m)})$$
$$= \frac{1}{2}\left[\frac{\delta_k^+(U_k^{(m+1)} + U_k^{(m)})}{2} \cdot \mu_k^+(U_k^{(m+1)} - U_k^{(m)})\right.$$
$$\left.+ \frac{\delta_k^-(U_k^{(m+1)} + U_k^{(m)})}{2} \cdot \mu_k^-(U_k^{(m+1)} - U_k^{(m)})\right]_0^N$$
$$= 0, \qquad (4.144)$$

$$B_{\mathrm{r},2}^{\langle 1 \rangle}(\boldsymbol{U}^{(m+1)}, \boldsymbol{U}^{(m)})$$
$$= \frac{1}{2}\left[\frac{\delta G_{\mathrm{d}}}{\delta(\boldsymbol{U}^{(m+1)}, \boldsymbol{U}^{(m)})_k}\left(s_k^{\langle 1 \rangle} \frac{\delta G_{\mathrm{d}}}{\delta(\boldsymbol{U}^{(m+1)}, \boldsymbol{U}^{(m)})_k}\right)\right]_0^N$$
$$= 0. \qquad (4.145)$$

Both conditions are satisfied if we discretize the periodic boundary condition as
$$U_k^{(m)} = U_{k \bmod N}^{(m)}, \qquad m = 0, 1, 2, \ldots. \qquad (4.146)$$

The KdV equation is a completely integrable equation which has infinitely many conservation laws. The scheme above follows another conservation law ("mass" conservation law):

$$\sum_{k=0}^{N}{}'' U_k^{(m)} \Delta x = \sum_{k=0}^{N}{}'' U_k^{(0)} \Delta x, \qquad m = 0, 1, 2, \ldots, \qquad (4.147)$$

which corresponds to
$$\frac{\mathrm{d}}{\mathrm{d}t}\int_0^L u \, \mathrm{d}x = 0,$$

in continuous context. This conservation is confirmed by

$$\frac{1}{\Delta t}\sum_{k=0}^{N}{}''\left(U_k^{(m+1)} - U_k^{(m)}\right)\Delta x = \sum_{k=0}^{N}{}''\left(\delta_k^{\langle 1 \rangle}\frac{\delta G_{\mathrm{d}}}{\delta(\boldsymbol{U}^{(m+1)}, \boldsymbol{U}^{(m)})_k}\right)\Delta x$$
$$= \left[\mu_k^{\langle 1 \rangle}\frac{\delta G_{\mathrm{d}}}{\delta(\boldsymbol{U}^{(m+1)}, \boldsymbol{U}^{(m)})_k}\right]_0^N = 0, \qquad (4.148)$$

under the periodic boundary condition (4.146).

REMARK 4.5 As noted in Remark 3.3, and demonstrated in Section 4.1.1, the discrete variational derivative can be found by directly considering the factorization (3.32) for the specific energy function (4.141). Let us see this again in the case of KdV as follows.

$$\sum_{k=0}^{N}{}''\left(G_{\mathrm{d},k}(\boldsymbol{U}^{(m+1)}) - G_{\mathrm{d},k}(\boldsymbol{U}^{(m)})\right)\Delta x =$$

$$= \sum_{k=0}^{N}{}''\left[\frac{1}{6}\left((U_k^{(m+1)})^3 - (U_k^{(m)})^3\right)\right.$$

$$\left. - \frac{1}{2}\left\{\left(\frac{(\delta_k^+ U_k^{(m+1)})^2 + (\delta_k^- U_k^{(m+1)})^2}{2}\right) - \left(\frac{(\delta_k^+ U_k^{(m)})^2 + (\delta_k^- U_k^{(m)})^2}{2}\right)\right\}\right]\Delta x$$

$$= \sum_{k=0}^{N}{}''\left[\frac{1}{2}\left(\frac{(U_k^{(m+1)})^2 + U_k^{(m+1)}U_k^{(m)} + (U_k^{(m)})^2}{3}\right)\left(U_k^{(m+1)} - U_k^{(m)}\right)\right.$$

$$\left. + \delta_k^{\langle 2 \rangle}\left(\frac{U_k^{(m+1)} + U_k^{(m)}}{2}\right)\left(U_k^{(m+1)} - U_k^{(m)}\right)\right]\Delta x + (\text{boundary terms}).$$

In the last equality the summation-by-parts formula (3.12a) is used. We here omit the concrete form of the boundary terms, which corresponds to the boundary term $B_{\mathrm{r},1}(\boldsymbol{U}^{(m+1)}, \boldsymbol{U}^{(m)})$ in (3.32), for brevity. □

4.2.1.3 Numerical Examples

An example of two-soliton propagation is shown in Figure 4.5.[1] The initial data is set to

$$u(x,0) = 48\operatorname{sech}^2(2(x-36)) + 12\operatorname{sech}^2(x-24) \tag{4.149}$$

and parameters are $L = 40$, $\Delta x = 0.05$ and $\Delta t = 0.0001$. Notice that the intensities of solitons are quite large. For such large solitons often numerical schemes tend to be unstable due to the nonlinearity of the equation. In the present example, however, the computation proceeds quite stably. This clearly shows the superiority of the *conservative* scheme. In Figure 4.6,[2] the evolutions of the discrete energy and mass are shown. Both are well conserved to the machine accuracy.

4.2.2 Zakharov–Kuznetsov Equation

4.2.2.1 Introduction to Problem

The Zakharov–Kuznetsov equation (2.31) (ZK):

$$\frac{\partial u}{\partial t} = \frac{\partial}{\partial x}\left(-\frac{u^2}{2} - \frac{\partial^2 u}{\partial x^2} - \frac{\partial^2 u}{\partial y^2}\right)$$

[1,2] Reprinted from *J. Comput. Phys.*, 156, D. Furihata, Finite difference schemes for $\frac{\partial u}{\partial t} = \left(\frac{\partial}{\partial x}\right)^\alpha \frac{\delta G}{\delta u}$ that inherit energy conservation or dissipation property, 181–205, Copyright (1999), with permission from Elsevier.

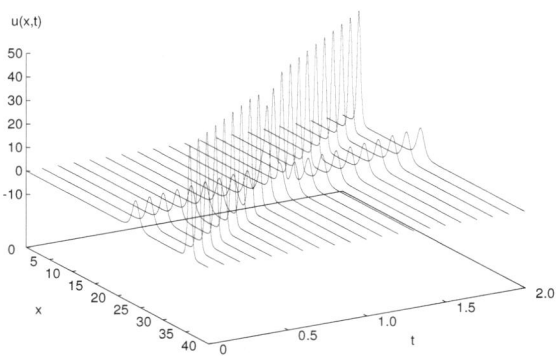

FIGURE 4.5: Two-soliton propagation dynamics for the KdV equation.

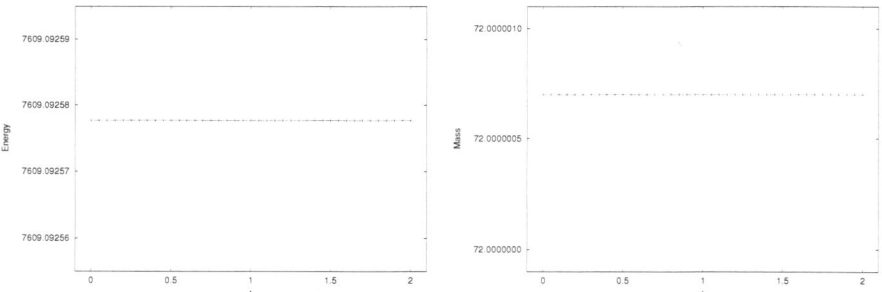

FIGURE 4.6: Time evolutions of discrete invariants for the KdV equation. The left shows the total energy and the right shows the mass.

is a two-dimensional PDE, and thus is not explicitly covered by the procedure in Chapter 3. However, as demonstrated in Section 4.1.1, it is easy to extend the procedure to the two-dimensional PDE as far as the domain is rectangular. If in this way we allow two-dimensional PDEs, the ZK can be regarded as an example of the target PDEs 2, with $s = 0$ and

$$G(u, \nabla u) = -\frac{u^3}{6} + \frac{(u_x)^2}{2} + \frac{(u_y)^2}{2}. \tag{4.150}$$

We take the periodic boundary conditions for this problem and assume that the space region Ω is rectangular. In addition to the standard energy:

$$J(u) \stackrel{\mathrm{d}}{\equiv} \int_\Omega G(u, \nabla u) \mathrm{d}\boldsymbol{x}, \tag{4.151}$$

the ZK has the "mass" as its invariant:

$$I(u) \stackrel{\mathrm{d}}{\equiv} \int_\Omega u \, \mathrm{d}\boldsymbol{x}. \tag{4.152}$$

4.2.2.2 Numerical Schemes

Let us define the two-dimensional discrete energy function by

$$\begin{aligned} G_{\mathrm{d},k,l}(\boldsymbol{U}) &\stackrel{\mathrm{d}}{\equiv} -\frac{1}{6}(U_{k,l})^3 \\ &+ \frac{(\delta_k^+ U_{k,l})^2 + (\delta_k^- U_{k,l})^2}{4} + \frac{(\delta_l^+ U_{k,l})^2 + (\delta_l^- U_{k,l})^2}{4}. \end{aligned} \tag{4.153}$$

Then by the two-dimensional version of Scheme 3.2, we obtain a nonlinear scheme:

$$\frac{U_{k,l}^{(m+1)} - U_{k,l}^{(m)}}{\Delta t} = \delta_k^{\langle 1 \rangle} \frac{\delta G_{\mathrm{d}}}{\delta(\boldsymbol{U}^{(m+1)}, \boldsymbol{U}^{(m)})_{k,l}} \quad 0 \le k \le N_x, \ 0 \le l \le N_y, \tag{4.154}$$

where

$$\frac{\delta G_{\mathrm{d}}}{\delta(\boldsymbol{U}^{(m+1)}, \boldsymbol{U}^{(m)})_{k,l}} = -\frac{(U_{k,l}^{(m+1)})^2 + (U_{k,l}^{(m+1)})(U_{k,l}^{(m)}) + (U_{k,l}^{(m)})^2}{6}$$

$$- \left(\delta_k^{\langle 2 \rangle} + \delta_l^{\langle 2 \rangle} \right) \left(\frac{U_{k,l}^{(m+1)} + U_{k,l}^{(m)}}{2} \right), \tag{4.155}$$

under the discrete periodic boundary condition.

This scheme has the following discrete invariants:

$$J_{\mathrm{d}}(\boldsymbol{U}^{(m)}) \stackrel{\mathrm{d}}{\equiv} \sum_{k=0}^{N_x} {}'' \sum_{l=0}^{N_y} {}'' G_{\mathrm{d},k,l}(\boldsymbol{U}^{(m)}) \Delta x \Delta y, \tag{4.156}$$

$$I_{\mathrm{d}}(\boldsymbol{U}^{(m)}) \stackrel{\mathrm{d}}{\equiv} \sum_{k=0}^{N_x} {}'' \sum_{l=0}^{N_y} {}'' U_{k,l}^{(m)} \Delta x \Delta y. \tag{4.157}$$

As in the Allen–Cahn case (Section 4.1.2), we are also able to design a linearly implicit scheme using the linearization technique discussed in Chapter 6. An example is (we omit the derivation detail):

$$\frac{U_{k,l}^{(m+1)} - U_{k,l}^{(m-1)}}{2\Delta t} = \delta_k^{\langle 1 \rangle} \frac{\delta G_\mathrm{d}}{\delta (\boldsymbol{U}^{(m+1)}, \boldsymbol{U}^{(m)}, \boldsymbol{U}^{(m-1)})_{k,l}} \qquad (4.158)$$

where

$$\frac{\delta G_\mathrm{d}}{\delta (\boldsymbol{U}^{(m+1)}, \boldsymbol{U}^{(m)}, \boldsymbol{U}^{(m-1)})_{k,l}} = -\frac{\left(U_{k,l}^{(m+1)} + U_{k,l}^{(m)} + U_{k,l}^{(m-1)}\right) U_{k,l}^{(m)}}{6}$$

$$- \left(\delta_k^{\langle 2 \rangle} + \delta_l^{\langle 2 \rangle}\right) \left(\frac{U_{k,l}^{(m+1)} + U_{k,l}^{(m-1)}}{2}\right). \quad (4.159)$$

This scheme can be derived out from

$$G_{\mathrm{d},k,l}(\boldsymbol{U}, \boldsymbol{V}) \stackrel{\mathrm{d}}{\equiv} -\frac{(U_{k,l} + V_{k,l}) U_{k,l} V_{k,l}}{12}$$

$$+ \frac{(\delta_k^+ U_{k,l})^2 + (\delta_k^- U_{k,l})^2 + (\delta_k^+ V_{k,l})^2 + (\delta_k^- V_{k,l})^2}{8}$$

$$+ \frac{(\delta_l^+ U_{k,l})^2 + (\delta_l^- U_{k,l})^2 + (\delta_l^+ V_{k,l})^2 + (\delta_l^- V_{k,l})^2}{8}. \quad (4.160)$$

This linearly implicit scheme also has two invariants. First,

$$J_\mathrm{d}(\boldsymbol{U}^{(m+1)}, \boldsymbol{U}^{(m)}) \stackrel{\mathrm{d}}{\equiv} \sum_{k=0}^{N_x} {}'' \sum_{l=0}^{N_y} {}'' G_{\mathrm{d},k,l}(\boldsymbol{U}^{(m+1)}, \boldsymbol{U}^{(m)}) \Delta x \Delta y \quad (4.161)$$

is preserved. Second,

$$I_\mathrm{d}(\boldsymbol{U}^{(m)}) = \begin{cases} I_\mathrm{d}(\boldsymbol{U}^{(0)}) & \text{if } m \text{ is even,} \\ I_\mathrm{d}(\boldsymbol{U}^{(1)}) & \text{if } m \text{ is odd,} \end{cases} \quad (4.162)$$

holds for any $m > 0$.

4.2.2.3 Numerical Examples

In Figure 4.7 we show numerical solutions of the *nonlinear* scheme, with $\Omega = [0, 32] \times [0, 32]$, $N_x = N_y = 100$ (i.e., $\Delta x = \Delta y = 0.32$), and $\Delta t = 0.01$. We first investigate the dynamics of solutions close to the 1D soliton, and thus take the initial state as

$$u_0(x, y) = 3 \, \mathrm{cosech}^2(0.5\sqrt{2}(x - 16)) + 0.05 \, \mathrm{rand}, \quad (4.163)$$

where "rand" is a random function with $0 \leq \mathrm{rand} \leq 1$.

Next, we show the dynamics of two 2D soliton-like solutions in Figure 4.8. Initial state profile is constructed by choosing appropriate two 2D soliton-like profiles obtained by random computations.

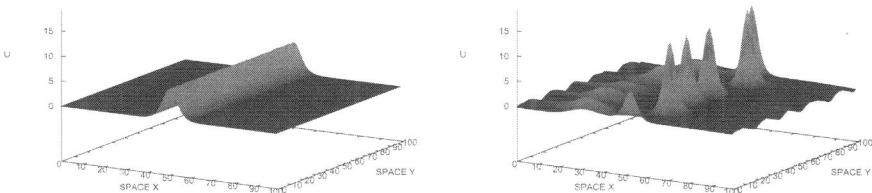

FIGURE 4.7: Time evolution of the traveling 1D soliton-like solutions by the nonlinear scheme. The left figure is the initial profile and the right is the profile of time $t = 15.0$.

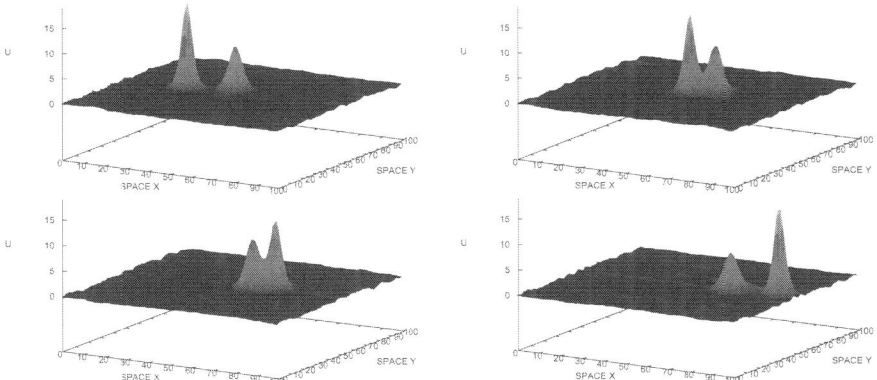

FIGURE 4.8: The time evolution of the two 2D soliton-like solutions. The left upper: the profile of time $t = 2.0$, the right upper: $t = 4.0$, the left bottom: $t = 5.0$, the right bottom: $t = 7.0$.

4.3 Target PDEs 3

In this section, examples for the target PDEs 3 (defined in Section 2.3; complex-valued, single, dissipative PDEs) are shown.

4.3.1 Complex-Valued Ginzburg–Landau Equation

4.3.1.1 Introduction to Problem

The complex-valued Ginzburg–Landau equation (CGL):

$$\frac{\partial u}{\partial t} = p\frac{\partial^2 u}{\partial x^2} + q|u|^2 u + ru, \qquad x \in (0, L), t > 0, p > 0, q < 0, r \in \mathbb{R}, \quad (4.164)$$

is an example of the dissipative PDEs 1. We consider the CGL under the periodic boundary condition:

$$u^{(j)}(0,t) = u^{(j)}(L,t), \qquad j = 0, 1, 2,\ t > 0. \quad (4.165)$$

The local energy $G(u, u_x)$ is given by

$$G(u, u_x) = p|u_x|^2 - \frac{q}{2}|u|^4 - r|u|^2. \quad (4.166)$$

4.3.1.2 Numerical Scheme

Following the procedure in Section 3.3, we firstly define the discrete local energy by

$$G_{\mathrm{d},k}(\boldsymbol{U}) \stackrel{\mathrm{d}}{=} p\left(\frac{|\delta_k^+ U_k|^2 + |\delta_k^- U_k|^2}{2}\right) - \frac{q}{2}|U_k|^4 - r|U_k|^2. \quad (4.167)$$

Then, by the definition of the complex discrete variational derivative (3.72b) (and the related definitions (3.70a)–(3.70c)), the discrete variational derivative can be calculated as

$$\frac{\delta G_\mathrm{d}}{\delta(\overline{\boldsymbol{U}^{(m+1)}}, \overline{\boldsymbol{U}^{(m)}})_k} =$$
$$-p\,\delta_k^{\langle 2\rangle}\left(\frac{U_k^{(m+1)} + U_k^{(m)}}{2}\right) - q\left(\frac{|U_k^{(m+1)}|^2 + |U_k^{(m)}|^2}{2}\right)\left(\frac{U_k^{(m+1)} + U_k^{(m)}}{2}\right)$$
$$- r\left(\frac{U_k^{(m+1)} + U_k^{(m)}}{2}\right). \quad (4.168)$$

Then from (3.76), we obtain a finite difference scheme:
$$\frac{U_k^{(m+1)} - U_k^{(m)}}{\Delta t} =$$
$$p\,\delta_k^{\langle 2 \rangle}\left(\frac{U_k^{(m+1)} + U_k^{(m)}}{2}\right) + q\left(\frac{|U_k^{(m+1)}|^2 + |U_k^{(m)}|^2}{2}\right)\left(\frac{U_k^{(m+1)} + U_k^{(m)}}{2}\right)$$
$$+ r\left(\frac{U_k^{(m+1)} + U_k^{(m)}}{2}\right). \qquad (4.169)$$

We here impose the discrete periodic boundary condition:
$$U_k^{(m)} = U_{k \bmod N}^{(m)}, \qquad m = 0, 1, 2, \ldots. \qquad (4.170)$$

Under the discrete periodic boundary condition, it is almost obvious that the condition (3.77) in Theorem 3.3 is satisfied (the periodic boundary condition almost automatically eliminates any boundary terms), and thus the dissipation property (3.78) holds.

4.3.2 Newell–Whitehead Equation

4.3.2.1 Introduction to Problem

Let us consider the Newell–Whitehead equation (NW):
$$\frac{\partial u}{\partial t}(t, x, y) = \mu u - |u|^2 u + \left(\frac{\partial}{\partial x} - \frac{i}{2k_c}\frac{\partial^2}{\partial y^2}\right)^2 u, \quad \begin{pmatrix} (x,y) \in [0, L_x] \times [0, L_y], \\ t > 0, \\ \mu, k_c \in \mathbb{R}. \end{pmatrix}. \qquad (4.171)$$

We assume for simplicity the periodic boundary condition in both directions. The local energy for the NW is
$$G(u, u_x, u_{yy}) = -\mu|u|^2 + \frac{1}{2}|u|^4 + \left|u_x - \frac{i}{2k_c}u_{yy}\right|^2. \qquad (4.172)$$

By integrating the local energy on the domain $[0, L_x] \times [0, L_y]$, we have the global energy for the NW accordingly:
$$J(u) = \int_0^{L_x} \int_0^{L_y} \left(-\mu|u|^2 + \frac{1}{2}|u|^4 + \left|\frac{\partial u}{\partial x} - \frac{i}{2k_c}\frac{\partial^2 u}{\partial y^2}\right|^2\right) dxdy. \qquad (4.173)$$

Note that this is a *two-dimensional* PDE. Still, the NW belongs to the target PDEs 3 (it is an easy exercise to check that the variational derivative of (4.172) coincides with the right hand side of the NW), and since now the domain is simply rectangular, by applying the procedure in Chapter 3 to x and y directions separately, we can construct a dissipative scheme. Below we demonstrate this.

4.3.2.2 Numerical Scheme

We define the discrete energy as

$$G_{\mathrm{d},k,l}(\boldsymbol{U}) \stackrel{\mathrm{d}}{=} -\mu|U_{k,l}|^2 + \frac{1}{2}|U_{k,l}|^4$$
$$+ \frac{1}{2}\left(\left|\delta_k^+ U_{k,l} - \frac{\mathrm{i}}{2k_c}\delta_l^{\langle 2\rangle} U_{k,l}\right|^2 + \left|\delta_k^- U_{k,l} - \frac{\mathrm{i}}{2k_c}\delta_l^{\langle 2\rangle} U_{k,l}\right|^2\right), \quad (4.174)$$

and accordingly the discrete global energy as

$$J_{\mathrm{d}}(\boldsymbol{U}) \stackrel{\mathrm{d}}{=} \sum_{k=0}^{N_x}{}''\sum_{l=0}^{N_y}{}'' G_{\mathrm{d},k,l}(\boldsymbol{U})\Delta x \Delta y, \quad (4.175)$$

where N_x and N_y are the number of grid points in x and y, $\Delta x \stackrel{\mathrm{d}}{=} L_x/N_x$, $\Delta y \stackrel{\mathrm{d}}{=} L_y/N_y$, and numerical solution $U_{k,l}^{(m)} \simeq u(m\Delta t, k\Delta x, l\Delta y)$ is now $\mathbb{C}^{(N_x+1)(N_y+1)}$ vector. The difference operators with the subscript l operate in l direction.

To define a discrete variational derivative we consider the difference $J_{\mathrm{d}}(\boldsymbol{U}) - J_{\mathrm{d}}(\boldsymbol{V})$ through analogy with the 1-dimensional case as follows.

$$J_{\mathrm{d}}(\boldsymbol{U}) - J_{\mathrm{d}}(\boldsymbol{V}) = \sum_{k=0}^{N_x}{}''\sum_{l=0}^{N_y}{}'' \left\{\frac{\delta G_{\mathrm{d}}}{\delta\left(\boldsymbol{U},\boldsymbol{V}\right)}\bigg|_{k,l}(U_{k,l} - V_{k,l}) \right.$$
$$\left. + \frac{\delta G_{\mathrm{d}}}{\delta\left(\overline{\boldsymbol{U}},\overline{\boldsymbol{V}}\right)}\bigg|_{k,l}(\overline{U_{k,l} - V_{k,l}})\right\}\Delta x \Delta y, \quad (4.176)$$

where

$$\frac{\delta G_{\mathrm{d}}}{\delta\left(\overline{\boldsymbol{U}},\overline{\boldsymbol{V}}\right)}\bigg|_{k,l} = -\mu\left(\frac{U_{k,l} + V_{k,l}}{2}\right) + \left(\frac{|U_{k,l}|^2 + |V_{k,l}|^2}{2}\right)\left(\frac{U_{k,l} + V_{k,l}}{2}\right)$$
$$- \left(\delta_k^{\langle 2\rangle} - \frac{\mathrm{i}}{k_c}\delta_k^{\langle 1\rangle}\delta_l^{\langle 2\rangle} - \frac{1}{4k_c^2}\delta_l^{\langle 4\rangle}\right)\left(\frac{U_{k,l} + V_{k,l}}{2}\right). \quad (4.177)$$

In the above calculation we used the summation-by-parts formula separately in k- and l-directions.

Then we have a finite difference scheme:
$$\frac{U_{k,l}^{(m+1)} - U_{k,l}^{(m)}}{\Delta t}$$
$$= -\frac{\delta G_{\mathrm{d}}}{\delta \left(\boldsymbol{U}^{(m+1)}, \overline{\boldsymbol{U}^{(m)}}\right)_{k,l}}$$
$$= \mu \left(\frac{U_{k,l}^{(m+1)} + U_{k,l}^{(m)}}{2}\right) - \left(\frac{\left|U_{k,l}^{(m+1)}\right|^2 + \left|U_{k,l}^{(m)}\right|^2}{2}\right)\left(\frac{U_{k,l}^{(m+1)} + U_{k,l}^{(m)}}{2}\right)$$
$$+ \left(\delta_k^{\langle 2\rangle} - \frac{\mathrm{i}}{k_c}\delta_k^{\langle 1\rangle}\delta_l^{\langle 2\rangle} - \frac{1}{4k_c^2}\delta_l^{\langle 4\rangle}\right)\left(\frac{U_{k,l}^{(m+1)} + U_{k,l}^{(m)}}{2}\right). \quad (4.178)$$

We impose the discrete periodic boundary condition in both directions as follows.
$$U_{k,l}^{(m)} = U_{(k \bmod N_x),l}^{(m)} = U_{k,(l \bmod N_y)}^{(m)}. \quad (4.179)$$
It is easy to see that under the discrete periodic boundary condition (4.179) the dissipation property holds for the scheme.

Since this problem is two-dimensional, the resulting scheme is relatively expensive as is. In Chapter 6 we will present a numerical example for the linearly implicit version of the above scheme.

4.4 Target PDEs 4

In this section, examples for the target PDEs 4 (defined in Section 2.3; complex-valued, single, conservative PDEs) are shown.

4.4.1 Nonlinear Schrödinger Equation

4.4.1.1 Introduction to Problem

We consider the nonlinear Schrödinger equation (NLS):
$$\mathrm{i}\frac{\partial u}{\partial t} = -\frac{\partial^2 u}{\partial x^2} - \gamma |u|^{p-1}u, \quad x \in (0,L), \ t > 0, \ \gamma \in \mathbb{R}, \ p = 3, 4, \ldots, \quad (4.180)$$
under the periodic boundary condition:
$$u^{(j)}(0,t) = u^{(j)}(L,t), \quad j = 0, 1, 2, \ t > 0. \quad (4.181)$$
This is an example of the conservative PDEs 4. The local energy $G(u, u_x)$ for NLS is given by
$$G(u, u_x) = -|u_x|^2 + \frac{2\gamma}{p+1}|u|^{p+1}. \quad (4.182)$$

We here also note that NLS has an additional invariant (which is often called "probability" in physical context):

$$P(u) = \int_0^L |u|^2 \mathrm{d}x, \qquad (4.183)$$

since

$$\frac{\mathrm{d}}{\mathrm{d}t} \int_{-L}^{L} |u(x,t)|^2 \mathrm{d}x = \int_{-L}^{L} \left(u \frac{\partial \overline{u}}{\partial t} + \frac{\partial u}{\partial t} \overline{u} \right) \mathrm{d}x = \mathrm{i} \int_{-L}^{L} \left(u \frac{\delta H}{\delta u} - \frac{\delta H}{\delta \overline{u}} \overline{u} \right) \mathrm{d}x$$

$$= \mathrm{i} \int_{-L}^{L} \left\{ u \left(-\overline{u_{xx}} - \gamma |u|^{p-1} \overline{u} \right) - \left(-u_{xx} - \gamma |u|^{p-1} u \right) \overline{u} \right\} \mathrm{d}x$$

$$= \mathrm{i} \int_{-L}^{L} (-u \overline{u_{xx}} + u_{xx} \overline{u}) \mathrm{d}x = \mathrm{i} \int_{-L}^{L} (u_x \overline{u_x} - u_x \overline{u_x}) \mathrm{d}x + \mathrm{i} \left[-u \overline{u_x} + u_x \overline{u} \right]_{-L}^{L}$$

$$= 0. \qquad (4.184)$$

The boundary term, $\mathrm{i} \left[-u \overline{u_x} + u_x \overline{u} \right]_{-L}^{L}$, vanishes in light of the periodic boundary condition.

4.4.1.2 Numerical Scheme

Let us construct a conservative scheme following the procedure in Section 3.3. We define the associated discrete local energy by

$$G_{\mathrm{d},k}(\boldsymbol{U}) \stackrel{\mathrm{d}}{=} -\frac{|\delta_k^+ U_k|^2 + |\delta_k^- U_k|^2}{2} + \frac{2\gamma}{p+1} |U_k|^{p+1}. \qquad (4.185)$$

Note that this G_d approximates $G(u, u_x)$ above, and can be decomposed as assumed in (3.66). Calculating mechanically the complex discrete variational derivatives by (3.72a), (3.72b), and (3.70a), (3.70b), (3.70c), we have

$$\frac{\delta G_\mathrm{d}}{\delta(\overline{\boldsymbol{U}^{(m+1)}}, \overline{\boldsymbol{U}^{(m)}})_k} = \delta_k^{\langle 2 \rangle} \left(\frac{U_k^{(m+1)} + U_k^{(m)}}{2} \right)$$
$$+ \gamma \left(\frac{|U_k^{(m+1)}|^{p+1} - |U_k^{(m)}|^{p+1}}{|U_k^{(m+1)}|^2 - |U_k^{(m)}|^2} \right) \left(\frac{U_k^{(m+1)} + U_k^{(m)}}{2} \right). \qquad (4.186)$$

(Here we like to stress again that, as repeatedly emphasized in Remark 3.3 and other related comments, it is much easier to directly consider the discrete variation process for (4.185).)

Then from (3.86) we obtain a finite difference scheme:

$$i\left(\frac{U_k^{(m+1)} - U_k^{(m)}}{\Delta t}\right) = -\delta_k^{\langle 2\rangle}\left(\frac{U_k^{(m+1)} + U_k^{(m)}}{2}\right)$$
$$-\gamma\left(\frac{|U_k^{(m+1)}|^{p+1} - |U_k^{(m)}|^{p+1}}{|U_k^{(m+1)}|^2 - |U_k^{(m)}|^2}\right)\left(\frac{U_k^{(m+1)} + U_k^{(m)}}{2}\right). \tag{4.187}$$

We employ the discrete periodic boundary condition:

$$U_k^{(m)} = U_{k \bmod N}^{(m)}, \qquad m = 0, 1, 2, \ldots. \tag{4.188}$$

As it satisfies (3.77), Theorem 3.4 holds; i.e., the discrete global energy is conserved. Moreover, the scheme preserves the discrete version of $P(u)$ as follows.

THEOREM 4.6
The solution of the scheme (4.187) satisfies

$$\sum_{k=0}^{N}{''}|U_k^{(m)}|^2 \Delta x = \text{const.}, \qquad m = 0, 1, 2, \cdots, \tag{4.189}$$

under the discrete periodic boundary condition (4.188).

PROOF The proof goes exactly the same as in the continuous case, but in order to avoid typesetting lengthy discrete formulas, below we split the discussion into parts. We firstly note that

$$\frac{1}{\Delta t}\sum_{k=0}^{N}{''}\left[|U_k^{(m+1)}|^2 - |U_k^{(m)}|^2\right]\Delta x$$
$$= \sum_{k=0}^{N}{''}\left[\left(\frac{U_k^{(m+1)} + U_k^{(m)}}{2}\right)\overline{\left(\frac{U_k^{(m+1)} - U_k^{(m)}}{\Delta t}\right)} + \text{(c.c.)}\right]\Delta x.$$
$$= i\sum_{k=0}^{N}{''}\left[\left(\frac{U_k^{(m+1)} + U_k^{(m)}}{2}\right)\frac{\delta G_d}{\delta(\boldsymbol{U}^{(m+1)}, \boldsymbol{U}^{(m)})_k} - \text{(c.c.)}\right]\Delta x.$$

If we substitute the concrete form of the discrete variational derivative (4.186) into the above, the second term in (4.186) obviously cancels out mutually in the complex conjugate pairs. For the first term of (4.186), we easily see by the summation-by-parts formula and the discrete periodic boundary condition,

$$\sum_{k=0}^{N}{}'' \left[\left(\frac{U_k^{(m+1)} + U_k^{(m)}}{2} \right) \left(-\delta_k^{\langle 2 \rangle} \overline{\frac{U_k^{(m+1)} + U_k^{(m)}}{2}} \right) \right] \Delta x$$
$$= \sum_{k=0}^{N}{}'' \left[\delta_k^+ \left(\frac{U_k^{(m+1)} + U_k^{(m)}}{2} \right) \cdot \delta_k^+ \left(\overline{\frac{U_k^{(m+1)} + U_k^{(m)}}{2}} \right) \right] \Delta x.$$

Thus it is canceled out as well. □

The scheme (4.187) coincides with the Delfour–Fortin–Payre scheme [35].

4.4.1.3 Numerical Examples

We here present some numerical examples (see also Section 5.2.4.2, where related examples are shown). The NLS is integrated in $0 \le t \le 100$, with the initial data:

$$u(x,0) = 4\text{sech}(2(x-10))e^{\mathrm{i}x} + 2\text{sech}(x-20)e^{\mathrm{i}x/2}.$$

Other parameters are set to $\gamma = 0.5$, $L = 30$, $N = 200$ (i.e. $\Delta x = 30/200$), and $\Delta t = 0.1$.

Figure 4.9 shows the time evolution of the solutions. The two-solitons propagate stably. In Figure 4.10, the evolution of the discrete energy and probability are shown. Both are well preserved.

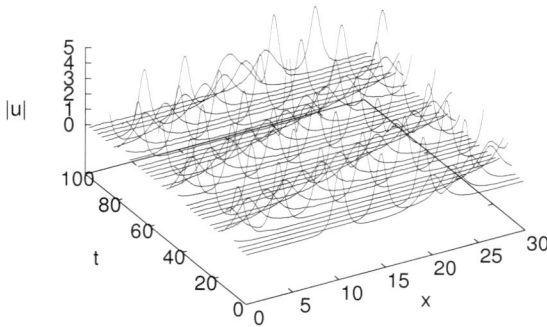

FIGURE 4.9: Evolution of the numerical solution.

Applications

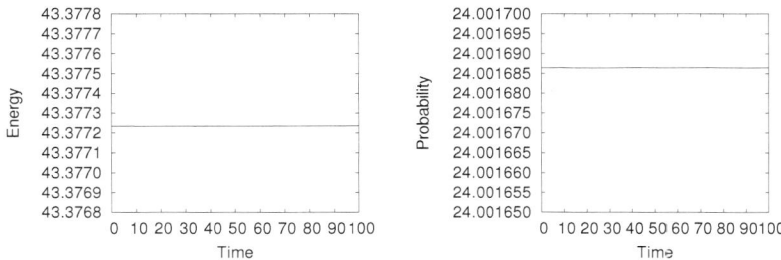

FIGURE 4.10: Evolution of the discrete invariants: (left) energy; (right) probability.

4.4.1.4 Analysis of Scheme

As mentioned above, the scheme has two discrete invariants. They serve as (discrete) a priori estimates, by which we can prove the stability and convergence of the numerical solution. Let us demonstrate it below.

Let us consider the case of $p = 3$, namely, when the NLS is *cubic*:

$$iu_t = -u_{xx} - \gamma |u|^2 u, \qquad x \in (0, L),\ t > 0,\ \gamma \in \mathbb{R}. \qquad (4.190)$$

In this case, it is known that there exists a global solution to the NLS (for larger p's, solutions can blow up, and it does not make sense to consider the "stability").

We first consider the existence and uniqueness of the solution to the nonlinear system (4.187), i.e., $\boldsymbol{U}^{(m+1)}$. We make use of the following well-known theorem.

THEOREM 4.7 Brouwer-type fixed-point theorem [8]
Let $(H, (\cdot, \cdot))$ be a finite dimensional inner product space and $\|\cdot\|$ the associated norm. Let $g : H \to H$ be a continuous function and assume that there exists $\alpha > 0$ such that for any $\boldsymbol{z} \in H$ that satisfies $\|\boldsymbol{z}\| = \alpha$,

$$\mathrm{Re}\,(g(\boldsymbol{z}), \boldsymbol{z}) > 0 \qquad (4.191)$$

holds. Then, there exists $\boldsymbol{z}^* \in H$ such that

$$g(\boldsymbol{z}^*) = 0 \quad and \quad \|\boldsymbol{z}^*\| \leq \alpha. \qquad (4.192)$$

THEOREM 4.8 Existence of a solution
The scheme (4.187) has at least one solution $\boldsymbol{U}^{(m+1)}$.

PROOF We use the notation of discrete functional analysis in Section 3.6. Rewriting the scheme with a new discrete quantity $V_k = \frac{1}{2}\left(U_k^{(m+1)} + U_k^{(m)}\right)$, we have

$$\frac{2\mathrm{i}}{\Delta t}\left(V_k - U_k^{(m)}\right) = -\delta_k^{\langle 2 \rangle} V_k - \gamma \left(|2V_k - U_k^{(m)}|^2 + |U_k^{(m)}|^2\right) V_k,$$

which can be simplified to

$$V_k = U_k^{(m)} + \frac{\mathrm{i}\Delta t}{2}\delta_k^{\langle 2 \rangle} V_k + \frac{\mathrm{i}\Delta t \gamma}{2}\left(|2V_k - U_k^{(m)}|^2 + |U_k^{(m)}|^2\right) V_k. \quad (4.193)$$

In order to prove the existence of a solution to the scheme, it suffices to show that there exists $\boldsymbol{V} \in \mathbb{S}_N$ satisfying (4.193).

Let us apply Brouwer's fixed-point theorem to the above equation. Let us define $g : \mathbb{S}_N \to \mathbb{S}_N$ by

$$(g(\boldsymbol{V}))_k = V_k - U_k^{(m)} - \frac{\mathrm{i}\Delta t}{2}\delta_k^{\langle 2 \rangle} V_k - \frac{\mathrm{i}\Delta t \gamma}{2}\left(|2V_k - U_k^{(m)}|^2 + |U_k^{(m)}|^2\right) V_k,$$

for $k = 0, \ldots, N-1$. Obviously, g is continuous. We also have

$$(g(\boldsymbol{V}), \boldsymbol{V}) = \|\boldsymbol{V}\|^2 - (\boldsymbol{U}^{(m)}, \boldsymbol{V}) - \frac{\mathrm{i}\Delta t}{2} \sum_{k=0}^{N-1} (\delta_k^{\langle 2 \rangle} \overline{V_k}) V_k \Delta x$$

$$- \frac{\mathrm{i}\Delta t \gamma}{2} \sum_{k=0}^{N-1} \left(|2V_k - U_k^{(m)}|^2 + |U_k^{(m)}|^2\right) |V_k|^2 \Delta x. \quad (4.194)$$

The last term on the right-hand side is a purely imaginary number. The third term on the right-hand side can be rewritten as, by the summation-by-parts formula,

$$\sum_{k=0}^{N-1} (\delta_k^{\langle 2 \rangle} \overline{V_k}) V_k \Delta x = -\sum_{k=0}^{N-1} |\delta_k^+ V_k|^2 \Delta x, \quad (4.195)$$

and hence becomes a purely imaginary number as well (the boundary term is canceled thanks to the discrete periodic boundary condition). Therefore, from (4.194) we have

$$\mathrm{Re}(g(\boldsymbol{V}), \boldsymbol{V}) = \|\boldsymbol{V}\|^2 - \mathrm{Re}(\boldsymbol{U}^{(m)}, \boldsymbol{V})$$
$$\geq \|\boldsymbol{V}\|(\|\boldsymbol{V}\| - \|\boldsymbol{U}^{(m)}\|).$$

By choosing $\alpha = \|\boldsymbol{U}^{(m)}\| + 1$, all the assumptions in Brouwer's fixed-point theorem are satisfied, and thus there exists $\boldsymbol{V}^* \in \mathbb{S}_N$ such that $g(\boldsymbol{V}^*) = 0$. This completes the proof. □

Next, we show that the solution is unique if Δt is chosen appropriately small. In order to accomplish this, we need some discrete norm estimates.

LEMMA 4.1

Let $\boldsymbol{V} \in \mathbb{S}_N$ be a vector that satisfies the following conditions (where c_1, c_2 are constants):

$$\left(\sum_{k=0}^{N-1} G_{\mathrm{d},k}(\boldsymbol{V})\Delta x =\right) - \|\boldsymbol{V}_x\|^2 + \frac{\gamma}{2}\|\boldsymbol{V}\|_4^4 = c_1, \qquad (4.196\mathrm{a})$$

$$\|\boldsymbol{V}\|^2 = c_2. \qquad (4.196\mathrm{b})$$

Then,
$$\|\boldsymbol{V}\|_4^4 \leq c(c_1, c_2, \gamma)$$

holds, where b is defined in Lemma 3.5, and

$$c(c_1, c_2, \gamma) = \begin{cases} bc_2^{\frac{3}{2}}\left(\dfrac{b\gamma c_2^{\frac{3}{2}}}{4} + \sqrt{\dfrac{b^2\gamma^2 c_2^3}{16} + c_1 + c_2}\right) & \text{when } \gamma \geq 0, \\ b\sqrt{-c_1 + c_2}\, c_2^{\frac{3}{2}} & \text{when } \gamma < 0. \end{cases} \qquad (4.197)$$

PROOF When $\gamma \geq 0$, from (4.196a) and (4.196b),

$$\frac{\gamma}{2}\|\boldsymbol{V}\|_4^4 = \|\boldsymbol{V}\|_{H^1}^2 - c_1 - c_2.$$

This, together with Lemma 3.5, shows

$$\|\boldsymbol{V}\|_{H^1}^2 \leq \frac{b\gamma}{2}\|\boldsymbol{V}\|_{H^1} \cdot c_2^{\frac{3}{2}} + c_1 + c_2.$$

Thus,
$$\|\boldsymbol{V}\|_{H^1}^2 \leq \frac{b\gamma c_2^{\frac{3}{2}}}{4} + \sqrt{\frac{b^2\gamma^2 c_2^3}{16} + c_1 + c_2}.$$

This, again together with Lemma 3.5, shows

$$\|\boldsymbol{V}\|_4^4 \leq bc_2^{\frac{3}{2}}\left(\frac{b\gamma c_2^{\frac{3}{2}}}{4} + \sqrt{\frac{b^2\gamma^2 c_2^3}{16} + c_1 + c_2}\right).$$

When $\gamma < 0$, from (4.196a) we immediately have $\|\boldsymbol{V}_x\|^2 \leq -c_1$ (note that, when $\gamma < 0$, $-c_1 > 0$ by (4.196a)). Thus, $\|\boldsymbol{V}\|_4^4 \leq b\sqrt{-c_1 + c_2}\, c_2^{\frac{3}{2}}$. □

Using the above lemmas, we can establish the uniqueness of the solution.

THEOREM 4.9 Uniqueness of the solution

Let c_1 be the initial discrete energy and c_2 be the initial discrete probability defined as follows.

$$c_1 = \sum_{k=0}^{N-1} \left(-|\delta_k^+ U_k^{(0)}|^2 + \frac{\gamma}{2} |U_k^{(0)}|^4 \right) \Delta x,$$

$$c_2 = \sum_{k=0}^{N-1} |U_k^{(0)}|^2 \Delta x,$$

and b be what was defined in Lemma 3.5, $c = c(c_1, c_2, \gamma)$ be what was defined in Lemma 4.1. Then, if Δt is sufficiently small such that

$$(\Delta t)^3 (18\Delta t + 37) < \frac{1}{18^3 b^2 c^2 \gamma^4}$$

holds (for example, when $\Delta t < \min\{(55 \times 18^3 b^2 c^2 \gamma^4)^{-1/3}, 1\}$), the solution of the scheme (4.187) is unique.

PROOF We show that if there exist two solutions to (4.193), say $\boldsymbol{V}, \boldsymbol{W} \in \mathbb{S}_N$, then they necessarily coincide: $\boldsymbol{V} = \boldsymbol{W}$. First, we have

$$\begin{aligned}
&\|\boldsymbol{V} - \boldsymbol{W}\|^2 \\
&= (\boldsymbol{V} - \boldsymbol{W}, \boldsymbol{V} - \boldsymbol{W}) \\
&= -\frac{i\Delta t}{2} \sum_{k=0}^{N-1} \left\{ \delta_k^{\langle 2 \rangle} (\overline{V_k - W_k}) \right\} (V_k - W_k) \Delta x \\
&\quad - \frac{i\Delta t \gamma}{2} \sum_{k=0}^{N-1} \left\{ |2V_k - U_k^{(m)}|^2 \overline{V_k} - |2W_k - U_k^{(m)}|^2 \overline{W_k} \right\} (V_k - W_k) \Delta x \\
&\quad - \frac{i\Delta t \gamma}{2} \sum_{k=0}^{N-1} |U_k^{(m)}|^2 |V_k - W_k|^2 \Delta x.
\end{aligned} \tag{4.198}$$

Applying the summation-by-parts formula, we have

$$\sum_{k=0}^{N-1} \left\{ \delta_k^{\langle 2 \rangle} (\overline{V_k - W_k}) \right\} (V_k - W_k) \Delta x = -\|(\boldsymbol{V} - \boldsymbol{W})_x\|^2.$$

From the real and imaginary parts of (4.198), we obtain

$$\|\boldsymbol{V} - \boldsymbol{W}\|^2$$
$$= \frac{\Delta t \gamma}{2} \operatorname{Im} \left[\sum_{k=0}^{N-1} \left\{ |2V_k - U_k^{(m)}|^2 \overline{V_k} - |2W_k - U_k^{(m)}|^2 \overline{W_k} \right\} (V_k - W_k) \Delta x \right]$$
$$\leq \frac{\Delta t \gamma}{2} \left| \sum_{k=0}^{N-1} \left\{ |2V_k - U_k^{(m)}|^2 \overline{V_k} - |2W_k - U_k^{(m)}|^2 \overline{W_k} \right\} (V_k - W_k) \Delta x \right|$$
$$\leq \frac{\Delta t \gamma}{2} \left(\sum_{k=0}^{N-1} \left| |2V_k - U_k^{(m)}|^2 \overline{V_k} - |2W_k - U_k^{(m)}|^2 \overline{W_k} \right|^{\frac{4}{3}} \Delta x \right)^{\frac{3}{4}} \|\boldsymbol{V} - \boldsymbol{W}\|_4,$$
(4.199)

$$\|(\boldsymbol{V} - \boldsymbol{W})_x\|^2$$
$$\leq \gamma \left| \sum_{k=0}^{N-1} \left\{ |2V_k - U_k^{(m)}|^2 \overline{V_k} - |2W_k - U_k^{(m)}|^2 \overline{W_k} \right\} (V_k - W_k) \Delta x \right|$$
$$+ \gamma \sum_{k=0}^{N-1} |U_k^{(m)}|^2 |V_k - W_k|^2 \Delta x$$
$$\leq \gamma \left(\sum_{k=0}^{N-1} \left| |2V_k - U_k^{(m)}|^2 \overline{V_k} - |2W_k - U_k^{(m)}|^2 \overline{W_k} \right|^{\frac{4}{3}} \Delta x \right)^{\frac{3}{4}} \|\boldsymbol{V} - \boldsymbol{W}\|_4$$
$$+ \gamma \sum_{k=0}^{N-1} |U_k^{(m)}|^2 |V_k - W_k|^2 \Delta x.$$
(4.200)

Since for any $z_1, z_2, z \in \mathbf{C}$, $\left| |2z_1 - z|^2 z_1 - |2z_2 - z|^2 z_2 \right| \leq 4 \left(|z_1| + |z_2| + \frac{1}{2}|z| \right)^2 |z_1 - z_2|$ holds, we have

$$\left(\sum_{k=0}^{N-1} \left| |2V_k - U_k^{(m)}|^2 \overline{V_k} - |2W_k - U_k^{(m)}|^2 \overline{W_k} \right|^{\frac{4}{3}} \Delta x \right)^{\frac{3}{4}}$$
$$\leq \left(\sum_{k=0}^{N-1} \left\{ 4(|V_k| + |W_k| + \frac{1}{2}|U_k^{(m)}|)^2 |V_k - W_k| \right\}^{\frac{4}{3}} \Delta x \right)^{\frac{3}{4}}$$
$$\leq 4 \left\{ \sum_{k=0}^{N-1} \left(|V_k| + |W_k| + \frac{1}{2}|U_k^{(m)}| \right)^4 \right\}^{\frac{1}{2}} \|\boldsymbol{V} - \boldsymbol{W}\|_4$$
$$\leq 36 \max \left\{ \|\boldsymbol{U}^{(m)}\|_4, \|\boldsymbol{V}\|_4, \|\boldsymbol{W}\|_4 \right\}^2 \|\boldsymbol{V} - \boldsymbol{W}\|_4.$$
(4.201)

Furthermore, the second term at the most right-hand side of (4.200) can be evaluated as

$$\sum_{k=0}^{N-1} |U_k^{(m)}|^2 |V_k - W_k|^2 \Delta x \leq \|\boldsymbol{U}^{(m)}\|_4^2 \|\boldsymbol{V} - \boldsymbol{W}\|_4^2. \quad (4.202)$$

Thus, substituting (4.201) and (4.202) into (4.199) and (4.200), we have

$$\|\boldsymbol{V} - \boldsymbol{W}\|^2 \leq 18\Delta t \gamma \max\left\{\|\boldsymbol{U}^{(m)}\|_4, \|\boldsymbol{V}\|_4, \|\boldsymbol{W}\|_4\right\}^2 \|\boldsymbol{V} - \boldsymbol{W}\|_4^2, \quad (4.203)$$

$$\|(\boldsymbol{V} - \boldsymbol{W})_x\|^2 \leq 37\gamma \max\left\{\|\boldsymbol{U}^{(m)}\|_4, \|\boldsymbol{V}\|_4, \|\boldsymbol{W}\|_4\right\}^2 \|\boldsymbol{V} - \boldsymbol{W}\|_4^2. \quad (4.204)$$

From the discrete energy and probability conservation properties, and Lemma 4.1, there exists a constant $\widetilde{c}(c_1, c_2, \gamma)$ such that $\|\boldsymbol{U}^{(m)}\|_4^4 \leq c(c_1, c_2, \gamma)$ holds. Since $\boldsymbol{V}$ and $\boldsymbol{W}$ are solutions of the form $\boldsymbol{V}^{(m+\frac{1}{2})} = (\boldsymbol{V}^{(m+1)} + \boldsymbol{V}^{(m)})/2$, there also exists a constant $c(c_1, c_2, \gamma)$ such that $\|\boldsymbol{V}\|_4^4, \|\boldsymbol{W}\|_4^4 \leq c(c_1, c_2, \gamma)$ holds. Thus,

$$\max\left\{\|\boldsymbol{U}^{(m)}\|_4, \|\boldsymbol{V}\|_4, \|\boldsymbol{W}\|_4\right\}^2 \leq c^{\frac{1}{2}}. \quad (4.205)$$

Thus, from (3.181) in Lemma 3.5, and from (4.203), (4.204), and (4.205), we have

$$\|\boldsymbol{V} - \boldsymbol{W}\|_4^4 \leq b\|\boldsymbol{V} - \boldsymbol{W}\|_{H^1} \|\boldsymbol{V} - \boldsymbol{W}\|^3 \quad (4.206)$$
$$\leq 18bc\gamma^2 \Delta t \sqrt{18\Delta t(18\Delta t + 37)} \|\boldsymbol{V} - \boldsymbol{W}\|_4^4.$$

If in (4.206) $18bc\gamma^2 \Delta t \sqrt{18\Delta t(18\Delta t + 37)} < 1$ holds, i.e.,

$$(\Delta t)^3 (18\Delta t + 37) < \frac{1}{18^3 b^2 c^2 \gamma^4}$$

holds, $\|\boldsymbol{V} - \boldsymbol{W}\|_4 = 0$. Hence $\boldsymbol{V} = \boldsymbol{W}$. □

From the discrete Sobolev inequality, and the discrete energy and probability conservation properties, we immediately obtain the boundedness of the the numerical solution.

THEOREM 4.10 Boundedness of numerical solution
The numerical solution is bounded:

$$\|\boldsymbol{U}^{(m)}\|_\infty < \infty, \qquad m = 0, 1, 2, \ldots.$$

Applications

PROOF We have $\|\boldsymbol{U}^{(m)}\|_2^2 = $ const., because of the discrete probability conservation property. From the proof of Lemma 4.1, $\|\boldsymbol{U}_x^{(m)}\|_2^2 < \infty$. Thus, from the discrete Sobolev inequality in Section 3.6.2.1, we have the claim of the theorem. □

In order to evaluate the convergence, let us first define the truncation error of the scheme, $F_k^{(m)}$, as follows.

$$\mathrm{i}\left(\frac{u(k\Delta x,(m+1)\Delta t)-u(k\Delta x,m\Delta t)}{\Delta t}\right)=$$
$$-\delta_k^{\langle 2\rangle}\left(\frac{u(k\Delta x,(m+1)\Delta t)+u(k\Delta x,m\Delta t)}{2}\right)$$
$$-\gamma\left(\frac{|u(k\Delta x,(m+1)\Delta t)|^2+|u(k\Delta x,m\Delta t)|^2}{2}\right)$$
$$\times\left(\frac{u(k\Delta x,(m+1)\Delta t)+u(k\Delta x,m\Delta t)}{2}\right)$$
$$+F_k^{(m)}. \qquad (4.207)$$

As to the truncation error $F_k^{(m)}$, the next lemma holds. Hereafter, $T = M\Delta t$ is the fixed "goal time" at which we measure the error.

LEMMA 4.2
Let $u \in C^2\left[[0,T],C^3\right]$. Then,

$$\Delta t \sum_{m=0}^{M} \|\boldsymbol{F}^{(m)}\|^2 \leq CT(\Delta t^4 + \Delta x^2),$$

where C is a constant that depends only on γ and the true solution u on $\Omega = [0,T] \times (-L,L)$. If $u \in C^2\left[[0,T],C^5\right]$, then the above estimate is improved to

$$\Delta t \sum_{m=0}^{M} \|\boldsymbol{F}^{(m)}\|^2 \leq C'T(\Delta t^4 + \Delta x^4),$$

where C' is another constant.

PROOF When $u \in C^2\left[[0,T],C^3\right]$, by considering the Taylor expansion of both sides of the scheme at $(x,t) = (k\Delta x,(m+\frac{1}{2})\Delta t)$, we evaluate the local truncation error as

$$|F_k^{(m)}| \leq c_3(\Delta t^2 + \Delta x),$$

where c_3 is a constant that only depends on γ and the true solution u on $\Omega = [0,T] \times (-L,L)$. Then by summing $|F_k^{(m)}|$ from $m=0$ to M, we obtain

the result. When $u \in C^2\left[[0,T], C^5\right]$, the local truncation error is replaced with
$$|F_k^{(m)}| \leq c_4(\Delta t^2 + \Delta x^2),$$
where c_4 is another constant, and the rest is the same. □

Now we are in a position to present a convergence theorem. Let us denote the error in the numerical solution by $e_k^{(m)} = u_k^{(m)} - U_k^{(m)}$, where $u_k^{(m)} \stackrel{\mathrm{d}}{=} u(k\Delta x, m\Delta t)$, and evaluate the error in the numerical solution at the "goal time": $\|\boldsymbol{e}^{(M)}\|$. Then, the following theorem holds.

THEOREM 4.11 Convergence of the scheme
Let $u \in C^2\left[[0,T], C^3\right]$ be the true solution to the cubic NLS, and C be the constant defined in Lemma 4.2. Then, there exists a constant c, which depends only on γ, c_1, c_2 (defined in Theorem 4.9) and the true solution u on $[0,T] \times (-L, L)$, such that the following estimate holds if Δt is chosen so that $1 - c\Delta t > 0$:
$$\|\boldsymbol{e}^{(M)}\|^2 \leq \frac{CT}{1 - c\Delta t}(\Delta t^4 + \Delta x^2)e^{\frac{cT}{1-c\Delta t}}.$$
Moreover, if $u \in C^2\left[[0,T], C^5\right]$, then the above estimate is improved to
$$\|\boldsymbol{e}^{(M)}\|^2 \leq \frac{C'T}{1 - c'\Delta t}(\Delta t^4 + \Delta x^4)e^{\frac{c'T}{1-c'\Delta t}},$$
where c' is a constant which depends only on γ, c_1, c_2 (as defined in Theorem 4.9) and the true solution u on $[0,T] \times (-L, L)$, and C' is the constant defined in Lemma 4.2.

PROOF Subtracting the scheme (4.187) from (4.207), we have
$$\mathrm{i}\left(\frac{e_k^{(m+1)} - e_k^{(m)}}{\Delta t}\right) =$$
$$-\delta_k^{\langle 2 \rangle}\left(\mu_m^+ e_k^{(m)}\right) - \gamma\left\{(\mu_m^+ u_k^{(m)})(\mu_m^+ |u_k^{(m)}|^2) - (\mu_m^+ U_k^{(m)})(\mu_m^+ |U_k^{(m)}|^2)\right\}$$
$$- F_k^{(m)}. \tag{4.208}$$

Let us apply $2\mathrm{Im}\sum_{k=0}^{N-1}(\cdot)\overline{(\mu_m^+ e_k^{(m)})}\Delta x$ to both sides of the above identity. The left-hand side becomes
$$\frac{1}{\Delta t}\mathrm{Im}\left\{\mathrm{i}\sum_{k=0}^{N-1}(e_k^{(m+1)} - e_k^{(m)})\overline{(e_k^{(m+1)} + e_k^{(m)})}\Delta x\right\}$$
$$= \frac{1}{\Delta t}\mathrm{Im}\left[\mathrm{i}\left\{\|\boldsymbol{e}^{(m+1)}\|^2 - \|\boldsymbol{e}^{(m)}\|^2 - (\boldsymbol{e}^{(m+1)}, \boldsymbol{e}^{(m)}) + (\boldsymbol{e}^{(m)}, \boldsymbol{e}^{(m+1)})\right\}\right]$$
$$= \frac{1}{\Delta t}\left\{\|\boldsymbol{e}^{(m+1)}\|^2 - \|\boldsymbol{e}^{(m)}\|^2\right\}. \tag{4.209}$$

The first term at the right-hand side of (4.208) becomes

$$-2\mathrm{Im}\left\{\sum_{k=0}^{N-1}(\delta_k^{\langle 2\rangle}\mu_m^+ e_k^{(m)})\overline{(\mu_m^+ e_k^{(m)})}\Delta x\right\}$$

$$= -2\mathrm{Im}\sum_{k=0}^{N-1}\left\{\frac{\mu_m^+ e_{k+1}^{(m)} - 2\mu_m^+ e_k^{(m)} + \mu_m^+ e_{k-1}^{(m)}}{(\Delta x)^2}\cdot\overline{(\mu_m^+ e_k^{(m)})}\Delta x\right\}$$

$$= -\frac{2}{(\Delta x)^2}\mathrm{Im}\left\{\sum_{k=0}^{N-1}(\mu_m^+ e_{k+1}^{(m)})\overline{(\mu_m^+ e_k^{(m)})}\Delta x + \sum_{k=0}^{N-1}(\mu_m^+ e_{k-1}^{(m)})\overline{(\mu_m^+ e_k^{(m)})}\Delta x\right\}$$

$$= 0. \qquad (4.210)$$

In the last equality, we used the discrete periodic boundary condition and rearranged the summation to find that inside $\{\cdot\}$ is a real number.

As to the second term at the right-hand side of (4.208),

$$(\mu_m^+ u_k^{(m)})(\mu_m^+|u_k^{(m)}|^2) - (\mu_m^+ U_k^{(m)})(\mu_m^+|U_k^{(m)}|^2) =$$
$$(\mu_m^+ u_k^{(m)})\left\{\mu_m^+(|u_k^{(m)}|^2 - |U_k^{(m)}|^2)\right\} + (\mu_m^+ e_k^{(m)})(\mu_m^+|U_k^{(m)}|^2)$$

holds. If we apply $2\mathrm{Im}\sum_{k=0}^{N-1}(\cdot)\overline{(\mu_m^+ e_k^{(m)})}\Delta x$, the second term at the right-hand side of the above identity vanishes. The first term is evaluated as

$$\left|2\mathrm{Im}\sum_{k=0}^{N-1}(\text{the first term})\overline{(\mu_m^+ e_k^{(m)})}\Delta x\right|$$

$$= \left|-2\gamma\mathrm{Im}\sum_{k=0}^{N-1}(\mu_m^+ u_k^{(m)})\left\{\mu_m^+(|u_k^{(m)}|^2 - |U_k^{(m)}|^2)\right\}\overline{(\mu_m^+ e_k^{(m)})}\Delta x\right|$$

$$\leq c|\gamma|\left|\sum_{k=0}^{N-1}\left[\mu_m^+\left\{(u_k^{(m)} + U_k^{(m)})\overline{e_k^{(m)}} + \overline{(u_k^{(m)} + U_k^{(m)})}e_k^{(m)}\right\}\right]\overline{(\mu_m^+ e_k^{(m)})}\Delta x\right|$$

$$\leq 4c^2|\gamma|\|\mu_m^+ e_k^{(m)}\|^2$$

$$\leq 8c^2|\gamma|\left(\|\boldsymbol{e}^{(m+1)}\|^2 + \|\boldsymbol{e}^{(m)}\|^2\right). \qquad (4.211)$$

In the above calculation, we used the fact that the numerical solution is bounded (Theorem 4.10).

Finally, the third term at the right-hand side of (4.208) is evaluated as

$$\left|2\mathrm{Im}\sum_{k=0}^{N-1}F_k^{(m)}\overline{(\mu_m^+ e_k^{(m)})}\Delta x\right| \leq 2\|\boldsymbol{F}^{(m)}\|\,\|\mu_m^+ \boldsymbol{e}^{(m)}\|$$

$$\leq \|\boldsymbol{F}^{(m)}\|^2 + 2\left(\|\boldsymbol{e}^{(m+1)}\|^2 + \|\boldsymbol{e}^{(m)}\|^2\right).\, (4.212)$$

Thus, from (4.209), (4.210), (4.211), and (4.212), we have

$$\frac{1}{\Delta t}\left(\|\boldsymbol{e}^{(m+1)}\|^2 - \|\boldsymbol{e}^{(m)}\|^2\right) \leq c\left(\|\boldsymbol{e}^{(m+1)}\|^2 + \|\boldsymbol{e}^{(m)}\|^2\right) + \|\boldsymbol{F}^{(m)}\|^2.$$

Therefore,

$$\|\boldsymbol{e}^{(m+1)}\|^2 \leq c\Delta t \sum_{l=0}^{m+1} \|\boldsymbol{e}^{(l)}\|^2 + \Delta t \sum_{l=0}^{m} \|\boldsymbol{F}^{(m)}\|^2.$$

If we choose Δt sufficiently small so that $1 - c\Delta t > 0$ holds, then we have

$$\|\boldsymbol{e}^{(m+1)}\|^2 \leq \frac{1}{1 - c\Delta t}\left(c\Delta t \sum_{l=0}^{m} \|\boldsymbol{e}^{(l)}\|^2 + \Delta t \sum_{l=0}^{m} \|\boldsymbol{F}^{(m)}\|^2\right). \qquad (4.213)$$

Taking $m = M - 1$, and from Lemma 3.6 and Lemma 4.2, we reach the claim of the theorem. $\square$

4.4.2 Gross–Pitaevskii Equation

4.4.2.1 Introduction to Problem

The Gross–Pitaevskii equation (2.41):

$$\mathrm{i}\frac{\partial u}{\partial t} = -\frac{\partial^2 u}{\partial x^2} - (|u|^2 - 1)u$$

is one of the target PDEs 4 in (2.39) with $G(u, u_x) = |u_x|^2 + \frac{1}{2}(1 - |u|^2)^2$. This equation is known as a mean field nonlinear Schrödinger equation and often used to investigate the dynamics of the Bose–Einstein condensation (BEC) phenomenon [79, 140]. This equation has the following invariants:

$$P(u) \stackrel{\mathrm{d}}{=} \int_\Omega |u|^2 \mathrm{d}x, \qquad (4.214)$$

$$J(u) = \int_\Omega G(u, u_x) \mathrm{d}x. \qquad (4.215)$$

The invariant $P(u)$ is called "charge."

4.4.2.2 Numerical Scheme

We derive a numerical scheme from the Scheme 3.4 in (3.86) and

$$G_{\mathrm{d},k}(\boldsymbol{U}) \stackrel{\mathrm{d}}{=} \frac{|\delta_k^+ U_k|^2 + |\delta_k^- U_k|^2}{2} + \frac{1}{2}\left(1 - |U_k|^2\right)^2. \qquad (4.216)$$

The scheme is

$$\mathrm{i}\left(\frac{U_k^{(m+1)} - U_k^{(m)}}{\Delta t}\right) = -\frac{\delta G_\mathrm{d}}{\delta(\overline{\boldsymbol{U}^{(m+1)}}, \overline{\boldsymbol{U}^{(m)}})_k}, \qquad k = 0, \ldots, N \qquad (4.217)$$

where
$$\frac{\delta G_\mathrm{d}}{\delta(\overline{\boldsymbol{U}^{(m+1)}, \boldsymbol{U}^{(m)}})_k} = -\delta_k^{\langle 2\rangle}(\mu_n^+ U_k^{(m)}) + \left(\mu_m^+ |U_k^{(m)}|^2 - 1\right)(\mu_m^+ U_k^{(m)}). \tag{4.218}$$

Under the discrete boundary conditions
$$\left. \delta_k^{\langle 1\rangle} U_k^{(m)} \right|_{k=0} = \left. \delta_k^{\langle 1\rangle} U_k^{(m)} \right|_{k=N} = 0, \qquad m = 0, 1, 2, \ldots \tag{4.219}$$

this scheme has the following discrete invariants:
$$P_\mathrm{d}(\boldsymbol{U}^{(m)}) \overset{\mathrm{d}}{\equiv} \sum_{k=0}^{N}{}'' |U_k^{(m)}|^2 \Delta x, \tag{4.220}$$

$$J_\mathrm{d}(\boldsymbol{U}^{(m)}) \overset{\mathrm{d}}{\equiv} \sum_{k=0}^{N}{}'' G_{\mathrm{d},k}(\boldsymbol{U}^{(m)}) \Delta x. \tag{4.221}$$

We are also able to design a linearly implicit scheme using the linearization technique discussed in the chapter 6. For example,
$$\mathrm{i}\left(\frac{U_k^{(m+1)} - U_k^{(m-1)}}{\Delta t}\right) = \frac{\delta G_\mathrm{d}}{\delta(\overline{\boldsymbol{U}^{(m+1)}, \boldsymbol{U}^{(m)}, \boldsymbol{U}^{(m-1)}})_k}, \tag{4.222}$$

where
$$\frac{\delta G_\mathrm{d}}{\delta(\overline{\boldsymbol{U}^{(m+1)}, \boldsymbol{U}^{(m)}, \boldsymbol{U}^{(m-1)}})_k} = -\delta_k^{\langle 2\rangle}(s_m^{\langle 1\rangle} U_k^{(m)}) + |U_k^{(m)}|^2 s_m^{\langle 1\rangle} U_k^{(m)} - U_k^{(m)} \tag{4.223}$$

is the linearly implicit for the Gross–Pitaevskii equation. This scheme has also two invariants $P_\mathrm{d}(\boldsymbol{U}^{(m+1)}, \boldsymbol{U}^{(m)})$ and $J_\mathrm{d}(\boldsymbol{U}^{(m+1)}, \boldsymbol{U}^{(m)})$. For the detail of this scheme and the related discussion, readers may refer to [152].

4.4.2.3 Numerical Examples

In Figure 4.11, we show numerical solutions with $L = 220$, $N = 1100$ ($\Delta x = 0.2$) and $\Delta t = 0.2$. We investigate the dynamics of a traveling wave solution whose initial state is
$$u_0(x) = \sqrt{1 - \frac{\gamma^2}{2\cosh(\gamma(x-20)/2)}} \times$$
$$\exp\left(\mathrm{i}\arctan\left(\frac{e^{\gamma(x-20)} + c^2 - 1}{-c\gamma}\right) - \mathrm{i}\arctan\left(\frac{-c}{\gamma}\right)\right) \tag{4.224}$$

where $\gamma = \sqrt{2-c^2}$, and $c > 0$ is the speed of the traveling wave.

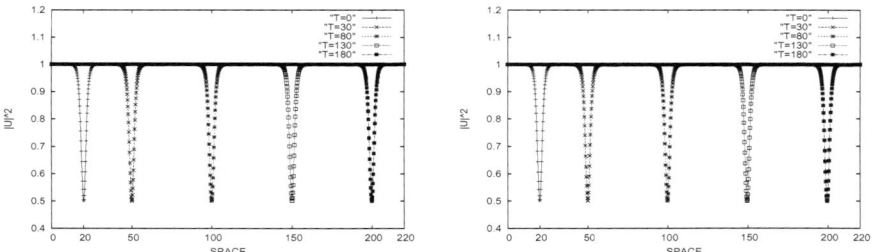

FIGURE 4.11: The time evolution profiles of the traveling wave solutions with the speed $c = 1.0$. The left figure: the nonlinear scheme (4.217), the right: the linearly implicit scheme (4.222).

4.4.2.4 Further Topic: Two-Dimensional Examples

The Gross–Pitaevskii equation has a variety of 2D-formulations. We here study the following equation described in Aftalion–Du [6].

$$\frac{\partial u}{\partial t} = \Delta u - g|u|^2 u - V_{\text{trap}}(\boldsymbol{x})u + \frac{\mu(u)}{\|u\|_2^2}u + \omega L_z u, \qquad (4.225)$$

where

$$\mu(u) \stackrel{\mathrm{d}}{=} \int_\Omega \left(|\nabla u|^2 + V_{\text{trap}}(\boldsymbol{x})|u|^2 + g|u|^4 - \omega \overline{u} L_z u \right) \mathrm{d}\boldsymbol{x},$$

and g is the parameter to describe the interaction between atoms in the condensation, $V_{\text{trap}}(\boldsymbol{x}) = ((w_x)^2 x^2 + (w_y)^2 y^2)/(2\min(w_x, w_y))$ is a trapped potential, w_x, w_y are the trap frequencies in x- and y-directions, and $\omega L_z = -i\omega\hbar(x\partial_y - y\partial_x)$ is the angular momentum at z-axis with frequency ω.

By applying (the two-dimensional version of) the discrete variational derivative method, we obtain the same scheme as Aftalion–Du [6]. Numerical examples are shown in Figure 4.12.

4.5 Target PDEs 5

In this section, an example for the target PDEs 5 (defined in Section 2.4; systems of conservative PDEs) is shown.

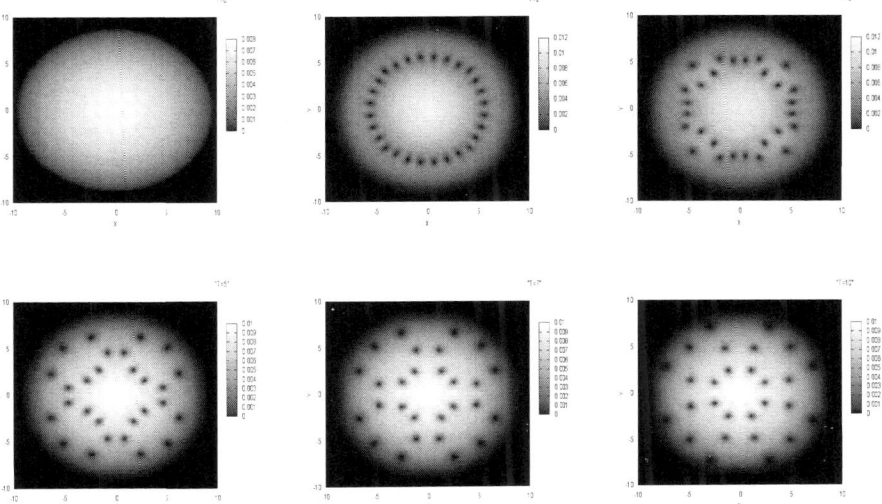

FIGURE 4.12: The time evolution contours of discrete density $|U_{k,l}|^2$ at time $t = 0, 2, 3, 5, 7, 10$.

4.5.1 Zakharov Equations

4.5.1.1 Introduction to Problem

Let us consider the Zakharov equations:

$$\begin{cases} iE_t + E_{xx} = nE, & t > 0, x \in (0, L), \\ n_{tt} - n_{xx} = (|E|^2)_{xx}, & t > 0, x \in (0, L), \\ E(0, x) = E_0(x), n(0, x) = n_0(x), n_t(0, x) = n_1(x), & x \in (0, L). \end{cases} \quad (4.226)$$

We assume the periodic boundary condition. The system of the Zakharov equations is an example of the PDEs 5, where $N_{\text{ex}} = 4$ and $(u_1, u_2, u_3, u_4) = (E, \overline{E}, n, v)$ (refer to (2.43)). In the classification of the target PDEs 5, the equations belong to Type C4 with two Type C2 subequations.

In what follows, we use the notation $(E, \overline{E}, n, v)$ instead of (u_1, u_2, u_3, u_4) for readability.

4.5.1.2 Numerical Scheme

Let us denote numerical solutions by $E_k^{(m)}$, $n_k^{(m)}$, $v_k^{(m)}$. We define the discrete local energy by

$$G_{\text{d},k}(\boldsymbol{E}^{(m)}, \boldsymbol{n}^{(m)}, \boldsymbol{v}^{(m)}) = |\delta_k^+ E_k^{(m)}|^2 + n_k^{(m)}|E_k^{(m)}|^2 + \frac{1}{2}\left(n_k^{(m)^2} + (\delta_k^+ v_k^{(m)})^2\right), \quad (4.227)$$

which complies with (3.93). We define the discrete global energy accordingly by

$$\sum_{k=0}^{N} {}''G_{\mathrm{d},k}(\boldsymbol{E}^{(m)}, \boldsymbol{n}^{(m)}, \boldsymbol{v}^{(m)})\Delta x. \qquad (4.228)$$

Taking discrete variation according to (3.96), we have

$$\sum_{k=0}^{N} {}''G_{\mathrm{d},k}(\boldsymbol{E}^{(m+1)}, \boldsymbol{n}^{(m+1)}, \boldsymbol{v}^{(m+1)}) - G_{\mathrm{d},k}(\boldsymbol{E}^{(m)}, \boldsymbol{n}^{(m)}, \boldsymbol{v}^{(m)})\Delta x$$

$$= \sum_{k=0}^{N-1} \left\{ \frac{\delta G_{\mathrm{d}}}{\delta(\boldsymbol{E}^{(m+1)}, \boldsymbol{E}^{(m)})_k} (E_k{}^{(m+1)} - E_k{}^{(m)}) \right.$$

$$+ \frac{\delta G_{\mathrm{d}}}{\delta(\overline{\boldsymbol{E}^{(m+1)}}, \overline{\boldsymbol{E}^{(m)}})_k} (\overline{E_k{}^{(m+1)} - E_k{}^{(m)}})$$

$$+ \frac{\delta G_{\mathrm{d}}}{\delta(\boldsymbol{n}^{(m+1)}, \boldsymbol{n}^{(m)})_k} (n_k{}^{(m+1)} - n_k{}^{(m)})$$

$$\left. + \frac{\delta G_{\mathrm{d}}}{\delta(\boldsymbol{v}^{(m+1)}, \boldsymbol{v}^{(m)})_k} (v_k^{(m+1)} - v_k^{(m)}) \right\} \Delta x, \qquad (4.229)$$

where

$$\frac{\delta G_{\mathrm{d}}}{\delta(\boldsymbol{E}^{(m+1)}, \boldsymbol{E}^{(m)})_k} = -\delta_k^{\langle 2 \rangle} \left(\frac{E_k^{(m+1)} + E_k^{(m)}}{2} \right)$$

$$+ \left(\frac{E_k^{(m)} + E_k^{(m)}}{2} \right) \left(\frac{n_k^{(m+1)} + n_k^{(m)}}{2} \right), \quad (4.230\mathrm{a})$$

$$\frac{\delta G_{\mathrm{d}}}{\delta(\overline{\boldsymbol{E}^{(m+1)}}, \overline{\boldsymbol{E}^{(m)}})_k} = \overline{\left(\frac{\delta G_{\mathrm{d}}}{\delta(\boldsymbol{E}^{(m+1)}, \boldsymbol{E}^{(m)})_k} \right)}, \qquad (4.230\mathrm{b})$$

$$\frac{\delta G_{\mathrm{d}}}{\delta(\boldsymbol{n}^{(m+1)}, \boldsymbol{n}^{(m)})_k} = \frac{n_k^{(m+1)} + n_k^{(m)}}{2} + \frac{|E_k^{(m+1)}|^2 + |E_k^{(m)}|^2}{2}, \quad (4.230\mathrm{c})$$

$$\frac{\delta G_{\mathrm{d}}}{\delta(\boldsymbol{v}^{(m+1)}, \boldsymbol{v}^{(m)})_k} = \delta_k^{\langle 2 \rangle} \left(\frac{v_k^{(m+1)} + v_k^{(m)}}{2} \right). \qquad (4.230\mathrm{d})$$

With the discrete variational derivatives, we define a numerical scheme according to (3.101).

$$\mathrm{i} \left(\frac{E_k^{(m+1)} - E_k^{(m)}}{\Delta t} \right) = \frac{\delta G_{\mathrm{d}}}{\delta(\boldsymbol{E}^{(m+1)}, \boldsymbol{E}^{(m)})_k}, \qquad (4.231\mathrm{a})$$

$$\frac{n_k^{(m+1)} - n_k^{(m)}}{\Delta t} = -\frac{\delta G_{\mathrm{d}}}{\delta(\boldsymbol{v}^{(m+1)}, \boldsymbol{v}^{(m)})_k}, \qquad (4.231\mathrm{b})$$

$$\frac{v_k^{(m+1)} - v_k^{(m)}}{\Delta t} = \frac{\delta G_{\mathrm{d}}}{\delta(\boldsymbol{n}^{(m+1)}, \boldsymbol{n}^{(m)})_k}. \qquad (4.231\mathrm{c})$$

We impose discrete periodic boundary conditions for all the variables. Since this scheme satisfies the assumptions in Scheme 3.5, the discrete global energy (4.228) is conserved.

4.6 Target PDEs 7

In this section, examples for the target PDEs 7 (defined in Section 2.5; second-order PDEs) are shown.

4.6.1 Nonlinear Klein–Gordon Equation

4.6.1.1 Introduction to Problem

We consider the nonlinear Klein–Gordon equation (2.58) as a specific example of the target equation (2.55) where

$$G(u, u_x) = \frac{1}{2}(u_x)^2 + \phi(u). \tag{4.232}$$

This is a well-known nonlinear equation with soliton solutions. This includes linear wave equation, the sine-Gordon equation, the double sine-Gordon equation and the phi-4 equation. Numerical studies regarding this equation are, for example, [2, 3, 4, 5, 15, 16, 20, 38, 42, 48, 50, 51, 53, 86, 99, 103, 107, 139, 155, 164, 173].

4.6.1.2 Numerical Schemes

In the numerical studies on this equation, such as the above, much effort has been devoted to "energy-preserving" computation. Most of the computations were, however, not completely discrete; for example, in some studies energies were defined by *integral*, not *summation*, whose "conservation" would necessarily be lost when the system was fully discretized. In the literature, we could find the following five schemes as fully-discrete energy-conserving schemes. Fortunately for us, all of them can be regarded as the special cases of Scheme (3.127) or (3.129).

Strauss scheme [155]

$$\delta_m^{\langle 2 \rangle} U_k^{(m)} = \delta_k^{\langle 2 \rangle} U_k^{(m)} - \frac{\mathrm{d}\phi}{\mathrm{d}\left(U_k^{(m+1)}, U_k^{(m-1)}\right)}, \tag{4.233}$$

where $\frac{\mathrm{d}\phi}{\mathrm{d}(a,b)} \stackrel{\mathrm{d}}{\equiv} \frac{\phi(a)-\phi(b)}{a-b}$. This is the implicit scheme (3.127) with

$$G_{\mathrm{d},k}(\boldsymbol{U}, \boldsymbol{V}) = \frac{1}{2}\left(\frac{\delta_k^+ U_k \delta_k^+ V_k + \delta_k^- U_k \delta_k^- V_k}{2}\right) + \frac{\phi(U_k) + \phi(V_k)}{2}. \tag{4.234}$$

Ben-Yu scheme [15]

$$\delta_m^{\langle 2 \rangle} U_k{}^{(m)} = \delta_k^{\langle 2 \rangle} s_m^{\langle 1 \rangle} U_k{}^{(m)} - \frac{\mathrm{d}\phi}{\mathrm{d}\left(U_k{}^{(m+1)}, U_k{}^{(m-1)}\right)}. \tag{4.235}$$

This scheme is a special case of the implicit scheme (3.127) with

$$G_{\mathrm{d},k}(\boldsymbol{U}, \boldsymbol{V}) = \frac{1}{2}\left(\frac{(\delta_k^+ U_k)^2 + (\delta_k^- U_k)^2 + (\delta_k^+ V_k)^2 + (\delta_k^- V_k)^2}{4}\right) + \frac{\phi(U_k) + \phi(V_k)}{2}. \tag{4.236}$$

Zhang implicit scheme [53]

$$\delta_m^{\langle 2 \rangle} U_k{}^{(m)} = \delta_k^{\langle 2 \rangle} U_k{}^{(m)} \\ - \frac{\mathrm{d}\phi}{\mathrm{d}\left(\frac{1}{2}(U_k{}^{(m+1)} + U_k{}^{(m)}), \frac{1}{2}(U_k{}^{(m)} + U_k{}^{(m-1)})\right)}. \tag{4.237}$$

This scheme is a special case of the implicit scheme (3.127) with

$$G_{\mathrm{d},k}(\boldsymbol{U}, \boldsymbol{V}) = \frac{1}{2}\left(\frac{\delta_k^+ U_k \delta_k^+ V_k + \delta_k^- U_k \delta_k^- V_k}{2}\right) + \phi\left(\frac{U_k + V_k}{2}\right). \tag{4.238}$$

Li scheme [107]

$$\delta_m^{\langle 2 \rangle} U_k{}^{(m)} = \delta_k^{\langle 2 \rangle} \mu_m^{\langle 1 \rangle} U_k{}^{(m)} - \frac{\mathrm{d}\phi}{\mathrm{d}\left(U_k{}^{(m+1)}, U_k{}^{(m-1)}\right)}. \tag{4.239}$$

This scheme is a special case of the implicit scheme (3.127) with

$$G_{\mathrm{d},k}(\boldsymbol{U}, \boldsymbol{V}) = \frac{1}{2}\left(\frac{\left(\delta_k^+\left(\frac{U_k+V_k}{2}\right)\right)^2 + \left(\delta_k^-\left(\frac{U_k+V_k}{2}\right)\right)^2}{2}\right) + \frac{\phi(U_k) + \phi(V_k)}{2}. \tag{4.240}$$

Zhang explicit scheme [53]

$$\delta_m^{\langle 2+ \rangle} U_k{}^{(m)} = \delta_k^{\langle 2 \rangle} \mu_m^+ U_k{}^{(m)} - \frac{\mathrm{d}\phi}{\mathrm{d}\left(U_k{}^{(m+1)}, U_k{}^{(m)}\right)}. \tag{4.241}$$

This scheme is a special case of the explicit scheme (3.129) with

$$G_{\mathrm{d},k}(\boldsymbol{U}) = \frac{1}{2}\left(\frac{(\delta_k^+ U_k)^2 + (\delta_k^- U_k)^2}{2}\right) + \phi(U_k). \tag{4.242}$$

As one can see, the variety comes from the degree of the freedom in the discrete energy function. By exploiting this feature, we can further construct other schemes based on (3.127) or (3.129) as below.

DVDM implicit scheme

$$\delta_m^{\langle 2 \rangle} U_k{}^{(m)} = (\delta_k^{\langle 1 \rangle})^2 \mu_m^{\langle 1 \rangle} U_k{}^{(m)} - \frac{\mathrm{d}\phi}{\mathrm{d}\left(U_k{}^{(m+1)}, U_k{}^{(m-1)}\right)} \tag{4.243}$$

is derived from (3.127) with

$$G_{\mathrm{d},k}(\boldsymbol{U},\boldsymbol{V}) = \frac{1}{2}\left(\delta_k^{\langle 1\rangle}\left(\frac{U_k+V_k}{2}\right)\right)^2 + \frac{\phi(U_k)+\phi(V_k)}{2}. \tag{4.244}$$

DVDM explicit scheme 1

$$\delta_m^{\langle 2+\rangle}{U_k}^{(m)} = s_k^{\langle 1\rangle}\delta_k^{\langle 2\rangle}\mu_m^+ {U_k}^{(m)} - \frac{\mathrm{d}\phi}{\mathrm{d}\left({U_k}^{(m+1)}, {U_k}^{(m)}\right)} \tag{4.245}$$

can be derived from (3.129) with

$$G_{\mathrm{d},k}(\boldsymbol{U}) = \frac{1}{2}(\delta_k^+ U_k)(\delta_k^- U_k) + \phi(U_k). \tag{4.246}$$

DVDM explicit scheme 2

$$\delta_m^{\langle 2+\rangle}{U_k}^{(m)} = (\delta_k^{\langle 1\rangle})^2 \mu_m^+ {U_k}^{(m)} - \frac{\mathrm{d}\phi}{\mathrm{d}\left({U_k}^{(m+1)}, {U_k}^{(m)}\right)} \tag{4.247}$$

is from (3.129), with

$$G_{\mathrm{d},k}(\boldsymbol{U}) = \frac{1}{2}(\delta_k^{\langle 1\rangle}U_k)^2 + \phi(U_k). \tag{4.248}$$

4.6.1.3 Numerical Examples

Let us test the above schemes numerically. We take the sine-Gordon equation as our example. The initial state is

$$u(x,0) = 4\arctan\left(\exp\left(\frac{x}{\sqrt{1-v^2}}\right)\right), \tag{4.249}$$

where $v = 0.2$. The exact solution for this initial state is

$$u(x,t) = 4\arctan\left(\exp\left(\frac{x-vt}{\sqrt{1-v^2}}\right)\right). \tag{4.250}$$

The energy $E^{\mathrm{TRUE}} \stackrel{\mathrm{d}}{=} \int \left\{\frac{1}{2}(u_t)^2 + G\right\}\mathrm{d}x$ for the exact solution is approximately

$$E^{\mathrm{TRUE}} \cong \frac{8}{\sqrt{1-v^2}} - L. \tag{4.251}$$

Now we set $L = 20$, which means $E^{\mathrm{TRUE}} \simeq -11.83503$. The momentum $M^{\mathrm{TRUE}} \stackrel{\mathrm{d}}{=} \int u_x u_t \mathrm{d}x$ for the exact solution is approximately

$$M^{\mathrm{TRUE}} \cong -\frac{8v}{\sqrt{1-v^2}} \simeq -1.632993. \tag{4.252}$$

Below we set $\Delta x = 0.5$, $\Delta t = 0.025$, and

$$u_{-j}^{(m)} \stackrel{\mathrm{d}}{\equiv} u_j^{(m)}, \qquad u_{N+j}^{(m)} \stackrel{\mathrm{d}}{\equiv} u_{N-j}^{(m)}, \qquad 1 \leq j \leq N \qquad (4.253)$$

for boundary conditions.

Among the schemes mentioned above, we employ the Strauss scheme (4.233), the Zhang implicit scheme (4.237), the Zhang explicit scheme (4.241), the DVDM implicit scheme (4.243) and the DVDM explicit scheme 2 (4.247) (below the DVDM schemes are just called "implicit scheme" and "explicit scheme"). The fourth order Runge–Kutta scheme is also employed for comparison. Numerical solutions obtained by those schemes agree quite well with the exact solution, while the former four schemes are slightly better than the latter schemes (4.243) and (4.247).

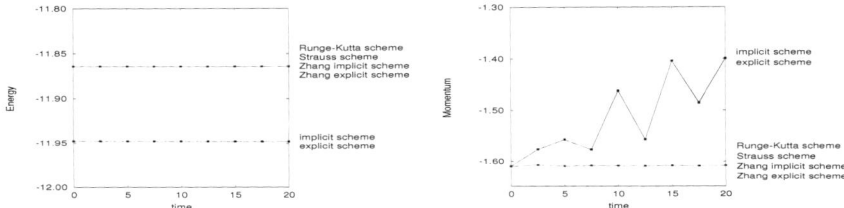

FIGURE 4.13: Evolution of the energies (left) and momenta (right).

Figure 4.13[3] shows the time evolution of the energies and momenta. We can see that both are well preserved in all of the tested schemes.

TABLE 4.1: Computation time and maximum Δt for each scheme

Scheme	Time(unit: second)	Max Δt
Runge–Kutta scheme	17.68	0.7
Strauss scheme	13.19	0.5
Zhang implicit scheme	14.36	0.5
Zhang explicit scheme	5.97	0.4
Implicit scheme (4.243)	48.37	0.1
Explicit scheme (4.247)	4.48	0.8

[3−6] Reprinted from *J. Comput. Appl. Math.*, 134, D. Furihata, Finite-difference schemes for nonlinear wave equation that inherit energy conservation property, 37–57, Copyright (2001), with permission from Elsevier.

Applications

Table 4.1[4] shows the computation time for each scheme using the SUN Ultra 1 model 170E (CPU: UltraSPARC, 167MHz). Computation times listed in the table represent the average of five calculations. The results indicate that the explicit schemes are much faster than the other schemes including the Runge–Kutta scheme.

The sensitivity of the schemes to the time mesh size Δt can be also judged from Table 4.1. The maximum mesh size for the implicit scheme (4.243) is smaller than the sizes for the other schemes. This is required for the convergence in the vector Newton method. The maximum time mesh size of the other schemes compares favorably with that of the Runge–Kutta scheme. This demonstrates the robustness of the other schemes.

4.6.2 Shimoji–Kawai Equation

4.6.2.1 Introduction to Problem

Here we consider the Shimoji–Kawai equation (2.59) as an example of the target equation (2.55) where

$$G(u, u_x) = \frac{1}{12}(u_x)^4. \tag{4.254}$$

This equation was first introduced in Shimoji–Kawai [154], where they showed multivalued exact solutions to the equation by a parametric equation.

4.6.2.2 Numerical Scheme

First we discretize the energy function as

$$G_{d,k}(\boldsymbol{U}) \stackrel{\mathrm{d}}{\equiv} \frac{1}{12} \left\{ \frac{(\delta_k^+ U_k)^4 + (\delta_k^- U_k)^4}{2} \right\}. \tag{4.255}$$

From this definition we obtain the following discrete variational derivative

$$\frac{\delta G_{\mathrm{d}}}{\delta(\boldsymbol{U},\boldsymbol{V})_k} = \frac{-1}{24\Delta x} \left(\sum_{j=0}^{3} (\delta_k^+ U_k^{(m)})^j (\delta_k^+ V_k^{(m)})^{3-j} - \sum_{j=0}^{3} (\delta_k^- U_k^{(m)})^j (\delta_k^- V_k^{(m)})^{3-j} \right), \tag{4.256}$$

and we can construct an explicit energy-conserving scheme:

$$\delta_m^{\langle 2+\rangle} U_k^{(m)} = -\frac{\delta G_{\mathrm{d}}}{\delta(\boldsymbol{U},\boldsymbol{V})_k} \tag{4.257}$$

from (3.129).

4.6.2.3 Numerical Examples

Parameters are set to $\Delta x = 0.05$, $\Delta t = 0.0001$, and the boundary conditions (4.253) are employed. The numerical investigation shows that the scheme (4.257) is quite promising as follows.

Figure 4.14[5] shows the numerical solutions for the initial state:
$$u(x,0) = e^{-(x-3)^2}, \tag{4.258}$$
$$u_t(x,0) = 2(x-3)^2 e^{-(x-3)^2}. \tag{4.259}$$

In the figures the numerical solutions are indicated by points and the exact solutions by lines. Below each graph, the energy values are also shown, which are well preserved. The exact energy value is $0.1384729571\cdots$. The difference comes from the spatial discretization in the numerical energy (i.e. discretization of the energy integral).

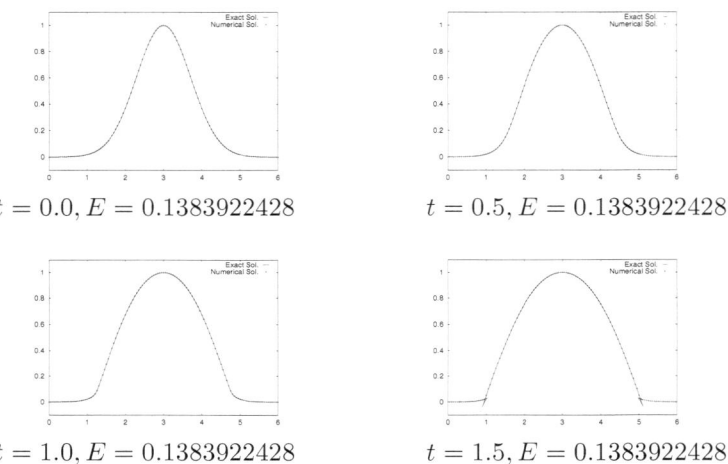

$t = 0.0, E = 0.1383922428$ $t = 0.5, E = 0.1383922428$

$t = 1.0, E = 0.1383922428$ $t = 1.5, E = 0.1383922428$

FIGURE 4.14: Numerical solutions by the scheme (4.257) for the Shimoji–Kawai equation with the initial state (4.258) and (4.259).

Let us consider another initial state. Figure 4.15[6] shows the numerical solutions for the initial state
$$u(x,0) = e^{-(x-3)^2}, \tag{4.260}$$
$$u_t(x,0) = -2(x-3)^2 e^{-(x-3)^2}. \tag{4.261}$$

The energy of exact solution is also $0.1384729571\cdots$. For this initial state we can find that the exact solution becomes multivalued, for example, when $t = 1.5$ in Figure 4.15. We can also find that the exact solution becomes slightly multivalued in Figure 4.14. After $t = 0.5$ the numerical solution deviates from the exact solution considerably, but energy of the numerical solution agrees with that of the exact solution. The well-posedness of the Shimoji–Kawai equation is still under investigation, and the numerical phenomenon should be carefully studied in connection with the theoretical understandings.

Applications 191

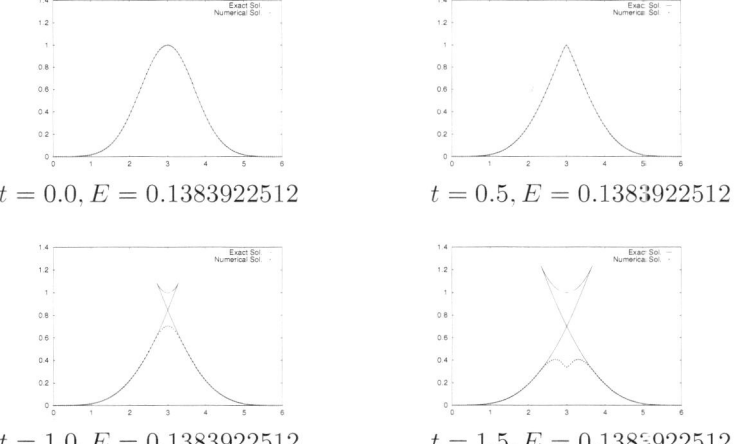

$t = 0.0, E = 0.1383922512$

$t = 0.5, E = 0.1383922512$

$t = 1.0, E = 0.1383922512$

$t = 1.5, E = 0.1383922512$

FIGURE 4.15: Numerical solutions by the scheme (4.257) for the Shimoji–Kawai equation with the initial state (4.260) and (4.261).

4.7 Other Equations

As mentioned in Chapter 2, there are PDEs that strictly speaking do not belong to the target PDEs 1–7, but are quite close to them, and still dissipative or conservative in some sense. In this section, we present such examples and demonstrate that by slightly modifying the procedure of the discrete variational method, we can still construct dissipative or conservative schemes for such PDEs.

4.7.1 Keller–Segel Equation

4.7.1.1 Introduction to Problem

As mentioned in Remark 2.3, the Keller–Segel equation (2.22):

$$\frac{\partial u}{\partial t} = \frac{\partial}{\partial x}\left(\frac{\partial u}{\partial x} - u\frac{\partial v}{\partial x}\right), \qquad (4.262)$$

$$0 = \frac{\partial^2 v}{\partial x^2} - a\,v + u, \qquad (4.263)$$

under the zero Neumann boundary conditions for u and v, is a kind of dissipative equation, although it formally does not belong to the target dissipative PDEs (1). The parameter a is a positive constant. This is due to the inverse of the Helmholtz operator included in the energy function G (see (2.23)). Below we slightly extend the procedure in Chapter 3 so that we can also handle the Helmholtz operator.

As the initial state we take $u_0(x) \geq 0$. This equation has the following important features in addition to the dissipation of the energy. The first one is the positiveness of the solutions:

$$u(x,t) \geq 0, \qquad x \in (0,L), \ t > 0.$$

The second one is mass conservation:

$$\frac{d}{dt} \int_0^L u(x,t)\, dx = 0, \qquad t > 0.$$

In order to handle the inverse of the Helmholtz operator: $(a - \partial^2/\partial x^2)^{-1}$, we consider the Green operator g defined by

$$(gu)(x,t) = \int_0^L \widehat{g}(x,y)\, u(y,t)\, dy \qquad (4.264)$$

where $\widehat{g}$ is the Green function of the Helmholtz operator $(a - \partial^2/\partial x^2)$ under the homogeneous Neumann boundary condition. With the aid of this operator, we rewrite the Keller–Segel equation as

$$\frac{\partial u}{\partial t} = \frac{\partial}{\partial x}\left(u \frac{\partial}{\partial x}\left(\frac{\delta G}{\delta u}\right)\right), \qquad (4.265)$$

where

$$G(u) = u \log u - u - \frac{1}{2} u\, gu. \qquad (4.266)$$

This expression can be easily confirmed, if we note that the variational derivative of G is

$$\frac{\delta G}{\delta u} = \log u - gu. \qquad (4.267)$$

From the variational formulation (4.265) and the positiveness of the solutions, the following important property holds.

$$\frac{d}{dt} J(u) \leq 0, \qquad (4.268)$$

where

$$J(u) \stackrel{d}{\equiv} \int_0^L G(u)\, dx. \qquad (4.269)$$

In this sense, the Keller–Segel equation is a dissipative equation.

4.7.1.2 Numerical Scheme

Below we show that by slightly modifying the procedure in Chapter 3, we can construct a dissipative scheme for the Keller–Segel equation.

Let us discretize the operator $(a - \partial^2/\partial x^2)$ by the standard central second-order difference operator, and denote it by a matrix H, which operates on the numerical solution vector $\boldsymbol{U}^{(m)}$. We note that the matrix H is nonsingular since it is strictly diagonally dominant.

Based on this, the discrete version of the operator g can be defined by $g_{\mathrm{d}} \stackrel{\mathrm{d}}{\equiv} H^{-1}$, and by which we define accordingly

$$G_{\mathrm{d},k}(\boldsymbol{U}) \stackrel{\mathrm{d}}{\equiv} U_k \log U_k - U_k - \frac{1}{2} U_k (g_{\mathrm{d}} \boldsymbol{U})_k, \tag{4.270}$$

and

$$I_{\mathrm{d}}(\boldsymbol{U}) \stackrel{\mathrm{d}}{\equiv} \sum_{k=0}^{N} {}'' G_{\mathrm{d},k}(\boldsymbol{U}) \Delta x. \tag{4.271}$$

We construct a discrete variational derivative scheme as

$$\frac{U_k^{(m+1)} - U_k^{(m)}}{\Delta t} = \delta_k^{\langle 1 \rangle} \left(U_k^{(m)} \delta_k^{\langle 1 \rangle} \left(\frac{\delta G_{\mathrm{d}}}{\delta(\boldsymbol{U}^{(m+1)}, \boldsymbol{U}^{(m)})_k} \right) \right), \tag{4.272}$$

where

$$\frac{\delta G_{\mathrm{d}}}{\delta(\boldsymbol{U}, \boldsymbol{V})_k} = \frac{\log U_k - \log V_k}{U_k - V_k} \cdot \frac{U_k + V_k}{2} - 1$$
$$+ \frac{\log U_k + \log V_k}{2} - \frac{(g_{\mathrm{d}} \boldsymbol{U})_k + (g_{\mathrm{d}} \boldsymbol{V})_k}{2}. \tag{4.273}$$

The discrete variational derivative can be obtained as follows. Since, as mentioned above, the energy function includes (the inverse of) the Helmholtz operator, the procedure in Chapter 3 does not apply as is. However, it is still possible to directly consider the discrete variation of the energy (4.270). The crucial part is the third term of the energy function, which goes as follows.

$$\sum_{k=0}^{N} {}'' \left(\frac{1}{2} U_k (g_{\mathrm{d}} \boldsymbol{U})_k - \frac{1}{2} V_k (g_{\mathrm{d}} \boldsymbol{V})_k \right) \Delta x$$
$$= \frac{1}{4} \sum_{k=0}^{N} {}'' \left\{ (U_k + V_k)(g_{\mathrm{d}}(\boldsymbol{U} - \boldsymbol{V}))_k + (U_k - V_k)(g_{\mathrm{d}}(\boldsymbol{U} + \boldsymbol{V}))_k \right\} \Delta x$$
$$= \frac{1}{4} \sum_{k=0}^{N} {}'' \left\{ (g_{\mathrm{d}}(\boldsymbol{U} + \boldsymbol{V}))_k (U_k - V_k) + (U_k - V_k)(g_{\mathrm{d}}(\boldsymbol{U} + \boldsymbol{V}))_k \right\} \Delta x$$
$$= \sum_{k=0}^{N} {}'' \left\{ \frac{(g_{\mathrm{d}}(\boldsymbol{U} + \boldsymbol{V}))_k}{2} (U_k - V_k) \right\} \Delta x. \tag{4.274}$$

The second equality is obtained since g_{d} is symmetric in the discrete function space $L^2(\Omega_N)$ (see Section 3.6 for the notation).

4.7.1.3 Numerical Examples

We show the dynamics of the numerical solutions under the periodic boundary conditions in Figure 4.16. Parameters were set to $L = 1.0$, $a = 10$, $\Delta x = 0.02$ and $\Delta t = 2 \times 10^{-5}$. The initial state was set to

$$u(x,0) = \begin{cases} 2\cos(8\pi x + \pi)/5 + 0.4 & (0 \leq x \leq 0.25, 0.75 \leq x \leq 1.0), \\ \cos(4\pi x)/2 + 0.5 & (\text{otherwise}). \end{cases} \quad (4.275)$$

Figure 4.17 confirms that the global energy successfully decreases monotonically as time evolves, and the global mass is almost conserved. Note that the latter—the conservation of the mass—is not guaranteed mathematically.

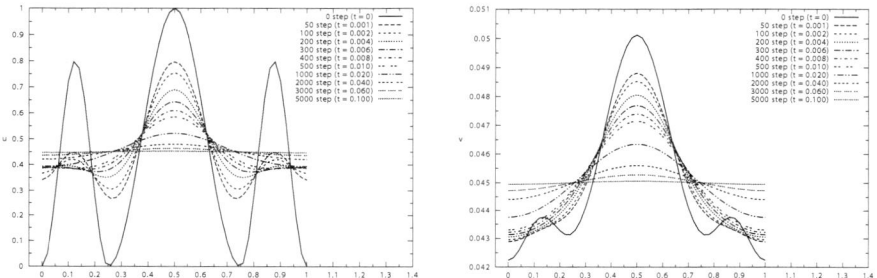

FIGURE 4.16: Numerical solutions to the Keller–Segel equation. Left: profiles of $u(x,t)$ at time $0 \leq t \leq 0.1$, right: profiles of v, which corresponds to $(a - \partial^2/\partial x^2)^{-1} u$.

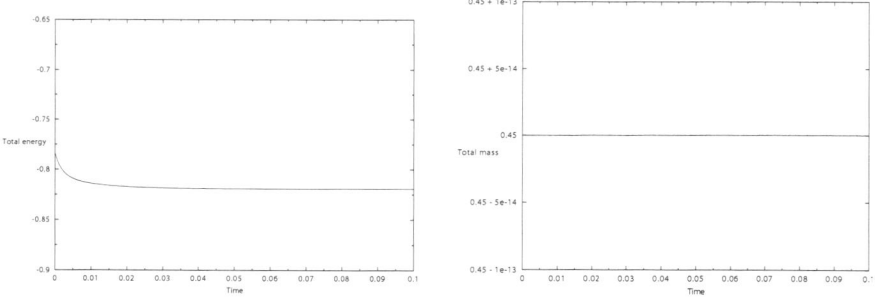

FIGURE 4.17: Left: the global energy, right: the global mass.

4.7.2 Camassa–Holm Equation

4.7.2.1 Introduction to Problem

As mentioned in Remark 2.5, the Camassa–Holm type equations [24] are conservative equations. Let us below consider the "limiting" Camassa–Holm (CH) equation, which reads

$$u_t - u_{xxt} = 2u_x u_{xx} + u u_{xxx} - 3u u_x, \quad x \in (0, L), t > 0. \tag{4.276}$$

This is a special case of (2.32) with $\kappa = 0$ and $\gamma = 1$. It is called the "limiting case," since in the original Camassa–Holm context, $\kappa > 0$ denotes the depth of shallow water, and the above case should be regarded as the extreme situation in the limit of $\kappa \to 0$. Due to the operator $(1 - \partial^2/\partial x^2)$ in the left hand side, it is not formally covered by the target PDEs 2.

It is well-known that by introducing an intermediate function $\omega = (1 - \partial^2/\partial x^2)u$, which is often called the "momentum variable,"[7] the CH can be written in two variational forms (the "bi-Hamiltonian form"):

$$\omega_t = -\left(\frac{\partial}{\partial x} - \frac{\partial^3}{\partial x^3}\right)\frac{\delta G}{\delta \omega}, \tag{4.277}$$

$$\omega_t = -\left(\frac{\partial}{\partial x}\omega + \omega\frac{\partial}{\partial x}\right)\frac{\delta \widetilde{G}}{\delta \omega}, \tag{4.278}$$

where

$$G \stackrel{\mathrm{d}}{\equiv} \frac{1}{2}\left(u^3 + u(u_x)^2\right), \tag{4.279}$$

$$\widetilde{G} \stackrel{\mathrm{d}}{\equiv} \frac{1}{2}\left(u^2 + (u_x)^2\right). \tag{4.280}$$

It is easy to see that

$$\frac{\delta G}{\delta u} = \frac{3}{2}u^2 + \frac{1}{2}(u_x)^2 - \frac{1}{2}\frac{\partial^2}{\partial x^2}(u^2), \tag{4.281}$$

$$\frac{\delta \widetilde{G}}{\delta u} = \left(1 - \frac{\partial^2}{\partial x^2}\right)u. \tag{4.282}$$

The related variational derivatives with respect to ω can be obtained via $\delta \bullet / \delta u = \left(1 - \frac{\partial^2}{\partial x^2}\right)(\delta \bullet / \delta \omega)$.

It is also an easy exercise to show that the Camassa–Holm equation has the following three invariants under the periodic boundary conditions:

$$I(u) \stackrel{\mathrm{d}}{\equiv} \int_0^L u \mathrm{d}x, \quad J(u) \stackrel{\mathrm{d}}{\equiv} \int_0^L G \mathrm{d}x, \text{ and } K(u) \stackrel{\mathrm{d}}{\equiv} \int_0^L \widetilde{G} \mathrm{d}x.$$

---

[7]In the standard notation of the Camassa–Holm studies, this is usually denoted by m, which is obviously for "momentum." In this book, however, we have already reserved m for the time index of numerical solutions. In order to avoid confusion, we denote it by ω.

We note that the Camassa–Holm equation has infinite invariants[24, 105], because it has a bi-Hamiltonian structure [136].

4.7.2.2 Numerical Schemes

We impose the discrete periodic boundary condition:

$$U_k^{(m)} = U_{k \bmod N}^{(m)}, \quad \forall k \in \mathbb{Z}, \quad m = 0, 1, \ldots, \tag{4.283}$$

which implies that we assume $\boldsymbol{U}^{(m)} \in \mathbb{S}_N$ (for this notation, refer to Section 3.6).

Now let us define discrete analogues of the invariants, I, J, K. For the purpose, we first discretize the energy functions $G = \left(u^3 + u(u_x)^2\right)/2$, and $\widetilde{G} = \left(u^2 + (u_x)^2\right)/2$ by, for $\boldsymbol{U}, \boldsymbol{V} \in \mathbb{S}_N$,

$$G_{\mathrm{d},k}(\boldsymbol{U}) \stackrel{\mathrm{d}}{\equiv} U_k \widetilde{G}_{\mathrm{d},k}(\boldsymbol{U}), \tag{4.284}$$

$$\widetilde{G}_{\mathrm{d},k}(\boldsymbol{U}) \stackrel{\mathrm{d}}{\equiv} \frac{1}{2}\left\{(U_k)^2 + \frac{(\delta_k^+ U_k)^2 + (\delta_k^- U_k)^2}{2}\right\}. \tag{4.285}$$

Then we define the discrete versions of the integral functionals $I(u), J(u), K(u)$, for $\boldsymbol{U}, \boldsymbol{V} \in \mathbb{S}_N$.

$$I_{\mathrm{d}}(\boldsymbol{U}) \stackrel{\mathrm{d}}{\equiv} \sum_{k=0}^{N}{}'' U_k \Delta x, \tag{4.286}$$

$$J_{\mathrm{d}}(\boldsymbol{U}) \stackrel{\mathrm{d}}{\equiv} \sum_{k=0}^{N}{}'' G_{\mathrm{d},k}(\boldsymbol{U}) \Delta x, \tag{4.287}$$

$$K_{\mathrm{d}}(\boldsymbol{U}) \stackrel{\mathrm{d}}{\equiv} \sum_{k=0}^{N}{}'' \widetilde{G}_{\mathrm{d},k}(\boldsymbol{U}) \Delta x. \tag{4.288}$$

For the discrete energy functions, we obtain the following discrete variational derivatives.

$$\frac{\delta G_{\mathrm{d}}}{\delta(\boldsymbol{U},\boldsymbol{V})_k} = \frac{3}{2}\frac{(U_k)^2 + (U_k)(V_k) + (V_k)^2}{3}$$
$$+ \frac{1}{2}\frac{(\delta_k^+ U_k)^2 + (\delta_k^- U_k)^2 + (\delta_k^+ V_k)^2 + (\delta_k^- V_k)^2}{4}$$
$$- \frac{1}{2}\delta_k^{\langle 2 \rangle}\left(\frac{U_k + V_k}{2}\right)^2, \tag{4.289}$$

$$\frac{\delta \widetilde{G}_{\mathrm{d}}}{\delta(\boldsymbol{U},\boldsymbol{V})_k} = (1 - \delta_k^{\langle 2 \rangle})\frac{U_k + V_k}{2}. \tag{4.290}$$

We are able to obtain these discrete variational derivatives by the formal approach, or the direct approach described in Section 4.1.1.

Applications

In view of the relation $\delta \bullet / \delta u = \left(1 - \frac{\partial^2}{\partial x^2}\right) (\delta \bullet / \delta w)$, it is straightforward to relate the above discrete derivatives to those on w:

$$\frac{\delta G_\mathrm{d}}{\delta w(\boldsymbol{U},\boldsymbol{V})_k} = (1 - \delta_k^{\langle 2 \rangle})^{-1} \frac{\delta G_\mathrm{d}}{\delta(\boldsymbol{U},\boldsymbol{V})_k}, \qquad (4.291)$$

$$\frac{\delta \widetilde{G}_\mathrm{d}}{\delta w(\boldsymbol{U},\boldsymbol{V})_k} = (1 - \delta_k^{\langle 2 \rangle})^{-1} \frac{\delta \widetilde{G}_\mathrm{d}}{\delta(\boldsymbol{U},\boldsymbol{V})_k} \quad \left(= \frac{U_k + V_k}{2} \right), \qquad (4.292)$$

where $(1-\delta_k^{\langle 2 \rangle})^{-1}$ is an inverse operator of $(1-\delta_k^{\langle 2 \rangle})$. We note that $(1-\delta_k^{\langle 2 \rangle})^{-1}$ is well-defined, since the matrix representation of $(1 - \delta_k^{\langle 2 \rangle})$ (considering the discrete periodic boundary condition) is regular.

Now we are in a position to define schemes [161, 162]. Based on (4.277), we construct the following scheme. We call the following scheme "IJ–NL" since it conserves I_d and J_d, and it is a nonlinear scheme.

$$\delta_m^+ V_k^{(m)} = -\delta_k^{\langle 1 \rangle} \left(1 - \delta_k^{\langle 2 \rangle}\right) \frac{\delta G_\mathrm{d}}{\delta w(\boldsymbol{U}^{(m+1)}, \boldsymbol{U}^{(m)})_k}, \qquad (4.293)$$

where

$$V_k^{(m)} \stackrel{\mathrm{d}}{\equiv} (1 - \delta_k^{\langle 2 \rangle}) U_k^{(m)}. \qquad (4.294)$$

For this scheme, the following properties hold.

$$I_\mathrm{d}(\boldsymbol{U}^{(m)}) = I_\mathrm{d}(\boldsymbol{U}^{(0)}), \qquad (4.295)$$
$$J_\mathrm{d}(\boldsymbol{U}^{(m)}) = J_\mathrm{d}(\boldsymbol{U}^{(0)}). \qquad (4.296)$$

Based on (4.278) we also construct another scheme. We call it "IK–NL" since it conserves I_d and K_d and it is a nonlinear scheme.

$$\delta_m^+ V_k^{(m)} = - \left(\delta_k^{\langle 1 \rangle} V_k^{(m+\frac{1}{2})} + V_k^{(m+\frac{1}{2})} \delta_k^{\langle 1 \rangle}\right) \frac{\delta \widetilde{G}_\mathrm{d}}{\delta w(\boldsymbol{U}^{(m+1)}, \boldsymbol{U}^{(m)})_k}, (4.297)$$

where

$$U_k^{(m+\frac{1}{2})} \stackrel{\mathrm{d}}{\equiv} \mu_m^+ U_k^{(m)} = \frac{(U_k^{(m+1)} + U_k^{(m)})}{2}. \qquad (4.298)$$

This scheme coincides with the one in [1]. For the solutions of this scheme, the following discrete conservation properties hold.

$$I_\mathrm{d}(\boldsymbol{U}^{(m)}) = I_\mathrm{d}(\boldsymbol{U}^{(0)}), \qquad (4.299)$$
$$K_\mathrm{d}(\boldsymbol{U}^{(m)}) = K_\mathrm{d}(\boldsymbol{U}^{(0)}). \qquad (4.300)$$

4.7.2.3 Linearly Implicit Schemes

In order to compare schemes in the next subsection, we here show two linearly implicit schemes. First we define the discrete energy functions,

$$G_{\mathrm{d}2,k}(\boldsymbol{U},\boldsymbol{V}) \stackrel{\mathrm{d}}{\equiv} \left(\frac{U_k + V_k}{2}\right)\widetilde{G}_{\mathrm{d}2,k}(\boldsymbol{U},\boldsymbol{V}),$$

$$\widetilde{G}_{\mathrm{d}2,k}(\boldsymbol{U},\boldsymbol{V}) \stackrel{\mathrm{d}}{\equiv} \frac{1}{2}\left\{U_k V_k + \frac{\left(\delta_k^+ U_k\right)\left(\delta_k^+ V_k\right) + \left(\delta_k^- U_k\right)\left(\delta_k^- V_k\right)}{2}\right\},$$

and accordingly the summations of them as:

$$J_{\mathrm{d}2}(\boldsymbol{U},\boldsymbol{V}) \stackrel{\mathrm{d}}{\equiv} \sum_{k=0}^{N}{}'' G_{\mathrm{d}2,k}(\boldsymbol{U},\boldsymbol{V})\Delta x,$$

$$K_{\mathrm{d}2}(\boldsymbol{U},\boldsymbol{V}) \stackrel{\mathrm{d}}{\equiv} \sum_{k=0}^{N}{}'' \widetilde{G}_{\mathrm{d}2,k}(\boldsymbol{U},\boldsymbol{V})\Delta x.$$

The subscript "2" is for distinguishing them from the previous discrete energy functions. The discrete variational derivatives of these discrete energy functions can be obtained using the technique in Chapter 6.

$$\frac{\delta G_{\mathrm{d}2}}{\delta(\boldsymbol{U},\boldsymbol{V},\boldsymbol{W})_k} = \frac{3}{2}\frac{V_k(U_k + V_k + W_k)}{3}$$
$$+ \frac{1}{4}\left\{(\delta_k^+ V_k)\left(\delta_k^+ \frac{U_k + W_k}{2}\right) + (\delta_k^- V_k)\left(\delta_k^- \frac{U_k + W_k}{2}\right)\right\}$$
$$- \left(\delta_k^{\langle 2 \rangle} V_k\right)\cdot \mu_k^{(1)}\left(\frac{U_k + 2V_k + W_k}{4}\right)$$
$$- \left(\delta_k^{\langle 1 \rangle} V_k\right)\cdot \delta_k^{\langle 1 \rangle}\left(\frac{U_k + 2V_k + W_k}{4}\right), \tag{4.301}$$

$$\frac{\delta \widetilde{G}_{\mathrm{d}2}}{\delta(\boldsymbol{U},\boldsymbol{V},\boldsymbol{W})_k} = (1 - \delta_k^{\langle 2 \rangle})V_k, \tag{4.302}$$

$$\frac{\delta G_{\mathrm{d}}}{\delta \omega(\boldsymbol{U},\boldsymbol{V},\boldsymbol{W})_k} = (1 - \delta_k^{\langle 2 \rangle})^{-1}\frac{\delta G_{\mathrm{d}}}{\delta(\boldsymbol{U},\boldsymbol{V},\boldsymbol{W})_k}, \tag{4.303}$$

$$\frac{\delta \widetilde{G}_{\mathrm{d}}}{\delta \omega(\boldsymbol{U},\boldsymbol{V},\boldsymbol{W})_k} = (1 - \delta_k^{\langle 2 \rangle})^{-1}\frac{\delta \widetilde{G}_{\mathrm{d}}}{\delta(\boldsymbol{U},\boldsymbol{V},\boldsymbol{W})_k} \quad (= V_k). \tag{4.304}$$

With these discrete variational derivatives we define two linearly implicit schemes. The first is called the "IJ–L" since it conserves I_{d} and $J_{\mathrm{d}2}$, and it is a linearly implicit scheme.

$$\delta_m^{\langle 1 \rangle} V_k^{(m)} = -\delta_k^{\langle 1 \rangle}\left(1 - \delta_k^{\langle 2 \rangle}\right)\frac{\delta G_{\mathrm{d}}}{\delta \omega(\boldsymbol{U}^{(m+1)},\boldsymbol{U}^{(m)},\boldsymbol{U}^{(m-1)})_k}. \tag{4.305}$$

We note that we have to prepare $\boldsymbol{U}^{(0)}$ and $\boldsymbol{U}^{(1)}$ for the linear scheme. The conservation properties of this scheme are

$$I_{\rm d}(\boldsymbol{U}^{(m)}) = \begin{cases} I_{\rm d}(\boldsymbol{U}^{(0)}) \text{ (n : even)}, \\ I_{\rm d}(\boldsymbol{U}^{(1)}) \text{ (n : odd)}, \end{cases} \quad (4.306)$$

$$J_{\rm d2}(\boldsymbol{U}^{(m+1)}, \boldsymbol{U}^{(m)}) = J_{\rm d2}(\boldsymbol{U}^{(1)}, \boldsymbol{U}^{(0)}). \quad (4.307)$$

The second is called the "IK–L" since it conserves $I_{\rm d}$ and $K_{\rm d2}$, and it is a linearly implicit scheme.

$$\delta_m^{\langle 1 \rangle} V_k^{(m)} = - \left(\delta_k^{\langle 1 \rangle} V_k^{(m)} + V_k^{(m)} \delta_k^{\langle 1 \rangle} \right) \frac{\delta \widetilde{G}_{\rm d}}{\delta \omega (\boldsymbol{U}^{(m+1)}, \boldsymbol{U}^{(m)}, \boldsymbol{U}^{(m-1)})_k}, \quad (4.308)$$

and we also need $\boldsymbol{U}^{(0)}$ and $\boldsymbol{U}^{(1)}$ as initial values. This scheme holds the following properties.

$$I_{\rm d}(\boldsymbol{U}^{(m)}) = \begin{cases} I_{\rm d}(\boldsymbol{U}^{(0)}) \text{ (n : even)}, \\ I_{\rm d}(\boldsymbol{U}^{(1)}) \text{ (n : odd)}, \end{cases} \quad (4.309)$$

$$K_{\rm d2}(\boldsymbol{U}^{(m+1)}, \boldsymbol{U}^{(m)}) = K_{\rm d2}(\boldsymbol{U}^{(1)}, \boldsymbol{U}^{(0)}). \quad (4.310)$$

4.7.2.4 Numerical Examples

In this subsection, we numerically demonstrate the presented schemes.

4.7.2.5 Comparison of the Schemes

We first compare the schemes above, the classical Runge–Kutta scheme (CRK), and the Heun scheme. The space and time ranges are set to $x \in [0, 100], t \in [0, 30]$, and discretized with the mesh sizes $\Delta x = 2^{-5}$, $\Delta t = 2^{-5}$. The initial profile is set to $u_0(x) = 0.8 \exp(-|x - 50|)$, so that we can simulate a traveling single-peakon solution.

Figures 4.20–4.22 show the profiles of the numerical solutions. The profile starts with a single peakon solution, and it should move to the right as time evolves. The results by the IK-NL, IK-L, CRK, and the Heun schemes seem fine. However, in the IJ–NL and IJ–L schemes, undesirable oscillation appears around the initial peakon position at $x = 50$. One explanation for this might be that the initial profile created from the exact single-peakon solution (which originally should be defined on the whole $\mathbb{R}$ domain) is not suitable for the schemes. On this issue, we have confirmed that the oscillation disappears as the space mesh size Δx decreases.

Next let us focus on the conservation properties. Figures 4.25, 4.27, and 4.26 show the evolution of the discrete invariants, $I_{\rm d}$, $J_{\rm d}$, and $K_{\rm d}$. According to Figure 4.25, $I_{\rm d}$ is well conserved by all the schemes. In Figure 4.27, $K_{\rm d}$ is well conserved by the IK–L, IK–NL, and CRK schemes, while on the contrary in Figure 4.26, $J_{\rm d}$ is well preserved by the IK–∗ schemes and the CRK scheme.

Thus we conclude that the IK–L, IK–NL and CRK schemes conserve those three quantities well.

Thirdly, we compare the computation times of the schemes. Table 4.2 shows the computation times. The linear schemes are generally faster than the nonlinear schemes. In particular, we like to point out that the IK–L scheme is the fastest, which is about four times as fast as the CRK, and twice as fast as the Heun scheme. The IJ–L scheme is also linear, but it falls behind the IK–L scheme, because the coefficient matrix is *constant* in IK–L scheme. More specifically, in the IK–L scheme, once we compute the inverse of the coefficient matrix by, for example, the LU decomposition at the beginning of the time evolution process, then we do not need to solve any linear systems ever after.

Finally, we mention the long time behaviors of the schemes. We continued the computation also for $t > 30$ to see the asymptotic behaviors. There we found that around $t = 34.5$, the numerical solution by the CRK scheme blows up; see Figure 4.24. The blowup occurs at the initial peak position mentioned above. We also observed that decreasing the time and space mesh sizes did not improve these instabilities very well, although the speed of the blowup was slightly relaxed. We have also observed that similar blowup occurs in the Heun scheme around $t = 34.0$. In contrast, no such explosions were found in the conservative schemes.

TABLE 4.2: Computation times

scheme	time (sec.)
IJ–NL scheme	2498
IK–NL scheme	1786
IJ–L scheme	622
IK–L scheme	103
Classical Runge–Kutta	397
Heun	199

4.7.2.5.1 Multi Peakon Solutions Next we try to capture the "multi peakon solutions," starting from the following initial profiles:

$$\text{(2-peakon)} \quad u_0(x) = 0.8\exp(-|x - 10|) + 0.2\exp(-|x - 40|), \quad (4.311)$$
$$\text{(3-peakon)} \quad u_0(x) = 0.8\exp(-|x - 10|) + 0.4\exp(-|x - 30|)$$
$$+ 0.2\exp(-|x - 40|). \quad (4.312)$$

We tested the IK–L scheme. Parameters were set to $x \in [0, 100]$, $t \in [0, 100]$, $\Delta x = 2^{-8}$ and $\Delta t = 2^{-9}$. Figure 4.28 shows the whole profile of the numerical solutions in the 2-peakon case, Figure 4.35 shows the 3-peakon

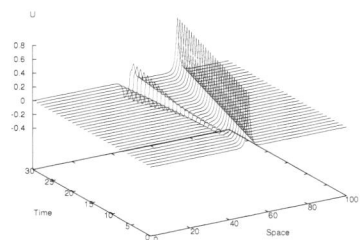

FIGURE 4.18: IJ–NL scheme.

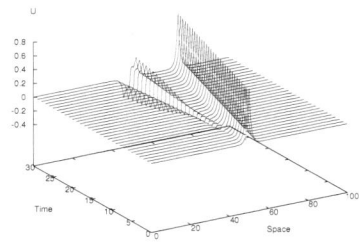

FIGURE 4.19: IJ–L scheme.

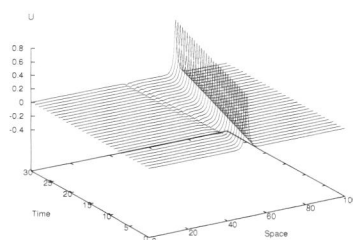

FIGURE 4.20: IK–NL scheme.

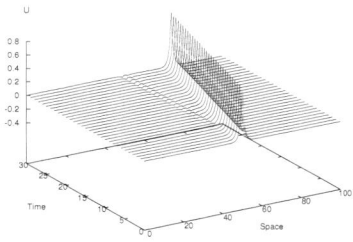

FIGURE 4.21: IK–L scheme.

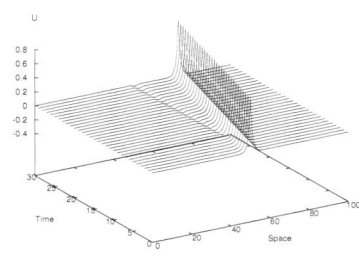

FIGURE 4.22: Classical Runge–Kutta scheme.

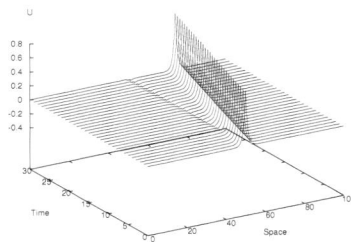

FIGURE 4.23: Heun scheme.

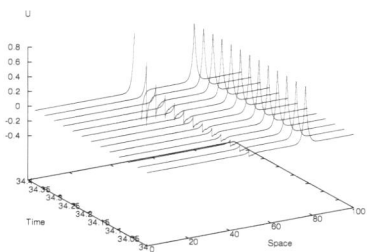

FIGURE 4.24: Blow-up solutions by the classical Runge–Kutta scheme ($t \in [34.0, 34.4]$).

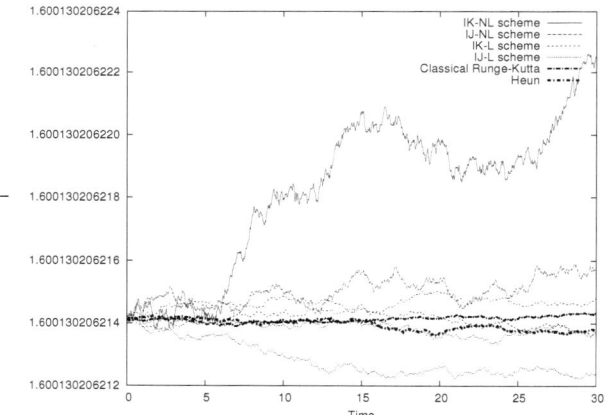

FIGURE 4.25: Evolutions of I_d ($t \in [0, 30]$).

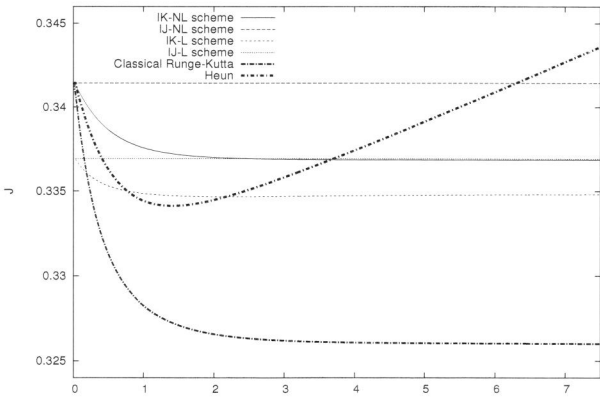

FIGURE 4.26: Evolutions of J_{d} and J_{d2} ($t \in [0, 7.5]$).

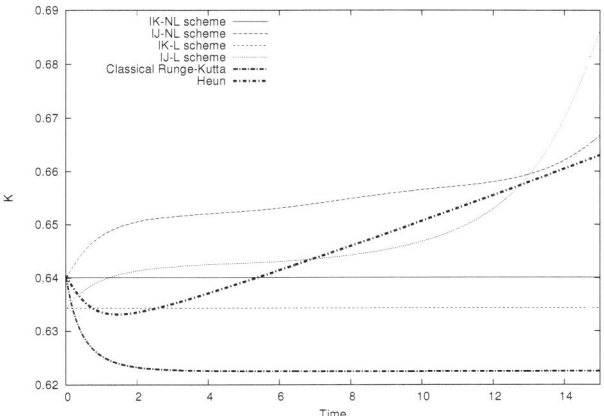

FIGURE 4.27: Evolutions of K_{d} and K_{d2} ($t \in [0, 15]$).

case. In Figures 4.29–4.34, the details of the 2-peakon profiles are shown at $t = 0, 20, 46, 50, 54$ and 100. Similarly, in Figures 4.36–4.41, the 3-peakon profiles are shown. From those figures we clearly see that the peakons recover their original profiles after collisions. In addition to that, we also observe some "phase shifts" after the collisions. These observations convince us that those peakons in fact behave like solitons, as widely believed.

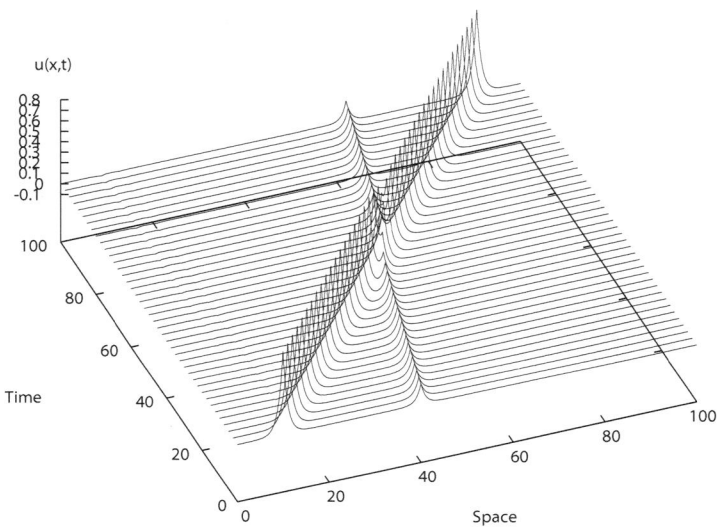

FIGURE 4.28: The 2-peakon case by the IK–L scheme.

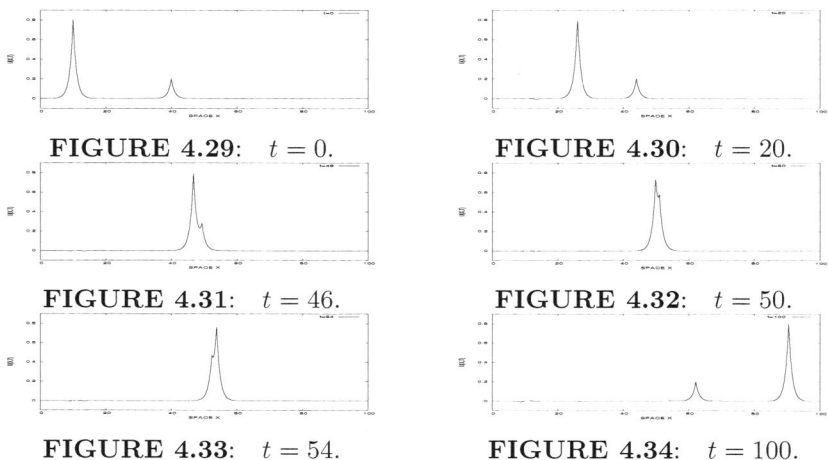

FIGURE 4.29: $t = 0$.　　**FIGURE 4.30**: $t = 20$.

FIGURE 4.31: $t = 46$.　　**FIGURE 4.32**: $t = 50$.

FIGURE 4.33: $t = 54$.　　**FIGURE 4.34**: $t = 100$.

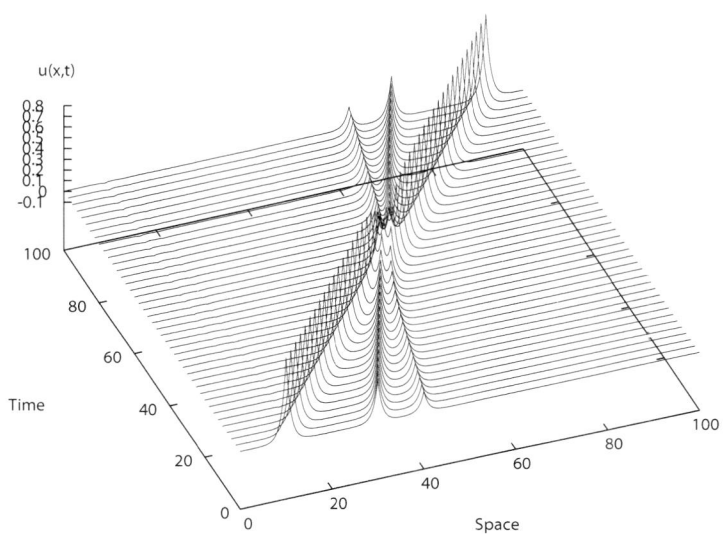

FIGURE 4.35: The 3-peakon case by the IK–L scheme.

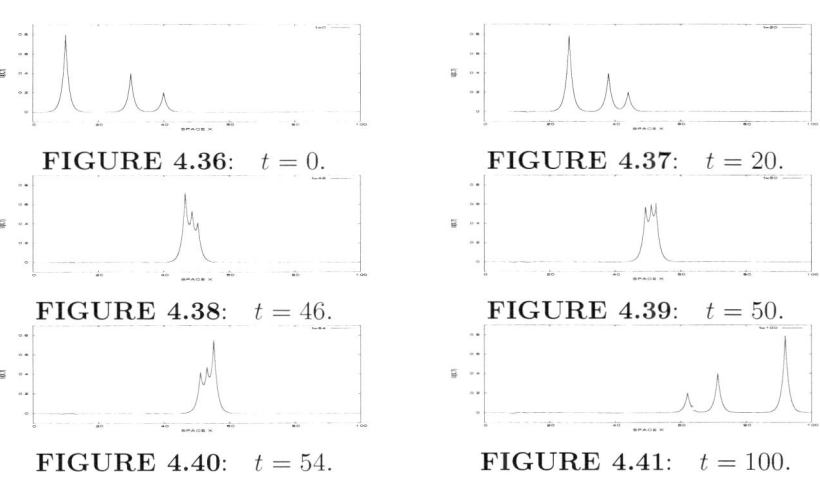

FIGURE 4.36: $t = 0$.

FIGURE 4.37: $t = 20$.

FIGURE 4.38: $t = 46$.

FIGURE 4.39: $t = 50$.

FIGURE 4.40: $t = 54$.

FIGURE 4.41: $t = 100$.

4.7.2.6 Analysis of Scheme

In this subsection, we note some theoretical properties of the IK–NL scheme; namely, the stability and the unique existence of numerical solutions.

4.7.2.6.1 Stability The IK–NL scheme enjoys the following stability estimate:

$$\|\boldsymbol{U}^{(m)}\|_\infty < C,$$

where $\boldsymbol{U}^{(m)}$ is the solution, and C is a constant independent of $n, \Delta x$ and Δt. This property has been already indicated in a similar context in Section 4.1.1 for the Cahn–Hilliard equation. We use the discrete Sobolev–Hilbert norm (4.23) and the discrete Sobolev lemma in Section 3.6.2.1. With this norm,

$$\|\boldsymbol{U}^{(m)}\|_{H^1} = 2J_\mathrm{d}(\boldsymbol{U}^{(m)}) = 2J_\mathrm{d}(\boldsymbol{U}^{(0)}) = \|\boldsymbol{U}^{(0)}\|_{H^1}$$

for the solutions of the IK–NL scheme. From this and the discrete Sobolev lemma, we obtain the following important evaluation.

$$\|\boldsymbol{U}^{(m)}\|_\infty \leq 2\max\left(\frac{1}{\sqrt{L}}, \sqrt{\frac{L}{2}}\right) \|\boldsymbol{U}^{(0)}\|_{H^1}.$$

This means stability in the supremum norm, aside from the effect of rounding errors.

4.7.2.6.2 Unique Existence of the Solution
Here we show that the IK–NL scheme and IK–L scheme are uniquely solvable under appropriate conditions at each time step.

First, let us consider the IK–NL scheme. Let us define

$$M_U := \sup_m \|\boldsymbol{U}^{(m)}\|, \quad M_V := \sup_m \|\boldsymbol{V}^{(m)}\|$$

for $\boldsymbol{V}^{(m)} = \left(I - D_k^{(2)}\right)\boldsymbol{U}^{(m)}$. With these definitions and the fact that $\|I - D_k^{(2)}\| \leq \|I\| + \|D_k^{(2)}\| \leq 1 + 4/(\Delta x)^2$, we obtain the following inequality: $M_V \leq (1 + 4/(\Delta x)^2)M_U$. Through some cumbersome computations, we obtain the following lemma.

Lemma 1 *If the condition:*

$$\Delta t \leq \frac{(\Delta x)^{3/2}}{6M_V} \tag{4.313}$$

is satisfied, then the IK–NL scheme has numerical solutions $\boldsymbol{U}^{(m+1)}$.

Furthermore, it is an easy exercise to show that the solution of the scheme is unique if $\Delta t \leq (\Delta x)^{3/2}/5M_V$. This, together with Lemma 1, proves the following existence theorem. The proof is by the standard contraction mapping theorem.

Theorem 1 *If the condition (4.313) is satisfied, the IK–NL scheme has a unique solution at the new time step.*

Here we consider the IK–L scheme (4.308), whose concrete form becomes

$$(1 - \delta_k^{\langle 2 \rangle})U_k^{(m+1)}$$
$$= (1 - \delta_k^{\langle 2 \rangle})U_k^{(m-1)} - 2\Delta t \left(\delta_k^{\langle 1 \rangle} V_k^{(m)} + V_k^{(m)} \delta_k^{\langle 1 \rangle} \right) U_k^{(m)}. \quad (4.314)$$

Since it is a linear scheme, it suffices to show that the coefficient matrix $(1 - \delta_k^{\langle 2 \rangle})$ is nonsingular. It is in fact clear, since it is strictly diagonally dominant. We also note that this proof is uniform in that it does not depend on Δx, Δt, nor the time step m.

4.7.2.7 Numerical Convergence Evaluations

Here, we numerically investigate the error convergence rates of the schemes above. Due to the spatial and temporal symmetries of the schemes, we expect $O(\Delta x^2 + \Delta t^2)$ convergence, at least for sufficiently smooth solutions.

We test two initial profiles; one is a singular peakon solution, and the other is a sufficiently smooth solution. We measure the errors in the discrete L_2 norm. When the exact solutions are not known, the numerical solutions with sufficiently fine meshes are used as their substitutes.

Let us first consider the following peakon solution:

$$u_0(x) = 0.8 \exp(-|x - 50|), \quad x \in [0, 100].$$

Since this is not an exact solution under the periodic boundary condition, we compute a fine solution with $\Delta x = 2^{-7}$ and $\Delta t = 2^{-8}$, and regard it as a substitute for the exact solution. We fix the time mesh to $\Delta t = 2^{-8}$, and compute numerical solutions with several space mesh sizes $\Delta x = 2^{-2}, 2^{-3}, 2^{-4}$ and 2^{-5}. The errors at $t = 5$ are shown in Figure 4.42. The lines in Figure 4.42 are drawn by the least-square approximations based on the data, whose gradients indicate the convergence rates. Table 4.3 summarizes the estimated convergence rates. We observe that numerical solutions by those schemes converge to the fine solution with the order $O(\Delta x^{0.8 \cong 0.9})$. Next, we fix the space mesh size to $\Delta x = 2^{-3}$ instead, and observe the convergence with respect to Δt. Figure 4.43 shows the errors at time $t = 5$ with the time mesh sizes $\Delta t = 2^{-3}, 2^{-4}, 2^{-5}$ and 2^{-6}. Again, Table 4.4 summarizes the estimated convergence rates. From the table, we can see that the convergence rates of the conservative schemes and the Heun scheme are around $O(\Delta t^2)$, while the CRK scheme achieves almost $O(\Delta t^4)$.

208 Discrete Variational Derivative Method

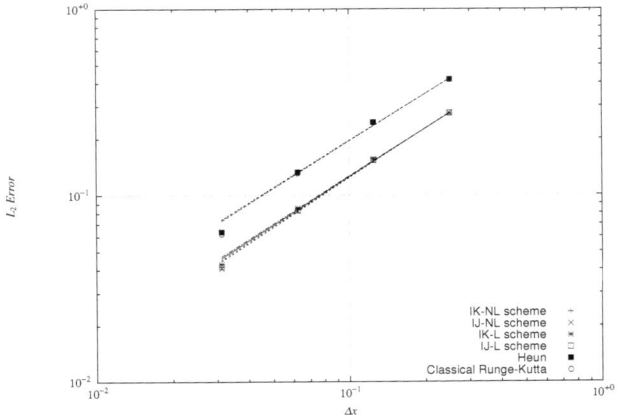

FIGURE 4.42: The error L_2 norm versus the space mesh size Δx at time $t = 5$ with the peakon initial profile.

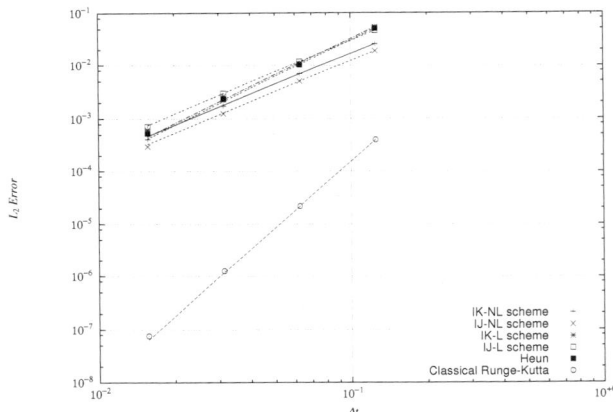

FIGURE 4.43: The error L_2 norm versus the time mesh size Δt at the time $t = 5$ with the peakon initial profile.

TABLE 4.3: The estimated convergence rates with respect to Δx

Scheme	Estimated error convergence rates
IJ–NL scheme	0.880103
IK–NL scheme	0.857593
IJ–L scheme	0.872815
IK–L scheme	0.849059
Heun scheme	0.835725
Classical Runge–Kutta scheme	0.842118

TABLE 4.4: The estimated convergence rates with respect to Δt

Scheme	Estimated error convergence rates
IJ–NL scheme	1.94711
IK–NL scheme	1.93010
IJ–L scheme	1.98482
IK–L scheme	2.29136
Heun scheme	2.30428
Classical Runge–Kutta scheme	4.20176

As the second experiment, let us take the initial profile to

$$u_0(x) = 0.6\,\text{sech}^2(\sqrt{2}(x-30)/4), \quad x \in [0, 100], \qquad (4.315)$$

which is a smooth function. We here compare the IK–L scheme, the Heun scheme, and the CRK scheme. We investigated the convergence rates with respect to Δx for various t in $1 \leq t \leq 90$, and plotted them in Figure 4.46. In the figure, we find a curious behavior in that the convergence rates considerably change during the time evolution. In the beginning, they are about $O(\Delta x^2)$, as expected for smooth solutions. But soon after that they quickly drop to $O(\Delta x^{0.8})$. In order to understand this strange behavior, we observed the numerical solutions carefully to find that the initial smooth profile had gradually peaked around $5 \leq t \leq 15$. Figure 4.44 shows the solution profiles by the IK–L scheme with $\Delta x = 2^{-7}$ and $\Delta t = 2^{-8}$. Figure 4.45 shows the snapshots at $t = 0, 5, 15, 60$ and 90. From these figures, we can clearly see the loss of regularity, and this should be the reason of the rate deficiency. In fact, in the first peakon experiment, we have already seen that the rate with respect to Δx is $O(\Delta x^{0.8})$, which completely agrees with this view.

Next, the convergence rates with respect to Δt are shown in Figure 4.47. We see that the rate of the CRK scheme is around $O(\Delta t^{4.0 \cong 4.8})$, and the IK–L scheme around $O(\Delta t^{2.0 \cong 2.3})$. Since the numerical solutions of the Heun scheme with $\Delta x = 2^{-3}$ blow up at $t = 64$, we show the estimated rate only before $t \leq 64$. Compared to the rate against Δx, the result with respect to Δt is quite natural.

To summarize, we observed that the convergence rates of the conservative schemes are $O(\Delta x^2 + \Delta t^2)$ for the sufficient smooth solutions, and $O(\Delta x + \Delta t^2)$ for non-smooth solutions, e.g., peakons.

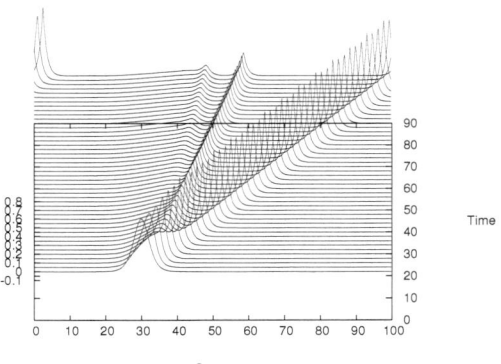

FIGURE 4.44: The numerical solutions by the IK–L scheme with $\Delta x = 2^{-7}$ and $\Delta t = 2^{-8}$ for the initial profile (4.315).

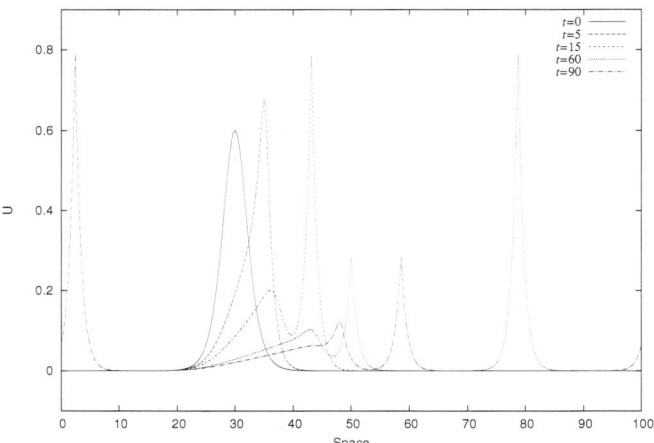

FIGURE 4.45: The numerical solutions in Figure 4.44 at time $t = 0, 5, 15, 60$ and 90.

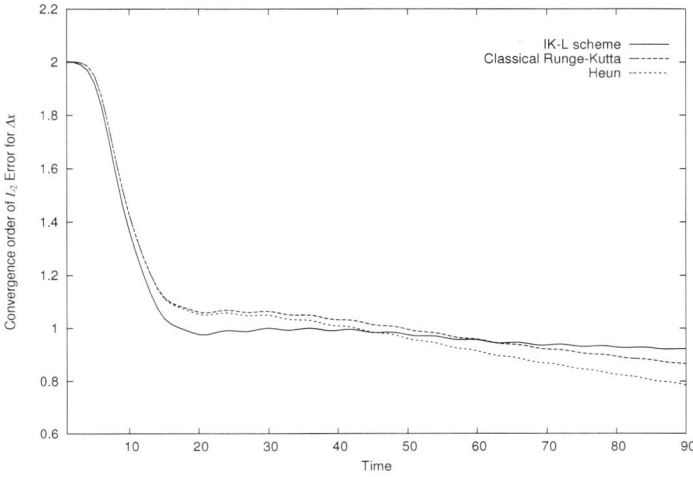

FIGURE 4.46: Time evolutions of the error convergence rate with the space mesh size Δx. The initial profile is (4.315).

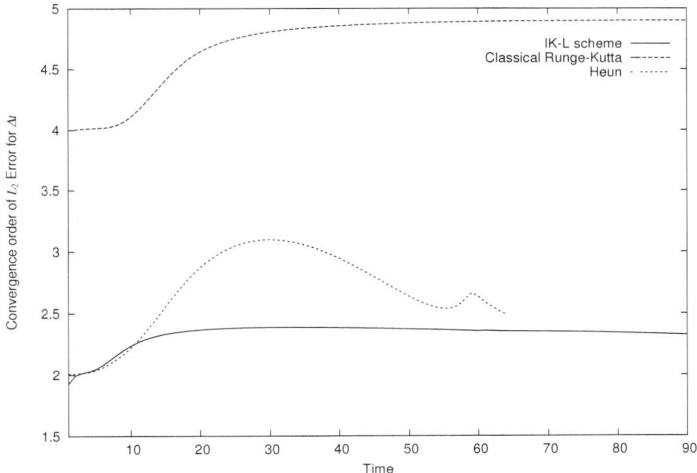

FIGURE 4.47: Time evolutions of the error convergence rate with the time mesh size Δt. The initial profile is (4.315).

4.7.3 Benjamin–Bona–Mahony Equation

4.7.3.1 Introduction to Problem

As mentioned in Remark 2.5, the Benjamin–Bona–Mahony equation (BBM; also known as the regularized long wave equation):

$$\left(1 - \frac{\partial^2}{\partial x^2}\right)\frac{\partial u}{\partial t} = -\frac{\partial}{\partial x}\left(\frac{\delta G}{\delta u}\right), \qquad x \in (0, L),\ t > 0, \tag{4.316}$$

where

$$G = \frac{1}{2}u^2 + \frac{1}{6}u^3. \tag{4.317}$$

is a conservative equation. We here impose the periodic boundary conditions of length $L > 0$,

$$\frac{\partial^l}{\partial x^l}\frac{\partial^{m'}}{\partial t^{m'}}u(0,t) = \frac{\partial^l}{\partial x^l}\frac{\partial^{m'}}{\partial t^{m'}}u(L,t), \quad l = 0, 1, 2,\ m' = 0, 1, \tag{4.318}$$

for $t > 0$.

This equation was proposed as a model for the undular bore problem by Peregrine [138]. Benjamin *et al.* [14] have investigated this equation as a regularized version of the Korteweg–de Vries (KdV) equation. This equation has solitary wave solutions similar to the KdV equation. However, there is a big difference that while the KdV equation has infinitely many invariants, it is proved by Olver [135] that the BBM equation admits only three independent invariants:

$$I(u) \stackrel{\mathrm{d}}{\equiv} \int_0^L u(x,t)\mathrm{d}x, \tag{4.319}$$

$$J(u) \stackrel{\mathrm{d}}{\equiv} \frac{1}{2}\int_0^L \left(u^2 + \frac{1}{3}u^3\right)\mathrm{d}x, \tag{4.320}$$

$$K(u) \stackrel{\mathrm{d}}{\equiv} \frac{1}{2}\int_0^L \left\{u^2 + \left(\frac{\partial u}{\partial x}\right)^2\right\}\mathrm{d}x. \tag{4.321}$$

Here, I, J, and K are called 'mass', 'energy', and 'momentum'. Since the number of conserved quantities is limited, we are not able to use the inverse-scattering technique, which is a powerful mathematical tool to obtain theoretical solutions of integrable equations such as the KdV equation. So far, large numbers of studies have been carried out for the numerical solutions of the BBM equation: for example, [14, 31, 32, 43, 46, 47, 138].

4.7.3.2 Numerical Schemes

As noted in Remark 2.5, the BBM equation is closely related to the Camassa–Holm equation, and thus we can follow exactly the same approach in Section 4.7.2. Here we like to leave this to the readers' exercise, and try another

Applications 213

approach to obtain four different finite difference schemes preserving either discrete momentum or discrete energy. Two of them are nonlinear, the rest are linear.

We impose the following discrete periodic boundary conditions:

$$U_k^{(m)} = U_{k \bmod N}^{(m)} \quad \text{for } {}^\forall k \in \mathbb{Z}. \tag{4.322}$$

In other words, we assume $\boldsymbol{U}^{(m)} \in \mathbb{S}_N$ (for the notation, see Section 3.6).

Let us begin by defining a discrete energy function as

$$G_{\mathrm{d},k}(\boldsymbol{U}) \stackrel{\mathrm{d}}{\equiv} \frac{1}{2}(U_k)^2 + \frac{1}{6}(U_k)^3, \tag{4.323}$$

from which we obtain the following discrete variational derivative by the standard procedure of the discrete variational derivative method:

$$\frac{\delta G_{\mathrm{d}}}{\delta(\boldsymbol{U},\boldsymbol{V})_k} = \left(\frac{U_k + V_k}{2}\right) + \frac{1}{2}\left(\frac{(U_k)^2 + U_k V_k + (V_k)^2}{3}\right). \tag{4.324}$$

Then we obtain the following scheme

$$\left(1 - (\delta_k^{\langle 1 \rangle})^2\right) \delta_m^+ U_k^{(m)} = -\delta_k^{\langle 1 \rangle} \frac{\delta G_{\mathrm{d}}}{\delta(\boldsymbol{U}^{(m+1)}, \boldsymbol{U}^{(m)})_k}. \tag{4.325}$$

This nonlinear scheme has the following conservation properties.

$$I_{\mathrm{d}}(\boldsymbol{U}^{(m)}) = I_{\mathrm{d}}(\boldsymbol{U}^{(0)}), \tag{4.326}$$

$$J_{\mathrm{d}}(\boldsymbol{U}^{(m)}) = J_{\mathrm{d}}(\boldsymbol{U}^{(0)}), \tag{4.327}$$

where

$$I_{\mathrm{d}}(\boldsymbol{U}) \stackrel{\mathrm{d}}{\equiv} \sum_{k=0}^{N} {}'' U_k \Delta x, \tag{4.328}$$

$$J_{\mathrm{d}}(\boldsymbol{U}) \stackrel{\mathrm{d}}{\equiv} \sum_{k=0}^{N} {}'' G_{\mathrm{d},k}(\boldsymbol{U}) \Delta x. \tag{4.329}$$

Below we call this scheme the $\underline{\text{N}}$onlinear $\underline{\text{E}}$nergy-conserving (NE) scheme.

With the same discrete variational derivative above, we can construct a slightly different scheme as follows.

$$\left(1 - \delta_k^{\langle 2 \rangle}\right) \delta_m^+ U_k^{(m)} = -\delta_k^{\langle 1 \rangle} \frac{\delta G_{\mathrm{d}}}{\delta(\boldsymbol{U}_+^{(m+1)/2}, \boldsymbol{U}_-^{(m+1)/2})_k}, \tag{4.330}$$

where

$$\left(\boldsymbol{U}_+^{(m+1)/2}\right)_k \stackrel{\mathrm{d}}{\equiv} s_k^+ \mu_m^+ U_k^{(m)}, \quad \left(\boldsymbol{U}_-^{(m+1)/2}\right)_k \stackrel{\mathrm{d}}{\equiv} s_k^- \mu_m^+ U_k^{(m)}. \tag{4.331}$$

That is, we utilize the discrete variational derivative *with shifted numerical solutions*. It is not so difficult to understand it because the mathematical key in the momentum-conservation property is

$$\int \frac{\partial u}{\partial x} \frac{\delta G}{\delta u} \mathrm{d}x = \int \frac{\partial}{\partial x} G(u, u_x) \mathrm{d}x \qquad (4.332)$$

since the invariance is shown as:

$$\frac{\mathrm{d}}{\mathrm{d}t} K(u) = \int (uu_t + u_x u_{xt}) \, \mathrm{d}x = \int u \, (u_t - u_{xxt}) \, \mathrm{d}x$$

$$= -\int u \frac{\partial}{\partial x} \frac{\delta G}{\delta u} \mathrm{d}x$$

$$= \int \frac{\partial u}{\partial x} \frac{\delta G}{\delta u} \mathrm{d}x = \int \frac{\partial}{\partial x} G(u, u_x) \mathrm{d}x = 0. \qquad (4.333)$$

The scheme has the following invariants:

$$I_{\mathrm{d}}(\boldsymbol{U}^{(m)}) = I_{\mathrm{d}}(\boldsymbol{U}^{(0)}), \qquad (4.334)$$

$$K_{\mathrm{d}}(\boldsymbol{U}^{(m)}) = K_{\mathrm{d}}(\boldsymbol{U}^{(0)}), \qquad (4.335)$$

where

$$K_{\mathrm{d}}(\boldsymbol{U}) \stackrel{\mathrm{d}}{=} \frac{1}{2} \sum_{k=0}^{N} {}'' \left\{ (U_k)^2 + \frac{(\delta_k^+ U_k)^2 + (\delta_k^- U_k)^2}{2} \right\} \Delta x. \qquad (4.336)$$

Due to the restriction of space, we here omit the detailed explanation on how the idea of shifted solutions in fact realizes the discrete conservation (readers may refer to [97]). Since this is a Nonlinear and Momentum-conserving scheme, we call it the NM scheme.

4.7.3.3 Linearly Implicit Schemes

Next let us construct linearly implicit schemes. First we define a discrete energy function:

$$G_{\mathrm{d}2,k}(\boldsymbol{U}, \boldsymbol{V}) \stackrel{\mathrm{d}}{=} \frac{1}{2} U_k V_k + \frac{1}{6} \left(\frac{(U_k)^2 V_k + U_k (V_k)^2}{2} \right). \qquad (4.337)$$

The subscript "2" is to distinguish it from the previous discrete energy function. From the energy we obtain the three-points discrete variational derivative:

$$\frac{\delta G_{\mathrm{d}2}}{\delta(\boldsymbol{U}, \boldsymbol{V}, \boldsymbol{W})_k} = V_k + \frac{1}{2} \left(\frac{(U_k + V_k + W_k) V_k}{3} \right). \qquad (4.338)$$

Then we obtain the following linearly implicit scheme:

$$\left(1 - (\delta_k^{\langle 1 \rangle})^2\right) \delta_m^{\langle 1 \rangle} U_k{}^{(m)} = -\delta_k^{\langle 1 \rangle} \frac{\delta G_{\mathrm{d}2}}{\delta(\boldsymbol{U}^{(m+1)}, \boldsymbol{U}^{(m)}, \boldsymbol{U}^{(m-1)})_k}. \qquad (4.339)$$

Applications

The associated conservation properties are

$$I_{\mathrm{d}}(\boldsymbol{U}^{(m)}) = \begin{cases} I_{\mathrm{d}}(\boldsymbol{U}^{(1)}), & \text{for odd } m > 0, \\ I_{\mathrm{d}}(\boldsymbol{U}^{(0)}), & \text{for even } m \geq 0, \end{cases} \quad (4.340)$$

$$J_{\mathrm{d}2}(\boldsymbol{U}^{(m+1)}, \boldsymbol{U}^{(m)}) = J_{\mathrm{d}2}(\boldsymbol{U}^{(1)}, \boldsymbol{U}^{(0)}), \quad \text{for } m \geq 0, \quad (4.341)$$

where

$$J_{\mathrm{d}2}(\boldsymbol{U}, \boldsymbol{V}) \stackrel{\mathrm{d}}{=} \sum_{k=0}^{N}{}'' G_{\mathrm{d}2,k}(\boldsymbol{U}, \boldsymbol{V}) \Delta x. \quad (4.342)$$

We call this scheme the <u>L</u>inear <u>E</u>nergy-conserving (LE) scheme.

As in the previous subsection, we can construct a slightly different scheme with the idea of shifted solutions as follows.

$$\left(1 - \delta_k^{\langle 2 \rangle}\right) \delta_m^{\langle 1 \rangle} U_k{}^{(m)} = -\delta_k^{\langle 1 \rangle} \frac{\delta G_{\mathrm{d}}}{\delta(\boldsymbol{U}_+^{(m)}, \boldsymbol{U}_-^{(m)})_k}, \quad (4.343)$$

where

$$\left(\boldsymbol{U}_+^{(m)}\right)_k \stackrel{\mathrm{d}}{=} s_k^+ U_k{}^{(m)}, \quad \left(\boldsymbol{U}_-^{(m)}\right)_k \stackrel{\mathrm{d}}{=} s_k^- U_k{}^{(m)}. \quad (4.344)$$

This keeps the following discrete invariants.

$$I_{\mathrm{d}}(\boldsymbol{U}^{(m)}) = \begin{cases} I_{\mathrm{d}}(\boldsymbol{U}^{(1)}), & \text{for odd } m > 0, \\ I_{\mathrm{d}}(\boldsymbol{U}^{(0)}), & \text{for even } m \geq 0, \end{cases} \quad (4.345)$$

$$K_{\mathrm{d}2}(\boldsymbol{U}^{(m+1)}, \boldsymbol{U}^{(m)}) = K_{\mathrm{d}2}(\boldsymbol{U}^{(1)}, \boldsymbol{U}^{(0)}), \quad \text{for } m \geq 0, \quad (4.346)$$

where

$$K_{\mathrm{d}2}(\boldsymbol{U}, \boldsymbol{V}) \stackrel{\mathrm{d}}{=} \frac{1}{2} \sum_{k=0}^{N}{}'' \left\{ U_k V_k + \frac{(\delta_k^+ U_k)(\delta_k^+ V_k) + (\delta_k^- U_k)(\delta_k^- V_k)}{2} \right\} \Delta x. \quad (4.347)$$

We call this scheme the <u>L</u>inear <u>M</u>omentum-conserving (LM) scheme.

4.7.3.4 Numerical Examples

In this subsection, we present several numerical examples.

4.7.3.4.1 One Solitary Wave The BBM equation has a one solitary wave solution,

$$u(x,t) = 3 \operatorname{sech}^2 \left(\frac{x - x_0 - 2t}{2\sqrt{2}} \right), \quad (4.348)$$

where $x_0 + 2t$ is the location of the solitary wave peak. In this subsection, we set x_0 to 20, and the initial state for the numerical computation to

$$u_0(x) = 3 \operatorname{sech}^2 \left(\frac{x - x_0}{2\sqrt{2}} \right). \quad (4.349)$$

For the linear schemes, i.e., the LM and LE schemes, we need another starting value: $\boldsymbol{U}^{(1)}$. We use the NM scheme for the LM scheme, and the NE scheme for the LE scheme.

In Table 4.5,[8] the relative errors in mass, energy, momentum, and the peak value of the numerical solutions and computation time obtained using the four proposed schemes and the Runge–Kutta scheme at $t = 40$ are listed. We note that the peak value of the exact solution (4.348) should be constant 3 (if the computation is exact). The computation parameters are set to $\Delta x = 1/4$, $\Delta t = 1/16$, and $L = 100$. For comparison, a Runge–Kutta scheme is constructed based on the ordinary differential equations of $\boldsymbol{U} : \mathbb{R} \to \mathbb{R}^N$:

$$\frac{\mathrm{d}}{\mathrm{d}t}\boldsymbol{U}(t) = -(I - D_2)^{-1} D_1 \left(\boldsymbol{U}(t) + \frac{1}{2}\boldsymbol{V}(t) \right), \qquad (4.350)$$

where $(\boldsymbol{V}(t))_k \stackrel{\mathrm{d}}{=} (U_k(t))^2$. In each time step of the nonlinear schemes (the NM and the NE schemes), we used the standard Newton method.

Energy fluctuations in the conservative schemes and the Runge–Kutta scheme are shown in Figure 4.48,[9] which shows that the discrete energies are well conserved in all the conservative schemes. In particular, they deserve attention so that even in the non-conservative schemes, i.e., the NM and the LM schemes, they are nearly conserved. On the other hand, in the Runge–Kutta scheme, the energy monotonically decreases.

Momentum fluctuations are shown in Figure 4.49,[10] where we find the same trend as the energy. The discrete momenta are well conserved by the conservative schemes, even by the NE and LE schemes. On the other hand, in the Runge–Kutta scheme, the energy monotonically decreases. In Figure 4.49, around $t \leq 4$, we observe oscillation in the LE scheme. This might be caused

TABLE 4.5: Relative errors in mass(Ms.), energy(E.), momentum(Mm.), and peak value(PV.) of numerical solutions and computation time(CPU) obtained using the proposed schemes and Runge–Kutta scheme at $t = 40$. Computation parameters are $\Delta x = 1/4$, $\Delta t = 1/16$, and $L = 100$.

Scheme	Ms.err.	E.err.	Mm.err.	PV.err.	CPU
NM	4.18691e–16	4.40752e–03	1.91544e–10	1.54169e–02	41m2s
NE	2.09345e–16	3.16232e–11	3.97717e–06	1.39999e–03	30m8s
LM	9.21120e–15	3.94501e–03	1.11896e–10	1.51433e–02	12m22s
LE	1.25607e–15	3.66348e–10	9.29801e–07	3.44311e–05	15m28s
RK	1.06307e–17	3.66779e–01	3.12369e–01	2.04904e–01	10m37s

[8–14] Reprinted from S. Koide and D. Furihata, Nonlinear and linear conservative finite difference schemes for regularized long wave equation, *Japan J. Indust. Appl. Math.*, 26, 15–40, Copyright (2009), with permission from JJIAM publishing committee.

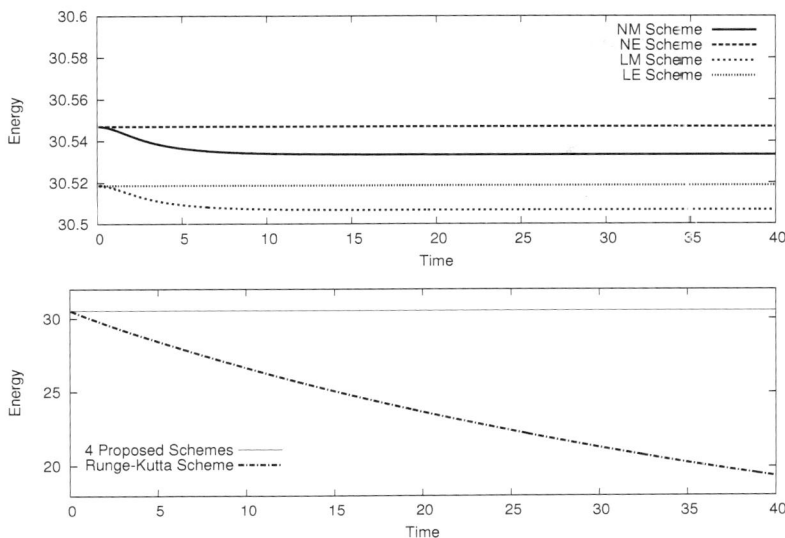

FIGURE 4.48: Energy fluctuations in the conservative schemes and the Runge–Kutta scheme.

by the fact that the starting value $\boldsymbol{U}^{(1)}$ is not necessarily appropriate for the LE scheme.

Peak value fluctuations are shown in Figure 4.50.[11] The conservative schemes preserve the peak value relatively well, while the Runge–Kutta scheme does not; this agrees with the view in the discussion above.

For the initial state (4.349), we are able to estimate the errors based on the exact solution (4.348). In Figures 4.51[12] and 4.52,[13] the errors in the four conservative schemes are shown. The left panel in Figure 4.51 shows the errors for the fixed time mesh size $\Delta t = 1/16$, where Δx was set to $1, 1/2, 1/4, 1/8$, and $1/16$. The right panel in Figure 4.51 shows the errors for the fixed space mesh size $\Delta x = 1/16$, where Δt was set to $1, 1/2, 1/4, 1/8$, and $1/16$. Figure 4.52 shows the errors for the case $\Delta x = \Delta t$. For the NM scheme, in particular, the max norm of the errors is estimated as $O(\Delta x^2 + \Delta t^2)$ in Theorem 4.14, and these figures confirm this fact. Although there is no similar theorem for the other three schemes, the figures suggest that the convergence rates are practically the same.

4.7.3.4.2 Two Solitary Waves

Let us next test another initial state:

$$u_0(x) = 3\operatorname{sech}^2\left(\frac{x - x_1}{\sqrt{2}}\right) + \frac{3}{2}\operatorname{sech}^2\left(\frac{x - x_2}{\sqrt{3}}\right), \qquad (4.351)$$

218 *Discrete Variational Derivative Method*

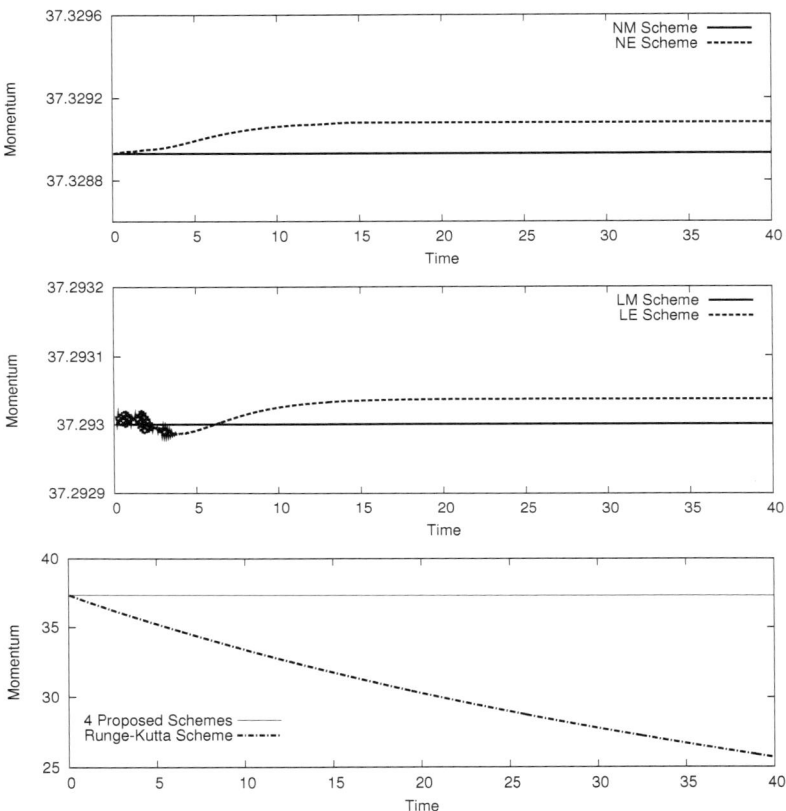

FIGURE 4.49: Momentum fluctuations in the conservative schemes and the Runge–Kutta scheme.

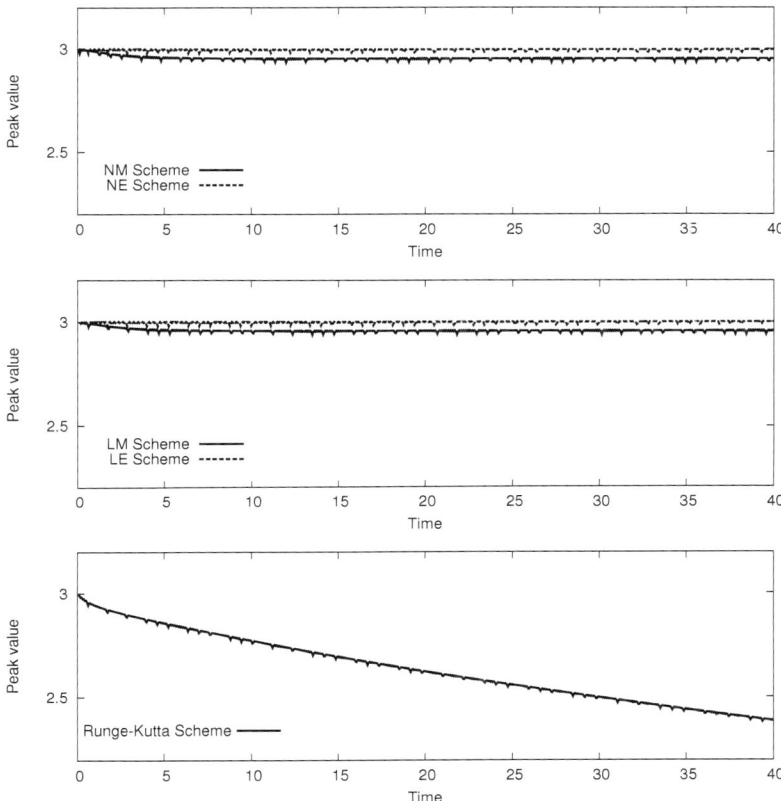

FIGURE 4.50: Peak value fluctuations in the conservative schemes and the Runge–Kutta scheme.

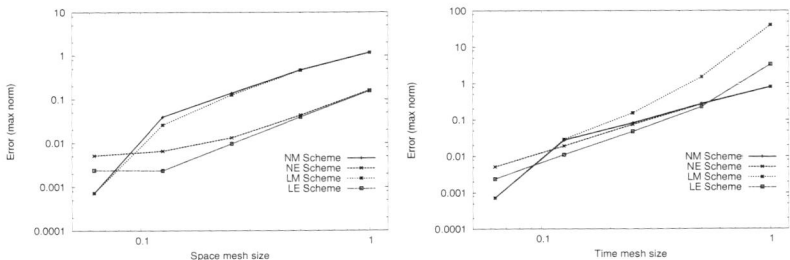

FIGURE 4.51: Numerical solution errors (max norm) at $t = 5$ in the conservative schemes; $L = 100$ and the initial state is given by (4.349). Left: errors with time mesh size $\Delta t = 1/16$; right: errors with $\Delta x = 1/16$.

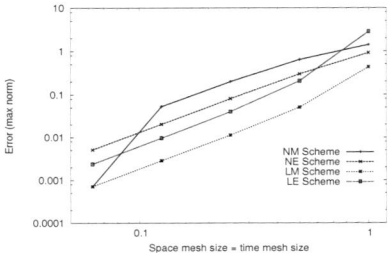

FIGURE 4.52: Numerical solution errors (max norm) at $t = 5$ for $\Delta x = \Delta t$. $L = 100$ and the initial state is given by (4.349).

where $x_1 = 20$ and $x_2 = 50$. This function approximates the wave with two peaks at x_1 and x_2. Figure 4.53[14] shows the profiles of the numerical

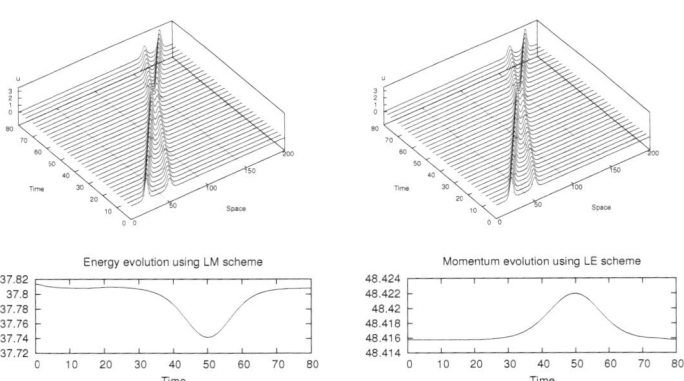

FIGURE 4.53: Computation for the two solitary waves initial state (4.351) with $\Delta x = 1/5$, $\Delta t = 1/32$, and $L = 200$. Top left: profiles of numerical solutions by the LM scheme, top right: by LE scheme, bottom left: energy evolution in the LM scheme, bottom right: in the LE scheme.

solutions, energy evolution, and momentum evolution obtained by the linear schemes with $\Delta x = 1/5$, $\Delta t = 1/32$, and $L = 200$. In the figures, we find that a phenomenon like 'phase shift' occurs when two peaks collide. The non-conserved quantities, i.e. energy in the LM scheme and momentum in the LE scheme, deviate to some extent when two peaks of a wave collide, but fortunately after that they attain their initial values.

4.7.3.5 Analysis of Scheme

Here we reveal several theoretical aspects of the NM scheme (4.330). First, applying the discrete Sobolev lemma (see Section 3.6.2.1) to (4.335), we immediately obtain the following inequality.

THEOREM 4.12

$$\|\boldsymbol{U}^{(m)}\|_\infty \leq 2\sqrt{2}\max\left(\frac{1}{\sqrt{L}}, \sqrt{\frac{L}{2}}\right)\sqrt{I_\mathrm{d}(\boldsymbol{U}^{(0)})}. \qquad (4.352)$$

This inequality implies that the NM scheme (4.330) is stable for any time step m as far as the numerical solutions of the scheme exist.

Next we exploit the conditions for the existence of numerical solutions.

THEOREM 4.13 Local Existence and Uniqueness of Solutions

If the condition:

$$\Delta t \leq \frac{2(\Delta x)^{3/2}}{\sqrt{\Delta x} + 2\|\boldsymbol{U}^{(m)}\|}, \qquad (4.353)$$

holds, the NM scheme (4.330) has a unique solution.

We leave the proof to [97]. The next estimate is easily obtained, in view of the conservation of the discrete momentum K_d.

$$\|\boldsymbol{U}^{(m)}\|^2 \leq \|\boldsymbol{U}^{(m)}\|_{H^1}^2 = 2K_\mathrm{d}(\boldsymbol{U}^{(m)}) = 2K_\mathrm{d}(\boldsymbol{U}^{(0)}). \qquad (4.354)$$

With this, and the local existence theorem, we reach the global existence result below.

COROLLARY 4.2 Global Existence and Uniqueness of Solutions

If the condition:

$$\Delta t \leq \frac{(\Delta x)^{3/2}}{\sqrt{\Delta x} + 2\left(2K_\mathrm{d}(\boldsymbol{U}^{(0)})\right)^{1/2}} \qquad (4.355)$$

holds, the NM scheme (4.330) has unique numerical solutions $\boldsymbol{U}^{(m)}$ for any $m \geq 0$.

Finally, we give an error estimate.

THEOREM 4.14
Assume that $T < \infty$ is given, and $N = L/\Delta x \geq 12$. If the solution of the BBM equation is sufficiently smooth such that

$$\left| \frac{\partial^l}{\partial x^l} \frac{\partial^{m'}}{\partial t^{m'}} u(x,t) \right| < \infty, \quad x \in (0,L),\ 0 \leq t \leq T,\ 0 \leq l \leq 5,\ 0 \leq m' \leq 3, \tag{4.356}$$

and the time mesh size Δt is sufficiently smooth such that

$$\Delta t < \frac{2}{3\lambda}, \tag{4.357}$$

then the error is evaluated as follows:

$$\max_{0 \leq k \leq N} \left| U_k^{(m)} - u(k\Delta x, m\Delta t) \right| \leq \sqrt{6T} \max(1, L) E_0\, e^{\frac{3}{4}\lambda T}, \quad \text{for } m \leq \frac{T}{\Delta t}, \tag{4.358}$$

where

$$\lambda \stackrel{\mathrm{d}}{=} 2^{9/8} \max\left(\left(\frac{4}{L}\right)^{1/4}, 1\right) \sqrt{K[u(\cdot,0)] + \frac{7}{4}L\widetilde{L}^2 \Delta x^2 + 1}, \tag{4.359}$$

$$E_0 \stackrel{\mathrm{d}}{=} \left(\Delta x^2 + \Delta t^2\right)\left(1 + \Delta x^2 + \Delta t^2 + \Delta x^4 + \Delta t^4\right)$$
$$\times \left(\frac{49}{32}\widetilde{L} + \frac{14245}{3456}\widetilde{L}^2\right), \tag{4.360}$$

$$\widetilde{L} \stackrel{\mathrm{d}}{=} \sup_{0 \leq t \leq T,\ x \in (0,L),\ 0 \leq m' \leq 3,\ l \in a(m')} \left\{ \left| \frac{\partial^l}{\partial x^l} \frac{\partial^{m'}}{\partial t^{m'}} u(x,t) \right| \right\}, \tag{4.361}$$

$$a(m') \stackrel{\mathrm{d}}{=} \begin{cases} \{0,1,2,3,4,5\}, & m' = 0, \\ \{3,4\}, & m' = 1, \\ \{0,1,2,3,4,5\}, & m' = 2, \\ \{0,1,2,3,4\}, & m' = 3. \end{cases} \tag{4.362}$$

The proof is, again, left to [97].

4.7.4 Feng Equation

4.7.4.1 Introduction to Problem

The Feng equation [55, 56] is

$$\frac{\partial^2 u}{\partial t^2} - \gamma \frac{\partial^4 u}{\partial x^2 \partial t^2} = -\Phi'(u), \tag{4.363}$$

where Φ is a function of u, with the appropriate periodic boundary conditions. As noted in Remark 2.8, this does not directly belong to the target PDEs

Applications

classified in Chapter 2, but can be regarded as a conservative PDE, if we rewrite this as follows.

$$\begin{cases} \dfrac{\delta G_1}{\delta u} + H(v) = 0, \\ \dfrac{\partial u}{\partial t} = v, \end{cases} \tag{4.364}$$

where

$$G_1(u) \stackrel{\mathrm{d}}{\equiv} \Phi(u), \tag{4.365}$$

$$G_2(v, v_x) \stackrel{\mathrm{d}}{\equiv} \frac{1}{2}v^2 + \frac{1}{2}\gamma(v_x)^2, \tag{4.366}$$

and

$$H(v) \stackrel{\mathrm{d}}{\equiv} v_t - \gamma v_{txx}. \tag{4.367}$$

Note that $H(v)$ satisfies the following identity.

$$\int_\Omega H(v)v \mathrm{d}x = \int_\Omega \frac{\delta G_2}{\delta v} v_t \mathrm{d}x. \tag{4.368}$$

This Feng equation has the following invariant.

$$J(u,v) \stackrel{\mathrm{d}}{\equiv} \int_\Omega \{G_1(u) + G_2(v, v_x)\} \, \mathrm{d}x. \tag{4.369}$$

4.7.4.2 Numerical Scheme

We impose the following discrete periodic boundary conditions:

$$U_k^{(m)} = U_{k \bmod N}^{(m)} \quad \text{for } {}^\forall k \in \mathbb{Z}. \tag{4.370}$$

In other words, we assume $\boldsymbol{U}^{(m)} \in \mathbb{S}_N$ (for the notation, see 3.6.) By appropriate discretizing of G_1 and G_2 we obtain the discrete energy functions

$$G_{1\mathrm{d},k}(\boldsymbol{U}) \stackrel{\mathrm{d}}{\equiv} \Phi(U_k), \tag{4.371}$$

$$G_{2\mathrm{d},k}(\boldsymbol{U}) \stackrel{\mathrm{d}}{\equiv} \frac{1}{2}(U_k)^2 + \frac{1}{2}\gamma\left(\frac{(\delta_k^+ U_k)^2 + (\delta_k^- U_k)^2}{2}\right) \tag{4.372}$$

and the discrete variational derivatives

$$\frac{\delta G_{1\mathrm{d}}}{\delta(\boldsymbol{U},\boldsymbol{V})_k} = \frac{\Phi(U_k) - \Phi(V_k)}{U_k - V_k}, \tag{4.373}$$

$$\frac{\delta G_{2\mathrm{d}}}{\delta(\boldsymbol{U},\boldsymbol{V})_k} = \left(1 - \gamma \delta_k^{\langle 2 \rangle}\right) \frac{U_k + V_k}{2}. \tag{4.374}$$

We are able to define the discrete H function as

$$H_{\mathrm{d},k}(\boldsymbol{U},\boldsymbol{V}) \stackrel{\mathrm{d}}{\equiv} \left(1 - \gamma \delta_k^{\langle 2 \rangle}\right) \frac{U_k - V_k}{\Delta t} \tag{4.375}$$

and they satisfy the following equality under the discrete periodic boundary conditions.

$$\sum_{k=0}^{N}{}'' H_{\mathrm{d},k}(\boldsymbol{U},\boldsymbol{V}) \frac{U_k + V_k}{2} \Delta x = \sum_{k=0}^{N}{}'' \frac{\delta G_{2\mathrm{d}}}{\delta(\boldsymbol{U},\boldsymbol{V})_k} \frac{U_k - V_k}{\Delta t} \Delta x. \quad (4.376)$$

So we obtain the following finite difference scheme

$$\begin{cases} \dfrac{\delta G_{1\mathrm{d}}}{\delta(\boldsymbol{U}^{(m+1)},\boldsymbol{U}^{(m)})_k} + H_{\mathrm{d},k}(\boldsymbol{V}^{(m+1)},\boldsymbol{V}^{(m)}) = 0, \\[2ex] \dfrac{U_k^{(m+1)} - U_k^{(m)}}{\Delta t} = \dfrac{V_k^{(m+1)} + V_k^{(m)}}{2}. \end{cases} \quad (4.377)$$

The solutions $\boldsymbol{U}, \boldsymbol{V}$ conserve the discrete invariant:

$$J_{\mathrm{d}}(\boldsymbol{U}^{(m)}, \boldsymbol{V}^{(m)}) \stackrel{\mathrm{d}}{=} \sum_{k=0}^{N}{}'' \Big(G_{1\mathrm{d},k}(\boldsymbol{U}^{(m)}) + G_{2\mathrm{d},k}(\boldsymbol{V}^{(m)}) \Big) \Delta x. \quad (4.378)$$

4.7.4.3 Numerical Examples

In Figure 4.54, we show numerical solutions with $\gamma = 1/4$ and $\Phi(u) = u^4$. Initial state was set to

$$u(x, 0) = A \operatorname{sech}(2x - 16), \quad (4.379\mathrm{a})$$
$$u_t(x, 0) = 0, \quad (4.379\mathrm{b})$$

where A is a positive constant.

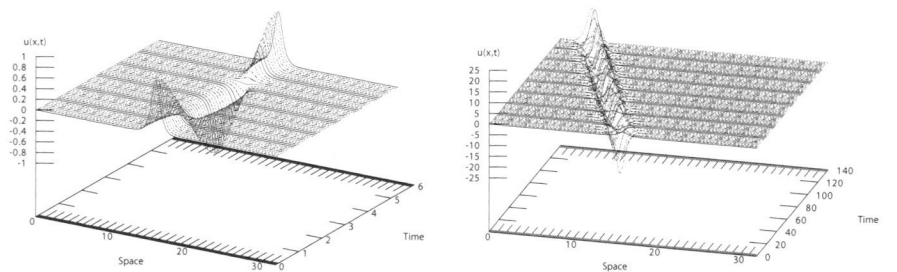

FIGURE 4.54: Numerical solutions to the Feng equation. Left: a stationary breather solution with $A = 1.0$, $\Delta x = 0.02$ and $\Delta t = 0.001$; right: a moving breather solution with $A = 2.0$, $\Delta x = 0.02$ and $\Delta t = 0.01$.

4.7.4.4 Analysis of Scheme

For some special $\Phi(u)$, we are able to obtain a stability result. Let us, for example, take Φ to

$$\Phi(u) \stackrel{\mathrm{d}}{=} \frac{1}{2}\alpha u^2 + \frac{1}{4}\beta u^4. \tag{4.380}$$

With this Φ, it is easy to see

$$\|\boldsymbol{V}^{(m)}\|_{H^1} \leq 8\left\{J_{\mathrm{d}}(\boldsymbol{U}^{(0)}, \boldsymbol{V}^{(0)}) + \frac{\alpha^2}{\beta}L\right\}, \tag{4.381}$$

and based on this and the discrete Sobolev lemma in Section 3.6.2.1, we can deduce the following theorem.

THEOREM 4.15
For the potential function (4.380), the following estimate holds under the discrete periodic boundary conditions.

$$\|\boldsymbol{U}^{(m)}\|_\infty \leq \|\boldsymbol{U}^{(0)}\|_\infty + C_1 T, \tag{4.382}$$

where $T = m\Delta t$, and

$$C_1 \stackrel{\mathrm{d}}{=} 4\max\left(\frac{1}{\sqrt{L}}, \sqrt{L}\right) \sqrt{J_{\mathrm{d}}(\boldsymbol{U}^{(0)}, \boldsymbol{V}^{(0)}) + \frac{\alpha^2}{\beta}L}. \tag{4.383}$$

An existence and uniqueness result can be obtained via a similar discussion as in Theorem 4.3.

THEOREM 4.16
Let us consider the potential function (4.380). If the condition:

$$(\Delta t)^2 < \min\left(\frac{\Delta x}{\sqrt{44\alpha^2(\Delta x)^2 + (7038/16)\beta^2 M^4}}, \frac{\Delta x}{\alpha \Delta x + (6\beta + 9/8)M^2}\right) \tag{4.384}$$

holds, then the solution $\boldsymbol{U}^{(m+1)}$ uniquely exists in the scheme (4.377) for any $m \geq 1$, where

$$M \stackrel{\mathrm{d}}{=} 2\|\boldsymbol{U}^{(m)}\|_2 + \|\boldsymbol{U}^{(m-1)}\|_2. \tag{4.385}$$

Chapter 5

Advanced Topic I: Design of High-Order Schemes

We have already glanced at the basic formulation of the discrete variational derivative method and its various applications. In this and subsequent chapters we will discuss more advanced topics. This chapter is devoted to the design of spatially and temporally higher-order schemes. The presented schemes so far were all second-order, both spatially and temporally. If more accuracy is demanded, we can increase the accuracy by the method presented here. This chapter is organized as follows. We first discuss the "orders of accuracy" in Section 5.1. Section 5.2 is devoted to the spatially high-order schemes. Section 5.3 and Section 5.4 are both for temporally high-order schemes; we have two different ways of designing such schemes. The first option is the use of the composition method; this issue will be discussed in Section 5.3. The second option is to consider the generalization of discrete variation process; this will be discussed in Section 5.4.

5.1 Orders of Accuracy of Schemes

Here we define the orders of accuracy of the schemes. Let u be the solution on $\Omega = [0, L] \times [0, T]$ with some prescribed $T > 0$, and $\boldsymbol{U}^{(m)}$ ($0 \leq m \leq M$, $M\Delta t = T$) be the corresponding numerical solution obtained by a numerical scheme. Then, the *global orders of accuracy* of the numerical solution (and accordingly of the scheme) are defined as follows. In the definition, we denote some vector norms as $\|\cdot\|$.

DEFINITION 5.1 Global orders of accuracy
Let $\boldsymbol{u}(T) = (u(0,T), u(\Delta x, T), \ldots, u(L,T))^\top$ *be the values of true solutions on the mesh points. If the error estimate:*

$$\|\boldsymbol{U}^{(M)} - \boldsymbol{u}(T)\| = O(\Delta x^{p'}, \Delta t^{q'}), \qquad \Delta x, \Delta t \to 0$$

holds, we say the numerical solution (and accordingly the scheme) is globally of p'th-order in space and globally of q'th-order in time. □

We cannot, however, always reach this kind of estimate; especially if the target PDE is nonlinear.

As a secondary estimate method, we employ the concept of *local orders of accuracy*. Let us formally write a numerical scheme as follows.

$$f_k(\boldsymbol{U}^{(m+1)}, \ldots, \boldsymbol{U}^{(m-l+2)}) = 0, \quad 0 \leq k \leq N-1, \ m = 0, 1, 2, \ldots,$$

where $l \geq 2$ is the number of steps that the scheme involves. For example, for Scheme 3.1 (page 80), we have $l = 2$, and

$$f_k(\boldsymbol{U}^{(m+1)}, \boldsymbol{U}^{(m)}) \stackrel{\mathrm{d}}{=} \frac{U_k^{(m+1)} - U_k^{(m)}}{\Delta t} - (-1)^{s+1} \delta_k^{\langle 2s \rangle} \frac{\delta G_\mathrm{d}}{\delta(\boldsymbol{U}^{(m+1)}, \boldsymbol{U}^{(m)})_k}.$$

Then, the *local truncation error* of the scheme is defined by

$$E_{\mathrm{local},k} \stackrel{\mathrm{d}}{=} f_k(u_k^{(m+1)}, \ldots, u_k^{(m-l+2)}),$$

where $u_k^{(m)} \stackrel{\mathrm{d}}{=} u(k\Delta x, m\Delta t)$ are the values of the true solution on the grid points. That is, the local truncation error shows how well the equation of the scheme is satisfied by the true solution u. Using Taylor expansion, we can always obtain an estimate:

$$E_{\mathrm{local},k} = f_k(u_k^{(m+1)}, \ldots, u_k^{(m-l+2)}) = O(\Delta x^p, \Delta t^q), \qquad \Delta x, \Delta t \to 0. \quad (5.1)$$

Then, the local orders of accuracy are defined as follows.

DEFINITION 5.2 Local orders of accuracy When the estimate (5.1) holds, the numerical solution (and accordingly the scheme) is locally of pth-order in space, and locally of qth-order in time. □

As opposed to the global orders of accuracy, the local orders of accuracy can be always obtained by considering Taylor expansion. Moreover, the local orders p, q often coincide with the global orders p', q', and thus can be used as the estimates of the global orders of accuracy. Thus, hereafter we employ the local orders as the measures of accuracy; namely, when we call a scheme pth-order in space and qth-order in time, we mean the *local* orders of accuracy. We denote such scheme a (p, q)*th-order scheme*.

Employing Taylor expansion, we can prove that Scheme 3.1 and Scheme 3.2 (pages 80 and 84), Scheme 3.3 and Scheme 3.4 (pages 96 and 99), and Scheme 3.5 and Scheme 3.6 (pages 105 and 109), presented in Chapter 3, are all temporally second-order (note that they are symmetric with respect to $\boldsymbol{U}^{(m+1)}$ and $\boldsymbol{U}^{(m)}$). The spatial order of the schemes is either one or two, depending on the concrete forms of the resulting schemes. The examples in Chapter 3 are all spatially second-order.

5.2 Spatially High-Order Schemes

In this section we design spatially high-order conservative or dissipative schemes. To this end, we first set one big assumption: in all the problems considered in this subsection,

$$\left.\frac{\partial^j u}{\partial x^j}\right|_{x=0} = \left.\frac{\partial^j u}{\partial x^j}\right|_{x=L}, \quad j = 0, 1, \ldots, J, \quad (5.2)$$

where $J \geq 1$ is specific to each problem is imposed.[1] We also limit ourselves to first-order real-valued and complex-valued PDEs for brevity. The extensions to other classes of target PDEs are rather straightforward.

5.2.1 Discrete Symbols and Formulas

The discrete symbols and the formulas used in this subsection are summarized.

Corresponding to the continuous periodic boundary condition (5.2), we apply the discrete periodic boundary condition:

$$U_k = U_{k+\omega N}, \quad 0 \leq k \leq N-1, \; \omega \in \mathbb{Z}. \quad (5.3)$$

Next we introduce the difference operator of $O(\Delta x^{2p})$ ($p = 1, 2, \ldots$) for the first derivative as follows.

$$\delta_k^{\langle 1 \rangle, 2p} U_k = \sum_{j=-p}^{p} \frac{\alpha_{p,j} U_{k+j}}{\Delta x}. \quad (5.4)$$

The coefficients $\alpha_{p,j}$ can be uniquely determined to gain accuracy of $O(\Delta x^{2p})$. The operators are skew-symmetric, i.e., $\alpha_{p,j} = -\alpha_{p,-j}$ for any p (see, for example, Fornberg [61]). When $p = 1$, it is the well-known central difference operator: $\delta_k^{\langle 1 \rangle} U_k = (U_{k+1} - U_{k-1})/2\Delta x$. And in the limit of $p \to \infty$, $\delta_k^{\langle 1 \rangle, 2p}$ becomes the so-called "spectral differentiation" operator [61]:

$$\delta_k^{\langle 1 \rangle, \infty} U_k = \left(F^{-1} \widetilde{D} F \boldsymbol{U} \right)_k, \quad (5.5)$$

where $\boldsymbol{U} = (U_0, U_1, \ldots, U_{N-1})^{\mathrm{T}}$ and

$$F_{jk} \stackrel{\mathrm{d}}{\equiv} \frac{1}{N} \exp\left(-\frac{2\mathrm{i}\pi j k}{N}\right), \quad (5.6)$$

$$\widetilde{D} \stackrel{\mathrm{d}}{\equiv} \mathrm{diag}\left\{0, \frac{2\pi\mathrm{i}}{L}, \ldots, \frac{2\pi\mathrm{i}(\frac{N}{2}-1)}{L}, 0, -(\frac{2\pi\mathrm{i}(\frac{N}{2}-1)}{L}), \ldots, -\frac{2\pi\mathrm{i}}{L}\right\}. \quad (5.7)$$

---

[1] Although we have already reserved "J" for the global energies, we use the same symbol for the degree of smoothness here, since it should not cause any confusion.

With regard to the differentiation operator $\delta_k^{\langle 1 \rangle, 2p}$, the following "summation-by-parts" formula holds; this is a generalization of the basic summation-by-parts formulas appearing in the preceding chapters.

LEMMA 5.1 Summation-by-parts with respect to $\delta_k^{\langle 1 \rangle, 2p}$
For any two N-periodic sequences U_k, V_k $(0 \leq k \leq N-1)$,

$$\sum_{k=0}^{N} {}'' (\delta_k^{\langle 1 \rangle, 2p} U_k) V_k \Delta x = -\sum_{k=0}^{N} {}'' U_k (\delta_k^{\langle 1 \rangle, 2p} V_k) \Delta x. \tag{5.8}$$

PROOF

$$\sum_{k=0}^{N} {}'' \left(\delta_k^{\langle 1 \rangle, 2p} U_k\right) V_k \Delta x = \sum_{k=0}^{N-1} \left(\delta_k^{\langle 1 \rangle, 2p} U_k\right) V_k \Delta x$$

$$= \sum_{k=0}^{N-1} \left(\sum_{j=-p}^{p} \alpha_{p,j} U_{k+j}\right) V_k$$

$$= \sum_{j=-p}^{p} \left(\sum_{k'=j}^{N-1+j} \alpha_{p,j} U_{k'} V_{k'-j}\right)$$

$$= \sum_{j=-p}^{p} \left(\sum_{k'=0}^{N-1} \alpha_{p,j} U_{k'} V_{k'-j}\right)$$

$$= -\sum_{j'=-p}^{p} \left(\sum_{k'=0}^{N-1} \alpha_{p,j'} U_{k'} V_{k'+j'}\right)$$

$$= -\sum_{k=0}^{N-1} U_k (\delta_k^{\langle 1 \rangle, 2p} V_k) \Delta x$$

$$= -\sum_{k=0}^{N} {}'' U_k (\delta_k^{\langle 1 \rangle, 2p} V_k) \Delta x. \tag{5.9}$$

The periodic boundary condition (5.2), and the skew-symmetry: $\alpha_{p,j} = -\alpha_{p,-j}$ are used. □

REMARK 5.1 Under the discrete periodic boundary condition (5.3) the trapezoidal rule $\sum_{k=0}^{N} {}''$ is completely equivalent to the rectangle rule $\sum_{k=0}^{N-1}$. We here choose the trapezoidal rule notation, simply because we have basically used it in the preceding chapters. □

5.2.2 Discrete Variational Derivative

In this subsection we define a *spatially high-order version* of the discrete variational derivative, in both real-valued and complex-valued cases. The key is to utilize the highly accurate spatial difference operator $\delta_k^{\langle 1 \rangle, 2p}$ instead of the central difference operator $\delta_k^{\langle 1 \rangle}$ used in the preceding chapters. This modification is realized by the high-order version of the summation-by-parts formula (5.8).

5.2.2.1 For the Real-Valued PDEs

We here consider the real dissipative PDEs 1 and conservative PDEs 2 under the periodic boundary condition (5.2). As in Chapter 3, for the sake of simplicity, we assume that $G(u, u_x)$ is of the form:

$$G(u, u_x) = \sum_{l=1}^{\widetilde{M}} f_l(u) g_l(u_x), \tag{5.10}$$

where $\widetilde{M} \in \mathbb{N}$, and $f_l, g_l : \mathbb{R} \to \mathbb{R}$ are differentiable functions. We then define the "discrete local energy" $G_{\mathrm{d}}(\boldsymbol{U})$ analogously to (5.10) as

$$G_{\mathrm{d},k}(\boldsymbol{U}) = \sum_{l=1}^{M} f_l(U_k) g_l(\delta_k^{\langle 1 \rangle, 2p} U_k), \tag{5.11}$$

where $M \, (\geq \widetilde{M})$ is an integer. We also define its associated global energy by

$$J_{\mathrm{d}}(\boldsymbol{U}^{(m)}) \stackrel{\mathrm{d}}{=} \sum_{k=0}^{N} {}'' G_{\mathrm{d},k}(\boldsymbol{U}^{(m)}) \Delta x. \tag{5.12}$$

For such a G_{d}, we consider the difference:

$$\sum_{k=0}^{N} {}'' G_{\mathrm{d},k}(\boldsymbol{U}) \Delta x - \sum_{k=0}^{N} {}'' G_{\mathrm{d},k}(\boldsymbol{V}) \Delta x$$

$$= \sum_{k=0}^{N} {}'' \left\{ \frac{G_{\mathrm{d}}}{\partial (\boldsymbol{U}, \boldsymbol{V})} \bigg|_k (U_k - V_k) + \frac{G_{\mathrm{d}}}{\partial \delta(\boldsymbol{U}, \boldsymbol{V})} \bigg|_k \left(\delta_k^{\langle 1 \rangle, 2p} U_k - \delta_k^{\langle 1 \rangle, 2p} V_k \right) \right\} \Delta x$$

$$= \sum_{k=0}^{N} {}'' \left\{ \frac{G_{\mathrm{d}}}{\partial (\boldsymbol{U}, \boldsymbol{V})} \bigg|_k - \delta_k^{\langle 1 \rangle, 2p} \frac{G_{\mathrm{d}}}{\partial \delta(\boldsymbol{U}, \boldsymbol{V})} \bigg|_k \right\} (U_k - V_k) \Delta x$$

$$= \sum_{k=0}^{N} {}'' \frac{\delta G_{\mathrm{d}}}{\delta (\boldsymbol{U}, \boldsymbol{V})} \bigg|_k (U_k - V_k) \Delta x, \tag{5.13}$$

where

$$\frac{\delta G_{\mathrm{d}}}{\delta (\boldsymbol{U}, \boldsymbol{V})} \bigg|_k \stackrel{\mathrm{d}}{=} \frac{\partial G_{\mathrm{d}}}{\partial (\boldsymbol{U}, \boldsymbol{V})} \bigg|_k - \delta_k^{\langle 1 \rangle, 2p} \left(\frac{\partial G_{\mathrm{d}}}{\partial \delta (\boldsymbol{U}, \boldsymbol{V})} \bigg|_k \right), \tag{5.14}$$

and

$$\frac{\partial G_{\mathrm{d}}}{\partial(\boldsymbol{U},\boldsymbol{V})_k} = \sum_{l=1}^{M}\left\{\left(\frac{g_l(\delta_k^{\langle 1\rangle,2p}U_k)+g_l(\delta_k^{\langle 1\rangle,2p}V_k)}{2}\right)\left(\frac{f_l(U_k)-f_l(V_k)}{(U_k-V_k)}\right)\right\},$$
(5.15a)

$$\frac{\partial G_{\mathrm{d}}}{\partial\delta(\boldsymbol{U},\boldsymbol{V})_k} = \sum_{l=1}^{M}\left\{\left(\frac{f_l(U_k)+f_l(V_k)}{2}\right)\left(\frac{g_l(\delta_k^{\langle 1\rangle,2p}U_k)-g_l(\delta_k^{\langle 1\rangle,2p}V_k)}{\delta_k^{\langle 1\rangle,2p}(U_k-V_k)}\right)\right\}.$$
(5.15b)

In the second equality of (5.13) we used the summation-by-parts formula (5.8). The discrete quantity
$$\frac{\delta G_{\mathrm{d}}}{\delta(\boldsymbol{U},\boldsymbol{V})_k}$$
is called the "*spatially high-order*" discrete variational derivative. Note that though we use the same symbol as in the second-order case, it utilizes the high-order spatial difference operator $\delta_k^{\langle 1\rangle,2p}$.

5.2.2.2 For the Complex-Valued PDEs

We here consider the complex dissipative PDEs 3 and conservative PDEs 4 under the periodic boundary condition (5.2). As in Chapter 3 we assume that the energy $G(u,u_x)$ takes the form

$$G(u,u_x) = \sum_{l=1}^{M} c_l |p_l(u)|^{N_l^P} |q_l(u_x)|^{N_l^Q} \tag{5.16}$$

where $c_l \in \mathbb{R}$, and $N_l^P, N_l^Q \in \{2,3,4,\ldots\}$, and $p_l, q_l : \mathbb{C} \to \mathbb{C}$ are assumed to be analytic functions which satisfy $p_l(\overline{u}) = \overline{p_l(u)}, q_l(\overline{u}) = \overline{q_l(u)}$ ($u \in \mathbb{C}$). Hereafter we abbreviate $|p_l(U_k)|^{N_l^P}$ as $P_l(U_k)$, and $|q_l(\delta_k^{\langle 1\rangle,2p}U_k)|^{N_l^Q}$ as $Q_l(U_k)$. We then define discrete energy $G_{\mathrm{d}}(\boldsymbol{U})$ analogously to (5.16) as

$$G_{\mathrm{d},k}(\boldsymbol{U}) \stackrel{\mathrm{d}}{\equiv} \sum_{l=1}^{M} c_l |p_l(U_k)|^{N_l^P} |q_l(\delta_k^{\langle 1\rangle,2p}U_k)|^{N_l^Q}, \tag{5.17}$$

and its associated global energy by

$$J_{\mathrm{d}}(\boldsymbol{U}^{(m)}) \stackrel{\mathrm{d}}{\equiv} \sum_{k=0}^{N}{''} G_{\mathrm{d},k}(\boldsymbol{U}^{(m)})\Delta x. \tag{5.18}$$

As in the real case, we consider the difference

$$\sum_{k=0}^{N}{''} G_{\mathrm{d},k}(\boldsymbol{U})\Delta x - \sum_{k=0}^{N}{''} G_{\mathrm{d},k}(\boldsymbol{V})\Delta x$$
$$= \sum_{k=0}^{N}{''}\left(\frac{\delta G_{\mathrm{d}}}{\delta(\boldsymbol{U},\boldsymbol{V})_k}(U_k-V_k) + \frac{\delta G_{\mathrm{d}}}{\delta(\overline{\boldsymbol{U}},\overline{\boldsymbol{V}})_k}(\overline{U_k-V_k})\right)\Delta x, \tag{5.19}$$

where

$$\frac{\delta G_{\rm d}}{\delta(\boldsymbol{U},\boldsymbol{V})_k} \stackrel{\rm d}{\equiv} \frac{\partial G_{\rm d}}{\partial(\boldsymbol{U},\boldsymbol{V})_k} - \delta_k^{\langle 1\rangle,2p}\left(\frac{\partial G_{\rm d}}{\partial\delta(\boldsymbol{U},\boldsymbol{V})_k}\right), \quad (5.20)$$

$$\frac{\delta G_{\rm d}}{\delta(\overline{\boldsymbol{U}},\overline{\boldsymbol{V}})_k} \stackrel{\rm d}{\equiv} \frac{\partial G_{\rm d}}{\partial(\overline{\boldsymbol{U}},\overline{\boldsymbol{V}})_k} - \delta_k^{\langle 1\rangle,2p}\left(\frac{\partial G_{\rm d}}{\partial\delta(\overline{\boldsymbol{U}},\overline{\boldsymbol{V}})_k}\right), \quad (5.21)$$

are the "*spatially high-order* complex discrete variational derivatives," whose concrete forms are

$$\frac{\partial G_{\rm d}}{\partial(\boldsymbol{U},\boldsymbol{V})_k} \stackrel{\rm d}{\equiv} \sum_{l=1}^{M} c_l \left(\frac{Q_l(U_k)+Q_l(V_k)}{2}\right)\left(\frac{p_l(U_k)-p_l(V_k)}{U_k-V_k}\right) \quad (5.22)$$
$$\times f\left(N_l^P; p_l(U_k), p_l(V_k)\right),$$

$$\frac{\partial G_{\rm d}}{\partial \delta(\boldsymbol{U},\boldsymbol{V})_k} \stackrel{\rm d}{\equiv} \sum_{l=1}^{M} c_l \left(\frac{P_l(U_k)+P_l(V_k)}{2}\right) \quad (5.23)$$
$$\times \left(\frac{q_l(\delta_k^{\langle 1\rangle,2p}U_k)-q_l(\delta_k^{\langle 1\rangle,2p}V_k)}{\delta_k^{\langle 1\rangle,2p}U_k - \delta_k^{\langle 1\rangle,2p}V_k}\right) f\left(N_l^Q; q_l(\delta_k^{\langle 1\rangle,2p}U_k), q_l(\delta_k^{\langle 1\rangle,2p}V_k)\right),$$

where

$$f(n;z_1,z_2) \stackrel{\rm d}{\equiv} \begin{cases} \dfrac{z_1+z_2}{2}(|z_1|^{n-2}+|z_1|^{n-4}|z_2|^2+\cdots+|z_2|^{n-2}), & n:\text{even},\\[2mm] \dfrac{z_1+z_2}{2}\dfrac{|z_1|^{n-1}+|z_1|^{n-2}|z_2|+\cdots+|z_2|^{n-1}}{|z_1|+|z_2|}, & n:\text{odd}. \end{cases}$$
(5.24)

In above calculation we used the summation-by-parts formula (5.8).

5.2.3 Design of Schemes

With the discrete variational derivative given above, we then define the finite difference schemes.

5.2.3.1 For the Real-Valued PDEs

We define a scheme for the real dissipative PDEs 1 as follows.

Scheme 5.1 (Spatially high-order dissipative scheme for the PDEs 1)
Let $U_k^{(0)} = u(k\Delta x, 0)$ *be initial values. Then, a spatially high-order scheme for the PDEs 1 is given by, for* $m = 0, 1, 2, \ldots$,

$$\frac{U_k^{(m+1)} - U_k^{(m)}}{\Delta t} = (-1)^{s+1}\left(\delta_k^{\langle 1\rangle,2p}\right)^{2s}\frac{\delta G_{\rm d}}{\delta(\boldsymbol{U}^{(m+1)},\boldsymbol{U}^{(m)})_k},$$
$$k = 0,\ldots,N-1. \quad (5.25)$$

THEOREM 5.1 Discrete dissipation property and orders of Scheme 5.1

Under the discrete periodic boundary condition (5.3), Scheme 5.1 is dissipative in the sense that the inequality

$$J_d(\boldsymbol{U}^{(m+1)}) \leq J_d(\boldsymbol{U}^{(m)}), \quad m = 0, 1, 2, \ldots \tag{5.26}$$

holds. Moreover, the scheme is $(2p, 2)$-th order.

PROOF The dissipation property is proved as follows.

$$\frac{1}{\Delta t}\left(\sum_{k=0}^{N}{}''G_{d,k}(\boldsymbol{U}^{(m+1)})\Delta x - \sum_{k=0}^{N}{}''G_{d,k}(\boldsymbol{U}^{(m)})\Delta x\right)$$

$$= \sum_{k=0}^{N}{}'' \frac{\delta G_d}{\delta(\boldsymbol{U}^{(m+1)}, \boldsymbol{U}^{(m)})_k} \left(\frac{U_k^{(m+1)} - U_k^{(m)}}{\Delta t}\right) \Delta x$$

$$= \sum_{k=0}^{N}{}'' \frac{\delta G_d}{\delta(\boldsymbol{U}^{(m+1)}, \boldsymbol{U}^{(m)})_k} (-1)^{s+1} \left(\delta_k^{\langle 1 \rangle, 2p}\right)^{2s} \left(\frac{\delta G_d}{\delta(\boldsymbol{U}^{(m+1)}, \boldsymbol{U}^{(m)})_k}\right) \Delta x$$

$$= -\sum_{k=0}^{N}{}'' \left\{\left(\delta_k^{\langle 1 \rangle, 2p}\right)^s \left(\frac{\delta G_d}{\delta(\boldsymbol{U}^{(m+1)}, \boldsymbol{U}^{(m)})_k}\right)\right\}^2 \Delta x$$

$$\leq 0. \tag{5.27}$$

The first equality is from (5.13), the second is from the definition of the scheme (5.25), and the third is from the repeated use of the summation-by-parts formula (5.8).

In order to prove the claim on the orders of accuracy, let us substitute the true solutions $u_k^{(m)} \stackrel{\mathrm{d}}{=} u(k\Delta x, m\Delta t)$ into $U_k^{(m)}$, and consider Taylor expansion of both sides of the scheme at $(x, t) = (k\Delta x, (m + \frac{1}{2})\Delta t)$. It is easy to see that

$$\frac{u_k^{(m+1)} - u_k^{(m)}}{\Delta t} = \frac{\mathrm{d}}{\mathrm{d}t} u(k\Delta x, (m+1/2)\Delta t) + O(\Delta t^2).$$

Moreover, the following estimate holds:

$$\frac{\delta G_d}{\delta(\boldsymbol{u}^{(m+1)}, \boldsymbol{u}^{(m)})_k} = \left.\frac{\delta G}{\delta u}\right|_{\substack{x=k\Delta x \\ t=(m-1/2)\Delta t}} + O(\Delta x^{2p}, \Delta t^2). \tag{5.28}$$

Evidently, both estimates yield

$$\frac{u_k^{(m+1)} - u_k^{(m)}}{\Delta t} = (-1)^{s+1} \left(\frac{\partial}{\partial x}\right)^{2s} \frac{\delta G_d}{\delta(\boldsymbol{u}^{(m+1)}, \boldsymbol{u}^{(m)})_k} + O(\Delta x^{2p}, \Delta t^2),$$

which means that the scheme is $(2p, 2)$th-order.

Advanced Topic I: Design of High-Order Schemes 235

We prove the estimate (5.28), by substituting $u_k^{(m)}$ into $U_k^{(m)}$ in the definition of the discrete variational derivative, and considering Taylor expansion of each term at $(x,t) = (k\Delta x, (m+\frac{1}{2})\Delta t)$. We use the following abbreviation:

$$(\cdot)|_{(k,m+1/2)} \stackrel{\mathrm{d}}{=} (\cdot)|_{x=k\Delta x, t=(m+1/2)\Delta t}.$$

We first consider (5.15a). Since $\delta_k^{\langle 1 \rangle, 2p} u_k^{(m)} = u_x|_{(k,m+1/2)} + O(\Delta x^{2p})$, we obtain

$$\frac{g_l\left(\delta_k^{\langle 1 \rangle, 2p} u_k^{(m+1)}\right) + g_l\left(\delta_k^{\langle 1 \rangle, 2p} u_k^{(m)}\right)}{2} = g_l(u_x)|_{(k,m+1/2)} + O(\Delta x^{2p}, \Delta t^2). \tag{5.29}$$

Since for any sufficiently smooth function $f(y)$, we have

$$\frac{f(y_1) - f(y_2)}{y_1 - y_2} = \left.\frac{\partial f}{\partial y}\right|_{y=(y_1+y_2)/2} + O\left((y_1 - y_2)^2\right), \tag{5.30}$$

the following estimate holds:

$$\frac{f_l(u_k^{(m+1)}) - f_l(u_k^{(m)})}{u_k^{(m+1)} - u_k^{(m)}} = \left.\frac{\partial f_l}{\partial u}\right|_{u=\left(u_k^{(m+1)}+u_k^{(m)}\right)/2} + O\left((u_k^{(m+1)} - u_k^{(m)})^2\right)$$

$$= \left.\frac{\partial f_l}{\partial u}\right|_{(k,m+1/2)} + O\left(\Delta t^2\right). \tag{5.31}$$

From (5.29) and (5.31) we have

$$\frac{\partial G_{\mathrm{d}}}{\partial (\boldsymbol{u}^{(m+1)}, \boldsymbol{u}^{(m)})_k} = \left.\frac{\partial G}{\partial u}\right|_{(k,m+1/2)} + O(\Delta x^{2p}, \Delta t^2). \tag{5.32}$$

In a similar manner, we obtain the following estimate as to (5.15b).

$$\frac{\partial G_{\mathrm{d}}}{\partial \delta(\boldsymbol{u}^{(m+1)}, \boldsymbol{u}^{(m)})_k} = \left.\frac{\partial G}{\partial u_x}\right|_{(k,m+1/2)} + O(\Delta x^{2p}, \Delta t^2). \tag{5.33}$$

Thus we obtain the following estimate:

$$\frac{\delta G_{\mathrm{d}}}{\delta(\boldsymbol{u}^{(m+1)}, \boldsymbol{u}^{(m)})_k} = \frac{\partial G_{\mathrm{d}}}{\partial(\boldsymbol{u}^{(m+1)}, \boldsymbol{u}^{(m)})_k} - \delta_k^{\langle 1 \rangle, 2p} \frac{\partial G_{\mathrm{d}}}{\partial \delta(\boldsymbol{u}^{(m+1)}, \boldsymbol{u}^{(m)})_k}$$

$$= \left.\frac{\partial G}{\partial u}\right|_{(k,m+1/2)} - \delta_k^{\langle 1 \rangle, 2p}\left(\left.\frac{\partial G}{\partial u_x}\right|_{(k,m+1/2)}\right) + O(\Delta x^{2p}, \Delta t^2)$$

$$= \left.\frac{\delta G}{\delta u}\right|_{(k,m+1/2)} + O(\Delta x^{2p}, \Delta t^2). \tag{5.34}$$

□

A scheme for the real conservative PDEs 2 is defined as follows.

Scheme 5.2 (Spatially high-order conservative scheme for PDEs 2)
Let $U_k^{(0)} = u(k\Delta x, 0)$ be initial values. Then, a spatially high-order scheme for the PDEs 2 is given by, for $m = 0, 1, 2, \ldots$,

$$\frac{U_k^{(m+1)} - U_k^{(m)}}{\Delta t} = \left(\delta_k^{\langle 1 \rangle, 2p}\right)^{2s+1} \frac{\delta G_d}{\delta(\boldsymbol{U}^{(m+1)}, \boldsymbol{U}^{(m)})_k},$$
$$k = 0, \ldots, N-1. \qquad (5.35)$$

THEOREM 5.2 Discrete conservation property and orders of Scheme 5.2

Under the discrete periodic boundary condition (5.3), Scheme 5.2 is conservative in the sense that the inequality

$$J_d(\boldsymbol{U}^{(m)}) = J_d(\boldsymbol{U}^{(0)}), \quad m = 1, 2, 3, \ldots \qquad (5.36)$$

holds. Moreover, the scheme is $(2p, 2)$-th order.

PROOF The conservation property is proved as follows.

$$\frac{1}{\Delta t}\left(\sum_{k=0}^{N}{}''G_d(\boldsymbol{U}^{(m+1)})\Delta x - \sum_{k=0}^{N}{}''G_d(\boldsymbol{U}^{(m)})\Delta x\right)$$

$$= \sum_{k=0}^{N}{}''\frac{\delta G_d}{\delta(\boldsymbol{U}^{(m+1)}, \boldsymbol{U}^{(m)})_k}\left(\frac{U_k^{(m+1)} - U_k^{(m)}}{\Delta t}\right)\Delta x$$

$$= \sum_{k=0}^{N}{}''\frac{\delta G_d}{\delta(\boldsymbol{U}^{(m+1)}, \boldsymbol{U}^{(m)})_k}\left(\delta_k^{\langle 1 \rangle, 2p}\right)^{2s+1}\left(\frac{\delta G_d}{\delta(\boldsymbol{U}^{(m+1)}, \boldsymbol{U}^{(m)})_k}\right)\Delta x$$

$$= (-1)^s \sum_{k=0}^{N}{}''\left\{\left(\delta_k^{\langle 1 \rangle, 2p}\right)^s\left(\frac{\delta G_d}{\delta(\boldsymbol{U}^{(m+1)}, \boldsymbol{U}^{(m)})_k}\right)\right.$$

$$\left.\times \left(\delta_k^{\langle 1 \rangle, 2p}\right)^{s+1}\left(\frac{\delta G_d}{\delta(\boldsymbol{U}^{(m+1)}, \boldsymbol{U}^{(m)})_k}\right)\right\}\Delta x$$

$$= 0. \qquad (5.37)$$

The first equality is from (5.13), the second is from the definition of the scheme (5.35). and the third is from the repeated use of the summation-by-parts formula (5.8). (Note that the equality $\sum_{k=0}^{N}{}''U_k \delta_k^{\langle 1 \rangle, 2p} U_k \Delta x = 0$ holds for any N-periodic sequence U_k, which can be easily obtained from (5.8) as a special case.)

The claim on the orders can be proved in the same way in Theorem 5.1. □

5.2.3.2 For the Complex-Valued PDEs

We define finite difference schemes for the complex-valued dissipative PDEs 3 as follows.

Scheme 5.3 (Spatially high-order dissipative scheme for the PDEs 3)

Let $U_k^{(0)} = u(k\Delta x, 0)$ be initial values. Then, a spatially high-order scheme for the PDEs 3 is given by, for $m = 0, 1, 2, \ldots$,

$$\frac{U_k^{(m+1)} - U_k^{(m)}}{\Delta t} = -\left(\frac{\delta G_d}{\delta(\boldsymbol{U}^{(m+1)}, \boldsymbol{U}^{(m)})}\right)_k,$$
$$k = 0, \ldots, N-1. \qquad (5.38)$$

THEOREM 5.3 Discrete dissipation property and orders of Scheme 5.3

Under the discrete periodic boundary condition (5.3), Scheme 5.3 is dissipative in the sense that the inequality

$$J_d(\boldsymbol{U}^{(m+1)}) \leq J_d(\boldsymbol{U}^{(m)}), \quad m = 0, 1, 2, \ldots \qquad (5.39)$$

holds. Moreover, the scheme is $(2p, 2)$-th order.

PROOF The dissipation property is proved as follows.

$$\frac{1}{\Delta t}\left(\sum_{k=0}^{N}{}''G_{d,k}(\boldsymbol{U}^{(m+1)})\Delta x - \sum_{k=0}^{N}{}''G_{d,k}(\boldsymbol{U}^{(m)})\Delta x\right)$$

$$= \sum_{k=0}^{N}{}''\left\{\left(\frac{\delta G_d}{\delta(\boldsymbol{U}^{(m+1)}, \boldsymbol{U}^{(m)})}\right)_k \left(\frac{U_k^{(m+1)} - U_k^{(m)}}{\Delta t}\right)\right.$$

$$+ \left(\frac{\delta G_d}{\delta(\overline{\boldsymbol{U}^{(m+1)}}, \overline{\boldsymbol{U}^{(m)}})}\right)_k \left(\overline{\frac{U_k^{(m+1)} - U_k^{(m)}}{\Delta t}}\right)\right\}\Delta x$$

$$= -2\sum_{k=0}^{N}{}''\left|\left(\frac{\delta G_d}{\delta(\boldsymbol{U}^{(m+1)}, \boldsymbol{U}^{(m)})}\right)_k\right|^2 \Delta x \leq 0. \qquad (5.40)$$

The first equality is from (5.19), and the second is from the definition of Scheme 5.3 and the fact that the complex (discrete) variational derivatives are complex conjugates of each other.

The claim on the orders can be proved in the same way in Theorem 5.1. □

We also define a scheme for the conservative PDEs 4 as follows.

Scheme 5.4 (Spatially high-order conservative scheme for the PDEs 4)

Let $U_k^{(0)} = u(k\Delta x, 0)$ be initial values. Then, a spatially high-order scheme for the PDEs 4 is given by, for $m = 0, 1, 2, \ldots$,

$$i\left(\frac{U_k^{(m+1)} - U_k^{(m)}}{\Delta t}\right) = -\left(\frac{\delta G_d}{\delta(\boldsymbol{U}^{(m+1)}, \boldsymbol{U}^{(m)})}\right)_k,$$
$$k = 0, \ldots, N-1. \qquad (5.41)$$

The accuracy of this scheme is $O(\Delta t^2, \Delta x^{2p})$.

THEOREM 5.4 Discrete conservation property and orders of Scheme 5.4

Under the discrete periodic boundary condition (5.3), Scheme 5.4 is dissipative in the sense that the inequality

$$J_{\mathrm{d}}(\boldsymbol{U}^{(m+1)}) = J_{\mathrm{d}}(\boldsymbol{U}^{(0)}), \quad m = 1, 2, 3, \ldots \tag{5.42}$$

holds. Moreover, the scheme is $(2p, 2)$-th order.

PROOF The conservation property is proved as follows.

$$\frac{1}{\Delta t}\left(\sum_{k=0}^{N}{}''G_{\mathrm{d},k}(\boldsymbol{U}^{(m+1)})\Delta x - \sum_{k=0}^{N}{}''G_{\mathrm{d},k}(\boldsymbol{U}^{(m)})\Delta x\right)$$
$$= \sum_{k=0}^{N}{}''\left\{\left(\frac{\delta G_{\mathrm{d}}}{\delta(\boldsymbol{U}^{(m+1)},\boldsymbol{U}^{(m)})}\right)_k \left(\frac{U_k^{(m+1)} - U_k^{(m)}}{\Delta t}\right)\right.$$
$$\left. + \left(\overline{\frac{\delta G_{\mathrm{d}}}{\delta(\boldsymbol{U}^{(m+1)},\boldsymbol{U}^{(m)})}}\right)_k \left(\overline{\frac{U_k^{(m+1)} - U_k^{(m)}}{\Delta t}}\right)\right\}\Delta x$$
$$= \sum_{k=0}^{N}{}''\left\{\mathrm{i}\left|\left(\frac{\delta G_{\mathrm{d}}}{\delta(\boldsymbol{U}^{(m+1)},\boldsymbol{U}^{(m)})}\right)_k\right|^2 - \mathrm{i}\left|\left(\frac{\delta G_{\mathrm{d}}}{\delta(\boldsymbol{U}^{(m+1)},\boldsymbol{U}^{(m)})}\right)_k\right|^2\right\}\Delta x = 0. \tag{5.43}$$

The claim on the orders can be proved in the same way in Theorem 5.1. □

REMARK 5.2 As we have repeatedly mentioned so far, the derived schemes here can be implemented with variable time step sizes (observe that the dissipation or conservation theorems hold even if the temporal mesh size is changed during the time evolution process).

The schemes here are nonlinearly implicit in general due to the nonlinearity in the PDEs. We can, however, further modify the method for designing *linearly* implicit schemes, using the idea described in Chapter 6. We omit the details, but will present several examples in the next subsection. □

5.2.4 Application Examples

In this section we show some examples.

5.2.4.1 Application to the Korteweg–de Vries Equation

Let us consider the Korteweg–de Vries equation (KdV):

$$\frac{\partial u}{\partial t} = \frac{\partial}{\partial x}\left(\frac{1}{2}u^2 + \frac{\partial^2 u}{\partial x^2}\right), \qquad x \in (0, L),\ t > 0, \tag{5.44}$$

under the periodic boundary condition (5.2). The energy function G for the KdV is defined by

$$G(u, u_x) = \frac{1}{6}u^3 - \frac{1}{2}\left(\frac{\partial u}{\partial x}\right)^2. \tag{5.45}$$

We have already seen an energy-conserving scheme whose accuracy is $O(\Delta t^2, \Delta x^2)$ in Section 4.2.1. Here we derive a $(2p, 2)$th-order conservative scheme. We start by defining the discrete local energy as

$$G_{\mathrm{d},k}(\boldsymbol{U}) \stackrel{\mathrm{d}}{\equiv} \frac{U_k{}^3}{6} - \frac{(\delta_k^{\langle 1 \rangle, 2p} U_k)^2}{2}. \tag{5.46}$$

Note that it can be decomposed as assumed in (5.11). Mechanically calculating the discrete variational derivative using (5.14), (5.15a), and (5.15b), we obtain

$$\frac{\delta G_{\mathrm{d}}}{\delta(\boldsymbol{U},\boldsymbol{V})}\bigg|_k = \frac{U_k{}^2 + U_k V_k + V_k{}^2}{6} + \left(\delta_k^{\langle 1 \rangle, 2p}\right)^2 \left(\frac{U_k + V_k}{2}\right). \tag{5.47}$$

Then Scheme 5.1 reads

$$\frac{U_k{}^{(m+1)} - U_k{}^{(m)}}{\Delta t} = \delta_k^{\langle 1 \rangle, 2p} \left\{ \frac{(U_k{}^{(m+1)})^2 + U_k{}^{(m+1)} U_k{}^{(m)} + (U_k{}^{(m)})^2}{6} \right.$$
$$\left. + \left(\delta_k^{\langle 1 \rangle, 2p}\right)^2 \left(\frac{U_k{}^{(m+1)} + U_k{}^{(m)}}{2}\right) \right\}. \tag{5.48}$$

This scheme keeps a discrete energy thanks to Theorem 5.1. It also should be noted that, as mentioned in Section 5.2, the scheme is pseudospectral-like in the limit of $p \to \infty$.

REMARK 5.3 We can also construct linearly implicit schemes. Below is an example.

$$\frac{U_k{}^{(m+3)} - U_k{}^{(m)}}{3\Delta t} =$$
$$-\delta_k^{\langle 1 \rangle, 2p} \left\{ \frac{U_k{}^{(m+2)} U_k{}^{(m+1)}}{2} + \left(\delta_k^{\langle 1 \rangle, 2p}\right)^2 \left(\frac{U_k{}^{(m+3)} + U_k{}^{(m)}}{2}\right) \right\}. \tag{5.49}$$

This scheme conserves the discrete global energy:

$$\sum_{k=0}^{N-1}\left(\frac{U_k^{(m+3)}U_k^{(m+2)}U_k^{(m+1)}}{6}\right.$$
$$\left.-\frac{\left|\delta_k^{\langle 1\rangle,2p}U_k^{(m+3)}\right|^2+\left|\delta_k^{\langle 1\rangle,2p}U_k^{(m+2)}\right|^2+\left|\delta_k^{\langle 1\rangle,2p}U_k^{(m+1)}\right|^2}{6}\right)\Delta x. \tag{5.50}$$

□

5.2.4.2 Application to the Cubic Nonlinear Schrödinger Equation

Let us consider the cubic nonlinear Schrödinger equation(NLS):

$$\mathrm{i}\frac{\partial u}{\partial t}=-u_{xx}-\gamma|u|^2 u,\quad x\in(0,L), t>0, \gamma\in\mathbb{R}, \tag{5.51}$$

under the periodic boundary condition (5.2). The energy function G for the NLS is defined by

$$G(u,u_x)=-|u_x|^2+\frac{\gamma}{2}|u|^4.$$

Let us derive a $(2p,2)$th-order conservative scheme for the NLS. We first define the discrete local energy as

$$G_{\mathrm{d},k}(\boldsymbol{U})\stackrel{\mathrm{d}}{=}-|\delta_k^{\langle 1\rangle,2p}U_k|^2+\frac{\gamma}{2}|U_k|^4. \tag{5.52}$$

Note that it can be decomposed as assumed in (5.16). By mechanically calculating the complex discrete variational derivatives using (5.20), (5.21), and (5.22),(5.23), we have

$$\left(\frac{\delta G_{\mathrm{d}}}{\delta(\overline{\boldsymbol{U}^{(m+1)}},\overline{\boldsymbol{U}^{(m)}})}\right)_k=(\delta_k^{\langle 1\rangle,2p})^2\left(\frac{U_k^{(m+1)}+U_k^{(m)}}{2}\right)$$
$$+\gamma\left(\frac{|U_k^{(m+1)}|^2+|U_k^{(m)}|^2}{2}\right)\left(\frac{U_k^{(m+1)}+U_k^{(m)}}{2}\right). \tag{5.53}$$

Then Scheme 5.3 reads

$$\mathrm{i}\left(\frac{U_k^{(m+1)}-U_k^{(m)}}{\Delta t}\right)=-(\delta_k^{\langle 1\rangle,2p})^2\left(\frac{U_k^{(m+1)}+U_k^{(m)}}{2}\right)$$
$$-\gamma\left(\frac{|U_k^{(m+1)}|^2+|U_k^{(m)}|^2}{2}\right)\left(\frac{U_k^{(m+1)}+U_k^{(m)}}{2}\right). \tag{5.54}$$

Advanced Topic I: Design of High-Order Schemes

The conservation Theorem 5.4 holds, namely,

$$\sum_{k=0}^{N} {}''G_{d,k}(\boldsymbol{U}^{(m)})\Delta x \qquad (5.55)$$

remains constant. Moreover, the scheme conserves the discrete "probability":

$$\sum_{k=0}^{N} {}''|U_k{}^{(m)}|^2 \Delta x. \qquad (5.56)$$

This conservation property is a discrete analogue of the continuous probability conservation law: $\int_0^L |u|^2 \mathrm{d}x = \mathrm{const}$. With these properties the scheme is shown to be stable and L^2-convergent.

REMARK 5.4 By employing the linearization technique in Chapter 6, we can also construct a linearly implicit version of the scheme as follows. We omit the detail of the derivation here.

$$\mathrm{i}\left(\frac{U_k{}^{(m+2)} - U_k{}^{(m)}}{2\Delta t}\right) =$$
$$- \left(\delta_k^{\langle 1 \rangle, 2p}\right)^2 \left(\frac{U_k{}^{(m+2)} + U_k{}^{(m)}}{2}\right) - \gamma |U_k{}^{(m+1)}|^2 \left(\frac{U_k{}^{(m+2)} + U_k{}^{(m)}}{2}\right). \qquad (5.57)$$

This scheme preserves the discrete global energy:

$$\sum_{k=0}^{N} {}'' \left\{ -\frac{\left|\delta_k^{\langle 1 \rangle, 2p} U_k{}^{(m+1)}\right|^2 + \left|\delta_k^{\langle 1 \rangle, 2p} U_k{}^{(m)}\right|^2}{2} + \frac{\gamma}{2} \left|U_k{}^{(m+1)}\right|^2 \left|U_k{}^{(m)}\right|^2 \right\} \Delta x, \qquad (5.58)$$

and the discrete probability:

$$\sum_{k=0}^{N} {}'' \frac{|U_k{}^{(m+1)}|^2 + |U_k{}^{(m)}|^2}{2} \Delta x. \qquad (5.59)$$

□

Now we present some numerical results. We compare the following four numerical schemes:

- (CPS): the scheme (5.54) with $p = \infty$ (i.e., a pseudospectral-like scheme)

TABLE 5.1: Characteristics of the tested numerical schemes

name	t	x	energy	prob.
(CPS)	$O(\Delta t^2)$	spectral	yes	yes
(CFD)	$O(\Delta t^2)$	$O(\Delta x^2)$	yes	yes
(CLI)	$O(\Delta t^2)$	spectral	yes	yes
(RK4)	$O(\Delta t^4)$	spectral	no	no

- (CFD): the second-order scheme shown in Section 4.4.1
- (CLI): the linearly-implicit scheme (5.57)
- (RK4): the standard pseudospectral scheme which uses the 4th order Runge–Kutta method

We summarize their characteristics in Table 5.1.[2] In (CPS) and (CFD), we solved the nonlinear schemes (5.54) by simple iterations; i.e., we solved (5.54) by the iteration:

$$U_k^{(m+1),j+1} = U_k^{(m)} + \frac{\Delta t}{\mathrm{i}} \left\{ -(\delta_k^{\langle 1 \rangle, 2p})^2 \left(\frac{U_k^{(m+1),j} + U_k^{(m)}}{2} \right) \right.$$
$$\left. -\gamma \left(\frac{|U_k^{(m+1),j}|^2 + |U_k^{(m)}|^2}{2} \right) \left(\frac{U_k^{(m+1),j} + U_k^{(m)}}{2} \right) \right\}, \quad (5.60)$$

where $j \in \{0, 1, 2, \ldots\}$ is the index of the iteration.

We take up the exact periodic solution (for $\gamma > 0$):

$$u(x,t) = \sqrt{\frac{\lambda + \alpha}{\gamma}} \, \mathrm{cn}[\sqrt{\alpha}(x - vt), k] \exp\left[\mathrm{i}\frac{v}{2}x - \mathrm{i}\left(\frac{v^2}{4} - \lambda \right) t \right]. \quad (5.61)$$

Here, $\mathrm{cn}(x,k)$ is the Jacobi elliptic function of modulus k, and λ, α, v, k, L (the length of the domain) are constants. The constants are chosen to satisfy the relations: $\alpha = \sqrt{2\gamma A + \lambda^2}$, $k = \sqrt{(\lambda + \alpha)/(2\alpha)}$, $L = 4K(k)/\sqrt{\alpha}$ and $v = 4m\pi/L$ (m: an integer), where $K(k)$ is the complete elliptic integral of the first kind of modulus k to meet the periodic boundary condition (5.2). Thus we have only three free constants $\lambda, A \in \mathbb{R}$ and $m \in \mathbb{Z}$. In the following experiments, we set $\gamma = 2$, $A = 1.0 \times 10^{-5}$, and $\lambda = 1$ (so that $\alpha \simeq 1.00002$, $k \simeq 0.999990$, $L \simeq 28.5708$). The constant m, which determines the "speed"

[2-6] Reprinted from T. Matsuo, M. Sugihara, D. Furihata and M. Mori, Spatially accurate dissipative or conservative finite difference schemes derived by the discrete variational method, *Japan J. Indust. Appl. Math.*, 19, 311–330, Copyright (2002), with permission from JJIAM publishing committee.

of the solution, is chosen in each experiment. As m increases, the solution becomes "rapid" and harder to capture by numerical calculation. We carried out all numerical computations with double precision.

Firstly, we compare the accuracies of the numerical solutions. In the experiment we choose $v = 1$ ("slow" solution; temporal period $T_p \simeq 64.9583$). Table 5.2[3] shows the maximal errors at $t = T = 10$ (i.e., at about 1/6 period). We denote the exact solution by $\boldsymbol{u} = (u(T, x_0), u(T, x_1), \ldots, u(T, x_{N-1}))^\mathrm{T}$ at $t = T$, and the numerical solution by $\boldsymbol{U}^{(M)}$ where $M = T/\Delta t$. The (*) mark means that the result is almost at the best order which can be attained with the Δx value (i.e., the order cannot be improved by decreasing Δt). According to the marked result, we can confirm that the scheme (CPS) is spatially highly accurate as expected; the error decreases almost exponentially with regard to N, while the error by the scheme (CFD) decreases only at the rate of $O(\Delta x^2)$. The superiority becomes far more significant as N grows. The temporal accuracy of the scheme (CPS) is, however, still of $O(\Delta t^2)$ and may discourage us from utilizing the scheme in view of, for example, higher order schemes such as the Runge–Kutta scheme (RK4). This difficulty can be overcome by the composition technique. The scheme (CLI) gives as good a result as (CPS).

TABLE 5.2: Maximal errors at $T = 10$

Scheme	Δx (N)	Δt	$\|\boldsymbol{U}^{(M)} - \boldsymbol{u}\|_\infty$
(CPS)	0.4464 (64)	0.001	3.85×10^{-4} (*)
	0.2232 (128)	0.01	1.40×10^{-4}
		0.001	1.40×10^{-6}
		0.0001	1.92×10^{-8}
		0.00001	6.58×10^{-9} (*)
	0.1116 (256)	0.001	1.40×10^{-6}
		0.0001	1.40×10^{-8}
		0.00001	1.41×10^{-10}
		0.000001	1.08×10^{-12} (*)
(CFD)	0.4464 (64)	0.001	5.00×10^{-1} (*)
	0.2232 (128)	0.001	1.14×10^{-1} (*)
	0.1116 (256)	0.001	2.80×10^{-2} (*)
(CLI)	0.4464 (64)	0.001	3.85×10^{-4} (*)
	0.2232 (128)	0.00001	8.05×10^{-9} (*)
	0.1116 (256)	0.000001	2.33×10^{-12} (*)
(RK4)	0.4464 (64)	0.01	3.84×10^{-4} (*)
	0.2232 (128)	0.001	8.01×10^{-9} (*)
	0.1116 (256)	0.0001	3.58×10^{-12} (*)

Secondly, we compare the qualitative aspects of the numerical solutions. For this purpose we use a "rapid" solution: $v = 100$ (i.e., the temporal period $T_p \simeq 0.6496$), and see if the schemes can correctly preserve the soliton-like waveform of the solution. We carried out numerical calculations using the above schemes for $0 \leq t \leq T = 10$ (i.e., about 16 periods) with $\Delta t = 0.001$ and $\Delta x = 0.1116$ ($N = 256$). Figure 5.1[4] shows the progression results. We plotted the absolute values of the numerical solutions, i.e., $|U_k^{(m)}|$ against (t, x). Note that the graphs are drawn only at selected time points ($t = 0, 2, 4, 6, 8, 10$) to increase readability, therefore they do not reflect the right wave speed; the exact solution goes around the interval about 16 times, as stated earlier. Apparently the results by (CFD) and (RK4) are "qualitatively" wrong; they fail to preserve the soliton-like waveform of the solution. In other words, the mesh sizes are too coarse for them to catch up with the solution. The result by (CLI) is better, but not satisfactory. The results by (CPS) successfully preserve the waveform, and in that sense we can say that they achieve a qualitatively right solution. The reason for the failure of (CFD) is the lack of spatial resolution (as compared to (CPS)). The reason in (RK4) is the lack of conservation properties; Figure 5.2[5] and Figure 5.3[6] tell this clearly. In Figure 5.2 we plot the calculated discrete energy (5.55) for (CPS), (CFD), (RK4), and the energy (5.58) for (CLI). In Figure 5.3 we plot the calculated discrete probability (5.56) for (CPS), (CFD), (RK4), and the probability (5.59) for (CLI). In both graphs, the result by (RK4) does not conserve the invariants, while the others conserve them strictly. The readers may notice that the discrete energy of (CFD) differs from that of the other conservative schemes; this is due to the difference in the expressions of the discrete energy (5.52) ($p = \infty$ in (CPS),(CLI) and $p = 1$ in (CFD)).

As a result, we can conclude that the energy-conserving pseudospectral scheme (CPS) is highly accurate in space as expected, and qualitatively reliable.

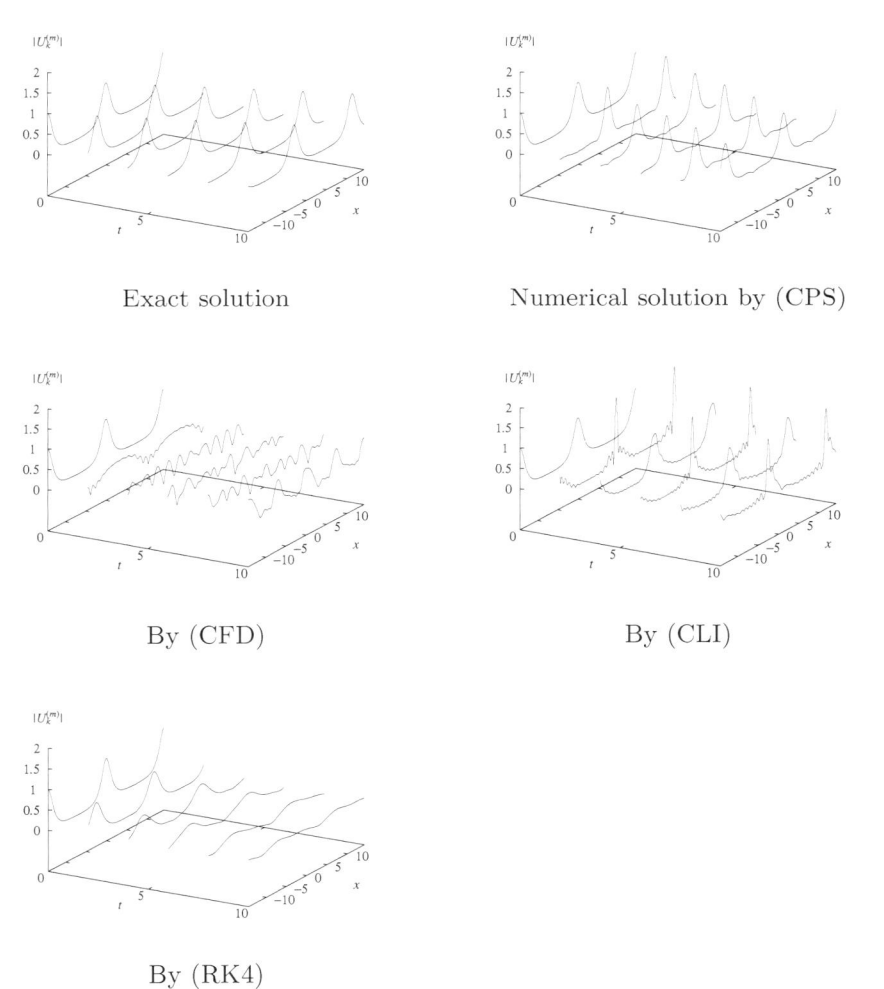

FIGURE 5.1: Evolution of numerical solutions.

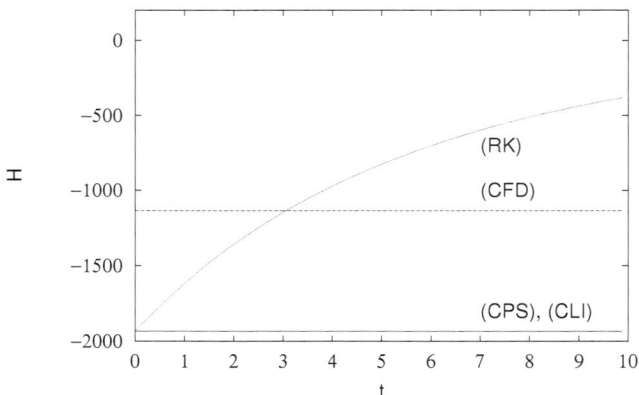

FIGURE 5.2: Conservation of the energy.

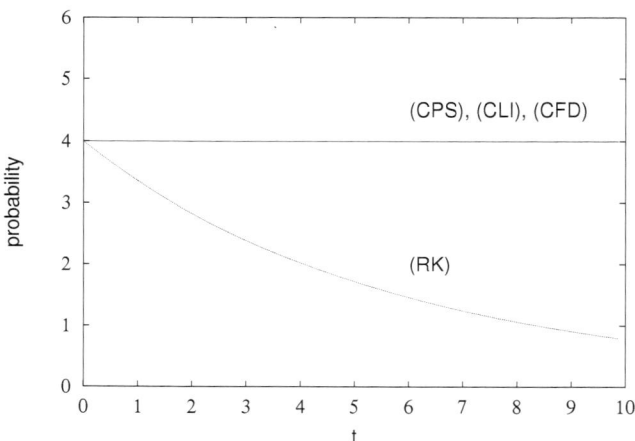

FIGURE 5.3: Conservation of the probability.

5.3 Temporally High-Order Schemes: Composition Method

The temporal accuracy of a conservative scheme can be increased by the the so-called "composition method." The method was originally proposed by Suzuki[157] and Yoshida[168], and then further developed by many researchers including Qin and Zhu[143]. Borrowing the description from Qin and Zhu[143], we summarize the most basic version of the composition method in the next proposition.

PROPOSITION 5.1 The composition method [143]
Let a system of ODEs

$$\frac{d}{dt}z(t) = f(z(t)), \quad t > 0,$$

be given, and $\phi(\Delta t)$ be a self-adjoint integrator of order $2n$, i.e.,

$$z(t + \Delta t) = \phi(\Delta t)z(t) + O(\Delta t^{2n+1}).$$

Let c_1, c_2 be constants which satisfy

$$2c_1^{2n+1} + c_2^{2n+1} = 0, \ 2c_1 + c_2 = 1.$$

Then the combination of integrators $\phi(c_1 \Delta t)\phi(c_2 \Delta t)\phi(c_1 \Delta t)$ is an integrator of order $2n + 2$.

By "self-adjoint integrator" we mean an integrator satisfying $\phi(-\Delta t) = \phi^{-1}(\Delta t)$; roughly-speaking, it means that the integrator is time-symmetric.

Suppose we have a second-order self-adjoint scheme. We can increase the order of the scheme to any order by repeatedly composing the scheme; from second-order to fourth-order, from fourth-order to sixth-order, and so on. We summarize the coefficients c_1, c_2 appearing in each step of composition in Table 5.3. In the table, N_2 is the number of second-order schemes in one step of the composed scheme; for example, in one step of the sixth-order scheme we use the second-order scheme nine times with different time steps. The quantity $\bar{c}$ will be discussed below.

The composition method is a very convenient and elegant way to obtain high-order schemes. Furthermore, obviously it maintains discrete conservation law.

PROPOSITION 5.2 Conservation property in composed schemes
Suppose a second-order self-adjoint scheme has a conservation law

$$I(\boldsymbol{U}^{(m)}) = I(\boldsymbol{U}^{(0)}), \quad m = 1, 2, \ldots, \tag{5.62}$$

TABLE 5.3: The coefficients in the standard composition method

stage	c_1	c_2	$\bar{c}$	N_2
2nd to 4th	1.351	−1.702	−1.702	3
4th to 6th	1.175	−1.349	−2.000	9
6th to 8th	1.116	−1.232	−2.829	27
8th to 10th	1.087	−1.174	−3.075	81

where $I(\boldsymbol{U}^{(m)})$ is a real-valued scalar function of $\boldsymbol{U}^{(m)}$. Then the high-order schemes obtained by composing the second-order scheme also keep $I(\boldsymbol{U}^{(m)})$ as their invariants.

Thus we can obtain temporally higher-order conservative schemes by the composition method. One difficulty in this approach may be the computational cost. By the standard composition method, we need to call the second-order scheme 3^{n-1} times to obtain a $2n$th-order scheme (refer N_2 in Table 5.3). This may be unbearably heavy if we count that the second-order scheme as nonlinear and we need iterative solvers in *every* second-order step required in one step of $2n$th-order scheme. Many efforts have been devoted to improve this situation. Now we have a wide variety of composition methods, some of which are designed to require as few second-order steps as possible. Interested readers may refer, for example, to McLachlan [125].

On the other hand, when we consider dissipative problems, things get even worse. It is proved that in *every* composition method there must be at least one step where the time step should be *negative*. In Table 5.3, $\bar{c}$ represents the width of the largest negative time step appearing in one step of the composed high-order scheme. For example, in the fourth-order scheme the largest negative time step is $-1.702\Delta t$, and in the sixth scheme $-2.000\Delta t$. The negative time step can destroy the overall dissipation property, even if we use a dissipative second-order scheme as the base scheme. So as far as we are interested in strictly dissipative schemes, we cannot adopt the composition method to obtain higher-order schemes.

5.4 Temporally High-Order Schemes: High-Order Discrete Variational Derivatives

The second approach to obtain temporally high-order schemes is to introduce a new concept, the "temporally high-order discrete variational derivative." Below, we discuss the following cases in order:

[A] $(1, q)$th- or $(2, q)$th-order schemes for the real-valued PDEs 1 and 2, i.e., temporally high-order versions of Scheme 3.1 and Scheme 3.2 (pages 80 and 84);

[B] (p, q)th-order schemes for the real-valued PDEs 1 and 2 (under the periodic boundary condition), i.e., temporally high-order versions of Scheme 5.1 and Scheme 5.2 (pages 233 and 236);

[C] $(1, q)$th- or $(2, q)$th-order schemes for the complex-valued PDEs 3 and 4, i.e., temporally high-order versions of Scheme 3.3 and Scheme 3.4 (pages 96 and 99);

[D] (p, q)th-order schemes for the complex-valued PDEs 3 and 4 (under the periodic boundary condition), i.e., temporally high-order versions of Scheme 5.3 and Scheme 5.4 (both on page 237).

5.4.1 Discrete Symbols

We use the following qth-order temporal difference operator throughout this section.

DEFINITION 5.3 q**th-order temporal difference operator**   Let $V^{(m)}$ ($m = 0, 1, 2, \ldots$) be a sequence. Let $\delta_{m;c}^{\langle 1 \rangle}$ be a temporal difference operator defined as

$$\delta_{m;c}^{\langle 1 \rangle} V^{(m)} \equiv \frac{1}{\Delta t} \sum_{i=-l_1}^{l_2} c_i V^{(m)}, \tag{5.63}$$

where $l_1, l_2 \in \{0, 1, 2, \ldots\}$, $c_i \in \mathbb{R}$, and $\mathbf{c} = (c_{-l_1}, \ldots, c_{l_2})$. We call $l = l_1 + l_2 - 1$ the number of the referenced points. If for any sufficiently smooth function $v(t) : \mathbb{R} \to \mathbb{R}$, there exists $\widetilde{t}(t, \Delta t, l_1, l_2)$ such that the estimate:

$$\frac{1}{\Delta t} \sum_{i=-l_1}^{l_2} c_i v(t + i\Delta t) = \left. \frac{d}{dt} v \right|_{t=\widetilde{t}} + O(\Delta t^q) \qquad \Delta t \to 0, \tag{5.64}$$

holds, $\delta_{m;c}^{\langle 1 \rangle}$ is called "qth-order at $\widetilde{t}$." When q is the highest order attained by varying $\widetilde{t}$, we call $\delta_{m;c}^{\langle 1 \rangle}$ a "qth-order difference operator," and denote it by $\delta_{m;c}^{\langle 1 \rangle, q}$. □

The subscript $\mathbf{c}$ is often omitted where no confusion occurs. The simplest example is

$$\delta_m^{\langle 1 \rangle, 2} V^{(m)} = \frac{V^{(m+1)} - V^{(m)}}{\Delta t},$$

i.e., the forward difference operator. Though it may seem a first-order difference operator, it is called "second-order" according to Definition 5.3. In fact, by substituting $v(m\Delta t)$ into $V^{(m)}$ and considering Taylor expansion at $\widetilde{t} = (m+\frac{1}{2})\Delta t$, we obtain an estimate:

$$\delta_m^{\langle 1 \rangle, 2} v(m\Delta t) = \left.\frac{\mathrm{d}}{\mathrm{d}t} v\right|_{t=(m+\frac{1}{2})\Delta t} + O(\Delta t^2).$$

Another example is

$$\delta_m^{\langle 1 \rangle, 4} v^{(m)} = \frac{1}{\Delta t}\left(-\frac{1}{24}v^{(m+2)} + \frac{9}{8}v^{(m+1)} - \frac{9}{8}v^{(m)} + \frac{1}{24}v^{(m-1)}\right),$$

which is of fourth-order (the fourth-order accuracy is attained at $\widetilde{t} = (m+1/2)\Delta t$).

5.4.2 Central Idea for High-Order Discrete Derivative

Since the notation will be quite complicated in the following sections, here we briefly summarize the central idea for temporally high-order schemes, so that the readers can get an idea of what is going on there.

Recall, for example, Scheme 3.1:

$$\frac{U_k^{(m+1)} - U_k^{(m)}}{\Delta t} = (-1)^{s+1} \delta_k^{\langle 2s \rangle} \frac{\delta G_\mathrm{d}}{\delta(\boldsymbol{U}^{(m+1)}, \boldsymbol{U}^{(m)})_k}. \qquad (5.65)$$

As in the proof of Theorem 5.1, we can prove that the scheme is temporally second-order (and spatially first- or second-order) by considering Taylor expansion of both sides of the scheme at $(x, t) = (k\Delta x, (m+\frac{1}{2})\Delta t)$. Note that the left-hand side is nothing but the temporally second-order difference operator $\delta_m^{\langle 1 \rangle, 2}$ in our new notation (see above), which implies that the resulting scheme is temporally second-order.

The idea for designing temporally high-order schemes is to use the qth-order temporal difference operators $\delta_m^{\langle 1 \rangle, q}$ instead of the second-order operator $\delta_m^{\langle 1 \rangle, 2}$, and search for the schemes of the form

$$\delta_m^{\langle 1 \rangle, q} U_k^{(m)} = (-1)^{s+1} \delta_k^{\langle 2s \rangle} \frac{\delta G_\mathrm{d}}{\delta(\boldsymbol{U}^{(m+l_2)}, \ldots, \boldsymbol{U}^{(m-l_1)})_k}. \qquad (5.66)$$

The complicated discrete symbol on the right-hand side is a discretization of $\delta G/\delta u$, which will be introduced in the next section. Because the left-hand side is a temporally qth-order approximation of $\frac{\mathrm{d}}{\mathrm{d}t} u(k\Delta x, \widetilde{t})$ ($\widetilde{t}$ is defined in Definition 5.3 as to $\delta_m^{\langle 1 \rangle, q}$), if the right-hand side is also a temporally qth-order approximation of the variational derivative at $(x, t) = (k\Delta x, \widetilde{t})$, the scheme must be temporally qth-order as a whole. In the following sections, we show that in fact we can find such an approximation.

The key tool there is the high-order operator $\delta_m^{\langle 1 \rangle, q}$ and its associated "discrete chain rule." They were first suggested in Matsuo [114] (see also [115]), where high-order schemes for dissipative or conservative ODEs had been considered. Before proceeding to the PDE cases, we briefly summarize the ODE case.

Suppose we hope to solve a system of ODEs:

$$\frac{\mathrm{d}}{\mathrm{d}t}\boldsymbol{z}(t) = A\nabla H(\boldsymbol{z}), \tag{5.67}$$

where $\boldsymbol{z} \in \mathbb{R}^N$, $H : \mathbb{R}^N \to \mathbb{R}$. Assume that A is skew-symmetric, i.e., for any $\boldsymbol{x}, \boldsymbol{y}$, $\boldsymbol{x} \cdot A\boldsymbol{y} = -(A\boldsymbol{x}) \cdot \boldsymbol{y}$. The symbol "·" denotes the inner product in $\mathbb{R}^N$. Then H is an invariant:

$$\frac{\mathrm{d}}{\mathrm{d}t}H(\boldsymbol{z}(t)) = (\nabla H(\boldsymbol{z})) \cdot \frac{\mathrm{d}\boldsymbol{z}}{\mathrm{d}t} = (\nabla H(\boldsymbol{z})) \cdot A(\nabla H(\boldsymbol{z})) = 0.$$

The last equality follows from the skew-symmetry of A.

Now let us introduce a "discrete derivative":

$$\nabla_\mathrm{d}^q H(\boldsymbol{z}^{(m)}) \stackrel{\mathrm{d}}{=} \nabla H(\widetilde{\boldsymbol{z}}) + \frac{\delta_m^{\langle 1 \rangle,q} H(\boldsymbol{z}^{(m)}) - (\nabla H(\widetilde{\boldsymbol{z}})) \cdot \delta_m^{\langle 1 \rangle,q} \boldsymbol{z}^{(m)}}{\|\delta_m^{\langle 1 \rangle,q} \boldsymbol{z}^{(m)}\|_2^2} \delta_m^{\langle 1 \rangle,q} \boldsymbol{z}^{(m)}, \tag{5.68}$$

where $\widetilde{\boldsymbol{z}}$ is some mean value of $\boldsymbol{z}^{(m)}$'s. Observe that the fraction is just a scalar. The fraction is, roughly speaking, of order $O(\Delta t^q)$ (provided that we choose $\widetilde{\boldsymbol{z}}$ appropriately), and the discrete derivative is a combination of ∇H, which is a "true" derivative at $\widetilde{\boldsymbol{z}}$, and the correction term of $O(\Delta t^q)$. The latter is needed for the following discrete chain rule.

$$\delta_m^{\langle 1 \rangle,q} H(\boldsymbol{z}^{(m)}) = (\nabla_\mathrm{d}^q H(\boldsymbol{z}^{(m)})) \cdot \delta_m^{\langle 1 \rangle,q} \boldsymbol{z}^{(m)}. \tag{5.69}$$

This can be easily verified by multiplying both sides of (5.68) by $\delta_m^{\langle 1 \rangle,q} \boldsymbol{z}^{(m)}$. This is a high-order generalization of Gonzalez's discrete gradient introduced in [74]. With the high-order discrete derivative, we can construct a high-order conservative scheme as follows.

$$\delta_m^{\langle 1 \rangle,q} \boldsymbol{z}^{(m)} = A \nabla_\mathrm{d}^q H(\boldsymbol{z}^{(m)}). \tag{5.70}$$

This is in fact $O(\Delta t^q)$, and enjoys the conservation:

$$\delta_m^{\langle 1 \rangle,q} H(\boldsymbol{z}^{(m)}) = 0.$$

We can do a similar thing also in the PDE cases. But we have to consider the summation-by-parts formula at the same time, which makes the discussion (and notation) cumbersome.

5.4.3 Temporally High-Order Discrete Variational Derivative and Design of Schemes

Now let us turn to the PDE cases. In what follows we define a temporally high-order discrete variational derivative, and present the method for designing temporally high-order dissipative or conservative schemes, for the cases [A] to [D], in this order.

5.4.3.1 [A] $(1,q)$th- or $(2,q)$th-Order Schemes for the Real-Valued PDEs

In this subsection, $(1,q)$th- or $(2,q)$th-order schemes for the real-valued PDEs 1 and 2 are presented.

5.4.3.1.1 Discrete variational derivative
For an energy function $G(u, u_x)$, we introduce a function $\widetilde{G}(u, v, w)$ by

$$\widetilde{G}(u, v, w) \stackrel{\mathrm{d}}{\equiv} \frac{1}{2}\left(G(u, v) + G(u, w)\right), \tag{5.71}$$

and define a discrete energy function $G_{\mathrm{d}}(\boldsymbol{U}^{(m)})$ by

$$G_{\mathrm{d},k}(\boldsymbol{U}^{(m)}) \stackrel{\mathrm{d}}{\equiv} \widetilde{G}(U_k^{(m)}, \delta_k^+ U_k^{(m)}, \delta_k^- U_k^{(m)}). \tag{5.72}$$

In this way the discrete energy function is related to the original energy function. This makes our analysis of accuracy easier. We also define an associated global energy by

$$J_{\mathrm{d}}(\boldsymbol{U}^{(m)}) \stackrel{\mathrm{d}}{\equiv} \sum_{k=0}^{N}{}'' G_{\mathrm{d},k}(\boldsymbol{U}^{(m)}) \Delta x. \tag{5.73}$$

Next, we consider the discrete variation using the qth-order difference operator:

$$\delta_m^{\langle 1 \rangle, q} J_{\mathrm{d}}(\boldsymbol{U}^{(m)}) = \sum_{k=0}^{N}{}'' \left(\frac{\delta G_{\mathrm{d}}}{\delta(\boldsymbol{U}^{(m+l_2)}, \ldots, \boldsymbol{U}^{(m-l_1)})_k} \cdot \delta_m^{\langle 1 \rangle, q} U_k^{(m)} \right) \Delta x$$
$$+ B_{\mathrm{tr},1}(\boldsymbol{U}^{(m+l_2)}, \ldots, \boldsymbol{U}^{(m-l_1)}), \tag{5.74}$$

where

$$\frac{\delta G_{\mathrm{d}}}{\delta(\boldsymbol{U}^{(m+l_2)}, \ldots, \boldsymbol{U}^{(m-l_1)})_k} \stackrel{\mathrm{d}}{\equiv} \frac{\partial G_{\mathrm{d}}^q}{\partial(\boldsymbol{U}^{(m+l_2)}, \ldots, \boldsymbol{U}^{(m-l_1)})_k}$$
$$- \delta_k^- \left(\frac{\partial G_{\mathrm{d}}^q}{\partial \delta^+ (\boldsymbol{U}^{(m+l_2)}, \ldots, \boldsymbol{U}^{(m-l_1)})_k} \right)$$
$$- \delta_k^+ \left(\frac{\partial G_{\mathrm{d}}^q}{\partial \delta^- (\boldsymbol{U}^{(m+l_2)}, \ldots, \boldsymbol{U}^{(m-l_1)})_k} \right), \tag{5.75a}$$

$$\frac{\partial G_{\mathrm{d}}^q}{\partial(\boldsymbol{U}^{(m+l_2)}, \ldots, \boldsymbol{U}^{(m-_1 l)})_k} \stackrel{\mathrm{d}}{\equiv} \left.\frac{\partial \widetilde{G}}{\partial u}\right|_{V_k} + r \delta_m^{\langle 1 \rangle, q} U_k^{(m)}, \tag{5.75b}$$

$$\frac{\partial G_{\mathrm{d}}^q}{\partial \delta^+ (\boldsymbol{U}^{(m+l_2)}, \ldots, \boldsymbol{U}^{(m-l_1)})_k} \stackrel{\mathrm{d}}{\equiv} \left.\frac{\partial \widetilde{G}}{\partial v}\right|_{V_k} + r \delta_m^{\langle 1 \rangle, q} (\delta_k^+ U_k^{(m)}), \tag{5.75c}$$

$$\frac{\partial G_{\mathrm{d}}^q}{\partial \delta^- (\boldsymbol{U}^{(m+l_2)}, \ldots, \boldsymbol{U}^{(m-l_1)})_k} \stackrel{\mathrm{d}}{\equiv} \left.\frac{\partial \widetilde{G}}{\partial w}\right|_{V_k} + r \delta_m^{\langle 1 \rangle, q} (\delta_k^- U_k^{(m)}), \tag{5.75d}$$

$$r \stackrel{\mathrm{d}}{\equiv} \frac{r_{\mathrm{n}}}{r_{\mathrm{d}}}, \tag{5.76a}$$

$$r_{\mathrm{n}} \stackrel{\mathrm{d}}{\equiv} \delta_m^{\langle 1 \rangle, q} J_{\mathrm{d}}(\boldsymbol{U}^{(m)}) - \sum_{k=0}^{N} {}'' \left.\frac{\partial \widetilde{G}}{\partial u}\right|_{V_k} \delta_m^{\langle 1 \rangle, q} U_k{}^{(m)} \Delta x$$

$$- \sum_{k=0}^{N} {}'' \left.\frac{\partial \widetilde{G}}{\partial v}\right|_{V_k} (\delta_m^{\langle 1 \rangle, q} \delta_k^+ U_k{}^{(m)}) \Delta x - \sum_{k=0}^{N} {}'' \left.\frac{\partial \widetilde{G}}{\partial w}\right|_{V_k} (\delta_m^{\langle 1 \rangle, q} \delta_k^+ U_k{}^{(m)}) \Delta x, \tag{5.76b}$$

$$r_{\mathrm{d}} \stackrel{\mathrm{d}}{\equiv} \sum_{k=0}^{N} {}'' \left\{ \left(\delta_m^{\langle 1 \rangle, q} U_k{}^{(m)}\right)^2 + \left(\delta_m^{\langle 1 \rangle, q} \delta_k^+ U_k{}^{(m)}\right)^2 + \left(\delta_m^{\langle 1 \rangle, q} \delta_k^- U_k{}^{(m)}\right)^2 \right\} \Delta x, \tag{5.76c}$$

$$B_{\mathrm{tr},1}(\boldsymbol{U}^{(m+l_2)}, \ldots, \boldsymbol{U}^{(m-l_1)}) =$$
$$\frac{1}{2} \left[\frac{\partial G_{\mathrm{d}}^q}{\partial \delta^+ (\boldsymbol{U}^{(m+l_2)}, \ldots, \boldsymbol{U}^{(m-l_1)})_k} (s_k^+ \delta_m^{\langle 1 \rangle, q} U_k{}^{(m)}) \right.$$
$$+ \left(s_k^- \frac{\partial G_{\mathrm{d}}^q}{\partial \delta^+ (\boldsymbol{U}^{(m+l_2)}, \ldots, \boldsymbol{U}^{(m-l_1)})_k} \right) \delta_m^{\langle 1 \rangle, q} U_k{}^{(m)}$$
$$+ \frac{\partial G_{\mathrm{d}}^q}{\partial \delta^- (\boldsymbol{U}^{(m+l_2)}, \ldots, \boldsymbol{U}^{(m-l_1)})_k} (s_k^- \delta_m^{\langle 1 \rangle, q} U_k{}^{(m)})$$
$$\left. + \left(s_k^+ \frac{\partial G_{\mathrm{d}}^q}{\partial \delta^- (\boldsymbol{U}^{(m+l_2)}, \ldots, \boldsymbol{U}^{(m-l_1)})_k} \right) \delta_m^{\langle 1 \rangle, q} U_k{}^{(m)} \right]_0^N. \tag{5.77}$$

In the above definitions

$$(\cdot)|_{V_k} \stackrel{\mathrm{d}}{\equiv} (\cdot)|_{u=V_k, \, v=\delta_k^+ V_k, \, w=\delta_k^- V_k},$$

where V_k is a function of $\{U_k{}^{(m)}\}$:

$$V_k \stackrel{\mathrm{d}}{\equiv} \sum_{i=-l_1}^{l_2} d_i U_k^{(m+i)}, \qquad d_i \in \mathbb{R}, \tag{5.78}$$

such that for any sufficiently smooth function $f(t)$ defined on $[(m-l_1)\Delta t, (m+l_2)\Delta t]$,

$$\sum_{i=-l_1}^{l_2} d_i f((m+i)\Delta t) = f|_{t=\tilde{t}} + O(\Delta t^q). \tag{5.79}$$

The discrete quantity

$$\frac{\delta G_{\mathrm{d}}}{\delta (\boldsymbol{U}^{(m+l_2)}, \ldots, \boldsymbol{U}^{(m-l_1)})_k},$$

defined in (5.75a) is an approximation of the original variational derivative. It can be checked that it in fact satisfies the discrete variation relation (5.74). In order to see this, let us introduce an extended vector $\boldsymbol{z}^{(m)}$ of length $3(N+1)$ by

$$\left(U_0^{(m)}, \ldots, U_N^{(m)},\ \delta_k^+ U_0^{(m)}, \ldots, \delta_k^+ U_N^{(m)},\ \delta_k^- U_0^{(m)}, \ldots, \delta_k^- U_N^{(m)} \right)^\top,$$

and regard $J_\mathrm{d}(\boldsymbol{U}^{(m)})$ as $J_\mathrm{d}(\boldsymbol{z}^{(m)})$ (i.e., as a function on the extended vector.) Then it is an easy exercise to check that the discrete derivative (5.68) on this expression coincides with

$$\nabla_\mathrm{d}^q J_\mathrm{d}(\boldsymbol{z}^{(m)}) =$$
$$\left(\frac{\partial G_\mathrm{d}^q}{\partial(\cdots)_0}, \ldots, \frac{\partial G_\mathrm{d}^q}{\partial(\cdots)_N},\ \frac{\partial G_\mathrm{d}^q}{\partial \delta^+(\cdots)_0}, \ldots, \frac{\partial G_\mathrm{d}^q}{\partial \delta^+(\cdots)_N},\ \frac{\partial G_\mathrm{d}^q}{\partial \delta^-(\cdots)_0}, \ldots, \frac{\partial G_\mathrm{d}^q}{\partial \delta^-(\cdots)_N} \right)^\top.$$

Here we simplified the expressions to save space. The right hand side vector should be understood to be an extended vector consisting of the right hand sides of (5.75b)–(5.75d) with $\boldsymbol{U}^{(m)}$ expanded in $\boldsymbol{z}^{(m)}$. The correspondences of other expressions are as follows.

$$\widetilde{\boldsymbol{z}} \Leftrightarrow (V_k, \delta_k^+ V_k, \delta_k^- V_k) \text{ with } \boldsymbol{V} \text{ of (5.79)}$$
$$\nabla H(\widetilde{\boldsymbol{z}}) \Leftrightarrow \left(\left.\frac{\partial \widetilde{G}}{\partial u}\right|_{V_k}, \left.\frac{\partial \widetilde{G}}{\partial v}\right|_{V_k}, \left.\frac{\partial \widetilde{G}}{\partial w}\right|_{V_k} \right)$$
$$\|\delta_m^{\langle 1 \rangle,1} \boldsymbol{z}^{(m)}\|_2^2 \Leftrightarrow r_\mathrm{d} \text{ (with additional } \Delta x \text{ in the norm)}$$
$$\delta_m^{\langle 1 \rangle,q} H(\boldsymbol{z}^{(m)}) - (\nabla H(\widetilde{\boldsymbol{z}}))^\top \delta_m^{\langle 1 \rangle,q} \boldsymbol{z}^{(m)} \Leftrightarrow r_\mathrm{n}$$
$$\nabla_\mathrm{d}^q H(\boldsymbol{z}^{(m)}) \Leftrightarrow \nabla_\mathrm{d}^q J_\mathrm{d}(\boldsymbol{z}^{(m)})$$

In the first two vectors, we mean extended vectors of length $3(N+1)$ with k varying from 0 to N in each term. Note also that now we define the inner product by $\boldsymbol{x} \cdot \boldsymbol{y} = \sum_{k=0}^{N} {}'' x_k y_k \Delta x$ for $\boldsymbol{x}, \boldsymbol{y} \in \mathbb{R}^{3(N+1)}$, and the norm $\|\cdot\|_2$ accordingly.

In view of above, we understand that

$$\delta_m^{\langle 1 \rangle,q} J_\mathrm{d}(\boldsymbol{z}^{(m)}) = \nabla_\mathrm{d}^q J_\mathrm{d}(\boldsymbol{z}^{(m)}) \cdot \delta_m^{\langle 1 \rangle,q} \boldsymbol{z}^{(m)},$$

which can be expanded as

$$\delta_m^{\langle 1 \rangle,q} J_\mathrm{d}(\boldsymbol{z}^{(m)}) =$$
$$\sum_{k=0}^{N} {}'' \Bigg[\frac{\partial G_\mathrm{d}^q}{\partial (\boldsymbol{U}^{(m+l_2)}, \ldots, \boldsymbol{U}^{(m-1\,l)})_k} \delta_m^{\langle 1 \rangle,q} U_k^{(m)}$$
$$+ \frac{\partial G_\mathrm{d}^q}{\partial \delta^+(\boldsymbol{U}^{(m+l_2)}, \ldots, \boldsymbol{U}^{(m-l_1)})_k} \delta_m^{\langle 1 \rangle,q} \delta_k^+ U_k^{(m)}$$
$$+ \frac{\partial G_\mathrm{d}^q}{\partial \delta^-(\boldsymbol{U}^{(m+l_2)}, \ldots, \boldsymbol{U}^{(m-l_1)})_k} \delta_m^{\langle 1 \rangle,q} \delta_k^- U_k^{(m)} \Bigg] \Delta x.$$

Advanced Topic I: Design of High-Order Schemes   255

At this point, we have a high-order version of (3.28) (put $\boldsymbol{U}^{(m+1)}$ into $\boldsymbol{U}$, and $\boldsymbol{U}^{(m)}$ into $\boldsymbol{V}$). Then by applying the summation-by-parts formula (3.12a), we obtain the desired variation relation, corresponding to (3.29). The expression (5.77) corresponds to the boundary term (3.31).

5.4.3.1.2 Design of schemes We design conservative or dissipative finite difference schemes with the discrete variational derivative. Let us here introduce another discrete quantity:

$$B_{\mathrm{tr},2}^{\langle s\rangle}(\boldsymbol{U}^{(m+l_2)},\ldots,\boldsymbol{U}^{(m-l_1)}) =$$

$$\left[-\sum_{\substack{1\leq l\leq s/2 \\ l:\mathrm{even}}} \frac{2\varphi_k^{\langle l-1\rangle}\varphi_k^{\langle s-l\rangle} + \left(\delta_k^+\varphi_k^{\langle l-2\rangle}\right)\left(s_k^+\varphi_k^{\langle s-l\rangle}\right) + \left(\delta_k^-\varphi_k^{\langle l-2\rangle}\right)\left(s_k^-\varphi_k^{\langle s-l\rangle}\right)}{4}\right.$$

$$\left.+\sum_{\substack{1\leq l\leq s/2 \\ l:\mathrm{odd}}} \frac{2\varphi_k^{\langle l-1\rangle}\varphi_k^{\langle s-l\rangle} + \left(s_k^+\varphi_k^{\langle l-1\rangle}\right)\left(\delta_k^+\varphi_k^{\langle s-l-1\rangle}\right) + \left(s_k^-\varphi_k^{\langle l-1\rangle}\right)\left(\delta_k^-\varphi_k^{\langle s-l-1\rangle}\right)}{4}\right]_0^N,$$

$$\text{if } s \text{ is even}, \quad (5.80\mathrm{a})$$

$$\left[-\sum_{\substack{1\leq l\leq (s-1)/2 \\ l:\mathrm{even}}} \frac{\left(\delta_k^+\varphi_k^{\langle l-2\rangle}\right)\left(\delta_k^+\varphi_k^{\langle s-l-1\rangle}\right) + \left(\delta_k^-\varphi_k^{\langle l-2\rangle}\right)\left(\delta_k^-\varphi_k^{\langle s-l-1\rangle}\right)}{2}\right.$$

$$\left.+\sum_{\substack{1\leq l\leq (s-1)/2 \\ l:\mathrm{odd}}} \frac{\varphi_k^{\langle l-1\rangle}\left(s_k^{\langle 1\rangle}\varphi_k^{\langle s-l\rangle}\right) + \left(s_k^{\langle 1\rangle}\varphi_k^{\langle l-1\rangle}\right)\varphi_k^{\langle s-l\rangle}}{2}\right.$$

$$\left.+\frac{1}{2}(-1)^{(s-1)/2}\Psi_k^{(s,(s-1)/2)}\right]_0^N, \quad \text{if } s \text{ is odd}, \quad (5.80\mathrm{b})$$

where $s = 1, 2, 3, \ldots$, and

$$\varphi_k^{\langle l\rangle} \stackrel{\mathrm{d}}{\equiv} \delta_k^{\langle l\rangle} \frac{\delta G_\mathrm{d}}{\delta(\boldsymbol{U}^{(m+l_2)},\ldots,\boldsymbol{U}^{(m-l_1)})_k}, \quad (5.81)$$

$$\Psi_k^{(l,l')} \stackrel{\mathrm{d}}{\equiv} \begin{cases} \varphi_k^{\langle l'\rangle}\left(s_k^{\langle l \bmod 2\rangle}\varphi_k^{\langle l'\rangle}\right), & \text{if } l' \text{ is even}, \\ \frac{1}{2}\left\{\left(\delta_k^+\varphi_k^{\langle l'-1\rangle}\right)^2 + \left(\delta_k^-\varphi_k^{\langle l'-1\rangle}\right)^2\right\}, & \text{if } l' \text{ is odd}, \end{cases} \quad (5.82)$$

for $l, l' \in \{0, 1, 2, \ldots\}$.

The $(1, q)$th- or $(2, q)$th-order schemes are defined as follows.

Scheme 5.5 (Temporally high-order scheme for the PDEs 1)
Let $U_k^{(0)} = u(k\Delta x, 0)$ be initial values, and $\boldsymbol{U}^{(m)}$ ($m = 1, 2, \ldots, l-2$) be given

starting values. Assume that a discrete boundary condition is imposed such that for $m = l_1, l_1 + 1, \ldots$

$$B_{\mathrm{tr},1}(\boldsymbol{U}^{(m+l_2)}, \ldots, \boldsymbol{U}^{(m-l_1)}) = 0,$$

and

$$B_{\mathrm{tr},2}^{\langle 2s \rangle}(\boldsymbol{U}^{(m+l_2)}, \ldots, \boldsymbol{U}^{(m-l_1)}) = 0,$$

are satisfied. Then, a scheme for the PDEs 1 is given by for $m = l_1, l_1 + 1, \ldots$

$$\delta_m^{\langle 1 \rangle, q} U_k^{(m)} = (-1)^{s+1} \delta_k^{\langle 2s \rangle} \frac{\delta G_\mathrm{d}}{\delta(\boldsymbol{U}^{(m+l_2)}, \ldots, \boldsymbol{U}^{(m-l_1)})_k}, \qquad k = 0, \ldots, N. \tag{5.83}$$

THEOREM 5.5 Discrete dissipation property and orders of Scheme 5.5

The scheme 5.5 is dissipative in the sense that it satisfies

$$\delta_m^{\langle 1 \rangle, q} J_\mathrm{d}(\boldsymbol{U}^{(m)}) \leq 0, \qquad m = l_1, l_1 + 1, \ldots, \tag{5.84}$$

and it is $(1, q)$th- or $(2, q)$th-order.

PROOF In view of (5.74), and by the assumptions on the boundary terms, the dissipation property is proved in the same way as in, for example, Theorem 3.1.

Because the left-hand side of the scheme is temporally qth-order and spatially exact approximation of $\frac{\mathrm{d}}{\mathrm{d}t} u(k \Delta x, \widetilde{t})$ ($\widetilde{t}$ is defined in Definition 5.3 as to $\delta_m^{\langle 1 \rangle, q}$), if it is proved that the right-hand side is also a temporally qth-order and spatially first- or second-order approximation of the variational derivative at $(x, t) = (k \Delta x, \widetilde{t})$, the claim on the orders is established. We prove this in the following three steps:

Step 1 We evaluate r, which appears in (5.75b), (5.75c), and (5.75d).

Step 2 We evaluate (5.75b), (5.75c), and (5.75d).

Step 3 Based on the above estimates, we evaluate (5.75a).

We use the following abbreviations throughout.

$$(\cdot)|_{(k, \widetilde{t})} \stackrel{\mathrm{d}}{\equiv} (\cdot)|_{(x, t) = (k \Delta x, \widetilde{t})}, \qquad (\cdot)|_{u(k, \widetilde{t})} \stackrel{\mathrm{d}}{\equiv} (\cdot)|_{u = (k \Delta x, \widetilde{t}), v = w = u_x(k \Delta x, \widetilde{t})}.$$

As in the previous sections, we denote the true solution by $u_k^{(m)} \stackrel{\mathrm{d}}{\equiv} u(k \Delta x, m \Delta t)$.

Step 1 Estimating r.

We first show that when we substitute the true solution $u_k^{(m)}$ into $U_k^{(m)}$,

$$r = O(\Delta x, \Delta t^q) \qquad \Delta x, \Delta t \to 0.$$

The following estimates are easily obtained.

$$\delta_m^{\langle 1 \rangle, q} u(k\Delta x, m\Delta t) = u_t|_{(k,\tilde{t})} + O(\Delta t^q), \qquad (5.85a)$$

$$\delta_m^{\langle 1 \rangle, q} \delta_k^+ u(k\Delta x, m\Delta t) = u_{tx}|_{(k,\tilde{t})} + O(\Delta x, \Delta t^q), \qquad (5.85b)$$

$$\delta_m^{\langle 1 \rangle, q} \delta_k^- u(k\Delta x, m\Delta t) = u_{tx}|_{(k,\tilde{t})} + O(\Delta x, \Delta t^q). \qquad (5.85c)$$

Hence, we obtain

$$r_{\mathrm{d}} = \sum_{k=0}^{N} {}'' \left((u_t)^2 + 2(u_{tx})^2 \right) \Delta x \bigg|_{(k,\tilde{t})} + O(\Delta x, \Delta t^q). \qquad (5.86)$$

As to r_{n}, we first evaluate the term $\delta_m^{\langle 1 \rangle, q} J_{\mathrm{d}}(\boldsymbol{U}^{(m)})$ as follows.

$$\delta_m^{\langle 1 \rangle, q} \sum_{k=0}^{N} {}'' \widetilde{G}(u(k\Delta x, m\Delta t), \delta_k^+ u(k\Delta x, m\Delta t), \delta_k^- u(k\Delta x, m\Delta t)) \Delta x$$

$$= \left[\frac{\mathrm{d}}{\mathrm{d}t} \sum_{k=0}^{N} {}'' \widetilde{G}(u(k\Delta x, t), \delta_k^+ u(k\Delta x, t), \delta_k^- u(k\Delta x, t)) \Delta x \right]_{t=\tilde{t}} + O(\Delta t^q)$$

$$= \left[\frac{\mathrm{d}}{\mathrm{d}t} \sum_{k=0}^{N} {}'' \widetilde{G}(u, u_x, u_x) \Delta x \right]_{(k,\tilde{t})} + O(\Delta x, \Delta t^q)$$

$$= \left[\sum_{k=0}^{N} {}'' \left\{ \frac{\partial \widetilde{G}}{\partial u} u_t + \frac{\partial \widetilde{G}}{\partial v} u_{tx} + \frac{\partial \widetilde{G}}{\partial w} u_{tx} \right\} \Delta x \right]_{(k,\tilde{t})} + O(\Delta x, \Delta t^q). \qquad (5.87)$$

In order to evaluate the remaining terms in r_{n}, i.e., the terms

$$\frac{\partial \widetilde{G}}{\partial u}\bigg|_{V_k}, \quad \frac{\partial \widetilde{G}}{\partial v}\bigg|_{V_k}, \quad \frac{\partial \widetilde{G}}{\partial w}\bigg|_{V_k},$$

let us introduce $\widetilde{u}(x,t)$ by

$$\widetilde{u}(x,t) \stackrel{\mathrm{d}}{=} \sum_{i=-l_1}^{l_2} d_i u(x, t + i\Delta t)$$

for which, by the definition of V_k, the following estimate holds.

$$\widetilde{u}(k\Delta x, m\Delta t) = \sum_{i=-l_1}^{l_2} d_i u(k\Delta x, (m+i)\Delta t) = u|_{(k,\tilde{t})} + O(\Delta t^q), \quad (5.88a)$$

$$\delta_k^+ \widetilde{u}(k\Delta x, m\Delta t) = u_x|_{(k,\tilde{t})} + O(\Delta x, \Delta t^q), \qquad (5.88b)$$

$$\delta_k^- \widetilde{u}(k\Delta x, m\Delta t) = u_x|_{(k,\tilde{t})} + O(\Delta x, \Delta t^q). \qquad (5.88c)$$

Thus, we have

$$\frac{\partial \widetilde{G}}{\partial u}(\widetilde{u}(k\Delta x, m\Delta t), \delta_k^+ \widetilde{u}(k\Delta x, m\Delta t), \delta_k^- \widetilde{u}(k\Delta x, m\Delta t)) = \left.\frac{\partial \widetilde{G}}{\partial u}\right|_{u(k,\widetilde{t})} + O(\Delta x, \Delta t^q),$$
(5.89a)

$$\frac{\partial \widetilde{G}}{\partial v}(\widetilde{u}(k\Delta x, m\Delta t), \delta_k^+ \widetilde{u}(k\Delta x, m\Delta t), \delta_k^- \widetilde{u}(k\Delta x, m\Delta t)) = \left.\frac{\partial \widetilde{G}}{\partial v}\right|_{u(k,\widetilde{t})} + O(\Delta x, \Delta t^q),$$
(5.89b)

$$\frac{\partial \widetilde{G}}{\partial w}(\widetilde{u}(k\Delta x, m\Delta t), \delta_k^+ \widetilde{u}(k\Delta x, m\Delta t), \delta_k^- \widetilde{u}(k\Delta x, m\Delta t)) = \left.\frac{\partial \widetilde{G}}{\partial w}\right|_{u(k,\widetilde{t})} + O(\Delta x, \Delta t^q).$$
(5.89c)

Taking the estimates (5.85a), (5.85b), (5.85c), (5.87), (5.89a), (5.89b), and (5.89c) into account, we obtain

$$r_\mathrm{n} = O(\Delta x, \Delta t^q).$$

This, together with (5.86), proves the aimed estimate

$$r = O(\Delta x, \Delta t^q).$$

Step 2 Evaluating (5.75b), (5.75c), and (5.75d).

By using (5.85a), (5.85b), (5.85c), (5.89a), (5.89b), (5.89c), and the result obtained in Step 1, we immediately obtain

$$\frac{\partial G_\mathrm{d}^q}{\partial (\boldsymbol{u}^{(m+l_2)}, \ldots, \boldsymbol{u}^{(m-l_1)})_k} = \left.\frac{\partial \widetilde{G}}{\partial u}\right|_{u(k,\widetilde{t})} + O(\Delta x, \Delta t^q), \qquad (5.90a)$$

$$\frac{\partial G_\mathrm{d}^q}{\partial \delta^+ (\boldsymbol{u}^{(m+l_2)}, \ldots, \boldsymbol{u}^{(m-l_1)})_k} = \left.\frac{\partial \widetilde{G}}{\partial v}\right|_{u(k,\widetilde{t})} + O(\Delta x, \Delta t^q), \qquad (5.90b)$$

$$\frac{\partial G_\mathrm{d}^q}{\partial \delta^- (\boldsymbol{u}^{(m+l_2)}, \ldots, \boldsymbol{u}^{(m-l_1)})_k} = \left.\frac{\partial \widetilde{G}}{\partial w}\right|_{u(k,\widetilde{t})} + O(\Delta x, \Delta t^q). \qquad (5.90c)$$

Step 3 Evaluating (5.75a).

From the results obtained in Step 2 and using (5.75a), we obtain

$$\frac{\delta G_\mathrm{d}^q}{\delta (\boldsymbol{u}^{(m+l_2)}, \ldots, \boldsymbol{u}^{(m-l_1)})_k} = \left.\frac{\partial \widetilde{G}}{\partial u}\right|_{u(k,\widetilde{t})} - \frac{\partial}{\partial x} \left.\frac{\partial \widetilde{G}}{\partial v}\right|_{u(k,\widetilde{t})} - \frac{\partial}{\partial x} \left.\frac{\partial \widetilde{G}}{\partial w}\right|_{u(k,\widetilde{t})}$$
$$+ O(\Delta x, \Delta t^q)$$
$$= \left.\frac{\delta G}{\delta u}\right|_{u(k,\widetilde{t})} + O(\Delta x, \Delta t^q). \qquad (5.91)$$

Advanced Topic I: Design of High-Order Schemes   259

In the last equality, we used the following trivial identity.

$$\frac{\delta G}{\delta u} = \left.\frac{\partial \widetilde{G}}{\partial u}\right|_{u=u,v=w=u_x} - \frac{\partial}{\partial x}\left(\left.\frac{\partial \widetilde{G}}{\partial v}\right|_{u=u,v=w=u_x} + \left.\frac{\partial \widetilde{G}}{\partial w}\right|_{u=u,v=w=u_x}\right).$$

This estimate is only first-order as to Δx. However, when $G_{\rm d}$ is defined so that it is symmetric with respect to $\delta_k^+ U_k^{(m)}$ and $\delta_k^- U_k^{(m)}$, then the discrete derivative (5.75a) is also symmetric with respect to them, and hence trivially second-order: $O(\Delta x^2)$.   □

REMARK 5.5 The accuracy and the dissipation property are guaranteed for *any* choice of high-order difference operator $\delta_m^{\langle 1\rangle,q}$ as the theorem claims. Actually, however, the choice is quite limited in view of stability of the resulting scheme. The high-order operator should be, for example, the so-called "backward-difference operator" (see, for example, the textbook by Hairer–Nørsett–Wanner [82]).   □

Scheme 5.6 (Scheme for the PDEs 2) Let $U_k^{(0)} = u(k\Delta x, 0)$ be initial values, and $\boldsymbol{U}^{(m)}$ ($m = 1, 2, \ldots, l-2$) be given starting values. Assume that a discrete boundary condition is imposed such that for $m = l_1, l_1+1,\ldots$

$$B_{\mathrm{tr},1}(\boldsymbol{U}^{(m+l_2)},\ldots,\boldsymbol{U}^{(m-l_1)}) = 0,$$

and

$$B_{\mathrm{tr},2}^{\langle 2s\rangle}(\boldsymbol{U}^{(m+l_2)},\ldots,\boldsymbol{U}^{(m-l_1)}) = 0$$

are satisfied. Then, a scheme for the PDEs 2 is given by for $m = l_1, l_1+1,\ldots$

$$\delta_m^{\langle 1\rangle,q} U_k^{(m)} = \delta_k^{\langle 2s+1\rangle} \frac{\delta G_{\rm d}}{\delta(\boldsymbol{U}^{(m+l_2)},\ldots,\boldsymbol{U}^{(m-l_1)})_k}, \qquad k = 0,\ldots,N. \quad (5.92)$$

THEOREM 5.6 Discrete conservation property and orders of Scheme 5.6

The scheme 5.6 is conservative in the sense that it satisfies

$$\delta_m^{\langle 1\rangle,q} J_{\rm d}(\boldsymbol{U}^{(m)}) = 0, \qquad m = l_1, l_1+1,\ldots, \quad (5.93)$$

and it is $(1,q)$th- or $(2,q)$th-order.

PROOF The proof goes in the same way as in the preceding theorem, and hence is omitted. (We will omit similar proofs in the subsequent dissipation or conservation theorems as well.)   □

The resulting schemes derived from Scheme 5.5 or Scheme 5.6 are nonlinearly implicit (see the classification of schemes in Chapter 1, page 41). It might seem even worse than the standard second-order nonlinearly implicit schemes in Chapter 3, since now it involves not only $U^{(m+l_2)}$ (the next solution) and $U^{(m+l_2-1)}$ (the previous solution) but also further information $U^{(m+l_2-2)}, \ldots, U^{(m-l_1)}$. One might think that this causes serious increases in computational cost.

The actual situation is, however, more optimistic; the actual computational cost would not increase in practice even if we increase the temporal accuracy q. This can be understood in the following way.

We first point out that although it is surely possible to utilize the Newton method or similar iterative solvers, often simple function iterations based on the scheme itself are sufficient. For Scheme 5.5, for example, we can use the simple iteration:

$$U_k^{(m+l_2),j+1} = \\ -\sum_{i=-l_1}^{l_2-1} c_i U_k^{(m+i)} + (-1)^{s+1}\delta_k^{\langle 2s \rangle} \frac{\delta G_{\mathrm{d}}}{\delta(U^{(m+l_2),j}, U^{(m+l_2-1)}, \ldots, U^{(m-l_1)})_k},$$
(5.94)

where $U^{(m+l_2),j}$ ($j = 0, 1, \ldots$) is the jth iterated solution.

In either case (Newton method or the function iteration), the computational cost depends on the following three factors.

The number of the unknowns: In the above schemes, the only unknown variable is always $U^{(m+l_2)}$, and thus the number of unknowns does not increase even when q is increased.

The number of required iterations: This is difficult to estimate theoretically, but in practice, the number is expected to remain at some satisfactory level even when q is increased, as far as we choose Δt appropriately small. See the example below.

The computational cost in each iteration: This remains almost the same even when q is increased. To see this, let us consider the case where we use the simple iteration (5.94). When q is increased, the number of known numerical solutions that appears at the right hand side of (5.94), i.e., $U_k^{(m+l_2-1)}, \ldots, U_k^{(m-l_1)}$, increases. This does not, however, mean the increase of the computational cost in each iteration at all, because the terms in (5.94) where the known values appear can be calculated in prior to the iterative calculation, and thus practically the increase due to this factor can be negligible.

Taking these three factors into account, we can conclude that computational cost does not increase in practice, even when the temporal order of accuracy q is increased. This note applies to all of the schemes presented hereafter.

As a demonstration, we here present a simple ODE example from [114]. Let us consider the Kepler problem with the Hamiltonian $H(z_1, z_2, z_3, z_4) = (z_1{}^2 + z_2{}^2)/2 - 1/\sqrt{z_3{}^2 + z_4{}^2}$. This is an example of the ordinary differential equation (5.67), where $N = 4$ and $A = \begin{pmatrix} 0 & -I \\ I & 0 \end{pmatrix}$. The initial value is set to $z(0) = (0, \sqrt{(1+\varepsilon)(1-\varepsilon)}, 1-\varepsilon, 0)^\top$ with the eccentricity $\varepsilon = 0.8$, and the problem is integrated in $0 \leq t \leq 10$. Let us then test the scheme (5.70) where $\delta_m^{\langle 1 \rangle, q}$ is chosen as the backward-difference operators of order 2, 4, and 6. The scheme is solved in the function iteration form like (5.94). For comparison, we also consider 4th- and 6th-order Hamiltonian-preserving schemes that are derived by the composition method (Section 5.3) based on the 2nd-order version of (5.70) (in this case, we choose the standard forward difference operator as $\delta_m^{\langle 1 \rangle, 2}$). The underlying 2nd-order scheme is solved in the same way as above.

Table 5.4 shows the number of function iterations, where N_t is the number of time mesh points (i.e. $\Delta t = 10/N_t$; N_t is chosen such that $\log N_t$'s distribute uniformly in $300 \leq N_t \leq 10000$), "COMPn" ($n = 2, 4, 6$) denotes the composition schemes, and "BDFn" represents the scheme (5.70). As is claimed above, the number of iterations in BDFn actually does not increase with q for wide range of Δt. In contrast to that, the number in COMPn significantly increases with q. This should be attributed to the fact that in the composition schemes, the underlying second-order scheme is repeatedly called within a single step of Δt.

TABLE 5.4: Total number of iterations for each method

N_t	COMP2	COMP4	COMP6	BDF2	BDF4	BDF6
300	3053	10196	19756	3083	2810	2748
425	3899	13042	26070	3995	3637	3489
604	5263	16784	33995	5140	4811	4676
858	6765	22235	45280	6816	6368	6145
1219	9200	29053	61532	8963	8397	8217
1732	12746	39363	84041	11844	11393	11220
2459	16573	54380	113179	16104	15707	15551
3492	22440	71828	150739	22256	20407	19880
4959	31025	95964	207648	28555	27352	27027
7042	43391	132413	289926	38557	37629	37289
10000	57804	185033	396610	53221	52377	52054

REMARK 5.6 When G_d is simple, the form of the discrete variational derivative (5.75a) can be simplified. Let us consider the case, for example, where G_d includes only U_k and not $\delta_k^+ U_k$ and $\delta_k^- U_k$. In this case, the presented

definitions can be modified to the following simple forms.

$$\delta_m^{\langle 1 \rangle, q} \sum_{k=0}^{N} {}''G_{\mathrm{d},k}(\boldsymbol{U}^{(m)}) \Delta x =$$
$$\sum_{k=0}^{N} {}'' \left(\frac{\delta G_{\mathrm{d}}}{\delta(\boldsymbol{U}^{(m+l_2)}, \ldots, \boldsymbol{U}^{(m-l_1)})_k} \cdot \delta_m^{\langle 1 \rangle, q} U_k^{(m)} \right) \Delta x, \quad (5.95)$$

where

$$\frac{\delta G_{\mathrm{d}}}{\delta(\boldsymbol{U}^{(m+l_2)}, \ldots, \boldsymbol{U}^{(m-l_1)})_k} \stackrel{\mathrm{d}}{\equiv} \frac{\partial G_{\mathrm{d}}^q}{\partial(\boldsymbol{U}^{(m+l_2)}, \ldots, \boldsymbol{U}^{(m-l_1)})_k}, \quad (5.96\mathrm{a})$$

$$\frac{\partial G_{\mathrm{d}}^q}{\partial(\boldsymbol{U}^{(m+l_2)}, \ldots, \boldsymbol{U}^{(m-l_1)})_k} \stackrel{\mathrm{d}}{\equiv} \left. \frac{\partial \widetilde{G}}{\partial u} \right|_{V_k} + r \delta_m^{\langle 1 \rangle, q} U_k^{(m)}, \quad (5.96\mathrm{b})$$

$$r \stackrel{\mathrm{d}}{\equiv} \frac{\delta_m^{\langle 1 \rangle, q} J_{\mathrm{d}}(\boldsymbol{U}^{(m)}) - \sum_{k=0}^{N} {}'' \left. \frac{\partial \widetilde{G}}{\partial u} \right|_{V_k} \delta_m^{\langle 1 \rangle, q} U_k^{(m)} \Delta x}{\sum_{k=0}^{N} {}'' \left(\delta_m^{\langle 1 \rangle, q} U_k^{(m)} \right)^2 \Delta x}. \quad (5.96\mathrm{c})$$

This remark applies also to the other cases ([B], [C], and [D]). □

5.4.3.2 [B]: (p,q)th-Order Schemes for the Real-Valued PDEs

For the real-valued PDEs 1 and 2, (p,q)th-order schemes are proposed. Suppose the discrete periodic boundary condition (5.3) is imposed.

5.4.3.2.1 Discrete variational derivative
We define the discrete energy function by

$$G_{\mathrm{d},k}(\boldsymbol{U}) \stackrel{\mathrm{d}}{\equiv} G(U_k, \delta_k^{\langle 1 \rangle, p} U_k), \quad (5.97)$$

and accordingly associated global energy by

$$J_{\mathrm{d}}(\boldsymbol{U}^{(m)}) \stackrel{\mathrm{d}}{\equiv} \sum_{k=0}^{N} {}''G_{\mathrm{d},k}(\boldsymbol{U}^{(m)}) \Delta x. \quad (5.98)$$

Note that since in this case we use only the high-order difference operator $\delta_k^{\langle 1 \rangle, p}$, we do not need to introduce $\widetilde{G}$ as the previous section. Then we can consider the discrete variation as:

$$\delta_m^{\langle 1 \rangle, q} \sum_{k=0}^{N} {}''G_{\mathrm{d},k}(\boldsymbol{U}^{(m)}) \Delta x =$$
$$\sum_{k=0}^{N} {}'' \left(\frac{\delta G_{\mathrm{d}}}{\delta(\boldsymbol{U}^{(m+l_2)}, \ldots, \boldsymbol{U}^{(m-l_1)})_k} \cdot \delta_m^{\langle 1 \rangle, q} U_k^{(m)} \right) \Delta x, \quad (5.99)$$

Advanced Topic I: Design of High-Order Schemes 263

where

$$\frac{\delta G_\mathrm{d}}{\delta (\boldsymbol{U}^{(m+l_2)}, \ldots, \boldsymbol{U}^{(m-l_1)})_k} \stackrel{\mathrm{d}}{\equiv} \frac{\partial G_\mathrm{d}^q}{\partial (\boldsymbol{U}^{(m+l_2)}, \ldots, \boldsymbol{U}^{(m-l_1)})_k}$$
$$- \delta_k^{\langle 1 \rangle, p} \left(\frac{\partial G_\mathrm{d}^q}{\partial \delta^+ (\boldsymbol{U}^{(m+l_2)}, \ldots, \boldsymbol{U}^{(m-l_1)})_k} \right), \tag{5.100a}$$

$$\frac{\partial G_\mathrm{d}^q}{\partial (\boldsymbol{U}^{(m+l_2)}, \ldots, \boldsymbol{U}^{(m-l_1)})_k} \stackrel{\mathrm{d}}{\equiv} \left. \frac{\partial G}{\partial u} \right|_{V_k} + r \delta_m^{\langle 1 \rangle, q} U_k^{(m)}, \tag{5.100b}$$

$$\frac{\partial G_\mathrm{d}^q}{\partial \delta (\boldsymbol{U}^{(m+l_2)}, \ldots, \boldsymbol{U}^{(m-l_1)})_k} \stackrel{\mathrm{d}}{\equiv} \left. \frac{\partial G}{\partial v} \right|_{V_k} + r \delta_m^{\langle 1 \rangle, q} (\delta_k^{\langle 1 \rangle, p} U_k^{(m)}), \tag{5.100c}$$

$$r \stackrel{\mathrm{d}}{\equiv} \frac{r_\mathrm{n}}{r_\mathrm{d}}, \tag{5.101a}$$

$$r_\mathrm{n} \stackrel{\mathrm{d}}{\equiv} \delta_m^{\langle 1 \rangle, q} J_\mathrm{d}(\boldsymbol{U}^{(m)}) - \sum_{k=0}^{N}{}'' \left. \frac{\partial G}{\partial u} \right|_{V_k} \delta_m^{\langle 1 \rangle, q} U_k^{(m)} \Delta x$$
$$- \sum_{k=0}^{N}{}'' \left. \frac{\partial G}{\partial v} \right|_{V_k} (\delta_m^{\langle 1 \rangle, q} \delta_k^{\langle 1 \rangle, p} U_k^{(m)}) \Delta x, \tag{5.101b}$$

$$r_\mathrm{d} \stackrel{\mathrm{d}}{\equiv} \sum_{k=0}^{N}{}'' \left\{ \left(\delta_m^{\langle 1 \rangle, q} U_k^{(m)} \right)^2 + \left(\delta_m^{\langle 1 \rangle, q} \delta_k^{\langle 1 \rangle, p} U_k^{(m)} \right)^2 \right\} \Delta x. \tag{5.101c}$$

5.4.3.2.2 Design of schemes With the discrete variational derivative defined as above, we can design conservative or dissipative finite difference schemes which are temporally qth-order.

Scheme 5.7 (Scheme for the PDEs 1) *Let $U_k^{(0)} = u(k\Delta x, 0)$ be initial values, and $\boldsymbol{U}^{(m)}$ ($m = 1, 2, \ldots, l-2$) be given starting values. Then, a scheme is given by, for $m = l_1, l_1+1, \ldots,$*

$$\delta_m^{\langle 1 \rangle, q} U_k^{(m)} = (-1)^{s+1} (\delta_k^{\langle 1 \rangle, p})^{2s} \frac{\delta G_\mathrm{d}}{\delta (\boldsymbol{U}^{(m+l_2)}, \ldots, \boldsymbol{U}^{(m-l_1)})_k}, \quad k = 0, \ldots, N. \tag{5.102}$$

□

THEOREM 5.7 Discrete dissipation property and orders of Scheme 5.7

The scheme 5.7 is dissipative in the sense that it satisfies

$$\delta_m^{\langle 1 \rangle, q} J_\mathrm{d}(\boldsymbol{U}^{(m)}) \leq 0, \quad m = l_1, l_1+1, \ldots, \tag{5.103}$$

and it is (p, q)th-order.

PROOF The same as in the preceding theorems. □

Scheme 5.8 (Scheme for the PDEs 2) *Let $U_k^{(0)} = u(k\Delta x, 0)$ be initial values, and $\boldsymbol{U}^{(m)}$ $(m = 1, 2, \ldots, l-2)$ be given starting values. Then, a scheme is given by, for $m = l_1, l_1 + 1, \ldots,$*

$$\delta_m^{\langle 1 \rangle, q} U_k^{(m)} = (\delta_k^{\langle 1 \rangle, p})^{2s+1} \frac{\delta G_\mathrm{d}}{\delta (\boldsymbol{U}^{(m+l_2)}, \ldots, \boldsymbol{U}^{(m-l_1)})_k}, \qquad k = 0, \ldots, N. \tag{5.104}$$
□

THEOREM 5.8 Discrete conservation property and orders of Scheme 5.8

The scheme 5.8 is conservative in the sense that it satisfies

$$\delta_m^{\langle 1 \rangle, q} J_\mathrm{d}(\boldsymbol{U}^{(m)}) = 0, \qquad m = l_1, l_1 + 1, \ldots, \tag{5.105}$$

and it is (p, q)th-order.

PROOF The same as in the preceding theorems. □

5.4.3.3 [C]: $(1, q)$th- or $(2, q)$-th Order Schemes for the Complex-Valued PDEs

For the complex-valued PDEs 3 and 4, $(1, q)$th- or $(2, q)$th-order schemes are presented.

5.4.3.3.1 Complex discrete variational derivative
For an energy function $G(u, u_x)$, we introduce another function $\widetilde{G}(u, v, w)$ by

$$\widetilde{G}(u, v, w) \stackrel{\mathrm{d}}{\equiv} \frac{1}{2}\left(G(u, v) + G(u, w)\right). \tag{5.106}$$

We then define the discrete energy function $G_\mathrm{d}(\boldsymbol{U}^{(m)})$ by

$$G_{\mathrm{d},k}(\boldsymbol{U}^{(m)}) \stackrel{\mathrm{d}}{\equiv} \widetilde{G}(U_k^{(m)}, \delta_k^+ U_k^{(m)}, \delta_k^- U_k^{(m)}), \tag{5.107}$$

and accordingly associated global energy by

$$J_\mathrm{d}(\boldsymbol{U}^{(m)}) \stackrel{\mathrm{d}}{\equiv} \sum_{k=0}^{N} {}''G_{\mathrm{d},k}(\boldsymbol{U}^{(m)}) \Delta x. \tag{5.108}$$

Using the qth-order difference operator, we consider the discrete variation as

$$\delta_m^{\langle 1 \rangle, q} \sum_{k=0}^{N} {}'' G_{\mathrm{d},k}(\boldsymbol{U}^{(m)}) \Delta x =$$

$$\sum_{k=0}^{N} {}'' \left(\frac{\delta G_\mathrm{d}}{\delta(\boldsymbol{U}^{(m+l_2)}, \ldots, \boldsymbol{U}^{(m-l_1)})_k} \cdot \delta_m^{\langle 1 \rangle, q} U_k^{(m)} \right)$$

$$+ \left(\frac{\delta G_\mathrm{d}}{\delta(\overline{\boldsymbol{U}^{(m+l_2)}}, \ldots, \overline{\boldsymbol{U}^{(m-l_1)}})_k} \cdot \delta_m^{\langle 1 \rangle, q} \overline{U_k^{(m)}} \right) \Delta x$$

$$+ B_{\mathrm{tr},1}(\boldsymbol{U}^{(m+l_2)}, \ldots, \boldsymbol{U}^{(m-l_1)}) + \overline{B_{\mathrm{tr},1}(\boldsymbol{U}^{(m+l_2)}, \ldots, \boldsymbol{U}^{(m-l_1)})},$$
(5.109)

where

$$\frac{\delta G_\mathrm{d}}{\delta(\boldsymbol{U}^{(m+l_2)}, \ldots, \boldsymbol{U}^{(m-l_1)})_k} \stackrel{\mathrm{d}}{\equiv} \frac{\partial G_\mathrm{d}^q}{\partial(\boldsymbol{U}^{(m+l_2)}, \ldots, \boldsymbol{U}^{(m-l_1)})_k}$$
$$- \delta_k^{-} \left(\frac{\partial G_\mathrm{d}^q}{\partial \delta^{+}(\boldsymbol{U}^{(m+l_2)}, \ldots, \boldsymbol{U}^{(m-l_1)})_k} \right)$$
$$- \delta_k^{+} \left(\frac{\partial G_\mathrm{d}^q}{\partial \delta^{-}(\boldsymbol{U}^{(m+l_2)}, \ldots, \boldsymbol{U}^{(m-l_1)})_k} \right),$$
(5.110a)

$$\frac{\partial G_\mathrm{d}^q}{\partial(\boldsymbol{U}^{(m+l_2)}, \ldots, \boldsymbol{U}^{(m-l_1)})_k} \stackrel{\mathrm{d}}{\equiv} \left. \frac{\partial \widetilde{G}}{\partial u} \right|_{V_k} + r\delta_m^{\langle 1 \rangle, q} U_k^{(m)}, \tag{5.110b}$$

$$\frac{\partial G_\mathrm{d}^q}{\partial(\overline{\boldsymbol{U}^{(m+l_2)}}, \ldots, \overline{\boldsymbol{U}^{(m-l_1)}})_k} \stackrel{\mathrm{d}}{\equiv} \left. \frac{\partial \widetilde{G}}{\partial \overline{u}} \right|_{V_k} + r\delta_m^{\langle 1 \rangle, q} \overline{U_k^{(m)}}, \tag{5.110c}$$

$$\frac{\partial G_\mathrm{d}^q}{\partial \delta^{+}(\boldsymbol{U}^{(m+l_2)}, \ldots, \boldsymbol{U}^{(m-l_1)})_k} \stackrel{\mathrm{d}}{\equiv} \left. \frac{\partial \widetilde{G}}{\partial v} \right|_{V_k} + r\delta_m^{\langle 1 \rangle, q}(\delta_k^{+} U_k^{(m)}), \tag{5.110d}$$

$$\frac{\partial G_\mathrm{d}^q}{\partial \delta^{+}(\overline{\boldsymbol{U}^{(m+l_2)}}, \ldots, \overline{\boldsymbol{U}^{(m-l_1)}})_k} \stackrel{\mathrm{d}}{\equiv} \left. \frac{\partial \widetilde{G}}{\partial \overline{v}} \right|_{V_k} + r\delta_m^{\langle 1 \rangle, q}(\delta_k^{+} \overline{U_k^{(m)}}), \tag{5.110e}$$

$$\frac{\partial G_\mathrm{d}^q}{\partial \delta^{-}(\boldsymbol{U}^{(m+l_2)}, \ldots, \boldsymbol{U}^{(m-l_1)})_k} \stackrel{\mathrm{d}}{\equiv} \left. \frac{\partial \widetilde{G}}{\partial w} \right|_{V_k} + r\delta_m^{\langle 1 \rangle, q}(\delta_k^{-} U_k^{(m)}), \tag{5.110f}$$

$$\frac{\partial G_\mathrm{d}^q}{\partial \delta^{-}(\overline{\boldsymbol{U}^{(m+l_2)}}, \ldots, \overline{\boldsymbol{U}^{(m-l_1)}})_k} \stackrel{\mathrm{d}}{\equiv} \left. \frac{\partial \widetilde{G}}{\partial \overline{w}} \right|_{V_k} + r\delta_m^{\langle 1 \rangle, q}(\delta_k^{-} \overline{U_k^{(m)}}), \tag{5.110g}$$

$$r \stackrel{\mathrm{d}}{\equiv} \frac{r_{\mathrm{n}}}{r_{\mathrm{d}}}, \tag{5.111a}$$

$$r_{\mathrm{n}} \stackrel{\mathrm{d}}{\equiv} \delta_m^{\langle 1 \rangle, q} J_{\mathrm{d}}(\boldsymbol{U}^{(m)})$$
$$- \sum_{k=0}^{N}{}'' \left.\frac{\partial \widetilde{G}}{\partial u}\right|_{V_k} \delta_m^{\langle 1 \rangle, q} U_k^{(m)} \Delta x - \sum_{k=0}^{N}{}'' \left.\frac{\partial \widetilde{G}}{\partial v}\right|_{V_k} (\delta_m^{\langle 1 \rangle, q} \delta_k^+ U_k^{(m)}) \Delta x$$
$$- \sum_{k=0}^{N}{}'' \left.\frac{\partial \widetilde{G}}{\partial w}\right|_{V_k} (\delta_m^{\langle 1 \rangle, q} \delta_k^+ U_k^{(m)}) \Delta x,$$
$$- \sum_{k=0}^{N}{}'' \left.\frac{\partial \widetilde{G}}{\partial \overline{u}}\right|_{V_k} \delta_m^{\langle 1 \rangle, q} \overline{U_k^{(m)}} \Delta x - \sum_{k=0}^{N}{}'' \left.\frac{\partial \widetilde{G}}{\partial \overline{v}}\right|_{V_k} (\delta_m^{\langle 1 \rangle, q} \delta_k^+ \overline{U_k^{(m)}}) \Delta x$$
$$- \sum_{k=0}^{N}{}'' \left.\frac{\partial \widetilde{G}}{\partial \overline{w}}\right|_{V_k} (\delta_m^{\langle 1 \rangle, q} \delta_k^+ \overline{U_k^{(m)}}) \Delta x \tag{5.111b}$$

$$r_{\mathrm{d}} \stackrel{\mathrm{d}}{\equiv} 2 \sum_{k=0}^{N}{}'' \left\{ \left|\delta_m^{\langle 1 \rangle, q} U_k^{(m)}\right|^2 + \left|\delta_m^{\langle 1 \rangle, q} \delta_k^+ U_k^{(m)}\right|^2 + \left|\delta_m^{\langle 1 \rangle, q} \delta_k^- U_k^{(m)}\right|^2 \right\} \Delta x, \tag{5.111c}$$

$$B_{\mathrm{tr},1}(\boldsymbol{U}^{(m+l_2)}, \ldots, \boldsymbol{U}^{(m-l_1)}) =$$
$$\frac{1}{2} \left[\frac{\partial G_{\mathrm{d}}^q}{\partial \delta^+ (\boldsymbol{U}^{(m+l_2)}, \ldots, \boldsymbol{U}^{(m-l_1)})_k} (s_k^+ \delta_m^{\langle 1 \rangle, q} U_k^{(m)}) \right.$$
$$+ \left(s_k^- \frac{\partial G_{\mathrm{d}}^q}{\partial \delta^+ (\boldsymbol{U}^{(m+l_2)}, \ldots, \boldsymbol{U}^{(m-l_1)})_k} \right) \delta_m^{\langle 1 \rangle, q} U_k^{(m)}$$
$$+ \frac{\partial G_{\mathrm{d}}^q}{\partial \delta^- (\boldsymbol{U}^{(m+l_2)}, \ldots, \boldsymbol{U}^{(m-l_1)})_k} (s_k^- \delta_m^{\langle 1 \rangle, q} U_k^{(m)})$$
$$+ \left. \left(s_k^+ \frac{\partial G_{\mathrm{d}}^q}{\partial \delta^- (\boldsymbol{U}^{(m+l_2)}, \ldots, \boldsymbol{U}^{(m-l_1)})_k} \right) \delta_m^{\langle 1 \rangle, q} U_k^{(m)} \right]_0^N. \tag{5.112}$$

The discrete quantities

$$\frac{\delta G_{\mathrm{d}}}{\delta(\boldsymbol{U}^{(m+l_2)}, \ldots, \boldsymbol{U}^{(m-l_1)})_k} \quad \text{and} \quad \frac{\delta G_{\mathrm{d}}}{\delta(\overline{\boldsymbol{U}^{(m+l_2)}}, \ldots, \overline{\boldsymbol{U}^{(m-l_1)}})_k}$$

which are defined in (5.110a), are approximations of the original variational derivatives.

Advanced Topic I: Design of High-Order Schemes

5.4.3.3.2 Design of schemes With the discrete variational derivative defined as above, we can design conservative or dissipative finite difference schemes which are temporally qth-order.

Scheme 5.9 (Scheme for the PDEs 3) Let $U_k^{(0)} = u(k\Delta x, 0)$ be initial values, and $\boldsymbol{U}^{(m)}$ ($m = 1, 2, \ldots, l-2$) be given starting values. Assume that a discrete boundary condition is imposed such that

$$B_{\mathrm{tr},1}(\boldsymbol{U}^{(m+l_2)}, \ldots, \boldsymbol{U}^{(m-l_1)}) = 0$$

is satisfied for $m = l_1, l_1 + 1, \ldots$. Then, a scheme for the PDEs 3 is given by, for $m = l_1, l_1 + 1, \ldots$,

$$\delta_m^{\langle 1 \rangle, q} U_k^{(m)} = -\frac{\delta G_\mathrm{d}}{\delta(\overline{\boldsymbol{U}^{(m+l_2)}}, \ldots, \overline{\boldsymbol{U}^{(m-l_1)}})_k}, \quad k = 0, \ldots, N. \quad (5.113)$$

THEOREM 5.9 Discrete dissipation property and orders of Scheme 5.9

The scheme 5.9 is dissipative in the sense that it satisfies

$$\delta_m^{\langle 1 \rangle, q} J_\mathrm{d}(\boldsymbol{U}^{(m)}) \leq 0, \quad m = l_1, l_1 + 1, \ldots, \quad (5.114)$$

and it is $(1, q)$th- or $(2, q)$th-order.

PROOF The same as in the preceding theorems. □

Scheme 5.10 (Scheme for the PDEs 4) Let $U_k^{(0)} = u(k\Delta x, 0)$ be initial values, and $\boldsymbol{U}^{(m)}$ ($m = 1, 2, \ldots, l-2$) be given starting values. Assume that a discrete boundary condition is imposed such that

$$B_{\mathrm{tr},1}(\boldsymbol{U}^{(m+l_2)}, \ldots, \boldsymbol{U}^{(m-l_1)}) = 0$$

is satisfied for $m = l_1, l_1 + 1, \ldots$. Then, a scheme for the PDEs 4 is given by, for $m = l_1, l_1 + 1, \ldots$,

$$\mathrm{i}\, \delta_m^{\langle 1 \rangle, q} U_k^{(m)} = -\frac{\delta G_\mathrm{d}}{\delta(\overline{\boldsymbol{U}^{(m+l_2)}}, \ldots, \overline{\boldsymbol{U}^{(m-l_1)}})_k} \quad k = 0, \ldots, N. \quad (5.115)$$

THEOREM 5.10 Discrete conservation property and orders of Scheme 5.10

The scheme 5.10 is conservative in the sense that it satisfies

$$\delta_m^{\langle 1 \rangle, q} J_\mathrm{d}(\boldsymbol{U}^{(m)}) = 0, \quad m = l_1, l_1 + 1, \ldots, \quad (5.116)$$

and it is $(1, q)$th- or $(2, q)$th-order.

PROOF The same as in the preceding theorems. □

5.4.3.4 [D]: (p,q)th-Order Schemes for the Complex-Valued PDEs

For the complex-valued PDEs 3 and 4 with the discrete periodic boundary condition (5.3), (p,q)th-order schemes are presented.

5.4.3.4.1 Complex discrete variational derivative
We define the discrete energy function by

$$G_{\mathrm{d},k}(\boldsymbol{U}) \stackrel{\mathrm{d}}{\equiv} G(U_k, \delta_k^{\langle 1 \rangle, p} U_k), \tag{5.117}$$

and accordingly associated global energy by

$$J_{\mathrm{d}}(\boldsymbol{U}^{(m)}) \stackrel{\mathrm{d}}{\equiv} \sum_{k=0}^{N} {}'' G_{\mathrm{d},k}(\boldsymbol{U}^{(m)}) \Delta x. \tag{5.118}$$

Then we can consider the discrete variation as:

$$\delta_m^{\langle 1 \rangle, q} \sum_{k=0}^{N} {}'' G_{\mathrm{d},k}(\boldsymbol{U}^{(m)}) \Delta x$$

$$= \sum_{k=0}^{N} {}'' \left(\frac{\delta G_{\mathrm{d}}}{\delta(\boldsymbol{U}^{(m+l_2)}, \ldots, \boldsymbol{U}^{(m-l_1)})_k} \cdot \delta_m^{\langle 1 \rangle, q} U_k^{(m)} \right.$$

$$\left. + \frac{\delta G_{\mathrm{d}}}{\delta(\overline{\boldsymbol{U}^{(m+l_2)}}, \ldots, \overline{\boldsymbol{U}^{(m-l_1)}})_k} \cdot \delta_m^{\langle 1 \rangle, q} \overline{U_k^{(m)}} \right) \Delta x, \tag{5.119}$$

where

$$\frac{\delta G_{\mathrm{d}}}{\delta(\boldsymbol{U}^{(m+l_2)}, \ldots, \boldsymbol{U}^{(m-l_1)})_k} \stackrel{\mathrm{d}}{\equiv} \frac{\partial G_{\mathrm{d}}^q}{\partial(\boldsymbol{U}^{(m+l_2)}, \ldots, \boldsymbol{U}^{(m-l_1)})_k}$$

$$- \delta_k^{\langle 1 \rangle, p} \left(\frac{\partial G_{\mathrm{d}}^q}{\partial \delta^+(\boldsymbol{U}^{(m+l_2)}, \ldots, \boldsymbol{U}^{(m-l_1)})_k} \right), \tag{5.120a}$$

$$\frac{\delta G_{\mathrm{d}}}{\delta(\overline{\boldsymbol{U}^{(m+l_2)}}, \ldots, \overline{\boldsymbol{U}^{(m-l_1)}})_k} \stackrel{\mathrm{d}}{\equiv} \frac{\partial G_{\mathrm{d}}^q}{\partial(\overline{\boldsymbol{U}^{(m+l_2)}}, \ldots, \overline{\boldsymbol{U}^{(m-l_1)}})_k}$$

$$- \delta_k^{\langle 1 \rangle, p} \left(\frac{\partial G_{\mathrm{d}}^q}{\partial \delta^+(\overline{\boldsymbol{U}^{(m+l_2)}}, \ldots, \overline{\boldsymbol{U}^{(m-l_1)}})_k} \right), \tag{5.120b}$$

Advanced Topic I: Design of High-Order Schemes 269

$$\frac{\partial G_d^q}{\partial (\boldsymbol{U}^{(m+l_2)}, \ldots, \boldsymbol{U}^{(m-l_1)})_k} \stackrel{\mathrm{d}}{\equiv} \left.\frac{\partial G}{\partial u}\right|_{V_k} + r\delta_m^{\langle 1\rangle,q} U_k^{(m)}, \tag{5.121a}$$

$$\frac{\partial G_d^q}{\partial (\overline{\boldsymbol{U}^{(m+l_2)}}, \ldots, \overline{\boldsymbol{U}^{(m-l_1)}})_k} \stackrel{\mathrm{d}}{\equiv} \left.\frac{\partial G}{\partial \overline{u}}\right|_{V_k} + r\delta_m^{\langle 1\rangle,q} \overline{U_k^{(m)}}, \tag{5.121b}$$

$$\frac{\partial G_d^q}{\partial \delta (\boldsymbol{U}^{(m+l_2)}, \ldots, \boldsymbol{U}^{(m-l_1)})_k} \stackrel{\mathrm{d}}{\equiv} \left.\frac{\partial G}{\partial v}\right|_{V_k} + r\delta_m^{\langle 1\rangle,q}(\delta_k^{\langle 1\rangle,p} U_k^{(m)}), \tag{5.121c}$$

$$\frac{\partial G_d^q}{\partial \delta (\overline{\boldsymbol{U}^{(m+l_2)}}, \ldots, \overline{\boldsymbol{U}^{(m-l_1)}})_k} \stackrel{\mathrm{d}}{\equiv} \left.\frac{\partial G}{\partial \overline{v}}\right|_{V_k} + r\delta_m^{\langle 1\rangle,q}(\delta_k^{\langle 1\rangle,p} \overline{U_k^{(m)}}), \tag{5.121d}$$

$$r \stackrel{\mathrm{d}}{\equiv} \frac{r_\mathrm{n}}{r_\mathrm{d}}, \tag{5.122a}$$

$$r_\mathrm{n} \stackrel{\mathrm{d}}{\equiv} \delta_m^{\langle 1\rangle,q} J_\mathrm{d}(\boldsymbol{U}^{(m)}) - \sum_{k=0}^{N}{}'' \left.\frac{\partial G}{\partial u}\right|_{V_k} \delta_m^{\langle 1\rangle,q} U_k^{(m)} \Delta x$$
$$- \sum_{k=0}^{N}{}'' \left.\frac{\partial G}{\partial v}\right|_{V_k} (\delta_m^{\langle 1\rangle,q} \delta_k^{\langle 1\rangle,p} U_k^{(m)}) \Delta x, \tag{5.122b}$$

$$r_\mathrm{d} \stackrel{\mathrm{d}}{\equiv} \sum_{k=0}^{N}{}'' \left\{ \left|\delta_m^{\langle 1\rangle,q} U_k^{(m)}\right|^2 + \left|\delta_m^{\langle 1\rangle,q} \delta_k^{\langle 1\rangle,p} U_k^{(m)}\right|^2 \right\} \Delta x. \tag{5.122c}$$

The discrete quantity (5.120a) is an approximation of the discrete derivative.

5.4.3.4.2 Design of schemes With the discrete variational derivative defined as above, we can design conservative or dissipative finite difference schemes which are temporally qth-order.

Scheme 5.11 (Scheme for the PDEs 3) *Let* $U_k^{(0)} = u(k\Delta x, 0)$ *be initial values, and* $\boldsymbol{U}^{(m)}$ $(m = 1, 2, \ldots, l-2)$ *be given starting values. Then, a scheme for the PDEs 3 is given by, for* $m = l_1, l_1+1, \ldots$,

$$\delta_m^{\langle 1\rangle,q} U_k^{(m)} = -\frac{\delta G_\mathrm{d}}{\delta (\overline{\boldsymbol{U}^{(m+l_2)}}, \ldots, \overline{\boldsymbol{U}^{(m-l_1)}})_k}, \qquad k = 0, \ldots, N. \tag{5.123}$$

THEOREM 5.11 Discrete dissipation property and orders of Scheme 5.11

The scheme 5.11 is dissipative in the sense that it satisfies

$$\delta_m^{\langle 1\rangle,q} J_\mathrm{d}(\boldsymbol{U}^{(m)}) \leq 0, \qquad m = l_1, l_1+1, \ldots, \tag{5.124}$$

and it is (p,q)*th-order.*

PROOF   The same as in the preceding theorems. □

Scheme 5.12 (Scheme for the PDEs 4) *Let $U_k^{(0)} = u(k\Delta x, 0)$ be initial values, and $\boldsymbol{U}^{(m)}$ ($m = 1, 2, \ldots, l-2$) be given starting values. Then, a (p,q)th-order scheme for the PDEs 4 is given by, for $m = l_1, l_1 + 1, \ldots$,*

$$\mathrm{i}\delta_m^{\langle 1 \rangle, q} U_k{}^{(m)} = -\frac{\delta G_\mathrm{d}}{\delta(\overline{\boldsymbol{U}^{(m+l_2)}}, \ldots, \overline{\boldsymbol{U}^{(m-l_1)}})_k}, \qquad k = 0, \ldots, N. \tag{5.125}$$

THEOREM 5.12 Discrete conservation property and orders of Scheme 5.12

The scheme 5.12 is conservative in the sense that it satisfies

$$\delta_m^{\langle 1 \rangle, q} J_\mathrm{d}(\boldsymbol{U}^{(m)}) = 0, \qquad m = l_1, l_1 + 1, \ldots, \tag{5.126}$$

and it is (p,q)th-order.

PROOF   The same as in the preceding theorems. □

Chapter 6

Advanced Topic II: Design of Linearly Implicit Schemes

For nonlinear partial differential equations, we usually obtain *nonlinear* schemes by the discrete variational derivative method, inheriting the nonlinearity. In some large problems the nonlinearity can be quite crucial. In this chapter, we present ways of relaxing this restriction by considering *linearly implicit* schemes. There the dissipation or conservation property still holds in a generalized sense.

6.1 Basic Idea for Constructing Linearly Implicit Schemes

The conservative or dissipative schemes presented in the previous chapters were all nonlinearly implicit with respect to the numerical solution at the next time step. Although they had their own beauty and superiority in that they keep the desired dissipation or conservation property, the nonlinearity becomes more difficult as the sizes of the target problems increase. One way of surviving this is to look for a better nonlinear solver. In fact, recently many practical solutions for large scale nonlinear equations have been devised. See Chapter 7 on this topic.

Another practical compromise for this difficulty is to consider *linearly implicit* versions of the dissipative or conservative schemes. Linearly implicit schemes are still implicit but *linear* with respect to the numerical solution at the next time step (recall Table 1.1 in Chapter 1). Obviously they are far cheaper than the nonlinearly implicit schemes. However, they generally lose the desired dissipation or conservation property in strict sense of the word; rather they hold in some more generalized senses, as will be described below.

We have already seen the basics of constructing linearly implicit schemes in Chapter 1, but let us review them again with a new example. Let us consider the cubic nonlinear Schrödinger equation (NLS):

$$\mathrm{i}\frac{\partial u}{\partial t} = -u_{xx} - \gamma |u|^2 u, \qquad \gamma \in \mathbb{R}, \tag{6.1}$$

under the periodic boundary condition:

$$u^{(j)}(t,0) = u^{(j)}(t,L), \quad j = 0,1,2. \tag{6.2}$$

As to the NLS, we have already presented a conservative scheme in Chapter 4, which was nonlinearly implicit with respect to $U_k^{(m+1)}$.

In order to obtain a linearly implicit scheme, it is essential to understand the mechanism of how the nonlinearity in the energy is passed down to the equation through the variation calculation. In the cubic NLS equation, the nonlinear term $|u|^4$ in the local energy:

$$G(u, u_x) = -|u_x|^2 + \frac{\gamma}{2}|u|^4 \tag{6.3}$$

is the source of the nonlinear term $|u|^2 u$ in the resulting equation (6.1). In general, the power in the nonlinear term in the energy is always one higher than that in the resulting nonlinear term. Hence we easily come to the conclusion that if we want the resulting scheme to be linear, we must restrict our discrete energy function to be *quadratic*, at most. In the cubic NLS, for example, this can be accomplished by breaking down $|U_k^{(m)}|^4$ to $|U_k^{(m+1)}|^2|U_k^{(m)}|^2$, with the aid of $U_k^{(m)}$ (i.e., we make the energy function *multistep*). Its discrete variation calculation becomes

$$|U_k^{(m+1)}|^2|U_k^{(m)}|^2 - |U_k^{(m)}|^2|U_k^{(m-1)}|^2 =$$

$$|U_k^{(m)}|^2 \left(\frac{U_k^{(m+1)} + U_k^{(m-1)}}{2}\right) \overline{\left(U_k^{(m+1)} - U_k^{(m-1)}\right)}$$

$$+ |U_k^{(m)}|^2 \overline{\left(\frac{U_k^{(m+1)} + U_k^{(m-1)}}{2}\right)} \left(U_k^{(m+1)} - U_k^{(m-1)}\right). \tag{6.4}$$

Now $|U_k^{(m)}|^2 (U_k^{(m+1)} + U_k^{(m-1)})/2$, which is the approximation of $|u|^2 u$, is successfully *linear* with regard to the unknown variable $U_k^{(m+1)}$.

Including the other terms as well, we can now construct a linearly implicit scheme for the cubic NLS equation as follows. We define a multistep discrete local energy by

$$G_{\mathrm{d},k}(\boldsymbol{U}^{(m+1)}, \boldsymbol{U}^{(m)}) \stackrel{\mathrm{d}}{\equiv}$$
$$\frac{|\delta_k^+ U_k^{(m+1)}|^2 + |\delta_k^- U_k^{(m+1)}|^2 + |\delta_k^+ U_k^{(m)}|^2 + |\delta_k^- U_k^{(m)}|^2}{4}$$
$$- \frac{\gamma}{2}|U_k^{(m+1)}|^2|U_k^{(m)}|^2, \tag{6.5}$$

and accordingly the discrete global energy by

$$J_{\mathrm{d}}(\boldsymbol{U}^{(m+1)}, \boldsymbol{U}^{(m)}) \stackrel{\mathrm{d}}{\equiv} \sum_{k=0}^{N}{}'' G_{\mathrm{d},k}(\boldsymbol{U}^{(m+1)}, \boldsymbol{U}^{(m)}) \Delta x. \tag{6.6}$$

Advanced Topic II: Design of Linearly Implicit Schemes 273

We here impose the discrete periodic boundary condition:
$$U_k = U_{k \bmod N}. \tag{6.7}$$

Taking its variation we have

$$\sum_{k=0}^{N}{}'' G_{\mathrm{d},k}(\boldsymbol{U}^{(m+1)}, \boldsymbol{U}^{(m)}) - G_{\mathrm{d},k}(\boldsymbol{U}^{(m)}, \boldsymbol{U}^{(m-1)}) \Delta x =$$

$$\frac{\delta G_{\mathrm{d}}}{\delta(\boldsymbol{U}^{(m+1)}, \boldsymbol{U}^{(m)}, \boldsymbol{U}^{(m-1)})_k} \frac{U_k^{(m+1)} - U_k^{(m-1)}}{2}$$

$$+ \frac{\delta G_{\mathrm{d}}}{\delta(\overline{\boldsymbol{U}^{(m+1)}, \boldsymbol{U}^{(m)}, \boldsymbol{U}^{(m-1)}})_k} \overline{\frac{U_k^{(m+1)} - U_k^{(m-1)}}{2}}, \tag{6.8}$$

where

$$\frac{\delta G_{\mathrm{d}}}{\delta(\boldsymbol{U}^{(m+1)}, \boldsymbol{U}^{(m)}, \boldsymbol{U}^{(m-1)})_k}$$
$$= -\frac{1}{2}\delta_k^{\langle 2 \rangle}(U_k^{(m+1)} + U_k^{(m-1)}) - \frac{\gamma}{2}|U_k^{(m)}|^2 \overline{(U_k^{(m+1)} + U_k^{(m-1)})}, \tag{6.9}$$

$$\frac{\delta G_{\mathrm{d}}}{\delta(\overline{\boldsymbol{U}^{(m+1)}, \boldsymbol{U}^{(m)}, \boldsymbol{U}^{(m-1)}})_k} = \overline{\frac{\delta G_{\mathrm{d}}}{\delta(\boldsymbol{U}^{(m+1)}, \boldsymbol{U}^{(m)}, \boldsymbol{U}^{(m-1)})_k}} \tag{6.10}$$

are "three points discrete variational derivatives," which have been already introduced in Section 3.5.

With them we can now define a linearly implicit finite difference scheme as

$$\mathrm{i}\left(\frac{U_k^{(m+1)} - U_k^{(m-1)}}{2\Delta t}\right)$$
$$= \frac{\delta G_{\mathrm{d}}}{\delta(\overline{\boldsymbol{U}^{(m+1)}, \boldsymbol{U}^{(m)}, \boldsymbol{U}^{(m-1)}})_k}$$
$$= -\frac{1}{2}\delta_k^{\langle 2 \rangle}(U_k^{(m+1)} + U_k^{(m-1)}) - \frac{\gamma}{2}|U_k^{(m)}|^2(U_k^{(m+1)} + U_k^{(m-1)}). \tag{6.11}$$

Because the scheme (6.11) is linear with respect to $U_k^{(m+1)}$, we only need to solve a linear system at each time step, and therefore it is much faster than the nonlinear scheme which needs heavy iterative calculations.

The scheme conserves the discrete energy under the periodic boundary condition (the proof is omitted because it is straightforward by the construction).

PROPOSITION 6.1 Discrete energy conservation
The solution of the linearly implicit scheme (6.11) *conserves the discrete energy* (6.6) *under the periodic boundary condition* (6.7). *Namely,*

$$J_{\mathrm{d}}(\boldsymbol{U}^{(m+1)}, \boldsymbol{U}^{(m)}) = J_{\mathrm{d}}(\boldsymbol{U}^{(1)}, \boldsymbol{U}^{(0)}), \qquad m = 1, 2, 3, \ldots. \tag{6.12}$$

The discrete probability conservation law and the stability and L^2-convergence of the solution can be also established, under the periodic or zero Dirichlet boundary condition.

Scheme (6.11) is the same as the one Zhang et al. [54] proposed on the entire spatial domain $(-\infty, \infty)$. They also proved that the scheme is energy- and probability-conserving, stable, and L^2-convergent on the entire spatial domain $(-\infty, \infty)$.

6.2 Multiple-Points Discrete Variational Derivative

The concept of the three-points discrete variational derivative is further generalized to the multiple-points discrete variational derivative in order to deal with stronger nonlinearity. We consider the real-valued PDEs 1 and 2, and the complex-valued PDEs 3 and 4 (for the definitions of these PDEs, see Chapter 2).

6.2.1 For Real-Valued PDEs

Let us consider the real-valued PDEs 1 and 2. In Chapter 3, the discrete variational derivative was defined so that it satisfies the following identity (see the equation (3.32) on page 79).

$$\sum_{k=0}^{N}{}'' G_{\mathrm{d},k}(\boldsymbol{U}^{(m+1)}) - G_{\mathrm{d},k}(\boldsymbol{U}^{(m)}) \Delta x =$$
$$\sum_{k=0}^{N}{}'' \left[\left(\frac{\delta G_{\mathrm{d}}}{\delta(\boldsymbol{U}^{(m+1)}, \boldsymbol{U}^{(m)})_k} \right) (U_k^{(m+1)} - U_k^{(m)}) \right] \Delta x + B_{\mathrm{r},1}(\boldsymbol{U}^{(m+1)}, \boldsymbol{U}^{(m)}).$$
(6.13)

In view of the assumptions in Scheme 3.1 or 3.2, we see the boundary term $B_{\mathrm{r},1}(\boldsymbol{U}^{(m+1)}, \boldsymbol{U}^{(m)})$ vanishes, and the identity is simplified to

$$\sum_{k=0}^{N}{}'' G_{\mathrm{d},k}(\boldsymbol{U}^{(m+1)}) - G_{\mathrm{d},k}(\boldsymbol{U}^{(m)}) \Delta x =$$
$$\sum_{k=0}^{N}{}'' \left[\left(\frac{\delta G_{\mathrm{d}}}{\delta(\boldsymbol{U}^{(m+1)}, \boldsymbol{U}^{(m)})_k} \right) (U_k^{(m+1)} - U_k^{(m)}) \right] \Delta x. \qquad (6.14)$$

In a similar manner, let us define the *multiple-points real discrete variational derivative* as follows. First, we introduce the concept of the multiple-points discrete energy function.

DEFINITION 6.1 Multiple-points (real) discrete energy function
We call

$$G_{\mathrm{d}}(\boldsymbol{U}^{(m)}, \ldots, \boldsymbol{U}^{(m-l+2)}) : \underbrace{\mathbb{R}^{N+1} \times \cdots \times \mathbb{R}^{N+1}}_{l-1} \to \mathbb{R}^{N+1}$$

a $(l-1)$-points (real) discrete energy function. □

We use the same notation G_{d} or $G_{\mathrm{d},k}$ for the multiple-points energy function. The number of the referenced points can be understood by (the number of) their arguments. When $l = 2$, this is just the discrete energy function appearing in Chapter 3. We have already seen an example of three-points discrete variational derivative, i.e., the case where $l = 3$, in Section 3.5.

As to the multiple-points discrete energy function, let us introduce the concept of the multiple-points discrete variational derivative as follows.

DEFINITION 6.2 Multiple-points (real) discrete variational derivative We call

$$\frac{\delta G_{\mathrm{d}}}{\delta(\boldsymbol{U}^{(m)}, \ldots, \boldsymbol{U}^{(m-l+1)})} : \underbrace{\mathbb{R}^{N+1} \times \cdots \times \mathbb{R}^{N+1}}_{l} \to \mathbb{R}^{N+1}$$

an *l-points discrete variational derivative* if it satisfies

$$\sum_{k=0}^{N}{}'' \left\{ G_{\mathrm{d},k}(\boldsymbol{U}^{(m+1)}, \ldots, \boldsymbol{U}^{(m-l+3)}) - G_{\mathrm{d},k}(\boldsymbol{U}^{(m)}, \ldots, \boldsymbol{U}^{(m-l+2)}) \right\} \Delta x =$$
$$\sum_{k=0}^{N}{}'' \left[\left(\frac{\delta G_{\mathrm{d}}}{\delta(\boldsymbol{U}^{(m+1)}, \ldots, \boldsymbol{U}^{(m-l+2)})_k} \right) \left(\frac{U_k^{(m+1)} - U_k^{(m-l+2)}}{l-1} \right) \right] \Delta x, \quad (6.15)$$

under some discrete boundary condition. □

We also call it a *multiple-points discrete variational derivative*. When $l = 2$, this is just the discrete variational derivative which appeared in Chapter 3. We have already seen an example of the three-points discrete variational derivative, i.e., the case where $l = 3$, in Section 3.5.

6.2.2 For Complex-Valued PDEs

Let us consider the complex-valued single PDEs 3 and 4. In Chapter 3, the complex discrete variational derivatives were defined to satisfy the following identity (see the equation (3.75) on page 96):

$$\sum_{k=0}^{N}{}''G_{\mathrm{d},k}(\boldsymbol{U}^{(m+1)}) - G_{\mathrm{d},k}(\boldsymbol{U}^{(m)})\Delta x =$$

$$\sum_{k=0}^{N}{}''\left[\left(\frac{\delta G_{\mathrm{d}}}{\delta(\boldsymbol{U},\boldsymbol{V})}\right)_k (U_k - V_k) + \left(\frac{\delta G_{\mathrm{d}}}{\delta(\overline{\boldsymbol{U}},\overline{\boldsymbol{V}})}\right)_k (\overline{U_k - V_k})\right]\Delta x, \quad (6.16)$$

under an appropriate discrete boundary condition. In a similar manner, here we define the *multiple-points complex discrete variational derivative* as follows.

As in the real-valued case, we commence by introducing the multiple-points complex discrete energy as follows.

DEFINITION 6.3 Multiple-points complex discrete energy function We call

$$G_{\mathrm{d}}(\boldsymbol{U}^{(m)},\ldots,\boldsymbol{U}^{(m-l+2)}) : \underbrace{\mathbb{C}^{N+1} \times \cdots \times \mathbb{C}^{N+1}}_{l-1} \to \mathbb{R}^{N+1}$$

a $(l-1)$-points complex discrete energy function. □

Next, we define the multiple-points complex discrete variational derivative as follows.

DEFINITION 6.4 Multiple-points complex discrete variational derivative We call

$$\frac{\delta G_{\mathrm{d}}}{\delta(\boldsymbol{U}^{(m)},\ldots,\boldsymbol{U}^{(m-l+1)})} : \underbrace{\mathbb{C}^{N+1} \times \cdots \times \mathbb{C}^{N+1}}_{l} \to \mathbb{R}^{N+1}$$

a l-points complex discrete variational derivative *if it satisfies*

$$\sum_{k=0}^{N}{}''\left\{G_{\mathrm{d},k}(\boldsymbol{U}^{(m+1)},\ldots,\boldsymbol{U}^{(m-l+3)}) - G_{\mathrm{d},k}(\boldsymbol{U}^{(m)},\ldots,\boldsymbol{U}^{(m-l+2)})\right\}\Delta x$$

$$= \sum_{k=0}^{N}{}''\left[\left(\frac{\delta G_{\mathrm{d}}}{\delta(\boldsymbol{U}^{(m+1)},\ldots,\boldsymbol{U}^{(m-l+2)})_k}\right)\left(\frac{U_k^{(m+1)} - U_k^{(m-l+2)}}{l-1}\right)\right.$$

$$\left. + \left(\frac{\delta G_{\mathrm{d}}}{\delta(\overline{\boldsymbol{U}^{(m+1)}},\ldots,\overline{\boldsymbol{U}^{(m-l+2)}})_k}\right)\left(\frac{\overline{U_k^{(m+1)}} - \overline{U_k^{(m-l+2)}}}{l-1}\right)\right]\Delta x, \quad (6.17)$$

under some discrete boundary condition. □

We also call it a *multiple-points complex discrete variational derivative*. When $l = 2$, this is just the complex discrete variational derivative which appeared in Chapter 3.

6.3 Design of Schemes

Dissipative or conservative finite difference schemes are defined with the multiple-points discrete variational derivatives. Again we consider real-valued and complex-valued cases in order.

6.3.1 For Real-Valued PDEs

Using the multiple-points real discrete variational derivative in Definition 6.2, we define numerical schemes for the PDEs 1 as follows.

Scheme 6.1 (Scheme for the PDEs 1) Let $U_k^{(0)} = u(k\Delta x, 0)$ be initial values, and $\boldsymbol{U}^{(m)}$ ($m = 1, 2, \ldots, l-2$) be given starting values. Then, a scheme for the PDE 1 is given by, for $m = l-2, l-1, \ldots$,

$$\frac{U_k^{(m+1)} - U_k^{(m-l+2)}}{(l-1)\Delta t} = (-1)^{s+1} \delta_k^{\langle 2s \rangle} \frac{\delta G_\mathrm{d}}{\delta(\boldsymbol{U}^{(m+1)}, \ldots, \boldsymbol{U}^{(m-l+2)})_k},$$
$$k = 0, \ldots, N-1. \quad (6.18)$$

THEOREM 6.1 Discrete dissipation property of Scheme 6.1
Assume that a discrete boundary condition, which satisfies the condition (6.15), is imposed on Scheme 6.1. Then the scheme is dissipative in the sense that the inequality:

$$\sum_{k=0}^{N}{}'' \left\{ G_{\mathrm{d},k}(\boldsymbol{U}^{(m+1)}, \ldots, \boldsymbol{U}^{(m-l+3)}) - G_{\mathrm{d},k}(\boldsymbol{U}^{(m)}, \ldots, \boldsymbol{U}^{(m-l+2)}) \right\} \Delta x \leq 0$$
$$(6.19)$$

holds for $m = l-2, l-1, \ldots$.

PROOF Trivial from the identity (6.15) and the summation-by-parts formula. □

Next we define numerical schemes for the PDEs 2 as follows.

Scheme 6.2 (Scheme for the PDEs 2) Let $U_k^{(0)} = u(k\Delta x, 0)$ be initial values, and $\boldsymbol{U}^{(m)}$ ($m = 1, 2, \ldots, l-2$) be given starting values. Then, a scheme for the PDE 2 is given by, for $m = l-2, l-1, \ldots$,

$$\frac{U_k^{(m+1)} - U_k^{(m-l+2)}}{(l-1)\Delta t} = \delta_k^{\langle 2s+1 \rangle} \frac{\delta G_\mathrm{d}}{\delta(\boldsymbol{U}^{(m+1)}, \ldots, \boldsymbol{U}^{(m-l+2)})_k}. \quad (6.20)$$

THEOREM 6.2 Discrete conservation property of Scheme 6.2
Assume that a discrete boundary condition, which satisfies the condition (6.15), is imposed on Scheme 6.2. Then the scheme is conservative in the sense that the inequality:

$$\sum_{k=0}^{N}{}'' \left\{ G_{\mathrm{d},k}(\boldsymbol{U}^{(m+1)}, \ldots, \boldsymbol{U}^{(m-l+3)}) - G_{\mathrm{d},k}(\boldsymbol{U}^{(m)}, \ldots, \boldsymbol{U}^{(m-l+2)}) \right\} \Delta x = 0 \tag{6.21}$$

holds for $m = l-2, l-1, \ldots$.

PROOF Trivial from the identity (6.15) and the summation-by-parts formula. □

Note that each scheme derived from Scheme 6.1 or Scheme 6.2 is not necessarily linearly implicit at this point. When the energy function $G(u, u_x)$ is of some special form, and the discrete energy function G_d is defined appropriately, the resulting schemes 6.1 and 6.2 become linearly implicit. We present several examples below.

(A) When $G(u, u_x)$ has the nonlinear term u^{2s} $(s = 2, 3, \ldots)$, use an s-points discrete energy, where the corresponding term is

$$(U_k^{(m)})^2 (U_k^{(m-1)})^2 \cdots (U_k^{(m-s+1)})^2.$$

Then, consider the corresponding $(s+1)$-points discrete variational derivative, and define a scheme with it. The resulting scheme becomes linearly implicit.

(B) When $G(u, u_x)$ has the nonlinear term u^{2s-1} $(m = 2, 3, \ldots)$, use an s-points discrete energy, where the corresponding term is

$$\frac{1}{s+1} \sum_{i=1}^{s+1} \left(U_k^{(m+2-i)} \prod_{j \neq i} (U_k^{(m+2-j)})^2 \right).$$

For example, when there is a nonlinear term u^5 in $G(u, u_x)$, then define a three-points discrete energy function:

$$G_{\mathrm{d},k}(\boldsymbol{U}^{(m+1)}, \boldsymbol{U}^{(m)}, \boldsymbol{U}^{(m-1)}) = \frac{(U_k^{(m+1)})^2 (U_k^{(m)})^2 U_k^{(m-1)}}{3}$$
$$+ \frac{(U_k^{(m+1)})^2 U_k^{(m)} (U_k^{(m-1)})^2}{3} + \frac{U_k^{(m+1)} (U_k^{(m)})^2 (U_k^{(m-1)})^2}{3}. \tag{6.22}$$

Note also that the resulting schemes need starting values other than the initial values. The starting values should be obtained by another numerical

6.3.2 For Complex-Valued PDEs

With the multiple-points complex discrete variational derivative defined in Definition 6.4, we define numerical schemes for the PDEs 3 as follows.

Scheme 6.3 (Scheme for the PDEs 3) *Let $U_k^{(0)} = u(k\Delta x, 0)$ be initial values, and $\boldsymbol{U}^{(m)}$ ($m = 1, 2, \ldots, l-2$) be given starting values. Then, a scheme for the PDE 3 is given by, for $m = l-2, l-1, \ldots,$*

$$\frac{U_k^{(m+1)} - U_k^{(m-l+2)}}{(l-1)\Delta t} = -\frac{\delta G_{\mathrm{d}}}{\delta(\boldsymbol{U}^{(m+1)}, \ldots, \overline{\boldsymbol{U}^{(m-l+2)}})_k}, \quad k = 0, \ldots, N. \tag{6.23}$$

□

THEOREM 6.3 Discrete dissipation property of Scheme 6.3
Assume that a discrete boundary condition, which satisfies the condition (6.17), is imposed on Scheme 6.3. Then the scheme is dissipative in the sense that the inequality:

$$\sum_{k=0}^{N} {}''\left\{ G_{\mathrm{d},k}(\boldsymbol{U}^{(m+1)}, \ldots, \boldsymbol{U}^{(m-l+3)}) - G_{\mathrm{d},k}(\boldsymbol{U}^{(m)}, \ldots, \boldsymbol{U}^{(m-l+2)}) \right\} \Delta x \leq 0 \tag{6.24}$$

holds for $m = l-2, l-1, \ldots.$

PROOF Trivial from the identity (6.17) and the summation-by-parts formula. □

We define numerical schemes for the PDEs 4 as follows.

Scheme 6.4 (Scheme for the PDEs 4) *Let $U_k^{(0)} = u(k\Delta x, 0)$ be initial values, and $\boldsymbol{U}^{(m)}$ ($m = 1, 2, \ldots, l-2$) be given starting values. Then, a scheme for the PDE 4 is given by, for $m = l-2, l-1, \ldots,$*

$$\mathrm{i}\left(\frac{U_k^{(m+1)} - U_k^{(m-l+2)}}{(l-1)\Delta t}\right) = -\frac{\delta G_{\mathrm{d}}}{\delta(\boldsymbol{U}^{(m+1)}, \ldots, \overline{\boldsymbol{U}^{(m-l+2)}})_k}, \quad k = 0, \ldots, N. \tag{6.25}$$

□

THEOREM 6.4 Discrete conservation property of Scheme 6.4
Assume that a discrete boundary condition, which satisfies the condition (6.17), *is imposed on Scheme 6.4. Then the scheme is conservative in the sense that the inequality:*

$$\sum_{k=0}^{N}{}'' \left\{ G_{\mathrm{d},k}(\boldsymbol{U}^{(m+1)}, \ldots, \boldsymbol{U}^{(m-l+3)}) - G_{\mathrm{d},k}(\boldsymbol{U}^{(m)}, \ldots, \boldsymbol{U}^{(m-l+2)}) \right\} \Delta x = 0 \tag{6.26}$$

holds for $m = l-2, l-1, \ldots$.

PROOF Trivial from the identity (6.17) and the summation-by-parts formula. □

As in the real-valued case, the resulting schemes are not necessarily linearly implicit. The following is an example where we can always construct linearly implicit schemes.

(C) When $G(u, u_x)$ has the nonlinearity $|u|^{2s}$ ($s = 2, 3, \ldots$), then use an s-points discrete energy, where the corresponding term is

$$|U_k^{(m)}|^2 |U_k^{(m-1)}|^2 \cdots |U_k^{(m-s+1)}|^2.$$

The odd order NLS has the nonlinearity of this form.

6.4 Applications

Here we show several examples of linearly implicit schemes.

6.4.1 Cahn–Hilliard Equation

To illustrate how the linearization works, a dissipative linearly implicit scheme for the Cahn–Hilliard equation:

$$\frac{\partial}{\partial t} u(x,t) = \frac{\partial^2}{\partial x^2} \left(pu + ru^3 + q u_{xx} \right), \qquad p < 0, \ q < 0, \ r > 0, \tag{6.27}$$

is derived based on Scheme 6.1. The local energy G for the Cahn–Hilliard equation is

$$G(u, u_x) = \frac{1}{2} pu^2 + \frac{1}{4} ru^4 - \frac{1}{2} q(u_x)^2. \tag{6.28}$$

The equation is dissipative under the following boundary conditions:

$$\frac{\partial u}{\partial x} = \frac{\partial}{\partial x} \frac{\delta G}{\delta u} = 0, \qquad x = 0, L. \tag{6.29}$$

According to the rule (A), we decompose the nonlinear term $ru^4/4$ into $r\left(U_k^{(m+1)}\right)^2 \left(U_k^{(m)}\right)^2/4$, and accordingly define a discrete local energy by

$$G_{\mathrm{d},k}(\boldsymbol{U}^{(m+1)}, \boldsymbol{U}^{(m)}) \stackrel{\mathrm{d}}{=}$$
$$\frac{1}{2}pU_k^{(m+1)}U_k^{(m)} + \frac{1}{4}r\left(U_k^{(m+1)}\right)^2\left(U_k^{(m)}\right)^2$$
$$-\frac{1}{2}q\left(\frac{(\delta_k^+ U_k^{(m+1)})^2 + (\delta_k^- U_k^{(m+1)})^2 + (\delta_k^+ U_k^{(m)})^2 + (\delta_k^- U_k^{(m)})^2}{4}\right). \tag{6.30}$$

For the discrete energy, we consider the discrete variation as follows.

$$\sum_{k=0}^{N}{}''\left\{G_{\mathrm{d},k}(\boldsymbol{U}^{(m+1)}, \boldsymbol{U}^{(m)}) - G_{\mathrm{d},k}(\boldsymbol{U}^{(m)}, \boldsymbol{U}^{(m-1)})\right\}\Delta x$$
$$= \left(\frac{\delta G_{\mathrm{d}}}{\delta(\boldsymbol{U}^{(m+1)}, \boldsymbol{U}^{(m)}, \boldsymbol{U}^{(m-1)})_k}\right)\left(\frac{U_k^{(m+1)} - U_k^{(m-1)}}{2}\right)\Delta x, \tag{6.31}$$

where

$$\frac{\delta G_{\mathrm{d}}}{\delta(\boldsymbol{U}^{(m+1)}, \boldsymbol{U}^{(m)}, \boldsymbol{U}^{(m-1)})_k} =$$
$$pU_k^{(m)} + r\left(\frac{U_k^{(m+1)} + U_k^{(m-1)}}{2}\right)\left(U_k^{(m)}\right)^2$$
$$+ q\delta_k^{\langle 2 \rangle}\left(\frac{U_k^{(m+1)} + U_k^{(m-1)}}{2}\right). \tag{6.32}$$

Then, a dissipative scheme is derived from Scheme 6.1 as follows.

$$\frac{U_k^{(m+1)} - U_k^{(m-1)}}{2\Delta t}$$
$$= \delta_k^{\langle 2 \rangle}\left(\frac{\delta G_{\mathrm{d}}}{\delta(\boldsymbol{U}^{(m+1)}, \boldsymbol{U}^{(m)}, \boldsymbol{U}^{(m-1)})_k}\right)$$
$$= \delta_k^{\langle 2 \rangle}\left\{pU_k^{(m)} + r\left(\frac{U_k^{(m+1)} + U_k^{(m-1)}}{2}\right)\left(U_k^{(m)}\right)^2\right.$$
$$\left. + q\delta_k^{\langle 2 \rangle}\left(\frac{U_k^{(m+1)} + U_k^{(m-1)}}{2}\right)\right\}. \tag{6.33}$$

We take the following discrete boundary conditions.

$$\delta_k^{\langle 1 \rangle} U_k^{(m)}\Big|_{k=0,N} = 0, \tag{6.34}$$

$$\delta_k^{\langle 3 \rangle} U_k^{(m)}\Big|_{k=0,N} = 0. \tag{6.35}$$

This scheme is linearly implicit as expected. The dissipation property is assured by Theorem 6.1 (page 277). In [68], it has been proved that the scheme has a unique numerical solution and is unconditionally stable and convergent, provided that Δt is small enough. The proof of the stability is different from the standard (nonlinear) version given in Theorem 4.2 (page 137). Below we briefly show the proof. First, the following inequality is obtained easily.

LEMMA 6.1
The solutions $\boldsymbol{U}^{(m)}$ ($m = 0, 1, \ldots$) of the scheme (6.33) under the boundary conditions (6.34) and (6.35) satisfy

$$\sum_{k=0}^{N-1} \left(\delta_k^+ U_k^{(m)}\right)^2 \Delta x \leq \left(\frac{4}{-q}\right) \left\{ \sum_{k=0}^{N} {}''G_{\mathrm{d},k}(\boldsymbol{U}^{(1)}, \boldsymbol{U}^{(0)}) \Delta x + \frac{p^2 L}{4r} \right\}. \quad (6.36)$$

Applying the discrete Poincaré–Wirtinger inequality in Lemma 3.3 (page 122) to (6.36), we obtain the following theorem. The inequality (6.37) in the theorem implies that the difference scheme is stable for any time step m since the constants U_C^+, U_C^- and ΔU are determined by the initial state.

THEOREM 6.5
The solutions $U_k^{(m)}$ ($m = 1, 2, \ldots$) of the scheme (6.33) under the boundary conditions (6.34) and (6.35) satisfy

$$U_C^- - \Delta U \leq U_k^{(m)} \leq U_C^+ + \Delta U, \quad (6.37)$$

where

$$U_C^+ \stackrel{\mathrm{d}}{\equiv} \frac{1}{L} \max\left(\sum_{k=0}^{N} {}''U_k^{(0)} \Delta x, \sum_{k=0}^{N} {}''U_k^{(1)} \Delta x \right), \quad (6.38\mathrm{a})$$

$$U_C^- \stackrel{\mathrm{d}}{\equiv} \frac{1}{L} \min\left(\sum_{k=0}^{N} {}''U_k^{(0)} \Delta x, \sum_{k=0}^{N} {}''U_k^{(1)} \Delta x \right), \quad (6.38\mathrm{b})$$

$$\Delta U \stackrel{\mathrm{d}}{\equiv} \left[\left(\frac{4L}{-q}\right) \left\{ \sum_{k=0}^{N} {}''G_{\mathrm{d},k}(\boldsymbol{U}^{(1)}, \boldsymbol{U}^{(0)}) \Delta x + \frac{p^2 L}{4r} \right\} \right]^{1/2}. \quad (6.38\mathrm{c})$$

REMARK 6.1 Theorem 6.5 is essentially independent of both Δx and Δt except for the dependence of the constants U_C^+, U_C^- and ΔU on Δx and Δt. This means that the numerical scheme (6.33) is unconditionally stable. □

Evaluation of the regularity condition of the coefficient matrix of the newest values $\boldsymbol{U}^{(m+1)}$ in the linearly implicit scheme (6.33) gives the following theorem.

THEOREM 6.6
When the following condition

$$\Delta t < \frac{-4q}{r^2 \max\left((U_C^- - \Delta U)^4, (U_C^+ + \Delta U)^4\right)} \quad (6.39)$$

holds, the linearly implicit scheme (6.33) has a unique solution $\boldsymbol{U}^{(m+1)}$.

6.4.2 Odd-Order Nonlinear Schrödinger Equation

Here we present a linearly implicit finite difference scheme for the odd-order nonlinear Schrödinger equation (NLS):

$$\mathrm{i}\frac{\partial u}{\partial t} = -u_{xx} - \gamma |u|^{2s} u, \qquad s = 1, 2, \cdots, \quad (6.40)$$

under the periodic boundary condition. The local energy G for the equation is

$$G(u, u_x) = -|u_x|^2 + \frac{\gamma}{s+1}|u|^{2s+2}. \quad (6.41)$$

When $s = 1$, this coincides with the cubic NLS (6.1). According to the rule (C), we define $(s+1)$-points discrete energy as follows.

$$G_{\mathrm{d},k}(\boldsymbol{U}^{(m+1)}, \ldots, \boldsymbol{U}^{(m-s+1)}) \stackrel{\mathrm{d}}{=}$$

$$\frac{|\delta_k^+ U_k^{(m+1)}|^2 + |\delta_k^+ U_k^{(m)}|^2 + \cdots + |\delta_k^+ U_k^{(m-s+1)}|^2}{2(s+1)}$$

$$+ \frac{|\delta_k^- U_k^{(m+1)}|^2 + |\delta_k^- U_k^{(m)}|^2 + \cdots + |\delta_k^- U_k^{(m-s+1)}|^2}{2(s+1)}$$

$$+ \frac{\gamma}{s+1}|U_k^{(m+1)}|^2 |U_k^{(m)}|^2 \cdots |U_k^{(m-s+1)}|^2. \quad (6.42)$$

Through the discrete variation calculation we have

$$\mathrm{i}\left(\frac{U_k^{(m+1)} - U_k^{(m-s)}}{(s+1)\Delta t}\right)$$

$$= \frac{\delta G_{\mathrm{d}}}{\delta(\overline{\boldsymbol{U}^{(m+1)}}, \overline{\boldsymbol{U}^{(m)}}, \cdots, \overline{\boldsymbol{U}^{(m-s)}})_k}$$

$$= -\frac{1}{2}\delta_k^{\langle 2 \rangle}\left(U_k^{(m+1)} + U_k^{(m-s)}\right)$$

$$- \frac{\gamma}{2}|U_k^{(m)}|^2 |U_k^{(m-1)}|^2 \cdots |U_k^{(m-s+1)}|^2 \left(U_k^{(m+1)} + U_k^{(m-s)}\right). \quad (6.43)$$

6.4.3 Ginzburg–Landau Equation

Let us consider the Ginzburg–Landau equation of the form

$$\frac{\partial u}{\partial t} = pu_{xx} + q|u|^2 u + ru, \qquad p > 0,\ q < 0,\ r \in \mathbb{R}. \quad (6.44)$$

We here impose the periodic boundary condition. The local energy $G(u, u_x)$ is given as

$$G(u, u_x) = p|u_x|^2 - \frac{q}{2}|u|^4 - r|u|^2. \tag{6.45}$$

According to the rule (C), we define the discrete local energy as

$$G_{d,k}(\boldsymbol{U}^{(m+1)}, \boldsymbol{U}^{(m)}) =$$
$$\frac{p}{4}\left(|\delta_k^+ U_k^{(m+1)}|^2 + |\delta_k^+ U_k^{(m)}|^2 + |\delta_k^- U_k^{(m+1)}|^2 + |\delta_k^- U_k^{(m)}|^2\right)$$
$$-\frac{q}{2}|U_k^{(m+1)}|^2|U_k^{(m)}|^2 - \frac{r}{2}\left(U_k^{(m+1)}\overline{U_k^{(m)}} + \overline{U_k^{(m+1)}}U_k^{(m)}\right). \tag{6.46}$$

From the discrete energy we obtain a dissipative linearly implicit finite difference scheme as

$$\frac{U_k^{(m+1)} - U_k^{(m-1)}}{2\Delta t} =$$
$$p\,\delta_k^{\langle 2 \rangle}\left(\frac{U_k^{(m+1)} + U_k^{(m-1)}}{2}\right) + q|U_k^{(m)}|^2\left(\frac{U_k^{(m+1)} + U_k^{(m-1)}}{2}\right) + rU_k^{(m)}. \tag{6.47}$$

6.4.4 Zakharov Equations

We present a conservative linearly implicit scheme for the Zakharov equations:

$$iE_t + E_{xx} = nE,$$
$$n_{tt} - n_{xx} = (|E|^2)_{xx}, \tag{6.48}$$
$$E(0, x) = E_0(x), \ n(0, x) = n_0(x), \ n_t(0, x) = n_1(x).$$

We assume the periodic boundary condition. The local energy G is defined as follows.

$$G = |E_x|^2 + n|E|^2 + \frac{1}{2}(n^2 + (v_x)^2), \tag{6.49}$$

where v is an intermediate variable such that $v_t = n + |E|^2$.

According to the rule (C), we define the discrete energy as

$$G_{d,k}(\boldsymbol{E}^{(m+1)}, \boldsymbol{E}^{(m)}, \boldsymbol{n}^{(m+1)}, \boldsymbol{n}^{(m)}, \boldsymbol{v}^{(m+1)}, \boldsymbol{v}^{(m)}) \stackrel{\mathrm{d}}{\equiv}$$
$$\frac{|\delta_k^+ E_k^{(m+1)}|^2 + |\delta_k^- E_k^{(m+1)}|^2 + |\delta_k^+ E_k^{(m)}|^2 + |\delta_k^- E_k^{(m)}|^2}{4}$$
$$+ \frac{n_k^{(m+1)}|E_k^{(m)}|^2 + n_k^{(m)}|E_k^{(m+1)}|^2}{2} + \frac{(n_k^{(m+1)})^2 + (n_k^{(m)})^2}{4}$$
$$+ \frac{(\delta_k^+ v_k^{(m+1)})^2 + (\delta_k^- v_k^{(m+1)})^2 + (\delta_k^+ v_k^{(m)})^2 + (\delta_k^- v_k^{(m)})^2}{4}. \tag{6.50}$$

Advanced Topic II: Design of Linearly Implicit Schemes 285

From the discrete energy, we obtain a conservative linearly implicit scheme as:

$$\mathrm{i}\left(\frac{E_k^{(m+1)} - E_k^{(m-1)}}{2\Delta t}\right) = -\delta_k^{\langle 2\rangle}\left(\frac{E_k^{(m+1)} + E_k^{(m-1)}}{2}\right)$$
$$+ n_k^{(m)}\left(\frac{E_k^{(m+1)} + E_k^{(m-1)}}{2}\right), \quad (6.51)$$

$$\frac{n_k^{(m+1)} - n_k^{(m-1)}}{2\Delta t} = \delta_k^{\langle 2\rangle}\left(\frac{v_k^{(m+1)} + v_k^{(m-1)}}{2}\right), \quad (6.52)$$

$$\frac{v_k^{(m+1)} - v_k^{(m-1)}}{2\Delta t} = \frac{n_k^{(m+1)} + n_k^{(m-1)}}{2} + |E_k^{(m)}|^2. \quad (6.53)$$

This scheme conserves the discrete energy.

6.4.5 Newell–Whitehead Equation

Let us consider the Newell–Whitehead (NW) equation:

$$\frac{\partial u}{\partial t}(t, x, y) = \mu u - |u|^2 u + \left(\frac{\partial}{\partial x} - \frac{\mathrm{i}}{2k_c}\frac{\partial^2}{\partial y^2}\right)^2 u, \quad \begin{pmatrix}(x,y)\in [0,L_x]\times [0,L_y],\\ t>0,\\ \mu, k_c \in \mathbb{R}\end{pmatrix}. \quad (6.54)$$

Note that this is a two-dimensional problem defined on a rectangular domain. We here impose the periodic boundary condition for both x, y-directions. The local energy G is defined as follows.

$$G(u, u_x, u_{yy}) = -\mu|u|^2 + \frac{1}{2}|u|^4 + \left|u_x - \frac{\mathrm{i}}{2k_c}u_{yy}\right|^2. \quad (6.55)$$

Note also that the energy function includes higher derivative u_{yy}, which was not basically assumed in the standard procedure in Chapter 3. But it is not difficult to see that the standard procedure can be naturally extended to such a case, by repeatedly using the integration-by-parts (and accordingly, the summation-by-parts) formula.

According to the rule (C), we define the discrete energy as

$$G_{\mathrm{d},k,l}(\boldsymbol{U}^{(m+1)}, \boldsymbol{U}^{(m)}) \stackrel{\mathrm{d}}{=}$$
$$-\frac{\mu}{2}\left(U_{k,l}^{(m+1)}\overline{U_{k,l}^{(m)}} + \overline{U_{k,l}^{(m+1)}}U_{k,l}^{(m)}\right) + \frac{1}{2}|U_{k,l}^{(m+1)}|^2|U_{k,l}^{(m)}|^2$$
$$+\frac{1}{4}\left(\left|\delta_k^+ U_{k,l}^{(m+1)} - \frac{\mathrm{i}}{2k_c}\delta_l^{\langle 2\rangle}U_{k,l}^{(m+1)}\right|^2 + \left|\delta_k^+ U_{k,l}^{(m)} - \frac{\mathrm{i}}{2k_c}\delta_l^{\langle 2\rangle}U_{k,l}^{(m)}\right|^2\right)$$
$$+\frac{1}{4}\left(\left|\delta_k^- U_{k,l}^{(m+1)} - \frac{\mathrm{i}}{2k_c}\delta_l^{\langle 2\rangle}U_{k,l}^{(m+1)}\right|^2 + \left|\delta_k^- U_{k,l}^{(m)} - \frac{\mathrm{i}}{2k_c}\delta_l^{\langle 2\rangle}U_{k,l}^{(m)}\right|^2\right). \quad (6.56)$$

From the discrete energy we obtain a dissipative linearly implicit finite difference scheme as

$$\frac{U_{k,l}^{(m+1)} - U_{k,l}^{(m-1)}}{2\Delta t} = \mu U_{k,l}^{(m)} - \left|U_{k,l}^{(m)}\right|^2 \left(\frac{U_{k,l}^{(m+1)} + U_{k,l}^{(m-1)}}{2}\right)$$
$$+ \left(\delta_k^{\langle 2 \rangle} - \frac{\mathrm{i}}{k_c}\delta_k^{\langle 1 \rangle}\delta_l^{\langle 2 \rangle} - \frac{1}{4k_c^2}\delta_l^{\langle 4 \rangle}\right)\left(\frac{U_{k,l}^{(m+1)} + U_{k,l}^{(m-1)}}{2}\right). \tag{6.57}$$

We present a simple numerical example of the scheme (6.57). We consider the problem in Sakaguchi [148], in which all numerical calculations are done by discretizing x, y by a finite difference method and integrating in time by the fourth order Runge–Kutta method. We call it simply the "Runge–Kutta scheme." The initial data $u_0(x, y)$ and the other parameters are chosen to be the same as those given in [148] (i.e. $k_c = \sqrt{\pi/2}$, $\mu = 27\pi^2/800$, $L_x = 40, L_y = 20, N_x = 120, N_y = 60$, and the initial state is $u(0, x, y) = \sqrt{\mu - 9\pi/400}e^{-3\mathrm{i}\pi x/20}(1 + \mathrm{i} \cdot 0.0105 e^{3\mathrm{i}\pi/10} + \mathrm{i} \cdot 0.0095 e^{-3\mathrm{i}\pi/10}))$. With these parameters the Eckhaus instability phenomena should occur, and the reconnection process of the roll pattern proceeds, until a stable oblique roll pattern finally emerges. In the scheme (6.57), the $\boldsymbol{U}^{(1)}$ needed to start computation is obtained by the Runge–Kutta scheme. We used the CG-type solver to solve the system of linear equations.

Figure 6.1[1] shows (a) the initial state ($t = 0$), (b) the final state ($t = 100$) obtained by the our scheme (6.57) with $\Delta t = 5$, and (c) the final state ($t = 100$) by the Runge–Kutta scheme with $\Delta t = 1/120$ which is ascertained to be the maximum step size allowed for that scheme. The real part of the pattern $u(t, x, y)$ is plotted in the figure. The scheme (6.57) successfully obtained the right final pattern in spite of extraordinarily coarse time step width (600 times larger than that of the Runge–Kutta scheme).

Figure 6.2[2] shows the evolution of the discrete energy. For the Runge–Kutta scheme, which is not strictly dissipative, we computed H_d as defined in (4.175) for comparison (the dashed line). The scheme is so sensitive to Δt that the energy suddenly blows up within a few steps when Δt exceeds the limit (i.e. $\Delta t > 1/120$; not shown in the figure). In our scheme H_d as defined in (6.56) is plotted for two different values of Δt, namely $\Delta t = 5$, and $5/6$. According to the result in the figure, $\Delta t = 5/6$ is enough in our scheme to obtain the same result as the one by the Runge–Kutta scheme, which is 100 times larger than that of the Runge–Kutta scheme. When Δt is chosen extraordinarily large ($\Delta t = 5$), the progress becomes quite slow. However, the scheme strictly dissipates the energy until it reaches the same final pattern as

---

[1–3] Reprinted from *J. Comput. Phys.*, 171, T. Matsuo and D. Furihata, Dissipative or conservative finite difference schemes for complex-valued nonlinear partial differential equations, 425–447, Copyright (2001), with permission from Elsevier.

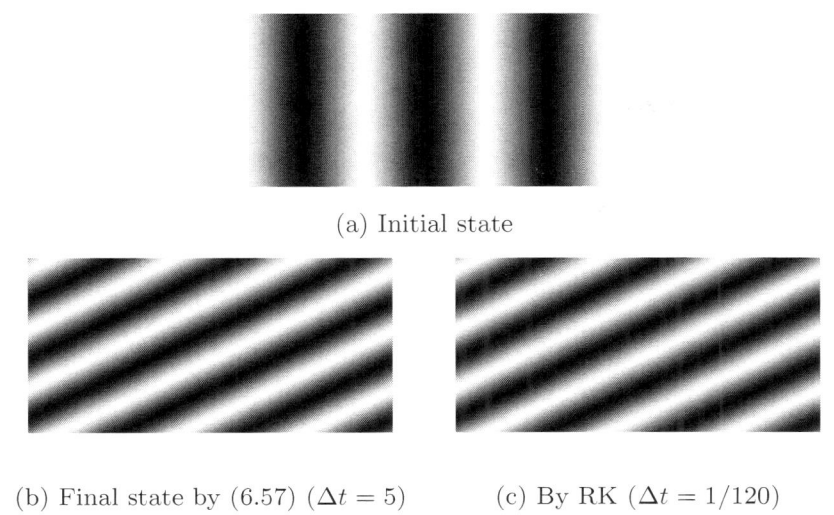

FIGURE 6.1: Initial and final states for the NW problem.

above, where the final energy is also the same as the one by the Runge–Kutta scheme (or by our scheme with fine mesh). The experiment assures us that our scheme is insensitive to Δt, i.e., numerically stable.

Table 6.1[3] shows the computation time for each scheme. We used a COMPAQ w AlphaStation XP1000 (CPU: Alpha 21264, 500 MHz) and DIGITAL Fortran 77 V5.2 compiler. Each scheme is tested several times and the mean computation time is listed in the table. According to the table, our scheme is much faster than the Runge–Kutta scheme by virtue of the large Δt and the linearity of the scheme itself.

TABLE 6.1: Computation time for each scheme (unit: seconds)

Runge–Kutta scheme	LI scheme ($\Delta t = 5$)	LI scheme ($\Delta t = 5/6$)
125	27.8	45.2

From the numerical experiment, we can conclude that the linearly implicit scheme is fast and stable.

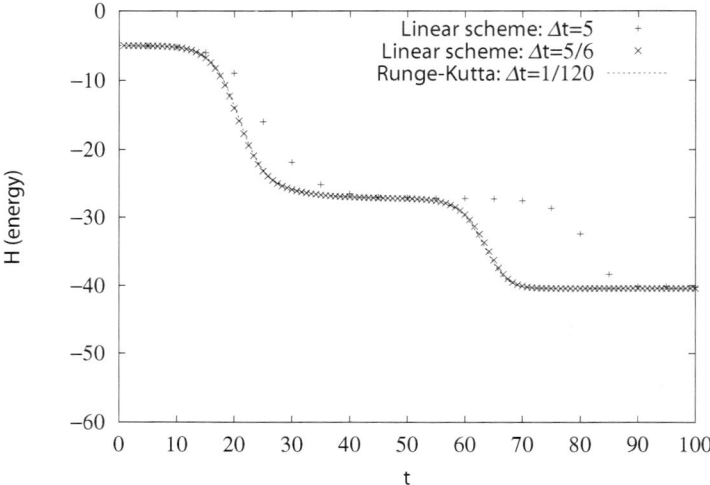

FIGURE 6.2: Dissipation of the discrete energies.

6.5 Remarks on the Stability of Linearly Implicit Schemes

As emphasized in Section 3.2.3 and other related places, in the discrete variational derivative method the choice of the discrete energy is left to each user, and the choice may severely affect the stability of the derived scheme. In this remark we present such an example, taking up the NW problem.

In order to clarify our point, let us first summarize the NW case again. The NW equation is of the form:

$$\frac{\partial u}{\partial t}(t, x, y) = \mu u - |u|^2 u + \left(\frac{\partial}{\partial x} - \frac{\mathrm{i}}{2k_c} \frac{\partial^2}{\partial y^2} \right)^2 u, \quad \begin{pmatrix} (x, y) \in [0, L_x] \times [0, L_y], \\ t > 0, \\ \mu, k_c \in \mathbb{R} \end{pmatrix}, \tag{6.54}$$

whose energy is

$$G(u, u_x, u_{yy}) = -\mu |u|^2 + \frac{1}{2} |u|^4 + \left| u_x - \frac{\mathrm{i}}{2k_c} u_{yy} \right|^2. \tag{6.55}$$

Advanced Topic II: Design of Linearly Implicit Schemes

To derive a linearly implicit scheme, we have defined a discrete energy as:

$$G_{d,k,l}(\boldsymbol{U}^{(m+1)}, \boldsymbol{U}^{(m)}) \stackrel{\mathrm{d}}{\equiv}$$
$$-\frac{\mu}{2}\left(U_{k,l}^{(m+1)}\overline{U_{k,l}^{(m)}} + \overline{U_{k,l}^{(m+1)}}U_{k,l}^{(m)}\right) + \frac{1}{2}|U_{k,l}^{(m+1)}|^2|U_{k,l}^{(m)}|^2$$
$$+\frac{1}{4}\left(\left|\delta_k^+ U_{k,l}^{(m+1)} - \frac{\mathrm{i}}{2k_c}\delta_l^{\langle 2\rangle}U_{k,l}^{(m+1)}\right|^2 + \left|\delta_k^+ U_{k,l}^{(m)} - \frac{\mathrm{i}}{2k_c}\delta_l^{\langle 2\rangle}U_{k,l}^{(m)}\right|^2\right)$$
$$+\frac{1}{4}\left(\left|\delta_k^- U_{k,l}^{(m+1)} - \frac{\mathrm{i}}{2k_c}\delta_l^{\langle 2\rangle}U_{k,l}^{(m+1)}\right|^2 + \left|\delta_k^- U_{k,l}^{(m)} - \frac{\mathrm{i}}{2k_c}\delta_l^{\langle 2\rangle}U_{k,l}^{(m)}\right|^2\right),$$
(6.58)

from which the linearly implicit scheme:

$$\frac{U_{k,l}^{(m+1)} - U_{k,l}^{(m-1)}}{2\Delta t} = \mu U_{k,l}^{(m)} - \left|U_{k,l}^{(m)}\right|^2 \left(\frac{U_{k,l}^{(m+1)} + U_{k,l}^{(m-1)}}{2}\right)$$
$$+ \left(\delta_k^{\langle 2\rangle} - \frac{\mathrm{i}}{k_c}\delta_k^{\langle 1\rangle}\delta_l^{\langle 2\rangle} - \frac{1}{4k_c^2}\delta_l^{\langle 4\rangle}\right)\left(\frac{U_{k,l}^{(m+1)} + U_{k,l}^{(m-1)}}{2}\right)$$
(6.59)

was derived. This scheme was shown to be stable by a numerical experiment.

The multiple-points discrete energy function is, however, *not unique* at all. In fact, for example, one can also define

$$G_{d,k,l}(\boldsymbol{U}^{(m+1)}, \boldsymbol{U}^{(m)}) \stackrel{\mathrm{d}}{\equiv}$$
$$-\frac{\mu}{2}\left(|U_{k,l}^{(m+1)}|^2 + |U_{k,l}^{(m)}|^2\right) + \frac{1}{2}|U_{k,l}^{(m+1)}|^2|U_{k,l}^{(m)}|^2$$
$$+\frac{1}{4}\left(\left|\delta_k^+ U_{k,l}^{(m+1)} - \frac{\mathrm{i}}{2k_c}\delta_l^{\langle 2\rangle}U_{k,l}^{(m+1)}\right|^2 + \left|\delta_k^+ U_{k,l}^{(m)} - \frac{\mathrm{i}}{2k_c}\delta_l^{\langle 2\rangle}U_{k,l}^{(m)}\right|^2\right)$$
$$+\frac{1}{4}\left(\left|\delta_k^- U_{k,l}^{(m+1)} - \frac{\mathrm{i}}{2k_c}\delta_l^{\langle 2\rangle}U_{k,l}^{(m+1)}\right|^2 + \left|\delta_k^- U_{k,l}^{(m)} - \frac{\mathrm{i}}{2k_c}\delta_l^{\langle 2\rangle}U_{k,l}^{(m)}\right|^2\right),$$
(6.60)

from which another linearly implicit scheme:

$$\frac{U_{k,l}^{(m+1)} - U_{k,l}^{(m-1)}}{2\Delta t} = \mu\left(\frac{U_{k,l}^{(m+1)} + U_{k,l}^{(m-1)}}{2}\right) - \left|U_{k,l}^{(m)}\right|^2\left(\frac{U_{k,l}^{(m+1)} + U_{k,l}^{(m-1)}}{2}\right)$$
$$+ \left(\delta_k^{\langle 2\rangle} - \frac{\mathrm{i}}{k_c}\delta_k^{\langle 1\rangle}\delta_l^{\langle 2\rangle} - \frac{1}{4k_c^2}\delta_l^{\langle 4\rangle}\right)\left(\frac{U_{k,l}^{(m+1)} + U_{k,l}^{(m-1)}}{2}\right),$$
(6.61)

is derived. The latter discrete energy (6.60) is different from the former energy (6.58) only in the first term (the terms correspond to the term $-\mu|u|^2$). Accordingly, the latter scheme (6.61) is different from the former scheme (6.59) only in the first term. At first glance, there seems no significant difference between these two choices. At least the local accuracies are more or less the same ($O(\Delta t^2, \Delta x^2)$). One might even think that the latter seem to be more natural than the former as an approximation to $\mu|u|^2$ (at least, the present authors think so!) The situation is, however, much more complicated than we simply expect.

Let us see what in fact happens if we try the scheme (6.61). The results are summarized in Figure 6.3 and Figure 6.4. Figure 6.3 shows the final pattern by the scheme (6.61). Obviously it does not match the correct pattern above. In Figure 6.4, the evolution of the discrete energy (6.60) is plotted (the points labeled as "Unstable linear scheme"). There, the other results already shown in Figure 6.2 are also drawn for comparison. As expected, the discrete energy (6.60) is strictly dissipated. This time, however, the discrete energy does not stay bounded from below but drops sharply to $-\infty$.

FIGURE 6.3: Final state by the unstable linearly implicit scheme (6.61).

These results reveal the fact that one must be very careful in choosing multiple-points energy function, in the design of linearly implicit schemes. The *generalized* dissipation (or conservation, respectively) property is different from the original dissipation (conservation) property after all, and the gap can cause severe instability in the resulting schemes.

Advanced Topic II: Design of Linearly Implicit Schemes

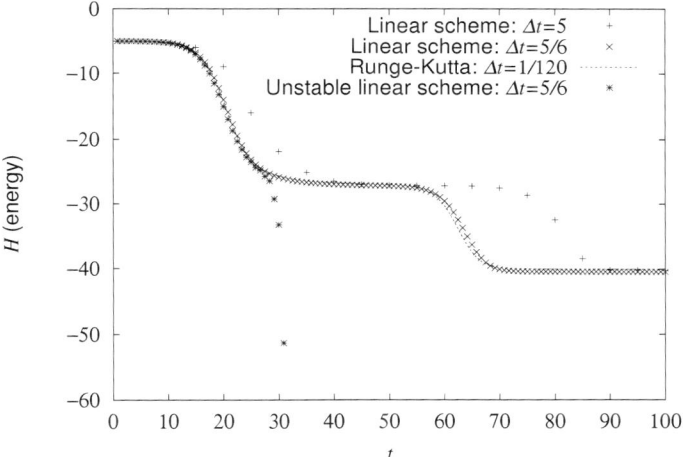

FIGURE 6.4: Dissipation of the discrete energies (with unstable case).

Chapter 7

Advanced Topic III: Further Remarks

In this chapter, other related remarks are presented. Specifically, here we consider the following three topics. First, in Section 7.1, we briefly mention the modern nonlinear solvers. Recall that, unless the linearization technique in Chapter 6 is utilized, the resulting schemes in the discrete variational derivative method are generally nonlinear, and good nonlinear solvers are inevitable. The next two sections are devoted to some basic techniques or ideas for handling spatially high-dimensional problems. In Section 7.2, we discuss how the discrete variational derivative method (DVDM) can be translated into the Galerkin framework. Finally in Section 7.3, we discuss another approach, where the DVDM is reconstructed on non-uniform grids.

7.1 Solving System of Nonlinear Equations

As we have seen in the previous chapters, when we simply apply the discrete variational derivative method (DVDM) to a nonlinear equation, the resulting scheme naturally inherits the nonlinearity, unless some linearization technique, such as the one discussed in Chapter 6, is utilized.

The issue of "either nonlinear or linear" seems to be quite a big problem, and no simple answer seems to be able to be found. From the perspective of speed, linear (linearly implicit) schemes are much more efficient; instead, they can lose stability and/or qualitative good behavior that nonlinear schemes generally have. In this respect, it seems that there are still many situations where nonlinear schemes should be willingly employed.

The biggest problem in nonlinear schemes is, of course, how we should solve them. We need some iterative solver, among which the most popular choice is the Newton method. As far as we can supply a good initial guess, the Newton method rapidly converges (if we observe the convergence per iteration), and quite reliable. The method has, however, several drawbacks when it comes to nonlinear equations arising from numerical schemes for nonlinear PDEs.

1. It requires explicit programming of Jacobian. When the target PDE is not so complicated, and furthermore we can limit ourselves to one-dimensional cases, this work is not so difficult. But as the problem grows, this rapidly becomes cumbersome. We would make many mistakes in the hand calculation of Jacobian, and even if we complete the calculation, next we have to code it without errors.

2. Even though the Newton method is surely "fast" in terms of the required iteration number, each iteration requires computational complexity of $O(N^3)$ (N is the number of spatial discretization), which is mainly consumed in the computation of the inverse of Jacobian. This is a lot compared to $O(N^2)$ cost in linearly implicit schemes. (In most finite difference schemes, Jacobian is sparse, often banded, and thus the effort can be much less. But in this case, one must write a code that reflects this sparsity. This might not be an easy task, in general, in particular when Jacobian has additional elements outside the band, as in the periodic boundary condition case.)

For these reasons, it is not practical to stick to the classical Newton method, and one may hope for a good alternative. Below several such alternatives are briefly listed.

7.1.1 Use of Numerical Newton Method Libraries

A good idea is to utilize a numerical Newton method routine, provided in various commercial and free libraries. Several examples are listed in Table 7.1. (Note that in some libraries there are several similar routines. In Table 7.1, only a typical example for each library is shown.)

TABLE 7.1: Numerical Newton Routines

Library name	Numerical Newton Method	Note
IMSL C	zero_sys_eqn	Visual Numerics, Inc.
IMSL Fortran	NEQNF	Visual Numerics, Inc.
MATLAB	fsolve	MathWorks
GNU Scientific Library	gsl_multiroot _fsolver_hybrids	free
minpack	hybrd.f	free

The numerical Newton methods above are more or less based on Powell's hybrid algorithm [141]. In most libraries, one can choose whether or not analytic Jacobian should be supplied by the user; when one refuses to supply analytic Jacobian, then in general it is substituted by finite difference Jacobian. In this case, one only has to supply the residual function $\boldsymbol{F}$ (i.e., we solve $\boldsymbol{F}(\boldsymbol{x}) = 0$), which can be easily written down based on numerical scheme. This greatly simplify the implementation of nonlinear schemes.

Although the stability can slightly decrease with finite-difference Jacobians, the convergence seem still satisfactory in most cases, as far as the authors have experienced. If one feels that the numerical Newton routines do not work well, then the initial guess at each time step can be improved by some less expensive integrators (see Section 7.1.4 below).

7.1.2 Variants of Newton Method

Mainly in the field of optimization, extensive effort has been continuously devoted to the Newton-like methods, such as the quasi-Newton method. Some of them have been translated for nonlinear equations. See, for example, [37] and references therein.

Among them, the so called "inexact Newton methods" [36] seem to deserve serious consideration. A rough outline of inexact Newton methods reads as follows.

1. Choose an initial guess $\boldsymbol{x}_0$ and $\eta_0 > 0$. Set $k = 0$.

2. Stop, if the stopping criterion is satisfied.

3. Solve $J(\boldsymbol{x}_k)\boldsymbol{d}_k = -\boldsymbol{F}(\boldsymbol{x}_k)$ *approximately*, such that $\|J(\boldsymbol{x}_k)\boldsymbol{d}_k + \boldsymbol{F}(\boldsymbol{x}_k)\| \leq \eta_k \|\boldsymbol{F}(\boldsymbol{x}_k)\|$.

4. Update $\boldsymbol{x}_{k+1} := \boldsymbol{x}_k + \boldsymbol{d}_k$, η_k (in some appropriate way), and set $k := k + 1$. Go to Step 2.

The parameter η_k is called the "forcing coefficient," and taken such that it becomes small near the convergence. For example, one can choose (see [49])

$$\eta_k = \gamma \left(\frac{\|\boldsymbol{F}(\boldsymbol{x}_k)\|}{\|\boldsymbol{F}(\boldsymbol{x}_{k-1})\|} \right)^2, \qquad \gamma \in (0, 1].$$

Notice that, in this particular choice, the forcing coefficient becomes quite small when $\boldsymbol{x}_k$ made great progress compared to $\boldsymbol{x}_{k-1}$, such that $\|\boldsymbol{F}(\boldsymbol{x}_k)\| \ll \|\boldsymbol{F}(\boldsymbol{x}_{k-1})\|$. In this case, in Step 3 the Newton direction $\boldsymbol{d}_k$ is computed relatively accurately, since it might be quite near the solution. On the other hand, when the condition is not satisfied, it is likely that the tentative solution is far from the solution, and thus the Newton direction can be less accurate. In this way, the computational cost is adaptively saved.

In order to approximately solve the system in Step 3, we can utilize efficient iterative methods such as the Krylov subspace methods. Here is another clever trick: in these iterative methods, often we only need to evaluate the matrix-vector product, say $J(\boldsymbol{x}_k)\boldsymbol{y}$ for a given vector $\boldsymbol{y}$. This can be quite efficiently computed in the present context by the finite-difference:

$$J(\boldsymbol{x}_k)\boldsymbol{y} \simeq \frac{\boldsymbol{F}(\boldsymbol{x}_k + h\boldsymbol{y}) - \boldsymbol{F}(\boldsymbol{x}_k)}{h}.$$

Supposing we already have $\boldsymbol{F}(\boldsymbol{x}_k)$ in testing the stopping criterion, we only need to evaluate $\boldsymbol{F}$ once for computing $J(\boldsymbol{x}_k)\boldsymbol{y}$! Note also that now we are interested in solving numerical schemes for PDEs; unless the target PDE has some nonlocal term, the computation of $\boldsymbol{F}$ representing the scheme is relatively fast. This means that inexact Newton methods can be quite efficient, and convenient (in that we do not have to supply the analytic Jacobian).

A possible drawback is that inexact Newton methods are relatively new, and to the best of authors' knowledge, so far they have not been implemented in major libraries such as those described above.

7.1.3 Spectral Residual Methods

Another newcomer is the "spectral residual method," which again appeared first in the context of optimization. It is a variant of the classical steepest descent method, but its characteristic is that it takes a special step length (in the line search), which is called "spectral step length." With this special choice, the line search is not necessarily monotonic, but often yields rapid convergence. The stepping was first devised in Barzilai–Borwein [13] (due to this, this group may be called "BB methods"). See also a review by Fletcher [58].

The spectral residual method was translated into nonlinear equations in La Cruz–Raydan [101]. There an algorithm called SANE (spectral approach for nonlinear equations) was proposed. Here we show the algorithm, for readers' convenience (consult the original paper for the detail).

1. Choose $\alpha_0 \in \mathbb{R}$, $\gamma > 0$, $0 < \sigma_1 < \sigma_2 < 1$, $0 < \varepsilon < 1$, M, and $\delta \in [\varepsilon, 1/\varepsilon]$. (Typical values are: $\alpha_0 = 1, \gamma = 10^{-4}, \varepsilon = 10^{-8}, M = 10, \sigma_1 = 0.1, \sigma_2 = 0.5$).

   Choose also $\boldsymbol{x}_0$ and set $k := 0$.

2. Stop, if the stopping criterion is satisfied.

   Also stop, if it stops improving: $|\boldsymbol{F}(\boldsymbol{x}_k)^\top J(\boldsymbol{x}_k)\boldsymbol{F}(\boldsymbol{x}_k))|/\|\boldsymbol{F}(\boldsymbol{x}_k)\|^2 < \varepsilon$.

3. If $\alpha_k \leq \varepsilon$ or $\alpha_k \geq 1/\varepsilon$, then set $\alpha_k := \delta$.

4. Set the direction: $\boldsymbol{d}_k := -\text{sgn}(\boldsymbol{F}(\boldsymbol{x}_k)^\top J(\boldsymbol{x}_k)\boldsymbol{F}(\boldsymbol{x}_k))\boldsymbol{F}(\boldsymbol{x}_k)$.

5. If $f(\boldsymbol{x}_k + \alpha_k \boldsymbol{d}_k) \leq \max_{0 \leq j \leq \min(k,M)} f(\boldsymbol{x}_{k-j}) + 2\gamma\alpha_k \boldsymbol{F}(\boldsymbol{x}_k)^\top J(\boldsymbol{x}_k)\boldsymbol{d}_k$, then proceed to Step 7.

6. Choose $\sigma \in [\sigma_1, \sigma_2]$, set $\alpha_k := \sigma\alpha_k$, and return to Step 5.

7. Set $\boldsymbol{x}_{k+1} := \boldsymbol{x}_k + \alpha_k \boldsymbol{d}_k$, and compute $\boldsymbol{y}_k := \boldsymbol{F}(\boldsymbol{x}_{k+1}) - \boldsymbol{F}(\boldsymbol{x}_k)$.

8. Set $\alpha_{k+1} := \text{sgn}(\boldsymbol{F}(\boldsymbol{x}_k)^\top J(\boldsymbol{x}_k)\boldsymbol{F}(\boldsymbol{x}_k)) \cdot \frac{\boldsymbol{d}_k^\top \boldsymbol{d}_k}{\boldsymbol{d}_k^\top \boldsymbol{y}_k} \cdot \alpha_k$.

   Set $k := k+1$, and return to Step 2.

The method can be summarized as follows.

- Generally speaking, an algorithm for optimization (minimization) can be translated for nonlinear equations $\boldsymbol{F}(\boldsymbol{x}) = 0$ by introducing the merit function $f(\boldsymbol{x}) = \|\boldsymbol{F}\|_2^2$. In this way, however, we need the computation of the gradient $\nabla f = J(\boldsymbol{x})^\top \boldsymbol{F}$, which cannot be efficiently computed (compare with the finite difference approximation of $J(\boldsymbol{x})\boldsymbol{F}$ above).

  To avoid this inconvenience, the SANE algorithm gives up the use of the gradient information, and instead simply employs the residual $\boldsymbol{F}(\boldsymbol{x})$ as the line search direction. More precisely, it considers the direction $-\mathrm{sgn}(\boldsymbol{F}(\boldsymbol{x}_k)^\top J(\boldsymbol{x}_k)\boldsymbol{F}(\boldsymbol{x}_k))\boldsymbol{F}(\boldsymbol{x}_k)$, which is expected to be a descent direction. It still needs Jacobian, but it can be approximated by the finite-difference.

- In order to compensate for the simplification, SANE is equipped with a nonmonotone line search algorithm (Step 5). It guarantees the global convergence of the algorithm, at least to some extent.

- SANE eventually ends in the following three patterns: (i) it successfully terminates with the stopping criterion satisfied ($\boldsymbol{F} \simeq 0$); (ii) it stops improving $|\boldsymbol{F}(\boldsymbol{x}_k)^\top J(\boldsymbol{x}_k)\boldsymbol{F}(\boldsymbol{x}_k))|/\|\boldsymbol{F}(\boldsymbol{x}_k)\|^2 < \varepsilon$, which means the residual vector $\boldsymbol{F}(\boldsymbol{x}_k)$ is orthogonal to the gradient $\nabla f(\boldsymbol{x}_k) = J(\boldsymbol{x}_k)\boldsymbol{F}(\boldsymbol{x}_k)$ and no further improvement is possible using the direction; (iii) too many iterations in the line search (Steps 5 and 6), or in the outer loop (Steps 2 through 8).

The SANE algorithm was then further extended to DF-SANE (derivative free SANE) algorithm [100], where even the Jacobian included in the sgn function is eliminated, so that the algorithm is totally derivative free. DF-SANE simply tries both $\pm \boldsymbol{F}(\boldsymbol{x})$ to find the descent direction. Interested readers can find the concrete algorithm in [100]. Below we simply call SANE and DF-SANE the "SANE algorithms" unless otherwise explicitly stated.

From a theoretical point of view, it seems that SANE algorithms are less understood compared to the original spectral residual methods for optimization. In particular, how the choice of $\boldsymbol{F}(\boldsymbol{x}_k)$ as the search direction affects the practical behavior of the algorithm near the solution is not well understood. It is chosen solely from the point of efficiency.

Still, it is quite interesting to observe that the practical behavior of the SANE algorithms is often surprisingly satisfactory. This was reported in the original papers on the SANE algorithms (for some test problems,) and also confirmed by the present authors in several numerical schemes appearing in this book. We observed the following.

- In general, the SANE algorithms can be quite unstable for bad initial guesses. By "unstable" we mean such cases that the algorithms do not terminate in prescribed maximum iterations (the case (iii) above). The necessity of a good initial guess is much stronger than for the Newton methods, including the numerical Newton methods.

- In the DVDM schemes, the SANE algorithms work well if the time step size Δt is chosen small enough. Note that by choosing Δt small enough, the numerical solution at step m should be a good initial guess for SANE algorithms. It can be further improved by utilizing the idea of predictor–corrector stepping (see Section 7.1.4,) if needed.

- When the SANE algorithms work, they are surprisingly fast, compared to other algorithms.

A FORTRAN implementation of DF-SANE can be found in the homepage of the authors of [100]. An R implementation is provided in the BB package of R. It seems standard libraries have not yet supported SANE algorithms.

7.1.4 Implementation as a Predictor–Corrector Method

In order to considerably minimize the computational effort in nonlinear solvers, an implementation as a predictor–corrector method often deserves consideration. That is, one step of a DVDM scheme can be combined with some other (cheap) numerical method in the following way.

1. Compute a numerical solution at time step $m + 1$ with some other numerical method (for example, the explicit Runge–Kutta methods).

2. Then solve the DVDM scheme by some iterative solver, with the solution above as its initial guess.

Whether or not it helps minimizing the computational effort seems to depend on the problem.

7.2 Switch to Galerkin Framework

Basically we are concerned with finite-difference DVDM schemes. As has been repeatedly emphasized, the most essential tool is the summation-by-parts formula. A consequent problem is that such formulas hold basically only in the simplest case, i.e., spatially one dimensional cases with equispaced meshes, and higher dimensional problems cannot be handled immediately unless the spatial domain is rectangular and the problem can be reduced to essentially one-dimensional problems. For example, in Section 4.1.1.5, we have considered the two-dimensional Cahn–Hilliard equation on a rectangular domain.

In order to adapt to more general cases, effort for extending the DVDM has been reported recently. In this book, a challenge on Voronoi mesh will be briefly shown in Section 7.3. Yaguchi–Matsuo–Sugihara [166] employed the concept of "computational space," and generalized the discussion of the DVDM to nonuniform grids. In a different study by the same authors, the combination of the discrete variational derivative method and the so-called "mimetic" spatial discretization has been considered [165] (for the "mimetic" discretization, see the references therein).

Another simple approach is to give up the finite difference method, and switch to the Galerkin framework [117]. In particular, we are interested in the finite element method. Below, we briefly summarize the method for the target PDEs 1, 2, and 4 described in Chapter 2. Extension to other cases is not so difficult. In order to keep notation simple, in what follows the discussion is given mainly in a one-dimensional setting. Still, the basic philosophy easily carries to multi-dimensional settings. An example of a two-dimensional problem is shown in the end of this section, which discusses several dissipative finite element schemes for the time-dependent Ginzburg–Landau equation for superconductivity.

In this project, we insist that the resulting schemes can be implemented only with cheap H^1-elements (i.e., we formulate the framework so that it does not necessarily require C^1 or any smoother elements). This is because when we hope to try two- or three-dimensional problems, smooth elements would be too expensive, and such schemes should be less attractive, even if they have advantageous conservation or dissipation properties.

In the rest of this section, the following notation is used. We denote the Galerkin approximate solution by $u^{(m)} \simeq u(x, m\Delta t)$ (similar expression will be used for other variables). Trial and test function spaces are denoted by S_j and W_j ($j = 1, 2, \ldots$). We use the standard notation on function spaces such as $L^2(\Omega)$ and $H^1(\Omega)$, and associated inner products. We denote the circle of length L by $\mathbb{S}$, which is meant to denote the space with the periodic boundary condition.

7.2.1 Design of Galerkin Schemes

7.2.1.1 Real-Valued Dissipative PDEs 1

We commence by introducing the concept of "discrete partial derivatives," which replaces the discrete *variational* derivatives in the previous chapters. We here suppose again that local energy is of the form

$$G(u, u_x) = \sum_{l=1}^{M} f_l(u) g_l(u_x), \qquad (7.1)$$

where $M \in \{1, 2, \ldots\}$, and f_l, g_l are real-valued functions (recall (3.22))[1]. Then "discrete partial derivatives" are defined as follows.

$$\frac{\partial G_d}{\partial(u^{(m+1)}, u^{(m)})} \stackrel{d}{\equiv} \sum_{l=1}^{M} \left(\frac{f_l(u^{(m+1)}) - f_l(u^{(m)})}{u^{(m+1)} - u^{(m)}} \right) \left(\frac{g_l(u_x^{(m+1)}) + g_l(u_x^{(m)})}{2} \right), \quad (7.2a)$$

$$\frac{\partial G_d}{\partial(u_x^{(m+1)}, u_x^{(m)})} \stackrel{d}{\equiv} \sum_{l=1}^{M} \left(\frac{f_l(u^{(m+1)}) + f_l(u^{(m)})}{2} \right) \left(\frac{g_l(u_x^{(m+1)}) - g_l(u_x^{(m)})}{u_x^{(m+1)} - u_x^{(m)}} \right). \quad (7.2b)$$

They correspond to $\partial G/\partial u$ and $\partial G/\partial u_x$, respectively. Compare them with the finite difference versions (3.27a)–(3.27c).

It can be easily verified that the following discrete chain rule holds (hereafter $G(u^{(m)}, u_x^{(m)})$ is abbreviated as $G(u^{(m)})$ to save space).

THEOREM 7.1 Discrete chain rule (real-valued case)
The discrete partial derivatives (7.2a) and (7.2b) satisfy the following identity.

$$\frac{1}{\Delta t} \int_0^L \left(G(u^{(m+1)}) - G(u^{(m)}) \right) dx$$
$$= \int_0^L \left\{ \frac{\partial G_d}{\partial(u^{(m+1)}, u^{(m)})} \left(\frac{u^{(m+1)} - u^{(m)}}{\Delta t} \right) \right.$$
$$\left. + \frac{\partial G_d}{\partial(u_x^{(m+1)}, u_x^{(m)})} \left(\frac{u_x^{(m+1)} - u_x^{(m)}}{\Delta t} \right) \right\} dx. \quad (7.3)$$

Now we are in a position to describe the schemes for the equation (2.14). The simplest case $s = 0$ and general cases $s = 1, 2, \ldots$ are treated separately. We use a set of trial and test function spaces S_1 and W_1.

Scheme 7.1 (Galerkin scheme for $s = 0$) Suppose $u^{(0)}(x)$ is given in S_1. Find $u^{(m)} \in S_1$ ($m = 1, 2, \ldots$) such that, for any $v \in W_1$,

$$\left(\frac{u^{(m+1)} - u^{(m)}}{\Delta t}, v \right) = -\left(\frac{\partial G_d}{\partial(u^{(m+1)}, u^{(m)})}, v \right) - \left(\frac{\partial G_d}{\partial(u_x^{(m+1)}, u_x^{(m)})}, v_x \right)$$
$$+ \left[\frac{\partial G_d}{\partial(u_x^{(m+1)}, u_x^{(m)})} v \right]_0^L. \quad (7.4)$$

---
[1] In the preceding chapters, we used $\widetilde{M}$ for original energy function, and M for its finite difference approximation. In the Galerkin framework, however, we do not have to distinguish these two, and we simply write both by M.

Because the discrete partial derivatives (7.2a) and (7.2b) do not include second derivatives, the scheme can be implemented using only H^1-elements, such as the standard piecewise linear function space. The scheme is dissipative in the following sense.

THEOREM 7.2 Dissipation property of Scheme 7.1
Assume that boundary conditions and the trial and test spaces are set such that

$$\left[\frac{\partial G_{\mathrm{d}}}{\partial(u_x^{(m+1)}, u_x^{(m)})} \left(\frac{u^{(m+1)} - u^{(m)}}{\Delta t} \right) \right]_0^L = 0, \qquad (7.5)$$

and $(u^{(m+1)} - u^{(m)})/\Delta t \in W_1$ holds. Then Scheme 7.1 is dissipative in the sense that

$$\frac{1}{\Delta t} \int_0^L \left(G(u^{(m+1)}) - G(u^{(m)}) \right) \mathrm{d}x \leq 0, \qquad m = 0, 1, 2, \ldots .$$

PROOF

$$\frac{1}{\Delta t} \int_0^L \left(G(u^{(m+1)}) - G(u^{(m)}) \right) \mathrm{d}x$$

$$= \left(\frac{\partial G_{\mathrm{d}}}{\partial(u^{(m+1)}, u^{(m)})}, \frac{u^{(m+1)} - u^{(m)}}{\Delta t} \right) + \left(\frac{\partial G_{\mathrm{d}}}{\partial(u_x^{(m+1)}, u_x^{(m)})}, \frac{u_x^{(m+1)} - u_x^{(m)}}{\Delta t} \right)$$

$$= - \left\| \frac{u^{(m+1)} - u^{(m)}}{\Delta t} \right\|_2^2 + \left[\frac{\partial G_{\mathrm{d}}}{\partial(u_x^{(m+1)}, u_x^{(m)})} \left(\frac{u^{(m+1)} - u^{(m)}}{\Delta t} \right) \right]_0^L \leq 0.$$

The first equality is by Theorem 7.1. The second one is shown by making use of expression (7.4) and the assumption $(u^{(m+1)} - u^{(m)})/\Delta t \in W_1$. The inequality is shown by the assumption (7.5). □

The assumption (7.5) corresponds to the condition (2.16). The assumption $(u^{(m+1)} - u^{(m)})/\Delta t \in W_1$ is an additional condition for the dissipation property, which can be usually satisfied with natural choices of S_1 and W_1. For example, when the Dirichlet boundary conditions $u(0) = a$, $u(L) = b$ are imposed, it is natural to take $S_1 = \{u \,|\, u(0) = a, u(L) = b\}$ and $W_1 = \{v \,|\, v(0) = 0, v(L) = 0\}$. In this setting the assumption is satisfied.

Next we proceed to the general case $s \geq 1$. We first observe that by recursively introducing intermediate variables: $p_1 = -(p_2)_{xx}$, ..., $p_{s-1} = -(p_s)_{xx}$, and $p_s = \delta G/\delta u$, the original equation (2.14) can be rewritten as a system of equations $u_t = (p_s)_{xx}$, $p_{j-1} = -(p_j)_{xx}$ ($j \in J$), and $p_s = \delta G/\delta u$, where the set $J = \{2, \ldots, s\}$ when $s \geq 2$ or $J = \emptyset$ when $s = 1$. This leads us to the following scheme. We assume that trial spaces $S_1, \ldots, S_{s+1}$, and test spaces $W_1, \ldots, W_{s+1}$ are prepared.

Scheme 7.2 (Galerkin scheme for $s \geq 1$) *Suppose that $u^{(0)}(x)$ is given in S_{s+1}. Find $u^{(m+1)} \in S_{s+1}$, $p_1^{(m+\frac{1}{2})} \in S_1$, ..., $p_s^{(m+\frac{1}{2})} \in S_s$ $(m = 0, 1, \ldots)$ such that, for any $v_1 \in W_1$, ..., $v_{s+1} \in W_{s+1}$,*

$$\left(\frac{u^{(m+1)} - u^{(m)}}{\Delta t}, v_1\right) = -\left((p_1^{(m+\frac{1}{2})})_x, (v_1)_x\right) + \left[(p_1^{(m+\frac{1}{2})})_x v_1\right]_0^L, \quad (7.6a)$$

$$\left(p_{j-1}^{(m+\frac{1}{2})}, v_j\right) = \left((p_j^{(m+\frac{1}{2})})_x, (v_j)_x\right) - \left[(p_j^{(m+\frac{1}{2})})_x v_j\right]_0^L, \quad (7.6b)$$

$$\left(p_s^{(m+\frac{1}{2})}, v_{s+1}\right) = \left(\frac{\partial G_d}{\partial(u^{(m+1)}, u^{(m)})}, v_{s+1}\right)$$
$$+ \left(\frac{\partial G_d}{\partial(u_x^{(m+1)}, u_x^{(m)})}, (v_{s+1})_x\right)$$
$$- \left[\frac{\partial G_d}{\partial(u_x^{(m+1)}, u_x^{(m)})} v_{s+1}\right]_0^L, \quad (7.6c)$$

where $j \in J$.

The equation (7.6b) is ignored when $J = \emptyset$. This scheme can be also implemented only with H^1-elements. The dissipation property is summarized in the next theorem.

THEOREM 7.3 Dissipation property of Scheme 7.2
Assume that boundary conditions and the trial and test spaces are set such that (i) the condition (7.5) is satisfied; (ii) $\left[(p_j^{(m+\frac{1}{2})})_x \cdot p_{s+1-j}^{(m+\frac{1}{2})}\right]_0^L = 0$ $(j = 1, 2, \ldots, s)$; (iii) $(u^{(m+1)} - u^{(m)})/\Delta t \in W_{s+1}$; and (iv) $W_j \supseteq S_{s+1-j}$ $(j = 1, 2, \ldots, s)$. Then Scheme 7.2 is dissipative in the sense that

$$\frac{1}{\Delta t} \int_0^L \left(G(u^{(m+1)}) - G(u^{(m)})\right) dx \leq 0, \quad m = 0, 1, 2, \ldots.$$

PROOF

$$\frac{1}{\Delta t} \int_0^L \left(G(u^{(m+1)}) - G(u^{(m)})\right) dx$$
$$= \left(\frac{\partial G_d}{\partial(u^{(m+1)}, u^{(m)})}, \frac{u^{(m+1)} - u^{(m)}}{\Delta t}\right) + \left(\frac{\partial G_d}{\partial(u_x^{(m+1)}, u_x^{(m)})}, \frac{u_x^{(m+1)} - u_x^{(m)}}{\Delta t}\right)$$
$$= \left(p_s^{(m+\frac{1}{2})}, \frac{u^{(m+1)} - u^{(m)}}{\Delta t}\right) + \left[\frac{\partial G_d}{\partial(u_x^{(m+1)}, u_x^{(m)})}\left(\frac{u^{(m+1)} - u^{(m)}}{\Delta t}\right)\right]_0^L$$
$$= -\left((p_1^{(m+\frac{1}{2})})_x, (p_s^{(m+\frac{1}{2})})_x\right) + \left[(p_1^{(m+\frac{1}{2})})_x p_s^{(m+\frac{1}{2})}\right]_0^L.$$

The second equality is shown by using equation (7.6c) with $v_{s+1} = (u^{(m+1)} - u^{(m)})/\Delta t$. The third equality is given by using equation (7.6a) with $v_1 = p_s^{(m+\frac{1}{2})}$ and the assumption $S_s \subseteq W_1$. By repeatedly making use of equation (7.6b) with $j = s, 2, s-1, 3, \ldots$ in this order, which is allowed by the assumption (iv), it can be seen that the right-hand side is equal to $-\|(p_{(s+1)/2}^{(m+\frac{1}{2})})_x\|_2^2$ when s is odd, or $-\|p_{s/2}^{(m+\frac{1}{2})}\|_2^2$ otherwise, and so the proof is complete. All the boundary terms vanish as a result of the boundary condition assumptions. □

REMARK 7.1 The assumption (ii) in Theorem 7.3 corresponds to the condition (2.17) (although the latter is written in weaker form). This can be checked as follows. Recall that by definition $p_s = \delta G/\delta u$, $p_{s-1} = -(\delta G/\delta u)^{(2)}$, ..., $p_1 = (-1)^{2s-1}(\delta G/\delta u)^{(2s-2)}$ (the superscripts denote the number of differentiations). Substituting them into $(p_j)_x \cdot p_{s+1-j}$, we understand that it means

$$\left[\left(\frac{\delta G}{\delta u}\right)^{(2s-2j+1)} \cdot \left(\frac{\delta G}{\delta u}\right)^{(2j-2)}\right]_0^L = 0, \qquad j = 1, 2, \ldots, s.$$

The superscripts cover $(2s-1, 0), (2s-3, 2), \ldots, (1, 2s-2)$. This exactly corresponds to (2.17). The assumptions (iii) and (iv) are purely additional conditions for the discrete dissipation property, which are likely to be satisfied in most settings of trial and test spaces. □

7.2.1.2 Real-Valued Conservative PDEs 2

Conservative schemes for the target PDEs 2 are presented using the discrete partial derivatives introduced in the previous section. The simplest case $s = 1$ and general cases $s = 2, 3, \ldots$ are treated separately. Let $S_1, \ldots, S_{s+1}$ be trial spaces, and $W_1, \ldots, W_{s+1}$ be test spaces.

Scheme 7.3 (Galerkin scheme for $s = 1$) Suppose that $u^{(0)}(x)$ is given in S_2. Find $u^{(m+1)} \in S_2$, $p_1^{(m+\frac{1}{2})} \in S_1$ such that, for any $v_1 \in W_1$, $v_2 \in W_2$,

$$\left(\frac{u^{(m+1)} - u^{(m)}}{\Delta t}, v_1\right) = \left((p_1^{(m+\frac{1}{2})})_x, v_1\right) \tag{7.7}$$

$$\left(p_1^{(m+\frac{1}{2})}, v_2\right) = \left(\frac{\partial G_d}{\partial(u^{(m+1)}, u^{(m)})}, v_2\right) + \left(\frac{\partial G_d}{\partial(u_x^{(m+1)}, u_x^{(m)})}, (v_2)_x\right)$$

$$- \left[\frac{\partial G_d}{\partial(u_x^{(m+1)}, u_x^{(m)})} v_2\right]_0^L. \tag{7.8}$$

THEOREM 7.4 Conservation property of Scheme 7.3
Assume that boundary conditions and the trial and test spaces are set such that (i) *the condition* (7.5) *is satisfied;* (ii) $\left[(p_1^{(m+\frac{1}{2})})^2\right]_0^L = 0$; (iii) $(u^{(m+1)} - u^{(m)})/\Delta t \in W_2$; *and* (iv) $S_1 \subseteq W_1$. *Then Scheme 7.3 is conservative in the sense that*

$$\frac{1}{\Delta t} \int_0^L \left(G(u^{(m+1)}) - G(u^{(m)}) \right) dx = 0, \qquad m = 0, 1, 2, \ldots.$$

PROOF

$$\frac{1}{\Delta t} \int_0^L \left(G(u^{(m+1)}) - G(u^{(m)}) \right) dx$$
$$= \left(p_1^{(m+\frac{1}{2})}, \frac{u^{(m+1)} - u^{(m)}}{\Delta t} \right) + \left[\frac{\partial G_d}{\partial (u_x^{(m+1)}, u_x^{(m)})} \left(\frac{u^{(m+1)} - u^{(m)}}{\Delta t} \right) \right]_0^L$$
$$= \left((p_1^{(m+\frac{1}{2})})_x, p_1^{(m+\frac{1}{2})} \right) = 0.$$

The first equality is shown by using equation (7.8) with $v_2 = (u^{(m+1)} - u^{(m)})/\Delta t$, while the second equality is given by using equation (7.7) with $v_1 = p_1^{(m+\frac{1}{2})}$ and the assumption $S_1 \subseteq W_1$. The last equality is from the assumption (ii). □

In order to describe the scheme for $s \geq 2$, let us define the set $J = \{2, \ldots, s\} \setminus \{n+1\}$ when $s = 2n$ ($n = 1, 2, \ldots$), or $J = \{2, \ldots, s\} \setminus \{n\}$ when $s = 2n - 1$ ($n = 2, 3, \ldots$).

Scheme 7.4 (Galerkin scheme for $s \geq 2$) *Suppose that $u^{(0)}(x)$ is given in S_{s+1}. Find $u^{(m+1)} \in S_{s+1}$, $p_1^{(m+\frac{1}{2})} \in S_1$, ..., $p_s^{(m+\frac{1}{2})} \in S_s$ ($m = 0, 1, \ldots$) such that, for any $v_1 \in W_1$, ..., $v_{s+1} \in W_{s+1}$,*

$$\left(\frac{u^{(m+1)} - u^{(m)}}{\Delta t}, v_1 \right) = -\left((p_1^{(m+\frac{1}{2})})_x, (v_1)_x \right) + \left[(p_1^{(m+\frac{1}{2})})_x v_1 \right]_0^L, \quad (7.9a)$$

$$\left(p_{j-1}^{(m+\frac{1}{2})}, v_j \right) = -\left((p_j^{(m+\frac{1}{2})})_x, (v_j)_x \right) + \left[(p_j^{(m+\frac{1}{2})})_x v_j \right]_0^L, \quad (7.9b)$$

$$\left(p_n^{(m+\frac{1}{2})}, (v_{n+1})_x \right) = \left((p_{n+1}^{(m+\frac{1}{2})})_x, (v_{n+1})_x \right) \quad (when\ s = 2n), \quad (7.9c)$$

$$\left(p_{n-1}^{(m+\frac{1}{2})}, v_n \right) = \left((p_n^{(m+\frac{1}{2})})_x, v_n \right) \quad (when\ s = 2n - 1), \quad (7.9d)$$

Advanced Topic III: Further Remarks

$$\left(p_s^{(m+\frac{1}{2})}, v_{s+1}\right) = \left(\frac{\partial G_{\mathrm{d}}}{\partial(u^{(m+1)}, u^{(m)})}, v_{s+1}\right)$$
$$+ \left(\frac{\partial G_{\mathrm{d}}}{\partial(u_x^{(m+1)}, u_x^{(m)})}, (v_{s+1})_x\right)$$
$$- \left[\frac{\partial G_{\mathrm{d}}}{\partial(u_x^{(m+1)}, u_x^{(m)})} v_{s+1}\right]_0^L \quad (7.9\mathrm{e})$$

where $j \in J$.

The equation (7.9b) is dropped when $J = \emptyset$. The conservation property is summarized in the next theorem.

THEOREM 7.5 Conservation property of Scheme 7.4

Assume that boundary conditions and trial and test spaces are set such that (i) *the condition* (7.5) *is satisfied;* (ii) $\left[(p_n^{(m+\frac{1}{2})})^2\right]_0^L = 0$ *and* $\left[(p_j^{(m+\frac{1}{2})})_x p_{s+1-j}^{(m+\frac{1}{2})}\right]_0^L = 0$ $(j \in J)$; (iii) $(u^{(m+1)} - u^{(m)})/\Delta t \in W_{s+1}$; *and* (iv) $W_j \supseteq S_{s+1-j}$ $(j = 1, \ldots, s)$. *Then Scheme 7.3 is conservative in the sense that*

$$\frac{1}{\Delta t}\int_0^L \left(G(u^{(m+1)}) - G(u^{(m)})\right)\mathrm{d}x = 0, \quad m = 0, 1, 2, \ldots.$$

PROOF The proof is similar to Theorem 7.3.

$$\frac{1}{\Delta t}\int_0^L \left(G(u^{(m+1)}) - G(u^{(m)})\right)\mathrm{d}x = -\left((p_1^{(m+\frac{1}{2})})_x, (p_s^{(m+\frac{1}{2})})_x\right)$$
$$= \begin{cases} -\left((p_n^{(m+\frac{1}{2})})_x, (p_{n+1}^{(m+\frac{1}{2})})_x\right) & \text{(when } s = 2n\text{),} \\ \left(p_{n-1}^{(m+\frac{1}{2})}, p_n^{(m+\frac{1}{2})}\right) & \text{(when } s = 2n - 1\text{),} \end{cases}$$
$$= (-1)^{s+1}\left(p_n^{(m+\frac{1}{2})}, (p_n^{(m+\frac{1}{2})})_x\right) = 0.$$

In the second equality the equation (7.9b) is repeatedly used. The third equality is either from (7.9c) or (7.9d). □

REMARK 7.2 The above schemes in this paper are more or less in mixed formulation (see, for example, [21]). The underlying weak forms are, however, carefully chosen for the targeted conservation/dissipation properties. Below we illustrate this using as an example the linear third-order dispersive equation $u_t = u_{xxx}$ under the periodic boundary condition. It is a special case of (2.28) with $s = 1$, $G = u^2/2$, and thus $\int_0^L (u^2/2)\mathrm{d}x$ is an invariant. Suppose that a grid and accordingly a periodic piecewise linear function space

$\mathbb{S}$ over the grid are appropriately given. Then the most straightforward mixed formulations of the problem might be to: find $u(\cdot,t), p(\cdot,t) \in \mathbb{S}$ such that

$$(u_t, v) = (p_x, v), \ (p, w) = -(u_x, w_x), \quad \forall v, w \in \mathbb{S}, \quad (7.10)$$

or

$$(u_t, v) = -(p_x, v_x), \ (p, w) = (u_x, w), \quad \forall v, w \in \mathbb{S}. \quad (7.11)$$

On the other hand, Scheme 7.4 suggests a slightly different form:

$$(u_t, v) = -(p_x, v_x), \ (p, w_x) = (u_x, w_x), \quad \forall v, w \in \mathbb{S}. \quad (7.12)$$

(Actually Scheme 7.4 literally suggests $(u_t, v) = -(p_x, v_x)$, $(p_1, w_x) = ((p_2)_x, w_x)$, $(p_2, z) = (u, z)$, $\forall v, w, z$, which immediately shrinks to the above.) The conservation property of scheme (7.12) is guaranteed by Theorem 7.5, but it can be also directly viewed as follows.

$$\frac{\mathrm{d}}{\mathrm{d}t} \int_0^L \frac{u^2}{2} \mathrm{d}x = (u_t, u) = -(p_x, u_x) = -(p, p_x) = 0,$$

where (7.12) with $v = u$, $w = p$ is used. The similar calculation with the straightforward schemes, (7.10) and (7.11), turns out to fail. Actually, unless the grid is completely uniform, the straightforward schemes are not conservative in general. Accordingly any full discrete schemes based on them cannot be conservative. This example illustrates that the conservation property is so "fragile" that it can easily be lost unless correct weak forms are carefully chosen. Only when these "correct" weak forms are integrated with the "correct" time-stepping using discrete partial derivatives, the conservation property is rigorously kept. Similar notice also applies to dissipative cases.

It is interesting to point out that (7.10) corresponds to Scheme 7.3 if we regard the original problem as the special case of (2.28) with $s = 0$ and $G = (u_x)^2/2$. Thus (7.10) conserves another invariant $\int_0^L ((u_x)^2/2) \mathrm{d}x$, but not $\int_0^L (u^2/2) \mathrm{d}x$. The other scheme (7.11) completely fails in preserving either of the invariants.

We tested three schemes based on the three weak forms, with a non-uniform grid. The numerical results are shown in Figure 7.1.[2] In the figure "H1" denotes the original invariant $\int_0^L (u^2/2) \mathrm{d}x$, and "H2" the other one $\int_0^L ((u_x)^2/2) \mathrm{d}x$. We can see that depending on the underlying weak forms, the conserved invariants differ. □

[2–7] Reprinted from *J. Comput. Appl. Math.*, 218, T. Matsuo, Dissipative/conservative Galerkin method using discrete partial derivatives for nonlinear evolution equations, 506–521, Copyright (2008), with permission from Elsevier.

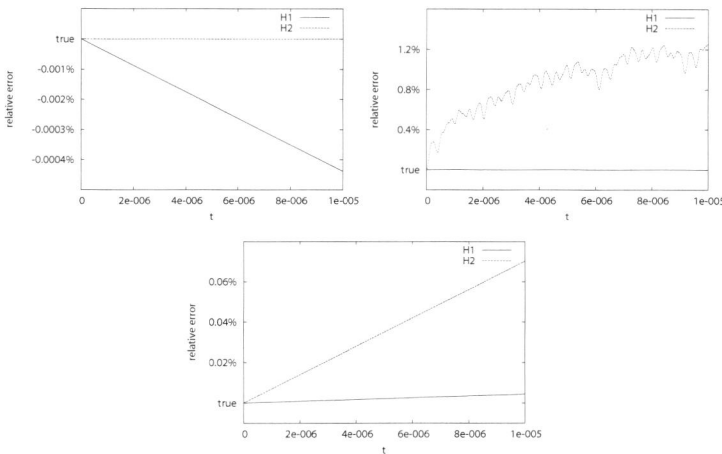

FIGURE 7.1: Comparison of invariants: (top left) scheme based on (7.10), (top right) on (7.12), (bottom) on (7.11).

7.2.1.3 Complex-Valued Conservative PDEs 4

We first introduce complex versions of the discrete partial derivatives. Suppose that the local energy is again of the form of equation (7.1), but that f_l and g_l are real-valued functions of a *complex-valued* function $u(x,t)$, which satisfy $f_l(u) = f_l(\bar{u})$, and $g_l(u_x) = g_l(\bar{u}_x)$. Throughout this section, we use the notation $(f,g) = \int_0^L \overline{f} g \, dx$. We call the discrete quantities

$$\frac{\partial G_\mathrm{d}}{\partial(u^{(m+1)}, u^{(m)})} \stackrel{\mathrm{d}}{\equiv} \sum_{l=1}^M \left(\frac{f_l(u^{(m+1)}) - f_l(u^{(m)})}{|u^{(m+1)}|^2 - |u^{(m)}|^2} \right) \left(\overline{\frac{u^{(m+1)} + u^{(m)}}{2}} \right)$$
$$\times \left(\frac{g_l(u_x^{(m+1)}) + g_l(u_x^{(m)})}{2} \right), \quad (7.13\mathrm{a})$$

$$\frac{\partial G_\mathrm{d}}{\partial(u_x^{(m+1)}, u_x^{(m)})} \stackrel{\mathrm{d}}{\equiv} \sum_{l=1}^M \left(\frac{f_l(u^{(m+1)}) + f_l(u^{(m)})}{2} \right) \left(\frac{g_l(u_x^{(m+1)}) - g_l(u_x^{(m)})}{|u_x^{(m+1)}|^2 - |u_x^{(m)}|^2} \right)$$
$$\times \left(\overline{\frac{u_x^{(m+1)} + u_x^{(m)}}{2}} \right), \quad (7.13\mathrm{b})$$

which correspond to $\partial G/\partial u$ and $\partial G/\partial u_x$ respectively, "complex discrete partial derivatives." Note that the complex discrete partial derivatives satisfy

$$\overline{\left(\frac{\partial G_{\mathrm{d}}}{\partial(u^{(m+1)}, u^{(m)})}\right)} = \frac{\partial G_{\mathrm{d}}}{\partial(\overline{u^{(m+1)}}, \overline{u^{(m)}})},$$

$$\overline{\left(\frac{\partial G_{\mathrm{d}}}{\partial(u_x^{(m+1)}, u_x^{(m)})}\right)} = \frac{\partial G_{\mathrm{d}}}{\partial(\overline{u_x^{(m+1)}}, \overline{u_x^{(m)}})}.$$

The following identity holds concerning the complex partial derivatives.

THEOREM 7.6 Discrete chain rule (complex-valued case)

$$\frac{1}{\Delta t}\int_0^L \left(G(u^{(m+1)}) - G(u^{(m)})\right) \mathrm{d}x$$
$$= \int_0^L \frac{\partial G_{\mathrm{d}}}{\partial(u^{(m+1)}, u^{(m)})} \left(\frac{u^{(m+1)} - u^{(m)}}{\Delta t}\right) \mathrm{d}x$$
$$+ \int_0^L \frac{\partial G_{\mathrm{d}}}{\partial(u_x^{(m+1)}, u_x^{(m)})} \left(\frac{u_x^{(m+1)} - u_x^{(m)}}{\Delta t}\right) \mathrm{d}x + (\mathrm{c.c.}), \quad (7.14)$$

where (c.c.) denotes the complex conjugates of the preceding terms.

Making use of the complex discrete partial derivatives, a conservative scheme for the target PDEs 4 is defined as follows:

Scheme 7.5 (Galerkin scheme for the PDEs 4) *Suppose that $u^{(0)}(x)$ is given in S_1. Find $u^{(m)} \in S_1$ ($m = 1, 2, \ldots$) such that, for any $v \in W_1$,*

$$\mathrm{i}\left(\frac{u^{(m+1)} - u^{(m)}}{\Delta t}, v\right) = -\left(\frac{\partial G_{\mathrm{d}}}{\partial(\overline{u^{(m+1)}}, \overline{u^{(m)}})}, v\right) - \left(\frac{\partial G_{\mathrm{d}}}{\partial(\overline{u_x^{(m+1)}}, \overline{u_x^{(m)}})}, v_x\right)$$
$$+ \left[\frac{\partial G_{\mathrm{d}}}{\partial(\overline{u_x^{(m+1)}}, \overline{u_x^{(m)}})} v\right]_0^L.$$

THEOREM 7.7 Conservation property of Scheme 7.5
Assume that boundary conditions are imposed so that

$$\left[\left(\frac{\partial G_{\mathrm{d}}}{\partial(u_x^{(m+1)}, u_x^{(m)})}\right)\left(\frac{\overline{u^{(m+1)} - u^{(m)}}}{\Delta t}\right) + (\mathrm{c.c.})\right]_0^L = 0,$$

and $(u^{(m+1)} - u^{(m)})/\Delta t \in W_1$. Then Scheme 7.5 is conservative in the sense that

$$\frac{1}{\Delta t}\int_0^L \left(G(u^{(m+1)}) - G(u^{(m)})\right) \mathrm{d}x = 0, \qquad m = 0, 1, 2, \ldots.$$

PROOF

$$\frac{1}{\Delta t}\int_0^L \left(G(u^{(m+1)}) - G(u^{(m)})\right) dx$$
$$= \left(\frac{\partial G_d}{\partial(u^{(m+1)}, u^{(m)})}, \frac{u^{(m+1)} - u^{(m)}}{\Delta t}\right) + \left(\frac{\partial G_d}{\partial(u_x^{(m+1)}, u_x^{(m)})}, \frac{u_x^{(m+1)} - u_x^{(m)}}{\Delta t}\right)$$
$$+ (\text{c.c.})$$
$$= -\mathrm{i}\left\|\frac{u^{(m+1)} - u^{(m)}}{\Delta t}\right\|_2^2 + \left[\frac{\partial G_d}{\partial(u_x^{(m+1)}, u_x^{(m)})}\left(\frac{u^{(m+1)} - u^{(m)}}{\Delta t}\right)\right]_0^L + (\text{c.c.})$$
$$= 0.$$

□

7.2.2 Application Examples

We show several examples. Below we suppose the one-dimensional region $[0, L]$ is divided into a mesh, and denote the piecewise linear function space on the mesh by S_h (other spaces can be also utilized, but we here use S_h for simplicity).

7.2.2.1 Cahn–Hilliard Equation

Let us consider the Cahn–Hilliard equation, which is an example of equation (2.14) with $s = 1$ and $G(u, u_x) = pu^2/2 + ru^4/4 - q(u_x)^2/2$. We assume the standard boundary conditions

$$u_x = 0 \quad \text{and} \quad \frac{\partial}{\partial x}\left(\frac{\delta G}{\delta u}\right) = 0 \quad \text{at } x = 0, L. \tag{7.15}$$

Motivated by nature of the boundary conditions, let us prepare the trial spaces as $S_1, S_2 = \{v \mid v \in S_h,\ v_x(0) = v_x(L) = 0\}$, and the test spaces as $W_1, W_2 = S_h$. Then Scheme 7.2 reads as follows: find $u^{(m)} \in S_2$ and $p_1^{(m+\frac{1}{2})} \in S_1$ such that, for all $v_1 \in W_1$ and $v_2 \in W_2$,

$$\left(\frac{u^{(m+1)} - u^{(m)}}{\Delta t}, v_1\right) = -\left((p_1^{(m+\frac{1}{2})})_x, (v_1)_x\right), \tag{7.16}$$

$$\left(p_1^{(m+\frac{1}{2})}, v_2\right) = \left(\frac{\partial G_d}{\partial(u^{(m+1)}, u^{(m)})}, v_2\right) + \left(\frac{\partial G_d}{\partial(u_x^{(m+1)}, u_x^{(m)})}, (v_2)_x\right). \tag{7.17}$$

The discrete partial derivatives are

$$\frac{\partial G_{\mathrm{d}}}{\partial (u^{(m+1)}, u^{(m)})} = p\left(\frac{u^{(m+1)} + u^{(m)}}{2}\right) +$$
$$r\left(\frac{(u^{(m+1)})^2 + (u^{(m)})^2}{2}\right)\left(\frac{u^{(m+1)} + u^{(m)}}{2}\right), \quad (7.18)$$

$$\frac{\partial G_{\mathrm{d}}}{\partial (u_x^{(m+1)}, u_x^{(m)})} = q\left(\frac{u_x^{(m+1)} + u_x^{(m)}}{2}\right), \quad (7.19)$$

which are obtained from (7.2a) and (7.2b). Note that the boundary terms $[(p_1^{(m+\frac{1}{2})})_x v_1]_0^L$ in (7.16) and $[\frac{\partial G_{\mathrm{d}}}{\partial (u_x^{(m+1)}, u_x^{(m)})} v_2]_0^L$ in (7.17) now vanish, because $(p_1^{(m+\frac{1}{2})})_x = u_x^{(m+1)} = u_x^{(m)} = 0$ at $x = 0, L$. It is easily checked that all the assumptions in Theorem 7.3 are satisfied, and thus the scheme is dissipative. This scheme coincides with the Du–Nicolaides scheme [39] (we like to note that Du–Nicolaides discussed this scheme only with (unphysical) zero Dirichlet boundary conditions).

REMARK 7.3 In practice, the trial spaces can be taken as $S_1 = S_2 = S_h$ as in the standard elliptic problems. Then the boundary conditions (7.15) are automatically recovered as the natural boundary conditions from the equations (7.16) and (7.17). □

REMARK 7.4 The scheme has an additional conservation law:

$$\frac{\mathrm{d}}{\mathrm{d}t}\int_0^L \frac{u^{(m+1)} - u^{(m)}}{\Delta t}\mathrm{d}x = 0, \quad m = 0, 1, 2, \ldots, \quad (7.20)$$

which can be easily seen from the equation (7.6a) with $v_1 = 1$. □

7.2.2.2 Korteweg-de Vries Equation

The KdV equation is an example of the target PDEs 2 with $s = 0$ and $G(u, u_x) = u^3/6 - (u_x)^2/2$. We suppose the periodic boundary condition.

Let us select the trial and test spaces as $S_1 = S_2 = W_1 = W_2 = S_h \cap H^1(\mathbb{S})$. Then Scheme 7.3 reads as follows: find $u^{(m)} \in S_2$ and $p_1^{(m+\frac{1}{2})} \in S_1$ such that, for all $v_1 \in W_1$ and $v_2 \in W_2$,

$$\left(\frac{u^{(m+1)} - u^{(m)}}{\Delta t}, v_1\right) = \left((p_1^{(m+\frac{1}{2})})_x, v_1\right), \quad (7.21)$$

$$\left(p_1^{(m+\frac{1}{2})}, v_2\right) = \left(\frac{\partial G_{\mathrm{d}}}{\partial (u^{(m+1)}, u^{(m)})}, v_2\right) + \left(\frac{\partial G_{\mathrm{d}}}{\partial (u_x^{(m+1)}, u_x^{(m)})}, (v_2)_x\right) \quad (7.22)$$

hold, where

$$\frac{\partial G_d}{\partial (u^{(m+1)}, u^{(m)})} = \frac{(u^{(m+1)})^2 + u^{(m+1)}u^{(m)} + (u^{(m)})^2}{6}, \quad (7.23)$$

$$\frac{\partial G_d}{\partial (u_x^{(m+1)}, u_x^{(m)})} = -\frac{u_x^{(m+1)} + u_x^{(m)}}{2}, \quad (7.24)$$

are obtained from definitions (7.2a) and (7.2b). The boundary term appearing in (7.8) vanishes due to the periodicity of S_1 and W_1. Due to the periodicity of S_1, the assumption $[(p_1^{(m+\frac{1}{2})})^2]_0^L = 0$ is satisfied. The periodicity also implies that condition (7.5) is satisfied, thus all the assumptions in Theorem 7.4 are satisfied, and hence the scheme is conservative. To the best of our knowledge, this scheme seems new.

REMARK 7.5 The scheme also has the additional conservation law (7.20). Set $v_1 = 1$ in the equation (7.7). □

Let us demonstrate the scheme numerically. The length of the spatial period is set to $L = 20$, and the period is divided into a non-uniform grid consisting of N points which concentrate at the center (see Figure 7.2³ for an example of $N = 201$). The approximation space $S_h \in H^1(0, L)$ is set to the standard piecewise linear function space over this grid. The initial condition is set to $u(x,0) = 48\text{sech}^2(2(x - 14)) + 24\text{sech}^2(x - 10)$ (soliton-like pulses). For comparison, a standard Crank–Nicolson type scheme:

$$\left(\frac{u^{(m+1)} - u^{(m)}}{\Delta t}, v_1\right) = \left((p_1^{(m+\frac{1}{2})})_x, v_1\right), \quad (7.25)$$

$$\left(p_1^{(m+\frac{1}{2})}, v_2\right) = \left(\frac{1}{2}\left(\frac{u^{(m+1)} + u^{(m)}}{2}\right)^2, v_2\right) - \left(\frac{u_x^{(m+1)} + u_x^{(m)}}{2}, (v_2)_x\right), \quad (7.26)$$

and a backward Euler scheme:

$$\left(\frac{u^{(m+1)} - u^{(m)}}{\Delta t}, v_1\right) = \left((p_1^{(m+\frac{1}{2})})_x, v_1\right), \quad (7.27)$$

$$\left(p_1^{(m+\frac{1}{2})}, v_2\right) = \left(\frac{(u^{(m+1)})^2}{2}, v_2\right) - \left(u_x^{(m+1)}, (v_2)_x\right), \quad (7.28)$$

are also tested.

First, the number of spatial mesh points is set to $N = 201$, the temporal mesh size $\Delta t = 0.025$, and the problem is integrated for $0 \leq t \leq 20$. Figure 7.3⁴ shows the evolution of the global energies. The conservative scheme strictly conserves the energy as constructed. In the Crank–Nicolson scheme

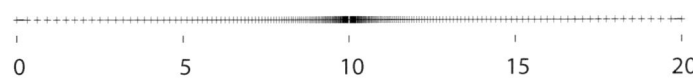

FIGURE 7.2: The non-uniform mesh ($N = 201$).

the energy is not conserved but goes down. This energy decrease is even more drastic in the backward Euler scheme, whereas the reason for the decrease is not quite the same as in the Crank–Nicolson case. This can be understood with Figure 7.4,[5] which shows the initial evolution of numerical solutions (for $0 \leq t \leq 3$). The solution by the backward Euler scheme collapses and eventually becomes flat; thus the energy decrease should be understood as the energy dissipation. On the other hand, the solution by the Crank–Nicolson scheme is not flattened; instead it develops undesired oscillations. The energy decrease should be attributed to this oscillations, which increases the term $\int_0^L u_x^2 \mathrm{d}x$ in the global energy. The oscillations can be also observed in the conservative scheme. The intensity is, however, smaller than the Crank–Nicolson case, and the solution is the best obtained among the three.

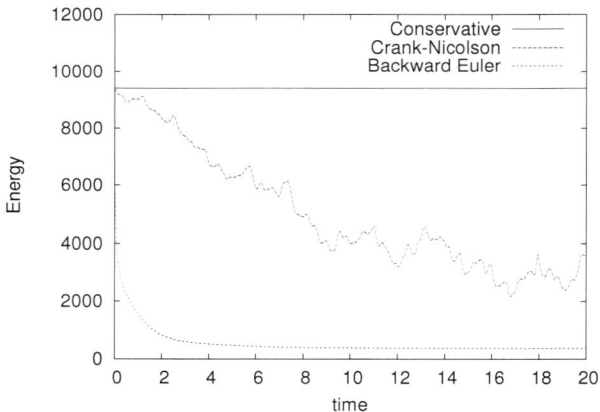

FIGURE 7.3: Evolution of the global energies ($N = 201$).

The undesired oscillations arise from the insufficiency in spatial accuracy. Next the conservative and Crank–Nicolson schemes are tested with the finer mesh $N = 401$, which is again non-uniform similar to the case of $N = 201$. The problem is then integrated for $0 \leq t \leq 100$ with $\Delta t = 0.01$. Figure 7.5[6] shows the evolution of the energies, and Figure 7.6[7] the solutions. The solutions are improved in both schemes. In the Crank–Nicolson scheme, however, the

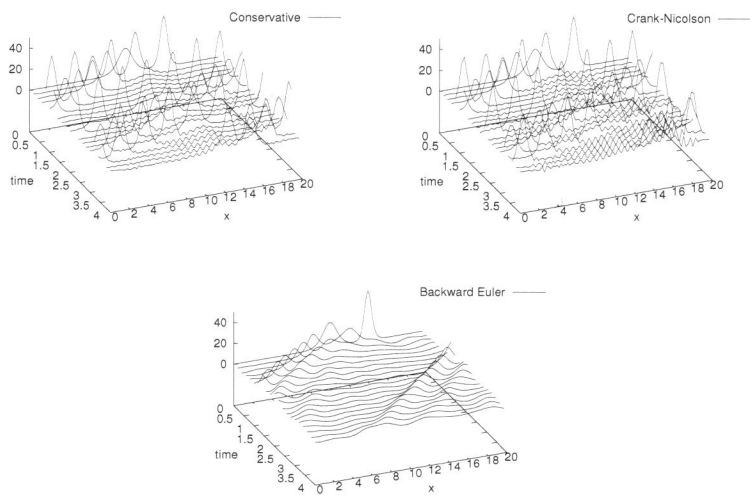

FIGURE 7.4: The numerical solutions ($N = 201$): (top left) the conservative scheme; (top right) the Crank–Nicolson scheme; (bottom) the backward Euler scheme.

drifting of the energy persists. On the other hand, the conservative scheme successfully conserves the energy, and thus is better.

7.2.2.3 Nonlinear Schrödinger Equation

Let us consider the nonlinear Schrödinger (NLS) equation under the periodic boundary condition. This is an example of the target PDEs 4 with $G(u, u_x) = -|u_x|^2 + 2\gamma |u|^{p+1}/(p+1)$. Let us select the trial and test spaces $S_1 = W_1 = S_h \cap H^1(\mathbb{S})$. Then Scheme 7.5 becomes: find $u \in S_1$ such that, for all $v \in W_1$,

$$\mathrm{i}\left(\frac{u^{(m+1)} - u^{(m)}}{\Delta t}, v\right) = -\left(\frac{\partial G_\mathrm{d}}{\partial(\overline{u^{(m+1)}}, \overline{u^{(m)}})}, v\right) - \left(\frac{\partial G_\mathrm{d}}{\partial(\overline{u_x^{(m+1)}}, \overline{u_x^{(m)}})}, v_x\right),$$

where the terms

$$\frac{\partial G_\mathrm{d}}{\partial(\overline{u^{(m+1)}}, \overline{u^{(m)}})} = \gamma \left(\frac{|u^{(m+1)}|^{p+1} - |u^{(m)}|^{p+1}}{|u^{(m+1)}|^2 - |u^{(m)}|^2}\right) \left(\frac{u^{(m+1)} + u^{(m)}}{2}\right), \tag{7.29}$$

$$\frac{\partial G_\mathrm{d}}{\partial(\overline{u_x^{(m+1)}}, \overline{u_x^{(m)}})} = -\frac{u_x^{(m+1)} + u_x^{(m)}}{2}, \tag{7.30}$$

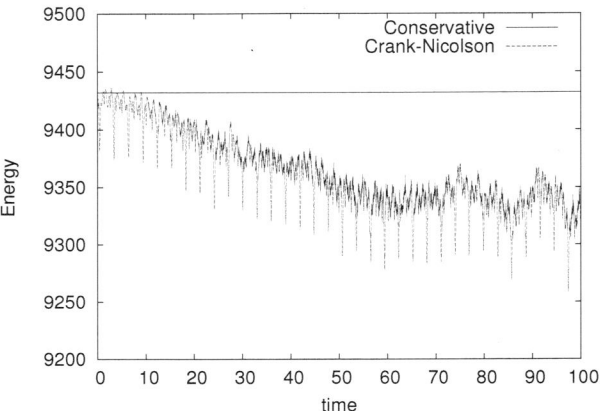

FIGURE 7.5: Evolution of the global energies ($N = 401$).

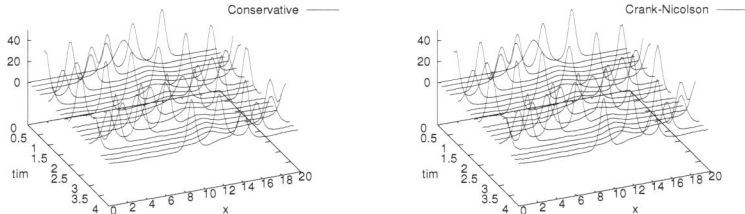

FIGURE 7.6: The numerical solutions ($N = 401$): (left) the conservative scheme; (right) the Crank–Nicolson scheme.

are obtained from definitions (7.13a) and (7.13b). The boundary term appearing in Scheme 7.5 vanishes due to the periodicity of S_1 and W_1. The periodicity also implies that condition (7.5) is satisfied, and thus the conservation property follows from Theorem 7.7. It may be noted that this scheme is simply the Akrivis–Dougalis–Karakashian scheme [8], whose stability and convergence are already guaranteed.

7.2.2.4 Camassa–Holm Type Equations

Next let us consider the Camassa–Holm type equations (2.32), which were briefly mentioned in Remark 2.5. These equations do not belong to the standard target PDEs, and thus are not immediately covered by the procedure described before. Still, it is not difficult to modify the procedure to accommodate them, and in fact, in Section 4.7.2 a way of working around was given in the finite difference context. We show a similar method below in the Galerkin framework. For the detail of this section, interested readers may refer to [123, 118].

This subsection is organized as follows. We first give a general discussion for the abstract equation (2.32). Then in Sections 7.2.2.4.1 through 7.2.2.4.3, we focus on the special cases: the limiting Camassa–Holm equation, the Dai equation, and the Benjamin–Bona–Mahony equation, respectively (for a brief introduction of these equations, see Remark 2.5). Finally, in Section 7.2.2.4.4, we present another formulation for the limiting Camassa–Holm equation, based on the bi-Hamiltonian structure.

We commence by recalling some mathematical properties of (2.32). It can be viewed as a gradient flow:

$$\left(1 - \frac{\partial^2}{\partial x^2}\right) u_t = \left(\frac{\delta G}{\delta u}\right)_x, \tag{7.31}$$

where

$$G(u, u_x) = -\frac{\kappa u^2 + u^3 + \gamma u (u_x)^2}{2}. \tag{7.32}$$

If we further introduce an operator $\mathcal{K} = (1 - \partial^2/\partial x^2)^{-1}$, which is a map $L^2(\mathbb{S}) \to H^2(\mathbb{S})$ [19] the equation can be rewritten as

$$u_t = \mathcal{K}\left(\frac{\delta G}{\delta u}\right)_x. \tag{7.33}$$

With the Green function:

$$k(x) = \frac{\cosh(x - L[x/L] - L/2)}{2 \sinh(L/2)}, \tag{7.34}$$

where the bracket $[x]$ means the largest integer which does not exceed x, the operator $\mathcal{K}$ can be expressed in terms of the convolution

$$(\mathcal{K}f)(x) = (k * f)(x) = \int_0^L k(x - \xi) f(\xi) \mathrm{d}\xi. \tag{7.35}$$

It is easy to see that $\mathcal{K}$ is symmetric, and thus due to the skew-symmetry of $\partial/\partial x$, the equation (7.33) is obviously conservative. But in order to clarify how the problem is now delicate, we explicitly present a conservation theorem below.

THEOREM 7.8 Conservation property of (7.33)
Suppose $u(\cdot, t) \in H^3(\mathbb{S}), u_t(\cdot, t) \in H^1(\mathbb{S})$, and $G(u, u_x)$ are sufficiently smooth with respect to their arguments. Then,

$$\frac{\mathrm{d}}{\mathrm{d}t} \int_0^L G \, \mathrm{d}x = 0. \qquad (7.36)$$

PROOF

$$\frac{\mathrm{d}}{\mathrm{d}t} \int_0^L G \, \mathrm{d}x = \int_0^L \left(\frac{\partial G}{\partial u} u_t + \frac{\partial G}{\partial u_x} u_{tx} \right) \mathrm{d}x = \int_0^L \frac{\delta G}{\delta u} u_t \, \mathrm{d}x + \left[\frac{\partial G}{\partial u_x} u_t \right]_0^L$$
$$= \int_0^L \frac{\delta G}{\delta u} \mathcal{K} \left(\frac{\delta G}{\delta u} \right)_x \mathrm{d}x = 0.$$

In the third equality, the boundary term is dropped due to the periodicity. In the last equality, an identity $(\mathcal{K} f_x, f) = 0$ which holds for any $f \in H^1(\mathbb{S})$ is used. ☐

The theorem is completely fine, as far as we deal with smooth solutions. The assumption $u(\cdot, t) \in H^3(\mathbb{S})$ is reasonable, since $(\delta G/\delta u)_x$ essentially includes u_{xxx}. However, the reason why the Camassa–Holm type equations have drawn much interest is that they can exhibit singular solutions like peakons. In order to justify peakons, an $H^1(\mathbb{S})$-formulation is inevitably required. In the critical CH ($\kappa = 0$) case, such a form is given in [28]:

$$u_t + \frac{1}{2} \left(u^2 + \mathcal{K} \left(u^2 + \frac{(u_x)^2}{2} \right) \right)_x = 0, \qquad (7.37)$$

which makes sense for $u(\cdot, t) \in H^1(\mathbb{S})$. This form is quite beautiful in that it not only accepts peakons but also clarifies the point that the equation is in fact a "conservation law" (in the terminology of fluid dynamics). However, since it seems that the conservation property of (7.37) cannot be directly established, (7.37) is not convenient in our project. Actually, in [29], the conservation property of (7.37) is established by expressing the target H^1-solution of (7.37) as the limit of a series of energy-conserving H^3-solutions of the original Camassa–Holm equation (with $\kappa = 0$). It seems difficult to do a similar thing in a discrete setting.

Instead we employ the following set of weak forms. Find $u(\cdot, t), p(\cdot, t) \in$

$H^1(\mathbb{S})$ such that for any $v_1, v_2 \in H^1(\mathbb{S})$,

$$(u_t, v_1) = (\mathcal{K} p_x, v_1), \tag{7.38}$$

$$(p, v_2) = \left(\frac{\partial G}{\partial u}, v_2\right) + \left(\frac{\partial G}{\partial u_x}, (v_2)_x\right). \tag{7.39}$$

It is obvious that the solution $u(\cdot, t) \in H^3(\mathbb{S})$ that solves (7.33) also solves this set of weak forms by setting $p = \delta G/\delta u$. From the weak forms (7.38), (7.39), the desired conservation property can be successfully deduced as shown in the next theorem. The deduction can be done completely in an abstract way as in Theorem 7.8; in this case, however, the key tools are the *partial* derivatives $\partial G/\partial u, \partial G/\partial u_x$, instead of the *variational* derivative $\delta G/\delta u$.

THEOREM 7.9 Conservation property of the weak forms (7.38), (7.39)

Suppose $u(\cdot, t), p(\cdot, t) \in H^1(\mathbb{S})$ are the solutions of the weak forms (7.38) and (7.39). Also assume that G is sufficiently smooth and $u_t(\cdot, t) \in H^1(\mathbb{S})$. Then the following holds

$$\frac{\mathrm{d}}{\mathrm{d}t} \int_0^L G \, \mathrm{d}x = 0. \tag{7.40}$$

PROOF

$$\frac{\mathrm{d}}{\mathrm{d}t} \int_0^L G \, \mathrm{d}x = \left(\frac{\partial G}{\partial u}, u_t\right) + \left(\frac{\partial G}{\partial u_x}, u_{tx}\right) = (p, u_t) = (\mathcal{K} p_x, p) = 0. \tag{7.41}$$

The first equality is just the chain rule. The second equality follows from (7.39) with $v_2 = u_t$, and the third one from (7.38) with $v_1 = p$. □

REMARK 7.6 Interested readers may compare the discussion here to the finite difference version in Section 4.7.2. In that context, all the derivatives are replaced with finite differences, and unless discrete functional analytic analysis is required, we usually do not pay attention to the "regularity" of the solution. That is, we simply assume any approximate solutions (i.e., solution vectors, such as in $\mathbb{S}_N$) are infinitely many times "differentiable" by finite differences. This is in sharp contrast to the Galerkin case, where regularity of approximate solutions should be explicitly kept in mind, and the complex discussion above to allow peakons in H^1 is inevitable. □

Now we proceed to the numerical schemes. For the function $G(u, u_x)$ defined in (7.32), the concrete forms of the discrete partial derivatives are

$$\frac{\partial G_\mathrm{d}}{\partial(u^{(m+1)},u^{(m)})} = -\kappa\left(\frac{u^{(m+1)} + u^{(m)}}{2}\right) - \frac{(u^{(m+1)})^2 + u^{(m+1)}u^{(m)} + (u^{(m)})^2}{2}$$
$$- \gamma\left(\frac{(u_x^{(m+1)})^2 + (u_x^{(m)})^2}{4}\right), \tag{7.42}$$

$$\frac{\partial G_\mathrm{d}}{\partial(u_x^{(m+1)},u_x^{(m)})} = -\gamma\left(\frac{u^{(m+1)} + u^{(m)}}{2}\right)\left(\frac{u_x^{(m+1)} + u_x^{(m)}}{2}\right). \tag{7.43}$$

With these discrete partial derivatives, we define an abstract scheme for the weak forms (7.38) and (7.39) as follows. Suppose S_1, S_2, W_1 and W_2 are appropriately chosen; for example, $S_1 = S_2 = W_1 = W_2 = S_h \cap H^1(\mathbb{S})$.

Scheme 7.6 (Abstract Galerkin scheme for (7.38), (7.39)**)** *Suppose that $u^{(0)}(x)$ is given in S_2. Find $u^{(m+1)} \in S_2$, $p^{(m+\frac{1}{2})} \in S_1$ ($m = 0, 1, 2, \ldots$) such that, for any $v_1 \in W_1$ and $v_2 \in W_2$,*

$$\left(\frac{u^{(m+1)} - u^{(m)}}{\Delta t}, v_1\right) = \left(\mathcal{K}(p^{(m+\frac{1}{2})})_x, v_1\right), \tag{7.44}$$

$$\left(p^{(m+\frac{1}{2})}, v_2\right) = \left(\frac{\partial G_\mathrm{d}}{\partial(u^{(m+1)},u^{(m)})}, v_2\right) + \left(\frac{\partial G_\mathrm{d}}{\partial(u_x^{(m+1)},u_x^{(m)})}, (v_2)_x\right). \tag{7.45}$$

The scheme enjoys the next conservation property. The proof can be done analogously to the continuous case.

THEOREM 7.10 Conservation property of Scheme 7.6
Assume the trial and test spaces S_1, S_2, W_1 and W_2 are set such that (i) $(u^{(m+1)} - u^{(m)})/\Delta t \in W_2$; *and* (ii) $S_1 \subseteq W_1$. *Scheme 7.6 is conservative in the sense that*

$$\frac{1}{\Delta t}\int_0^L \left(G(u^{(m+1)}) - G(u^{(m)})\right) \mathrm{d}x = 0, \qquad m = 0, 1, 2, \ldots.$$

PROOF
$$\frac{1}{\Delta t}\int_0^L \left(G(u^{(m+1)}) - G(u^{(m)})\right) \mathrm{d}x$$
$$= \left(\frac{\partial G_\mathrm{d}}{\partial(u^{(m+1)},u^{(m)})}, \frac{u^{(m+1)} - u^{(m)}}{\Delta t}\right) + \left(\frac{\partial G_\mathrm{d}}{\partial(u_x^{(m+1)},u_x^{(m)})}, \frac{u_x^{(m+1)} - u_x^{(m)}}{\Delta t}\right)$$
$$= \left(p^{(m+\frac{1}{2})}, \frac{u^{(m+1)} - u^{(m)}}{\Delta t}\right)$$
$$= \left(\mathcal{K}(p^{(m+\frac{1}{2})})_x, p^{(m+\frac{1}{2})}\right) = 0.$$

The first equality follows from the discrete chain rule (Theorem 7.1). The second one is shown by using the equation (7.45) with $v_2 = (u^{(m+1)} - u^{(m)})/\Delta t$ (the substitution is allowed by the assumption (i)), while the third one is given by using the equation (7.44) with $v_1 = p^{(m+\frac{1}{2})}$ (allowed by the assumption (ii)). □

REMARK 7.7 The equation (7.33) can also be viewed as a conservation law:
$$u_t - \left(\mathcal{K}\frac{\delta G}{\delta u}\right)_x = 0, \tag{7.46}$$

(note that for $f \in H^1(\mathbb{S})$ it holds $(\mathcal{K}f)_x = \mathcal{K}(f_x)$), and there is another invariant:
$$\frac{\mathrm{d}}{\mathrm{d}t}\int_0^L u\,\mathrm{d}x = \left(\mathcal{K}\left(\frac{\delta G}{\delta u}\right)_x, 1\right) = 0. \tag{7.47}$$

The final equality follows from an identity $((\mathcal{K}f)_x, 1) = 0$ which holds for any $f \in H^1(\mathbb{S})$. Scheme 7.6 also conserves this invariant:
$$\frac{1}{\Delta t}\int_0^L \left(u^{(m+1)} - u^{(m)}\right)\mathrm{d}x = (\mathcal{K}(p^{(m+\frac{1}{2})})_x, 1) = 0, \quad m = 0, 1, 2, \ldots. \tag{7.48}$$
□

Let us test how Scheme 7.6 in fact works. We set $S_1 = S_2 = W_1 = W_2 = S_h \cap H^1(\mathbb{S})$, where meshes are either equispaced or non-uniform depending on the problems. Given the approximation space, the concrete form of Scheme 7.6 is
$$A\left(\frac{\boldsymbol{u}^{(m+1)} - \boldsymbol{u}^{(m)}}{\Delta t}\right) = K\boldsymbol{p}^{(m+\frac{1}{2})}, \tag{7.49}$$
$$A\boldsymbol{p}^{(m+\frac{1}{2})} = \boldsymbol{f}(\boldsymbol{u}^{(m+1)}, \boldsymbol{u}^{(m)}), \tag{7.50}$$

where
$$\boldsymbol{u}^{(m)} = (u^{(m)}(x_0), \ldots, u^{(m)}(x_{N-1}))^\top,$$
$$\boldsymbol{p}^{(m+\frac{1}{2})} = (p^{(m+\frac{1}{2})}(x_0), \ldots, p^{(m+\frac{1}{2})}(x_{N-1}))^\top,$$

and $\boldsymbol{f}(\boldsymbol{u}^{(m+1)}, \boldsymbol{u}^{(m)})$ is the vector arising from the right hand side of (7.45) which in general nonlinearly include $\boldsymbol{u}^{(m+1)}$ and $\boldsymbol{u}^{(m)}$. The matrix A is the standard mass matrix whose elements are $A_{ij} = (\phi_i, \phi_j)$, where ϕ_i ($i = 0, \ldots, N-1$) are the standard basis functions of S_p, and $K_{ij} = (\mathcal{K}(\phi_i)_x, \phi_j)$. Note that the matrices A and K depend only on the approximate space (i.e., the mesh), and can be prepared in prior to the time evolution process. The preparation of the matrix K involves the computation of convolutions, which can be done by hand in the case of S_p. When more general approximate spaces are required, it is also possible to employ some numerical integrator with sufficient accuracy. Since the matrix A is invertible, the equations (7.49) and (7.50) immediately reduce to

$$A\left(\frac{\boldsymbol{u}^{(m+1)} - \boldsymbol{u}^{(m)}}{\Delta t}\right) = KA^{-1}\boldsymbol{f}(\boldsymbol{u}^{(m+1)}, \boldsymbol{u}^{(m)}). \tag{7.51}$$

That is, the computation of the intermediate variable $\boldsymbol{p}^{(m+\frac{1}{2})}$ can be skipped, and the dimension of the system to be solved is N, instead of $2N$. In what follows, the numerical calculations are based on this expression.

The nonlinear equations (7.51) should be solved by some iterative method. A convenient way is to use some reliable numerical Newton library. In the experiments below, the routine `imsl_d_zeros_sys_eqn` in the IMSL was used.

7.2.2.4.1 Limiting Camassa–Holm equation Let us consider the "limiting" Camassa–Holm equation, which is obtained by setting $\kappa = 0, \gamma = 1$. Originally, the Camassa–Holm (CH) equation only makes sense for $\kappa > 0$ in physical context, since κ corresponds to the critical shallow water speed that should be strictly positive (see [24]). Mathematically, however, main interest is usually on the limiting case $\kappa = 0$, where solitons become peaked. Below we consider this case. The concrete form of Scheme 7.6 then becomes as follows. With the function

$$G(u, u_x) = -\frac{u^3 + u(u_x)^2}{2}, \tag{7.52}$$

which is obtained by setting $\kappa = 0$, $\gamma = 1$ in (7.32), the discrete partial derivatives (7.42) and (7.43) become

$$\frac{\partial G_d}{\partial(u^{(m+1)}, u^{(m)})} = -\frac{(u^{(m+1)})^2 + u^{(m+1)}u^{(m)} + (u^{(m)})^2}{2} \\ - \left(\frac{(u_x^{(m+1)})^2 + (u_x^{(m)})^2}{4}\right), \tag{7.53}$$

$$\frac{\partial G_d}{\partial(u_x^{(m+1)}, u_x^{(m)})} = -\left(\frac{u^{(m+1)} + u^{(m)}}{2}\right)\left(\frac{u_x^{(m+1)} + u_x^{(m)}}{2}\right). \tag{7.54}$$

Note that for the energy function (7.52) the (continuous) partial derivatives are

$$\frac{\partial G}{\partial u} = -\frac{3}{2}u^2 - \frac{1}{2}(u_x)^2 \quad \text{and} \quad \frac{\partial G}{\partial u_x} = -uu_x,$$

and we can see the correspondence between the continuous and discrete ones. Substituting the discrete partial derivatives into Scheme 7.6, we obtain the concrete form of the scheme, which is then implemented as described above, i.e., as (7.51).

REMARK 7.8   Interested readers may compare the above discrete *partial* derivatives with the discrete *variational* derivative in the finite difference context in Section 4.7.2.   □

For comparison, the following two implicit schemes have been also tested; The Crank–Nicolson scheme:
$$\left(\frac{u^{(m+1)} - u^{(m)}}{\Delta t}, v_1\right) = \left(\mathcal{K}(p^{(m+\frac{1}{2})})_x, v_1\right) \tag{7.55}$$

$$\left(p^{(m+\frac{1}{2})}, v_2\right) =$$
$$-\left(\frac{3}{2}\left(\frac{u^{(m+1)} + u^{(m)}}{2}\right)^2 + \frac{1}{2}\left(\frac{u_x^{(m+1)} + u_x^{(m)}}{2}\right)^2, v_2\right)$$
$$-\left(\left(\frac{u^{(m+1)} + u^{(m)}}{2}\right)\left(\frac{u_x^{(m+1)} + u_x^{(m)}}{2}\right), (v_2)_x\right), \tag{7.56}$$

and the implicit Euler scheme:
$$\left(\frac{u^{(m+1)} - u^{(m)}}{\Delta t}, v_1\right) = \left(\mathcal{K}(p^{(m+\frac{1}{2})})_x, v_1\right) \tag{7.57}$$

$$\left(p^{(m+\frac{1}{2})}, v_2\right) = -\left(\frac{3}{2}\left(u^{(m+1)}\right)^2 + \frac{1}{2}\left(u_x^{(m+1)}\right)^2, v_2\right)$$
$$-\left(u^{(m+1)} u_x^{(m+1)}, (v_2)_x\right). \tag{7.58}$$

Note that since all of these schemes are based on the same weak forms (7.38) and (7.39), the spatial discretization is exactly the same, and only the temporal discretizations are different.

First, the three schemes are tested on the equispaced mesh with $L = 40$ and $N = 200$ (thus $\Delta x = 0.2$). The initial data is set to $u(x,0) = 5e^{-|x-x_a|} + 2e^{-|x-x_b|}$, where $x_a = 13.43$ and $x_b = 26.77$, and the problem is integrated in $0 \leq t \leq 10$ with $\Delta t = 0.1$. Since larger peakons are faster, the larger peakon initially centered at x_a overtakes the smaller one at x_b as time passes. Figure 7.7[8] shows the numerical results obtained by the three schemes. According to the figure, both the conservative scheme and the Crank–Nicolson scheme seem to correctly track the overtaking phenomenon (note that since now the periodic boundary condition is applied, the outgoing peakons come back to the interval from the left boundary). On the other hand, in the implicit Euler case, although the computation itself is stable, the peakons gradually become flattened. Figure 7.8[9] shows the evolution of the energy $\int_0^L G(u^{(m)})\mathrm{d}x$;

[8–20] Reprinted from *J. Comput. Phys.*, 228, T. Matsuo and H. Yamaguchi, An energy-conserving Galerkin scheme for a class of nonlinear dispersive equations, 4346–4358, Copyright (2009), with permission from Elsevier.

the left figure shows the evolution near the starting time, and the right figure the overall profile. As suggested in the wave pattern in Figure 7.7, we observe strong energy dissipation in the case of the implicit Euler scheme (see left figure); the energy rapidly tends to zero. Although in Figure 7.7 the results by the conservative scheme and the Crank–Nicolson scheme look quite similar, the energy profiles are considerably different (see right figure). In the conservative scheme, the energy is strictly conserved to the machine accuracy, while in the Crank–Nicolson scheme it drifts.

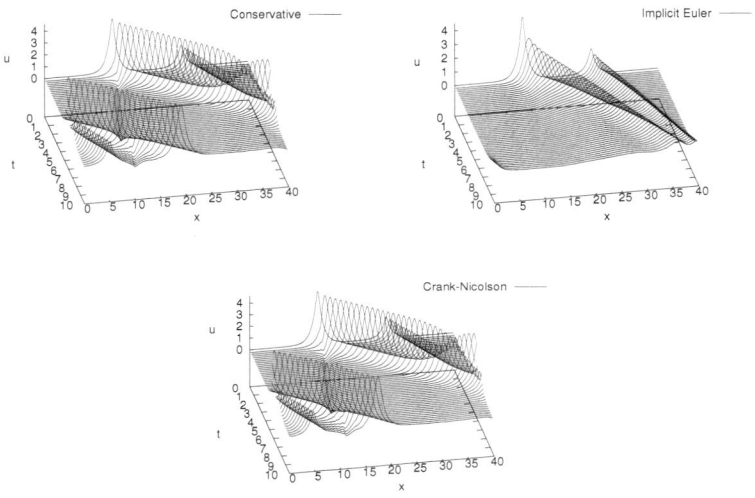

FIGURE 7.7: Evolution of the two peakons; (top left) the conservative scheme, (top right) the implicit Euler scheme, (bottom) the Crank–Nicolson scheme.

Next, in order to check the long-time stability, the problem is solved for $0 \leq t \leq 70$ with the time mesh size $\Delta t = 0.02$ and the number of spatial grid points $N = 400$ ($\Delta x = 0.1$). With these parameters, the larger peakon goes around the spatial interval about ten times. The conservative scheme successfully integrates the problem with the energy strictly kept (Figure 7.9[10] (left) and Figure 7.10[11]). In the Crank–Nicolson case, the energy is nearly conserved in the early stage $0 \leq t \leq 20$; the energy periodically oscillates and stays around the exact value. As time passes, however, the oscillation becomes irregular ($20 \leq t \leq 50$), and then completely unstable ($50 \leq t \leq 70$). This instability can be observed in the wave pattern in Figure 7.9, where the peakons in the Crank–Nicolson case are completely broken at $t = 70$. For

Advanced Topic III: Further Remarks

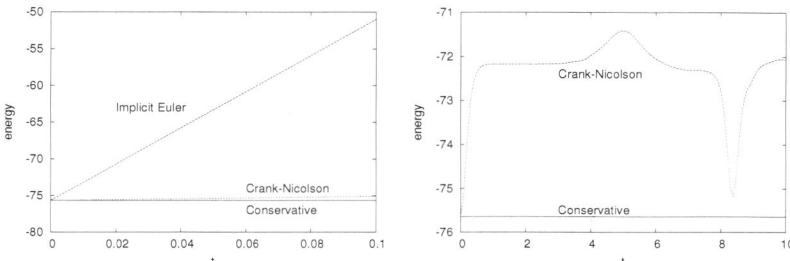

FIGURE 7.8: Evolution of the energies (two peakons case); (left) detailed profile near the starting time, (right) global profile.

$t \geq 70$, it turns out that the numerical Newton solver does not work in the Crank–Nicolson scheme, and it is impossible to continue the computation. This result strongly suggests that the conservative scheme is in fact more reliable than the standard Crank–Nicolson scheme.

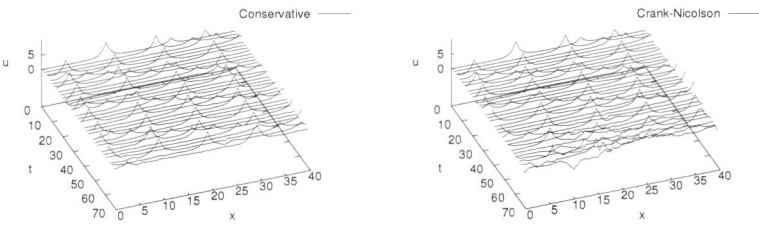

FIGURE 7.9: Long-time computation of the two-peakons problem; (left) the conservative scheme, (right) the Crank–Nicolson scheme.

The third experiment is to check whether the presented scheme works on non-equispaced grids as well. To this end, the Camassa–Holm is solved on the spatial interval $[0, 200]$ with the grid shown in Figure 7.11[12] ($N = 200$), and with the triangle shaped initial data

$$u(x, 0) = \begin{cases} x - x_c + 20 & \text{if } x \in [x_c - 20, x_c), \\ -(x - x_c) + 20 & \text{if } x \in [x_c, 20), \\ 0 & \text{otherwise,} \end{cases}$$

where $x_c = 80.5$.

Figure 7.12[13] shows the numerical results by the three schemes, where the time mesh width is set to $\Delta t = 0.05$. For comparison, a result by a stan-

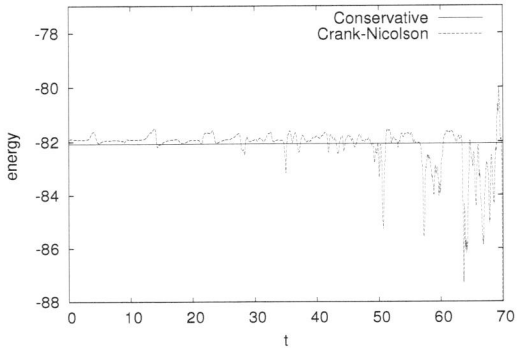

FIGURE 7.10: Evolution of the energies in the long-time computation.

FIGURE 7.11: The non-uniform mesh ($N = 200$).

dard numerical method on finer equispaced mesh ($N = 2000$), marked as "FD/RK" in the figures, is also presented. The scheme is obtained by discretizing space variable by the standard central finite differences (with second-order accuracy), and then by discretizing time stepping by the standard 4th-order Runge–Kutta method. The time-stepping width is chosen considerably small ($\Delta t = 0.0005$) such that the result is accurate enough as a substitute for the unknown exact solution. As the solution suggests (Figure 7.12, bottom right), in this problem setting the initial triangle shaped data soon splits into a number of peakons. The splitting mainly occurs at the center of the interval, which is the reason why the grid is chosen to be dense at the center. The result by the implicit Euler scheme (top right) again exhibits strong dissipation, which can be also observed in the energy profile (Figure 7.13[14]). The result by the conservative scheme (Figure 7.12, top left) is similar to the accurate result by FD/RK, with the excellent energy conservation profile (Figure 7.13). Compared to this result, even with considerably fine mesh sizes, the energy in FD/RK scheme monotonically moves apart from the exact value; this means that however mesh is refined the FD/RK method is not so reliable that it can be used as an integrator for long-time computations. The shape of the peakons in Crank–Nicolson case seems to be quite similar to the conservative and FD/RK cases (Figure 7.12, bottom left). The energy profile, however, behaves dreadfully, where the error exceeds 10% in magnitude. In this example, the peakons are quite sharp and high, and the slight error in the shapes of peakons is magnified as the big error in the energy.

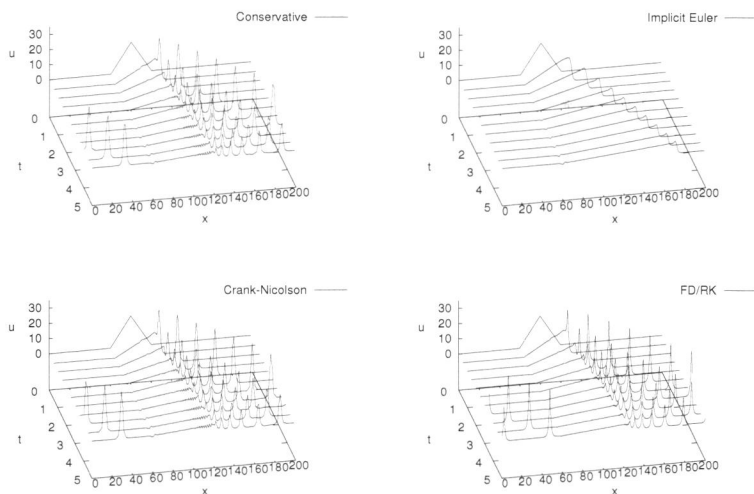

FIGURE 7.12: Generation of peakons; (top left) the conservative scheme, (top right) the implicit Euler scheme, (bottom left) the Crank–Nicolson scheme, (bottom right) the FD/RK solution on the finer mesh.

7.2.2.4.2 Dai equation The Dai equation is obtained by setting $\kappa = 0$, $\gamma \in \mathbb{R}$. This is quite similar to the limiting CH case, but now we have a freedom in the choice of γ. As described before, soliton solutions are expected to be smooth when $\gamma < 1$ and become "cusped" solutions when $\gamma > 1$. Below we have tested two cases: $\gamma = 0.5$ and $\gamma = 1.4$. The energy function is

$$G(u, u_x) = -\frac{u^3 + \gamma u u_x^2}{2}. \tag{7.59}$$

and accordingly the discrete partial derivatives are

$$\frac{\partial G_d}{\partial(u^{(m+1)}, u^{(m)})} = -\frac{(u^{(m+1)})^2 + u^{(m+1)}u^{(m)} + (u^{(m)})^2}{2}$$
$$- \gamma \left(\frac{(u_x^{(m+1)})^2 + (u_x^{(m)})^2}{4}\right), \tag{7.60}$$

$$\frac{\partial G_d}{\partial(u_x^{(m+1)}, u_x^{(m)})} = -\gamma \left(\frac{u^{(m+1)} + u^{(m)}}{2}\right)\left(\frac{u_x^{(m+1)} + u_x^{(m)}}{2}\right). \tag{7.61}$$

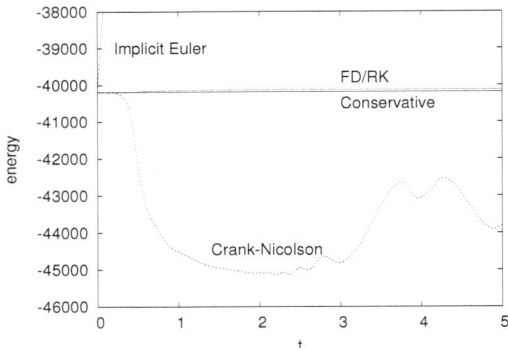

FIGURE 7.13: Evolution of the energies (peakon train case).

We test Scheme 7.6 and the Crank–Nicolson scheme. The latter is constructed in a similar manner as in the previous section.

First, the case of $\gamma = 0.5$ is considered. With this parameter, solitons are smooth and the computation is rather easy. In order to check the long-time stability of the schemes, the problem is solved in a long interval $0 \leq t \leq 500$ with the temporal mesh size $\Delta t = 0.1$. The initial data is $u(x,0) = 5\,\mathrm{sech}(x - 5) + 2\,\mathrm{sech}(x - 15)$. The length of the spatial interval L is set to 40, for which the equispaced grid with $N = 200$ is employed (i.e., $\Delta x = 0.2$). Figure 7.14[15] shows the evolution of the numerical solutions. The computation proceeds quite stably as expected, and the shapes of the solitons are successfully preserved in both schemes, although the phase speeds of the solitons are different. Figure 7.15[16] shows the evolution of the energies. In the conservative scheme, the energy is strictly kept. In the Crank–Nicolson scheme, the energy oscillates, but stays near the exact value.

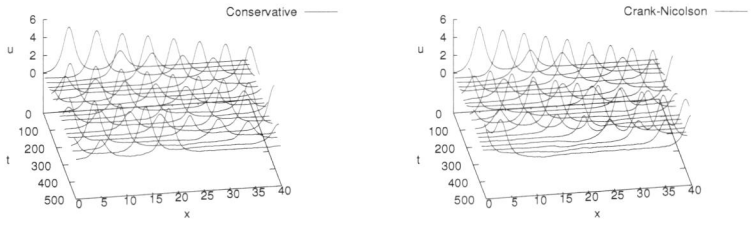

FIGURE 7.14: Smooth solitons in the Dai equation ($\gamma = 0.5$); (left) the conservative scheme, (right) the Crank–Nicolson scheme.

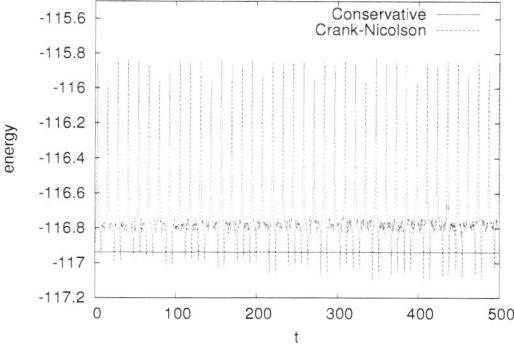

FIGURE 7.15: Evolution of the energies (the Dai equation, $\gamma = 0.5$).

Next, the results with $\gamma = 1.4$ are presented. The equispaced grid on the spatial interval $[0, 40]$ with $N = 200$ or 400 is used, and the problem is solved in $0 \leq t \leq 10$ with the time mesh size $\Delta t = 0.1$. The initial data is set to the same one as in the limiting CH case, i.e., $u(x,0) = 5e^{-|x-x_a|} + 2e^{-|x-x_b|}$ with $x_a = (200/3 + 1/2)\Delta x$ and $x_b = (400/3 + 1/2)\Delta x$. Figure 7.16[17] shows the numerical solutions of $N = 400$, and Figure 7.17[18] the evolution of the energies in both $N = 200$ and 400 cases. From Figure 7.16, both schemes succeed in catching the peaked solutions (although numerically it is difficult to judge whether the solutions are really "cusped" rather than "peaked"). Comparing Figure 7.17 (left) and Figure 7.15, we notice that with the same mesh ($N = 200$) the energy deviation in the Crank–Nicolson scheme becomes much larger when the solutions become singular, although it can be improved by refining the spatial mesh (Figure 7.17, right). In any case, the conservative scheme seems to be safer when we achieve such singular solutions.

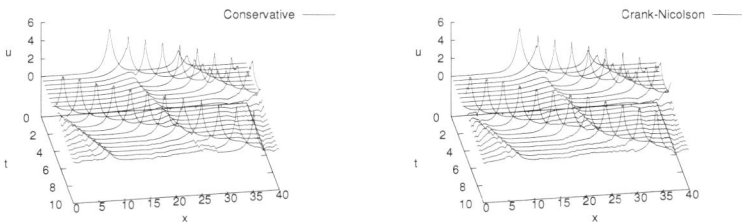

FIGURE 7.16: Cusped solutions in the Dai equation ($\gamma = 1.4, N = 400$); (left) the conservative scheme, (right) the Crank–Nicolson scheme.

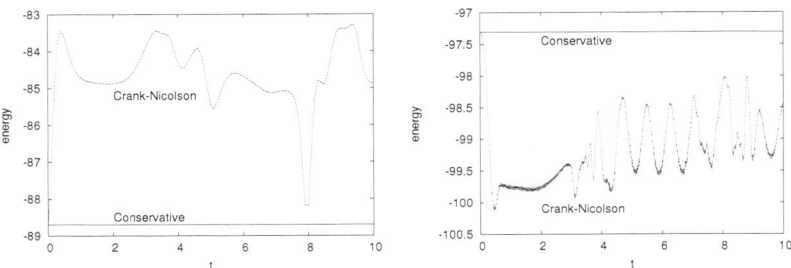

FIGURE 7.17: Evolution of the energies (the Dai equation, $\gamma = 1.4$); (left) $N = 200$, (right) $N = 400$.

7.2.2.4.3 Benjamin–Bona–Mahony equation The BBM equation is obtained by setting $\kappa > 0$ and $\gamma = 0$ and we set $\kappa = 1$ in this paragraph. The equation is considered over the spatial domain $[0, 40]$ using the equispaced mesh with the number of grid points $N = 100$. Then the problem is integrated in $0 \leq t \leq 20$ with the time mesh size $\Delta t = 0.25$. The initial data is set to $u(x,0) = c_1 \text{sech}^2(0.35(x - 15)) + c_2 \text{sech}^2(0.25(x - 25))$, where $c_1 = 9 \times 0.7^2/(1 - 0.7^2), c_2 = 9 \times 0.5^2/(1 - 0.5^2)$ (see [47] for this initial data). The conservative scheme and the Crank–Nicolson scheme are tested. Figure 7.18[19] shows the numerical solutions, and Figure 7.19[20] the evolution of the energies. Both schemes successfully capture the propagation of the two-soliton. The conservative scheme strictly preserves the energy, while in the Crank–Nicolson scheme the energy oscillates around the exact value.

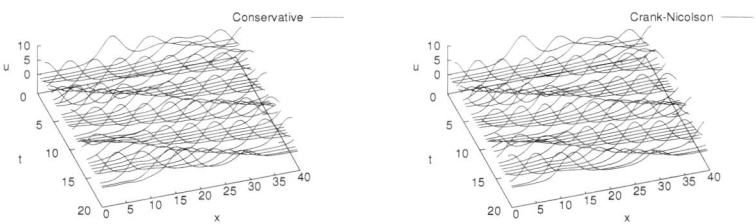

FIGURE 7.18: Two-soliton in the BBM equation; (left) the conservative scheme, (right) the Crank–Nicolson scheme.

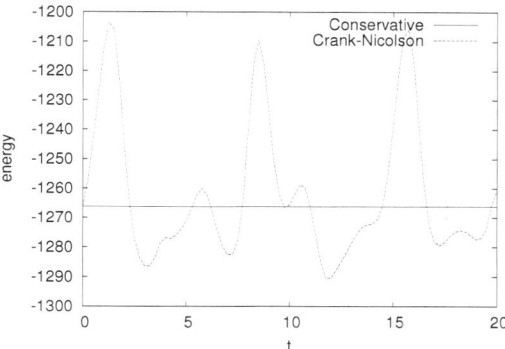

FIGURE 7.19: Evolution of the energies (the BBM equation).

7.2.2.4.4   Limiting Camassa–Holm Equation: Second Approach   So far we have glanced through the conservative schemes for the Camassa–Holm type equations (2.32). For a special case, the limiting Camassa–Holm equation with $\kappa = 0, \gamma = 1$, we can take another approach. The key is that the limiting Camassa–Holm equation has another invariant:

$$\widetilde{G}(u, u_x) = \frac{u^2 + (u_x)^2}{2}, \qquad \frac{\mathrm{d}}{\mathrm{d}t} \int_0^L \widetilde{G}(u, u_x)\mathrm{d}x = 0. \qquad (7.62)$$

(Actually, it is completely integrable, and has infinitely many invariants.) Then by introducing a new variable $\omega = (1 - \partial^2/\partial x^2)u$, it can be written in another variational form[21]:

$$\omega_t = -\left(\frac{\partial}{\partial x}\omega + \omega\frac{\partial}{\partial x}\right)\left(\frac{\delta \widetilde{G}}{\delta \omega}\right). \qquad (7.63)$$

With the help of the operator $\mathcal{K} = (1 - \partial^2/\partial x^2)^{-1}$, the energy function (7.62) can be represented in ω as

$$\widetilde{G}(\mathcal{K}\omega, \mathcal{K}\omega_x) = \frac{(\mathcal{K}\omega)^2 + (\mathcal{K}\omega_x)^2}{2}, \qquad (7.64)$$

and its variational derivative with respect to ω can be defined (note that for $f \in H^1(\mathbb{S})$, $\mathcal{K}f_x = (\mathcal{K}f)_x$ holds). Throughout this section, ω is assumed to

---

[21] For the same reason described in Section 4.7.2, we use the symbol ω in place of m.

be sufficiently smooth. By simply differentiating we obtain

$$\frac{d}{dt}\int_0^L \widetilde{G}(\mathcal{K}\omega, \mathcal{K}\omega_x)dx$$

$$= \int_0^L \left(\frac{\partial \widetilde{G}}{\partial(\mathcal{K}\omega)}\cdot\mathcal{K}\omega_t + \frac{\partial \widetilde{G}}{\partial(\mathcal{K}\omega_x)}\cdot\mathcal{K}\omega_{xt}\right)dx$$

$$= \int_0^L \left(\mathcal{K}\frac{\partial \widetilde{G}}{\partial(\mathcal{K}\omega)} - \left(\mathcal{K}\frac{\partial \widetilde{G}}{\partial(\mathcal{K}\omega_x)}\right)_x\right)\omega_t\,dx. \qquad (7.65)$$

Boundary terms are dropped due to the periodicity of ω and its derivatives. In light of the equation above, we define the variational derivative by

$$\frac{\delta \widetilde{G}}{\delta \omega} \stackrel{d}{=} \mathcal{K}\frac{\partial \widetilde{G}}{\partial(\mathcal{K}\omega)} - \left(\mathcal{K}\frac{\partial \widetilde{G}}{\partial(\mathcal{K}\omega_x)}\right)_x. \qquad (7.66)$$

It is easy to see that with this particular choice the variational equation (7.63) coincides with the limiting Camassa–Holm equation. In fact, since

$$\frac{\partial \widetilde{G}}{\partial(\mathcal{K}\omega)} = \mathcal{K}\omega \quad \text{and} \quad \frac{\partial \widetilde{G}}{\partial(\mathcal{K}\omega_x)} = \mathcal{K}\omega_x, \qquad (7.67)$$

the concrete form of (7.66) is

$$\frac{\delta \widetilde{G}}{\delta \omega} = \mathcal{K}(\mathcal{K}\omega) - (\mathcal{K}(\mathcal{K}\omega_x))_x = \mathcal{K}\left(1 - \frac{\partial^2}{\partial x^2}\right)\mathcal{K}\omega = \mathcal{K}\omega. \qquad (7.68)$$

Here the trivial identity $(1 - \partial^2/\partial x^2)\mathcal{K} = 1$ (the identity map) is used. Substituting this into (7.63) and using $\omega = (1 - \partial^2/\partial x^2)u$, we recover the limiting Camassa–Holm equation.

The conservation law (7.62) directly follows from the skew-symmetry of the operator $\mathcal{J} \stackrel{d}{=} -((\partial/\partial x)\omega + \omega(\partial/\partial x))$: for any $f, g \in H^1(\mathbb{S})$,

$$\int_0^L f\mathcal{J}g\,dx = -\int_0^L (\mathcal{J}f)g\,dx. \qquad (7.69)$$

This means that the variational PDE (7.63) is a Hamiltonian PDE, and the corresponding Hamiltonian is conserved. In fact, from (7.65) we immediately obtain

$$\frac{d}{dt}\int_0^L \widetilde{G}(\mathcal{K}\omega, \mathcal{K}\omega_x)dx = \int_0^L \frac{\delta \widetilde{G}}{\delta \omega}\omega_t\,dx = \int_0^L \frac{\delta \widetilde{G}}{\delta \omega}\cdot\mathcal{J}\frac{\delta \widetilde{G}}{\delta \omega}dx = 0. \qquad (7.70)$$

Observe that the conservation law solely comes from the skew-symmetry of $\mathcal{J}$ and the variational form defined with variational derivative (7.63), and the concrete form of $\widetilde{G}$ is not essential. This enables us to employ the strategy described in the next section.

We present a $\widetilde{G}$-conserving Galerkin scheme. To that end, we commence by defining a new set of weak forms introducing a new intermediate variable p: find $\omega(\cdot, t), p \in H^1(\mathbb{S})$ such that for any $v_1, v_2 \in H^1(\mathbb{S})$

$$(\omega_t, v_1) = (\mathcal{J}p, v_1), \tag{7.71}$$

$$(p, v_2) = \left(\frac{\partial \widetilde{G}}{\partial(\mathcal{K}\omega)}, \mathcal{K}v_2 \right) + \left(\frac{\partial \widetilde{G}}{\partial(\mathcal{K}\omega_x)}, \mathcal{K}(v_2)_x \right) \tag{7.72}$$

hold. This set of weak forms happily keeps the conservation law as follows.

THEOREM 7.11
Suppose $\omega_t(\cdot, t), p \in H^1(\mathbb{S})$. Then the solution ω of the weak forms (7.71), (7.72) satisfies the conservation law (7.62).

PROOF   From (7.65), we see

$$\frac{\mathrm{d}}{\mathrm{d}t} \int_0^L \widetilde{G}(\mathcal{K}\omega, \mathcal{K}\omega_x) \mathrm{d}x = \left(\frac{\partial \widetilde{G}}{\partial(\mathcal{K}\omega)}, \mathcal{K}\omega_t \right) + \left(\frac{\partial \widetilde{G}}{\partial(\mathcal{K}\omega_x)}, \mathcal{K}\omega_{xt} \right)$$

$$= (p, \omega_t) = (\mathcal{J}p, p) = 0. \tag{7.73}$$

The first equality is from (7.65), the second is from (7.72) with the assumption $\omega_t(\cdot, t) \in H^1(\mathbb{S})$, and the third is from (7.72) with the assumption $p \in H^1(\mathbb{S})$.
□

We denote the approximate solutions by $\omega^{(m)} \simeq \omega(\cdot, m\Delta t)$ and $p^{(m+\frac{1}{2})} \simeq p(\cdot, (m+\frac{1}{2})\Delta t)$ $(m = 0, 1, 2, \ldots)$. To mimic the variational weak forms (7.71), (7.72), we first define a discrete version of $\widetilde{G}$ by

$$G_\mathrm{d}(\mathcal{K}\omega^{(m)}, \mathcal{K}\omega_x^{(m)}) \stackrel{\mathrm{d}}{=} \frac{(\mathcal{K}\omega^{(m)})^2 + (\mathcal{K}\omega_x^{(m)})^2}{2}. \tag{7.74}$$

Below this will be often abbreviated as $G_\mathrm{d}^{(m)}$ for saving space. We then define associated discrete partial derivatives by

$$\frac{\partial G_\mathrm{d}}{\partial(\mathcal{K}\omega^{(m+1)}, \mathcal{K}\omega^{(m)})} \stackrel{\mathrm{d}}{=} \mathcal{K}\omega^{(m+\frac{1}{2})}, \quad \frac{\partial G_\mathrm{d}}{\partial(\mathcal{K}\omega_x^{(m+1)}, \mathcal{K}\omega_x^{(m)})} \stackrel{\mathrm{d}}{=} \mathcal{K}\omega_x^{(m+\frac{1}{2})}, \tag{7.75}$$

where $\omega^{(m+\frac{1}{2})} \stackrel{\mathrm{d}}{=} (\omega^{(m+1)} + \omega^{(m)})/2$. They apparently approximate the continuous case (7.67), and it is easy to check that they satisfy the following discrete chain rule corresponding to (7.65).

$$\frac{1}{\Delta t}\int_0^L \left(G_d^{(n+1)} - G_d^{(m)}\right)dx$$
$$= \left(\frac{\partial G_d}{\partial(\mathcal{K}w^{(m+1)}, \mathcal{K}w^{(m)})}, \mathcal{K}\left(\frac{w^{(m+1)} - w^{(m)}}{\Delta t}\right)\right)$$
$$+ \left(\frac{\partial G_d}{\partial(\mathcal{K}w_x^{(m+1)}, \mathcal{K}w_x^{(m)})}, \mathcal{K}\left(\frac{w_x^{(m+1)} - w_x^{(m)}}{\Delta t}\right)\right). \quad (7.76)$$

With the discrete partial derivatives, a scheme is defined as follows. Let S_1, S_2 be some appropriate trial spaces, and W_1, W_2 test spaces. We define an operator

$$\mathcal{J}^{(m+\frac{1}{2})} \stackrel{d}{=} -\left(\frac{\partial}{\partial x}w^{(m+\frac{1}{2})} + w^{(m+\frac{1}{2})}\frac{\partial}{\partial x}\right), \quad (7.77)$$

which approximates $\mathcal{J}$, and is skew-symmetric.

Scheme 7.7 ($\widetilde{G}$-conserving scheme) Find $w^{(m)} \in S_2$ and $p^{(m+\frac{1}{2})} \in S_1$ ($m = 0, 1, 2, \ldots$) such that for any $v_1 \in W_1$ and $v_2 \in W_2$,

$$\left(\frac{w^{(m+1)} - w^{(m)}}{\Delta t}, v_1\right) = \left(\mathcal{J}^{(m+\frac{1}{2})}p^{(m+\frac{1}{2})}, v_1\right), \quad (7.78)$$

$$\left(p^{(m+\frac{1}{2})}, v_2\right) = \left(\frac{\partial G_d}{\partial(\mathcal{K}w^{(m+1)}, \mathcal{K}w^{(m)})}, \mathcal{K}v_2\right)$$
$$+ \left(\frac{\partial G_d}{\partial(\mathcal{K}w_x^{(m+1)}, \mathcal{K}w_x^{(m)})}, \mathcal{K}(v_2)_x\right) \quad (7.79)$$

hold.

Then the scheme enjoys the following conservation property. Observe that the proof goes exactly the same way as in the continuous case.

THEOREM 7.12 Discrete $\widetilde{G}$ conservation law
Suppose $(w^{(m+1)} - w^{(m)})/\Delta t \in W_2$ ($m = 0, 1, 2, \ldots$) and $S_1 \subseteq W_1$. Then Scheme 7.7 is conservative in the sense that

$$\frac{1}{\Delta t}\int_0^L \left(G_d^{(m+1)} - G_d^{(m)}\right)dx = 0 \quad (m = 0, 1, 2, \ldots) \quad (7.80)$$

holds.

PROOF   From the discrete chain rule (7.76),

$$\frac{1}{\Delta t}\int_0^L \left(G_{\mathrm d}^{(m+1)} - G_{\mathrm d}^{(m)}\right) \mathrm dx$$
$$= \left(\frac{\partial G_{\mathrm d}}{\partial(\mathcal K\omega^{(m+1)}, \mathcal K\omega^{(m)})}, \mathcal K\left(\frac{\omega^{(m+1)} - \omega^{(m)}}{\Delta t}\right)\right) \quad (7.81)$$
$$+ \left(\frac{\partial G_{\mathrm d}}{\partial(\mathcal K\omega_x^{(m+1)}, \mathcal K\omega_x^{(m)})}, \mathcal K\left(\frac{\omega_x^{(m+1)} - \omega_x^{(m)}}{\Delta t}\right)\right)$$
$$= \left(p^{(m+\frac{1}{2})}, \frac{\omega_x^{(m+1)} - \omega_x^{(m)}}{\Delta t}\right)$$
$$= \left(\mathcal J^{(m+\frac{1}{2})} p^{(m+\frac{1}{2})}, p^{(m+\frac{1}{2})}\right)$$
$$= 0. \quad (7.82)$$

In the second and third equality, the assumptions are used. The last equality follows from the skew-symmetry of $\mathcal J^{(m+\frac{1}{2})}$. □

The trial and test function spaces can be set to various standard ones such as the standard finite-dimensional Fourier space or the finite element spaces, depending on the users' preferences. The theorem above clarifies the conditions for the scheme to be successfully conservative. The simplest and most useful choice would be the use of the standard periodic piecewise linear function space on some fixed grid for all of S_1, S_2, W_1 and W_2; in that case, the assumptions in the theorem are trivially satisfied.

An important outcome of preserving the $\widetilde G$ conservation law is that Scheme 7.7 gains the following stability property. Let us denote the approximate solution of u by $u^{(m)} \stackrel{\mathrm d}{\equiv} \mathcal K\omega^{(m)}$.

THEOREM 7.13 Stability of Scheme 7.7
Scheme 7.7 is stable in the sense that (in exact arithmetic) $\|u^{(m)}\|_\infty < \infty$ *($m = 0, 1, 2, \ldots$).*

PROOF   From the $\widetilde G$-conservation, we readily see that there exists a constant c such that

$$\|u^{(m)}\|_\infty \le c\|u^{(m)}\|_{H^1(\mathbb S)} = \mathrm{const.}, \quad (7.83)$$

by the Sobolev lemma. □

REMARK 7.9   This stability property is the Galerkin version of the stability discussed in Section 4.7.2.6.1. □

Let us demonstrate the scheme. In what follows the equispaced spatial mesh of N grid points ($x_0 = 0, x_N = L$) is assumed, and the standard periodic piecewise linear function space on the mesh, denoted as S_p, is used as the trial and test spaces. The basis functions are denoted by $\phi_k(x)$ ($k = 0, \ldots, N-1$).

In actual computation, the inverse operator $\mathcal{K} = (1 - \partial^2/\partial x^2)^{-1}$ is realized as the convolution

$$(\mathcal{K}f)(x) = (k * f)(x) = \int_0^L k(x - \xi) f(\xi) \mathrm{d}\xi, \qquad (7.84)$$

with the Green function:

$$k(x) = \frac{\cosh(x - L[x/L] + L/2)}{2\sinh(L/2)}. \qquad (7.85)$$

The operator appears in the second equation of Scheme 7.7, which reads

$$\left(p^{(m+\frac{1}{2})}, v_2\right) = \left(\mathcal{K}\omega^{(m+\frac{1}{2})}, \mathcal{K}v_2\right) + \left(\mathcal{K}\omega_x^{(m+\frac{1}{2})}, \mathcal{K}(v_2)_x\right). \qquad (7.86)$$

If we introduce the matrices

$$A_{ij} \stackrel{\mathrm{d}}{=} (\phi_i, \phi_j), \quad (K_1)_{ij} \stackrel{\mathrm{d}}{=} (\mathcal{K}\phi_i, \mathcal{K}\phi_j), \quad \text{and} \quad (K_2)_{ij} \stackrel{\mathrm{d}}{=} (\mathcal{K}(\phi_i)_x, \mathcal{K}(\phi_j)_x), \qquad (7.87)$$

then the concrete form of Scheme 7.7 becomes

$$A\left(\frac{\boldsymbol{\omega}^{(m+1)} - \boldsymbol{\omega}^{(m)}}{\Delta t}\right) = \boldsymbol{g}(\boldsymbol{\omega}^{(m+\frac{1}{2})}, \boldsymbol{p}^{(m+\frac{1}{2})}), \qquad (7.88)$$

$$A\boldsymbol{p}^{(m+\frac{1}{2})} = K_1 \boldsymbol{\omega}^{(m+\frac{1}{2})} + K_2 \boldsymbol{\omega}^{(m+\frac{1}{2})}. \qquad (7.89)$$

The vectors are defined as $\boldsymbol{\omega}^{(m)} \stackrel{\mathrm{d}}{=} (m_0^{(n)}, \ldots, m_{N-1}^{(n)})^\mathrm{T}$, $\boldsymbol{\omega}^{(m+\frac{1}{2})} \stackrel{\mathrm{d}}{=} (\boldsymbol{\omega}^{(m+1)} + \boldsymbol{\omega}^{(m)})/2$, $\boldsymbol{p}^{(m+\frac{1}{2})} \stackrel{\mathrm{d}}{=} (p_0^{(n)}, \ldots, p_{N-1}^{(n)})^\mathrm{T}$, and $\boldsymbol{g}$ is the vector function that represents the nonlinear part in the first equation (we omit the concrete form of $\boldsymbol{g}$ here, since it is straightforward and not important for the discussion here). The equations above represent a system of nonlinear and linear equations of dimension $2N$, but can be readily reduced to

$$A\left(\frac{\boldsymbol{\omega}^{(m+1)} - \boldsymbol{\omega}^{(m)}}{\Delta t}\right) = \boldsymbol{g}\left(\boldsymbol{\omega}^{(m+\frac{1}{2})}, A^{-1}(K_1 + K_2)\boldsymbol{\omega}^{(m+\frac{1}{2})}\right), \qquad (7.90)$$

which is of dimension N; that is, the intermediate variable $\boldsymbol{p}^{(m+\frac{1}{2})}$ can be erased in actual computation. Note that the matrices A, K_1 and K_2 depend only on the grid and basis functions, and can be computed *in prior to* the time evolution process; heavy convolutions are not required during the main computation. In the numerical experiment below, at each time step the equation (7.90) is solved by the hybrid Newton algorithm imsl_d_zeros_sys_eqn in the IMSL library.

Since the time integration is solely carried out in w space, we have to switch from/to the original variable u as pre- and post-processes. Let N_t be the number of temporal time steps. Then the overall integration procedure is as follows.

1. For a given initial data $u(x, 0)$, compute $w(x, 0) = (1 - \partial^2/\partial x^2)u(x, 0)$.

2. Time integration: repeat $w^{(m+1)} \leftarrow w^{(m)}$ ($m = 0, 1, 2, \ldots$) by (7.90).

3. For the obtained final data $w^{(N_t)}(x)$, compute $u^{(N_t)} = \mathcal{K}w^{(N_t)}$ as the solution.

Note that when we need the approximate solution in the form of u, we have to compute the convolution $\mathcal{K}w^{(m)}$, which is relatively time consuming. Usually, however, we need u itself at relatively few time steps compared to the whole number of computation steps, and the additional cost is considered to be acceptable in practical situations.

We consider the collision of two soliton-like solutions as an illustrative example. We set $L = 40$, which is divided into $N = 100$ grids. The initial data is set to $u(x, 0) = 0.2\,\mathrm{sech}(x - 403/15) + 0.5\,\mathrm{sech}(x - 203/15)$. Then the problem is integrated in the time interval $[0, 200]$, with the time mesh size $\Delta t = 0.1$ (i.e., the number of temporal grids $N_t = 2000$). In addition to the scheme 7.7, we also tested for comparison an implicit Euler scheme:

$$\left(\frac{w^{(m+1)} - w^{(m)}}{\Delta t}, v_1\right) = \left(-\left(\frac{\partial}{\partial x}w^{(m+1)} + w^{(m+1)}\frac{\partial}{\partial x}\right)p^{(m+\frac{1}{2})}, v_1\right),$$
$$\forall v_1 \in S_p, \quad (7.91)$$

$$\left(p^{(m+\frac{1}{2})}, v_2\right) = \left(\mathcal{K}w^{(m+1)}, \mathcal{K}v_2\right) + \left(\mathcal{K}w_x^{(m+1)}, \mathcal{K}(v_2)_x\right),$$
$$\forall v_2 \in S_p, \quad (7.92)$$

an explicit Euler scheme:

$$\left(\frac{w^{(m+1)} - w^{(m)}}{\Delta t}, v_1\right) = \left(-\left(\frac{\partial}{\partial x}w^{(m)} + w^{(m)}\frac{\partial}{\partial x}\right)p^{(m+\frac{1}{2})}, v_1\right),$$
$$\forall v_1 \in S_p, \quad (7.93)$$

$$\left(p^{(m+\frac{1}{2})}, v_2\right) = \left(\mathcal{K}w^{(m)}, \mathcal{K}v_2\right) + \left(\mathcal{K}w_x^{(m)}, \mathcal{K}(v_2)_x\right),$$
$$\forall v_2 \in S_p \quad (7.94)$$

and the G-conserving scheme presented in Section 7.2.2.4.1. Note that the implicit Euler and explicit Euler schemes above are also based on the conservative weak forms (7.71), (7.72), but all $w^{(m+\frac{1}{2})}$'s in Scheme 7.7 are replaced with $\omega^{(m+1)}$ or $\omega^{(m)}$, and thus the conservation law is lost in those cases. Note also that the computational complexity of the implicit Euler scheme is almost the same as that of Scheme 7.7.

Figure 7.20[22] shows the evolution of the approximate solutions. In the result by the conservative scheme (top left figure), the collisions of the two soliton-like solutions are rightly captured; the larger (thus faster) soliton-like solution overtakes the smaller (slower) one as expected. The computation proceeded quite stably. The implicit Euler scheme (bottom left) is favorably stable as well, but the stability rather comes from the strong dissipation property that is often observed in general implicit Euler schemes; in fact, the solution rapidly gets flattened. As a consequence, the soliton-like solutions get slower (recall that the speed of a soliton-like solution depends on its size), and the larger solution goes around the interval only once, instead of three times originally expected. Thus the implicit Euler scheme should be rejected, when the qualitative behavior of the problem is of our interest. The explicit Euler scheme (bottom right) is quite unstable as expected, and the solution blows up soon after the start of computation. This scheme does not deserve further consideration. On the result from the G-conserving scheme (top right), some careful discussion is required. In the early phase of computation (more precisely speaking, at least until around $t = 100$), it happily captures the collision process and the qualitative behavior agrees with that of $\widetilde{G}$-conserving scheme. After that, however, the solution shows instability. The difference between the $\widetilde{G}$- and G-conserving schemes in terms of stability should be attributed to the additional stability property of the G-conserving scheme stated in Theorem 7.13. In this sense, we can say that the property is of practical importance. (Note that the result here does not immediately imply that the G-conserving scheme is unstable; it has been confirmed in the preceding section that the G-conserving scheme is actually stabler than several generic schemes. The result just claims the $\widetilde{G}$-conserving one is better.)

The evolutions of the invariants $\widetilde{G}$ and G in each scheme (except the explicit Euler scheme) are shown in Figures 7.21[23] and 7.22.[24] In Figure 7.21, we can see that the $\widetilde{G}$-conserving scheme rightly conserves $\widetilde{G}$, while the other two schemes fail. In the implicit Euler scheme, $\widetilde{G}$ is steadily dissipated. In the G-conserving scheme, $\widetilde{G}$ stays around the exact value in the early phase of evolution, but finally it nearly blows up; this corresponds to the instability observed in Figure 7.20. The graphs in Figure 7.22 show the evolution of G; the left figure shows the overall profile, and the right shows its detail around the true G value, which is to clarify the difference between the $\widetilde{G}$- and G-conserving schemes. According to the graphs, in the implicit Euler scheme G is again soon dissipated. The G-conserving scheme strictly conserves

---

[22-25] Reprinted from *J. Comput. Appl. Math.*, 234, T. Matsuo, A Hamiltonian-conserving Galerkin scheme for the Camassa–Holm equation, 1258–1266, Copyright (2010), with permission from Elsevier.

the invariant as the theory suggests, while the $\widetilde{G}$-conserving scheme *nearly* conserves it.

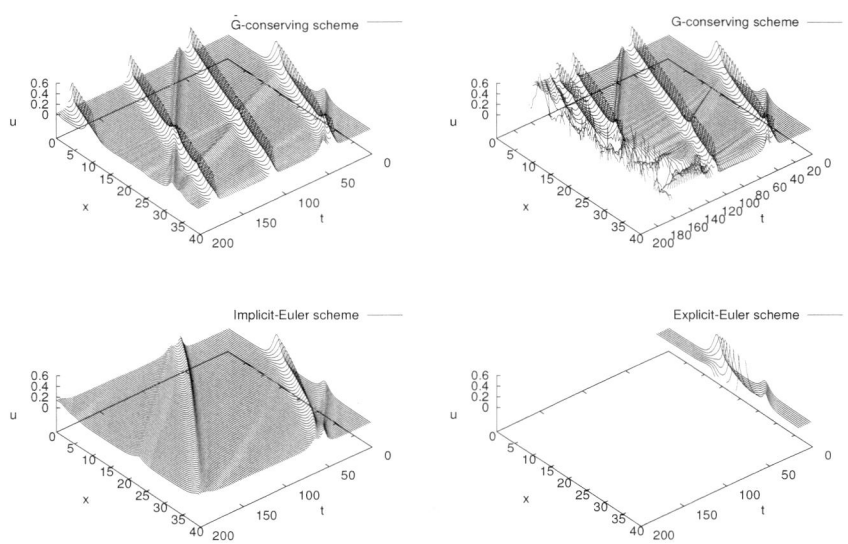

FIGURE 7.20: Evolution of the numerical solutions: (top left) the $\widetilde{G}$-conserving scheme, (top right) the G-conserving scheme, (bottom left) the implicit Euler scheme, (bottom right) the explicit Euler scheme.

Finally, the $\widetilde{G}$-conserving scheme is checked on coarser meshes $N = 20$ (i.e., $\Delta x = 2$) and $N = 40$ ($\Delta x = 1$), in order to check whether the scheme is stable with respect to the spatial discretization. The time mesh size is kept the same ($\Delta t = 0.1$). Figure 7.23[25] shows the results, which suggest that the scheme is stable even with very coarse mesh.

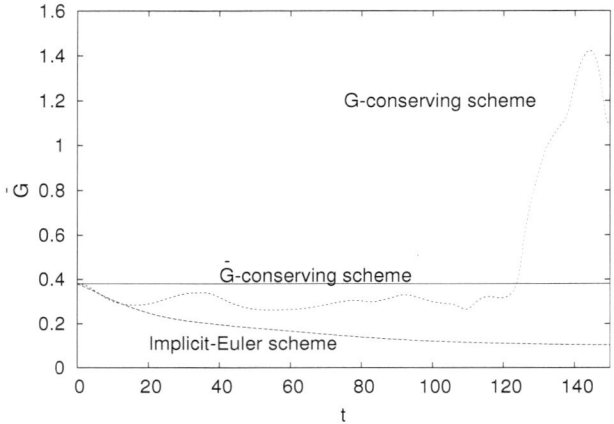

FIGURE 7.21: Evolution of $\widetilde{G}$.

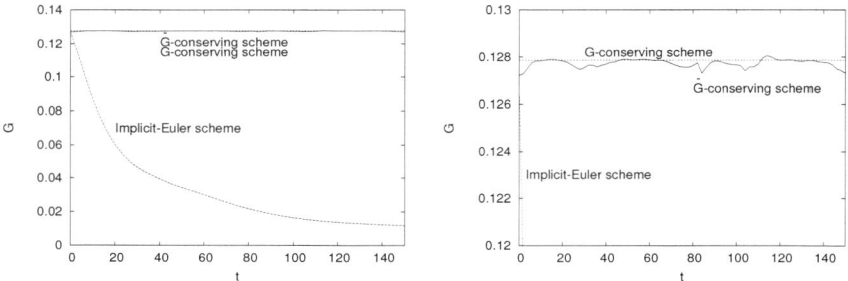

FIGURE 7.22: Evolution of G: (left) overall profile, (right) detail.

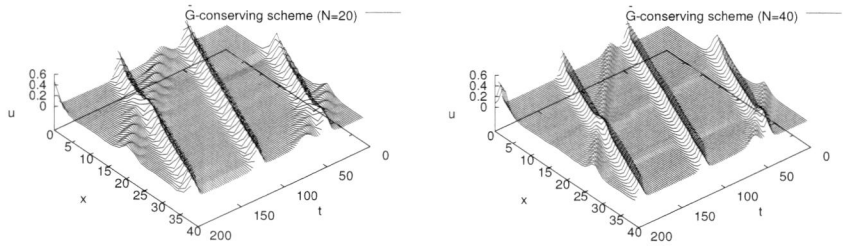

FIGURE 7.23: Evolution of the numerical solutions on coarser meshes with the $\widetilde{G}$-conserving scheme: (left) $N = 20$, (right) $N = 40$.

7.2.2.5 Ginzburg–Landau Equation for Superconductivity

This is a sample section for essentially two-dimensional computation. The phenomenological behavior of superconductivity is governed by the so-called Ginzburg–Landau model. The model in the so-called "zero electric potential gauge" is described as the following time-dependent Ginzburg–Landau (TDGL) equations:

$$\eta \frac{\partial \psi}{\partial t} + \frac{1}{2}\left\{\left(\frac{\mathrm{i}}{\kappa}\nabla + \boldsymbol{A}\right)^2 \psi + \left(|\psi|^2 - 1\right)\psi\right\} = 0 \quad \text{in } \Omega, \quad (7.95\mathrm{a})$$

$$\frac{\partial \boldsymbol{A}}{\partial t} + \mathrm{Re}\left[\overline{\psi}\left(\frac{\mathrm{i}}{\kappa}\nabla + \boldsymbol{A}\right)\psi\right] + \nabla \times (\nabla \times \boldsymbol{A} - \boldsymbol{H}) = 0 \quad \text{in } \Omega, \quad (7.95\mathrm{b})$$

where $\Omega \subset \mathbb{R}^d$ is a bounded subdomain with smooth boundary, $\kappa > 0$ is the material constant called the Ginzburg–Landau parameter, $\eta > 0$ is the friction coefficient, $\boldsymbol{H} \in \mathbb{R}^d$ is the applied magnetic field, $\psi : \Omega \times [0,T] \to \mathbb{C}$ is the complex-valued order parameter which denotes the conducting state of the material, and $\boldsymbol{A} : \Omega \times [0,T] \to \mathbb{R}^d$ is the magnetic potential. By $\overline{\psi}$ we mean the complex conjugate of ψ. The associated boundary conditions are:

$$\nabla \psi \cdot \boldsymbol{n} = 0, \quad \boldsymbol{A} \cdot \boldsymbol{n} = 0, \quad \boldsymbol{n} \times (\nabla \times \boldsymbol{A} - \boldsymbol{H}) = 0 \quad \text{on } \partial\Omega \quad (7.96)$$

where $\boldsymbol{n}$ is the exterior unit normal of the boundary $\partial\Omega$. For this gauge choice and the well-posedness of the associated Cauchy problem, see [40].

The advantage of this particular gauge choice is that the problem can be viewed as a gradient flow of the Ginzburg–Landau energy functional:

$$G(\psi, \boldsymbol{A}) = \frac{1}{2}\left|\left(\frac{\mathrm{i}}{\kappa}\nabla + \boldsymbol{A}\right)\psi\right|^2 + \frac{1}{4}(1 - |\psi|^2)^2 + \frac{1}{2}|\nabla \times \boldsymbol{A} - \boldsymbol{H}|^2, \quad (7.97\mathrm{a})$$

$$J = \int_\Omega G(\psi, \boldsymbol{A})\mathrm{d}\boldsymbol{x}, \quad (7.97\mathrm{b})$$

$$\eta \frac{\partial \psi}{\partial t} = -\frac{\delta G}{\delta \overline{\psi}}, \quad \frac{\partial \boldsymbol{A}}{\partial t} = -\frac{\delta G}{\delta \boldsymbol{A}}, \quad (7.98)$$

where $\delta G/\delta \overline{\psi}$ and $\delta G/\delta \boldsymbol{A}$ denote variational derivatives. This energy in other words serves as a Lyapunov functional of the system, and this suggests us to employ numerical schemes having some discrete counterpart of this property for stability and correct asymptotic behavior (see, for example, Lord [109], Montagne et al. [131], and Mu [133] for related discussions and numerical schemes). Below we show that by (extending) the Galerkin version of the discrete variational derivative method, we can deduce fully implicit and linearly implicit schemes for the TDGL that preserve discrete versions of the Lyapunov functional.

Note that since the Ginzburg–Landau energy and the related equations include the symbols of vector calculus, and now the domain Ω is generally not rectangular, this is an essentially high-dimensional problem. In the rest of this subsection, we briefly demonstrate that the Galerkin framework shown in this section can be naturally applied to this problem. The problem usually makes sense for $d = 3$, but if we assume the material is cylinder-shaped and the external magnetic field $\boldsymbol{H}$ is constant, the problem can be viewed as two-dimensional $d = 2$; below we assume this.

In order to simplify the discussion, in what follows we limit ourselves to the simplified model, ignoring all the magnetic effects:

$$\eta \frac{\partial \psi}{\partial t} = \frac{1}{2} \left\{ \frac{\Delta \psi}{\kappa^2} + (1 - |\psi|^2)\psi \right\} \quad \text{in } \Omega, \qquad \nabla \psi \cdot \boldsymbol{n} = 0 \quad \text{on } \partial \Omega. \qquad (7.99)$$

This still deserves investigation since it involves interesting physical solutions such as vortices and a Lyapunov functional:

$$J = \int_\Omega G(\psi) \mathrm{d}\boldsymbol{x}, \quad \text{where } G(\psi) = \frac{1}{2}\left|\frac{\nabla \psi}{\kappa}\right|^2 + \frac{1}{4}(1 - |\psi|^2)^2. \qquad (7.100)$$

The simplified equation (7.99) is formally a gradient flow with respect to the energy:

$$\eta \frac{\partial \psi}{\partial t} = -\frac{\delta G}{\delta \overline{\psi}}. \qquad (7.101)$$

We also assume $d = 2$ for brevity (we consider, say, a unit disk). Let $H_\mathrm{c}^1(\Omega)$ be the standard Sobolev space of complex-valued functions and $(\cdot, \cdot)$ be its associated inner product. Let S_1 and W_1 be the finite-dimensional subspaces in $H_\mathrm{c}^1(\Omega)$ for trial and test functions satisfying $S_1 \subseteq W_1$ (in most cases we simply take $S_1 = W_1$, in particular to the standard piecewise linear function space).

7.2.2.5.1 Fully Implicit Schemes for the Simplified GL Equation

By a natural extension of the Galerkin framework shown before, we reach the following fully implicit scheme. We denote the numerical solution by $\psi^{(m)}(\boldsymbol{x}) \simeq \psi(m\Delta t, \boldsymbol{x})$.

Scheme 7.8 (Fully Implicit Scheme ([132])) *Suppose an initial data $\psi^{(0)} \in S_1$ is given. Find $\psi^{(m)} \in S_1$ ($m = 1, 2, \ldots$) such that for any $\phi \in W_1$*

$$\eta \left(\frac{\psi^{(m+1)} - \psi^{(m)}}{\Delta t}, \phi \right)$$
$$= -\left(\frac{\partial G_\mathrm{d}}{\partial(\nabla \overline{\psi}^{(m+1)}, \nabla \overline{\psi}^{(m)})}, \nabla \phi \right) - \left(\frac{\partial G_\mathrm{d}}{\partial(\overline{\psi}^{(m+1)}, \overline{\psi}^{(m)})}, \phi \right),$$

where

$$\frac{\partial G_\mathrm{d}}{\partial(\nabla\overline{\psi}^{(m+1)},\nabla\overline{\psi}^{(m)})} = \frac{1}{2\kappa^2}\left(\frac{\nabla\psi^{(m+1)}+\nabla\psi^{(m)}}{2}\right),$$

$$\frac{\partial G_\mathrm{d}}{\partial(\overline{\psi}^{(m+1)},\overline{\psi}^{(m)})} = -\frac{1}{2}\left(1 - \frac{|\psi^{(m+1)}|^2 + |\psi^{(m)}|^2}{2}\right)\left(\frac{\psi^{(m+1)}+\psi^{(m)}}{2}\right).$$

This scheme has a desired dissipation property.

PROPOSITION 7.1 Dissipation Property of Scheme 7.8 ([132])
Let $\psi^{(m)}$ ($m = 1, 2, \ldots$) be the solutions of Scheme 7.8. Then the following discrete dissipation property holds:

$$\frac{1}{\Delta t}\int_\Omega G(\psi^{(m+1)}) - G(\psi^{(m)})\mathrm{d}\boldsymbol{x} = -2\eta\int_\Omega \left|\frac{\psi^{(m+1)} - \psi^{(m)}}{\Delta t}\right|^2 \mathrm{d}\boldsymbol{x} \leq 0.$$

This means that in Scheme 7.8 the original energy G dissipates as in the continuous case. This implies that the asymptotic behavior of the approximate solutions must be quite similar to that of the original TDGL (strictly speaking, to that of the corresponding ODE derived by discretizing the space variable).

In [41], an implicit Euler type scheme is derived from the energy functional based on minimization theory. Here only the resulting scheme is shown.

Scheme 7.9 (Fully Implicit Scheme ([41])) *Suppose an initial data $\psi^{(0)} \in S_1$ is given. Find $\psi^{(m)} \in S_1$ ($m = 1, 2, \ldots$) such that for any $\phi \in W_1$*

$$\eta\left(\frac{\psi^{(m+1)} - \psi^{(m)}}{\Delta t}, \phi\right) = -\frac{1}{2\kappa^2}\left(\nabla\psi^{(m+1)}, \nabla\phi\right) - \frac{1}{2}\left((|\psi^{(m+1)}|^2 - 1)\psi^{(m+1)}, \phi\right).$$

PROPOSITION 7.2 Dissipation property of Scheme 7.9 ([41])
Let $\psi^{(m)}$ ($m = 1, 2, \ldots$) be the solutions of Scheme 7.9. Then the following discrete dissipation property holds:

$$\frac{1}{\Delta t}\int_\Omega G(\psi^{(m+1)}) - G(\psi^{(m)})\mathrm{d}\boldsymbol{x} \leq -2\eta\int_\Omega \left|\frac{\psi^{(m+1)} - \psi^{(m)}}{\Delta t}\right|^2 \mathrm{d}\boldsymbol{x} \leq 0.$$

Thus the scheme should have similar asymptotic behavior as above; in fact, in [41], a detailed discussion on the asymptotic behavior is given for the full TDGL (7.95).

In these two similar schemes, however, we find several essential differences. First, notice that the first equality in Proposition 7.1 is replaced with an inequality in Proposition 7.2, whose equality does not hold in general (this can be understood by carefully inspecting its proof; interested readers may refer to [41]). Since in the continuous case, the equality holds: $(\mathrm{d}/\mathrm{d}t)\int G\mathrm{d}\boldsymbol{x} =$

$-2\eta \int |\psi_t|^2 d\boldsymbol{x}$, we can say that Scheme 7.8 is closer to the original TDGL. Although the implicit Euler scheme happily keeps the Lyapunov functional, the dissipation (how the energy is dissipated) is slightly stronger there than it should be. Second, Scheme 7.8 should be second order with respect to Δt due to its temporal symmetry, while Scheme 7.9 is only first order.

Both schemes have an unwelcome feature in common: they are fully implicit, and require time-consuming iterative solvers. This disadvantage becomes even more crucial, if we consider the full TDGL, or proceed to the $d = 3$ cases. In the next subsection, we consider a linearly implicit scheme in order to overcome this disadvantage.

7.2.2.5.2 A Linearly Implicit Scheme for the Simplified GL Equation
By utilizing the linearization technique in Chapter 6, we can derive the following linearly implicit scheme.

Scheme 7.10 (Linearly Implicit Scheme) *Suppose an initial data $\psi^{(0)} \in S_1$ and a starting value $\psi^{(1)}$ are given. Find $\psi^{(m)} \in S_1$ ($m = 2, 3, \ldots$) such that for any $\phi \in W_1$*

$$\eta\left(\frac{\psi^{(m+1)} - \psi^{(m-1)}}{2\Delta t}, \phi\right) = -\left(\frac{\partial G_{\mathrm{d}}}{\partial(\nabla\overline{\psi}^{(m+1)}, \nabla\overline{\psi}^{(m)}, \nabla\overline{\psi}^{(m-1)})}, \nabla\phi\right)$$
$$- \left(\frac{\partial G_{\mathrm{d}}}{\partial(\overline{\psi}^{(m+1)}, \overline{\psi}^{(m)}, \overline{\psi}^{(m-1)})}, \phi\right),$$

where

$$\frac{\partial G_{\mathrm{d}}}{\partial(\nabla\overline{\psi}^{(m+1)}, \nabla\overline{\psi}^{(m)}, \nabla\overline{\psi}^{(m-1)})} = \frac{1}{2\kappa^2}\left\{b\nabla\psi^{(m)} + (1-b)\frac{\nabla\psi^{(m+1)} + \nabla\psi^{(m-1)}}{2}\right\},$$

$$\frac{\partial G_{\mathrm{d}}}{\partial(\overline{\psi}^{(m+1)}, \overline{\psi}^{(m)}, \overline{\psi}^{(m-1)})} = \frac{a}{2}\left(-1 + \frac{\psi^{(m+1)} + \psi^{(m-1)}}{2}\overline{\psi}^{(m)}\right)\psi^{(m)}$$
$$+ \frac{1-a}{2}(-1 + |\psi^{(m)}|^2)\left(\frac{\psi^{(m+1)} + \psi^{(m-1)}}{2}\right),$$

and $a, b \in \mathbb{R}$ are scheme parameters.

The scheme parameters a, b should be chosen carefully, since they severely affect the stability of the resulting scheme as will be shown below. Observe that the scheme is linear with respect to the latest value $\psi^{(m+1)}$. This scheme enjoys the following dissipation property.

Advanced Topic III: Further Remarks

THEOREM 7.14 Dissipation Property of Scheme 7.10
Let $\psi^{(m)}$ $(m = 2, 3, \ldots)$ be the solutions of Scheme 7.10. Then the following discrete dissipation property holds:

$$\int_\Omega G_d(\psi^{(m+1)}, \psi^{(m)}) - G_d(\psi^{(m)}, \psi^{(m-1)}) d\boldsymbol{x}$$
$$= -2\eta \int_\Omega \left| \frac{\psi^{(m+1)} - \psi^{(m-1)}}{2\Delta t} \right|^2 d\boldsymbol{x} \leq 0,$$

where

$$G_d(\psi^{(m+1)}, \psi^{(m)})$$
$$= \frac{1}{4} \left\{ a(1 - \psi^{(m+1)}\overline{\psi}^{(m)})(1 - \overline{\psi}^{(m+1)}\psi^{(m)}) \right.$$
$$\left. + (1-a)(1 - |\psi^{(m+1)}|^2)(1 - |\psi^{(m)}|^2) \right\}$$
$$+ \frac{1}{2\kappa^2} \left\{ b \left(\frac{\nabla \psi^{(m)} \cdot \nabla \overline{\psi}^{(m+1)} + \nabla \overline{\psi}^{(m)} \cdot \nabla \psi^{(m)}}{2} \right) \right.$$
$$\left. + (1-b) \left(\frac{|\nabla \psi^{(m+1)}|^2 + |\nabla \psi^{(m)}|^2}{2} \right) \right\}. \quad (7.102)$$

Note that now the discrete energy function (7.102) depends on two consecutive numerical solutions (i.e., it is "multistep"), and is quadratic with respect to the latest value $\psi^{(m+1)}$; this is the key for the linearization. The scheme parameters a, b appear as the coefficients of the linear combination of the quadratic approximations. The theorem states that for any choice of a, b, the discrete dissipation property holds in the above sense. The discrete energy function (7.102) is, however, totally different from the original one (7.100), and as a consequence the discrete dissipation property does not immediately imply the correct asymptotic behavior, as was the case in the fully implicit schemes.

Still, the discrete energy function gives us useful information for designing good (stable) schemes; more specifically, for the choice of appropriate scheme parameters a, b. Below we demonstrate this. The first step is to rewrite the energy function as follows.

$$G_d(\psi^{(m+1)}, \psi^{(m)}) = \frac{1}{4} \left\{ |1 - \psi^{(m+1)}\overline{\psi}^{(m)}|^2 + (a-1)|\psi^{(m+1)} - \psi^{(m)}|^2 \right\}$$
$$\frac{1}{2\kappa^2} \left\{ \left| \frac{\nabla \psi^{(m+1)} + \nabla \psi^{(m)}}{2} \right|^2 + (1-2b) \left| \frac{\nabla \psi^{(m+1)} - \nabla \psi^{(m)}}{2} \right|^2 \right\}.$$
$$(7.103)$$

Let us then consider a "doubled" phase space $(\psi^{(m+1)}, \psi^{(m)})$, and consider that Scheme 7.10 defines a discrete map from the doubled space to itself:

$(\psi^{(m-1)}, \psi^{(m-2)}) \mapsto (\psi^{(m+1)}, \psi^{(m)})$. We then observe that depending on the parameters a, b the dynamical system can behave in the following three ways.

1. When $a < 1$ or $b > 1/2$, $G_\mathrm{d}(\psi^{(m+1)}, \psi^{(m)})$ obviously is not bounded from below, and thus it can never serve as Lyapunov functional. In this case, by losing the Lyapunov property the system can be unstable.

2. When $a = 1$ and $b = 1/2$, which here we call the "critical" case, the energy function is bounded and can serve as Lyapunov functional. By the Lyapunov theory, the dynamical system it governs asymptotically tends to the minimizers. But notice that the dynamics is a bit different from the original one. Let us consider the global minimizers $\int G_\mathrm{d}(\psi^{(m+1)}, \psi^{(m)}) \mathrm{d}\boldsymbol{x} = 0$. In view of (7.103), we see that the global minimizers are such points that $\psi^{(m+1)}\overline{\psi}^{(m)} = 1$ and $\nabla(\psi^{(m+1)} + \psi^{(m)}) = 0$. This allows an oscillatory "steady state" solution $\psi^{(m)} = c$, $\psi^{(m+1)} = 1/\overline{c}$ where $c \in \mathbb{C}$ is an arbitrary constant. This is in fact "steady state" in that in the doubled phase space, it corresponds to a fixed point $(c, 1/\overline{c})$ of the dynamical system; in the original undoubled space, however, it represents an oscillatory solution $c \to 1/\overline{c} \to c \to 1/\overline{c} \to \cdots$. Thus we conclude that in the critical case, the system is equipped with a Lyapunov functional, but the dynamics is different such that it allows spurious fixed points (in the doubled space).

3. When $a > 1$ and $b \leq 1/2$, the spurious fixed points vanish, and the Lyapunov functional allows only original steady state solutions as its fixed points.

In the last case, the dynamical system is expected to behave the same way as the fully implicit cases, although the corresponding linearly implicit scheme is far cheaper. We like to generalize the above observation as follows: as an unavoidable consequence of the linearization, the resulting scheme should be necessarily multistep, and the associated dynamical system should be understood in the doubled (or more higher) phase space. There are often degrees of freedom in the definition of multistep energy functions that crucially determine the dynamics observed as its (numerical) stability. In some happy cases, such as the above, by carefully choosing the free (scheme) parameters we can enforce the scheme (the dynamical system) to behave the same as the original system. A question, however, concerns in which circumstances we can find such "happy" cases. In particular, whether or not we can do that for any PDEs is an important open problem to be answered.

7.2.2.5.3 Numerical Examples In this section we present numerical examples that illustrate the discussion in the previous section. We here test Scheme 3, with two parameter sets $(a, b) = (0.9, 0.5)$ and $(2, -0.5)$, each of which corresponds to the first and third patterns described above. For comparison, we also test the standard semi-implicit scheme, where the diffusion term is discretized in time by the implicit Euler, and the nonlinear term by the explicit Euler. We set the TDGL parameters to be $\eta = 1$, $\kappa = 15$, and solved the simplified TDGL on the unit disk with a triangulation of 9,375 elements by FreeFEM. As the initial data, we set the two vortices of indices $+1$ and -1. With this setting, it is known that the annihilation (disappearing by merging) of vortices should occur.

First we show a result with a fine time mesh $\Delta t = 0.1$. We tested the semi-implicit scheme and Scheme 7.10 with $(a, b) = (2, -0.5)$, and found no difference; both schemes ran quite happily in this case. We show the result in Figure 7.24.

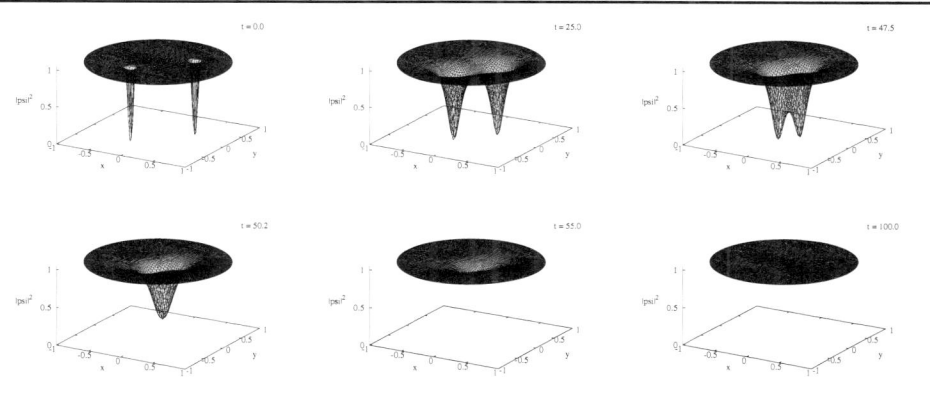

FIGURE 7.24: Evolution of the solution with $\Delta t = 0.1$: the semi-implicit scheme and Scheme 7.10 with $(a, b) = (2, -0.5)$.

The corresponding energy profiles are shown in Figure 7.25. For Scheme 7.10, we calculated the summation of G_d (the multistep energy function (7.102)) and G (the original energy function (7.100)). For the semi-implicit scheme, we calculated only the latter. In this setting, all the three lines well agree.

The semi-implicit scheme, however, becomes unstable as Δt increases. We demonstrate it by setting $\Delta t = 1.1$ in Figure 7.26 which shows snapshots of four consecutive time steps around $t = 50$. We can observe severe numerical oscillation there. In contrast, Scheme 3 holds out with the same coarse time step as shown in Figure 7.27. The energy profiles are shown in Figure 7.28, where we can observe oscillation in the semi-implicit scheme.

Finally we test Scheme 3 with the parameters $(a, b) = (0, 9, 0.5)$ with $\Delta t = 0.5$. As shown in Figure 7.29, the result is catastrophic. This agrees with the discussion in the previous section.

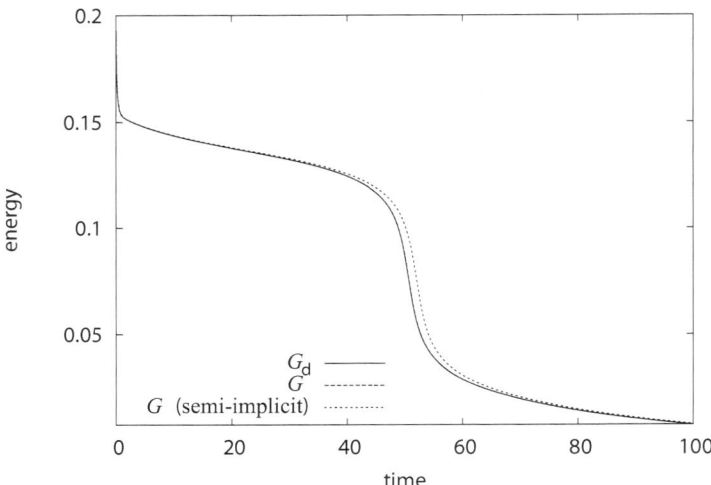

FIGURE 7.25: Evolution of the energies with $\Delta t = 0.1$: the semi-implicit scheme and Scheme 7.10 with $(a, b) = (2, -0.5)$.

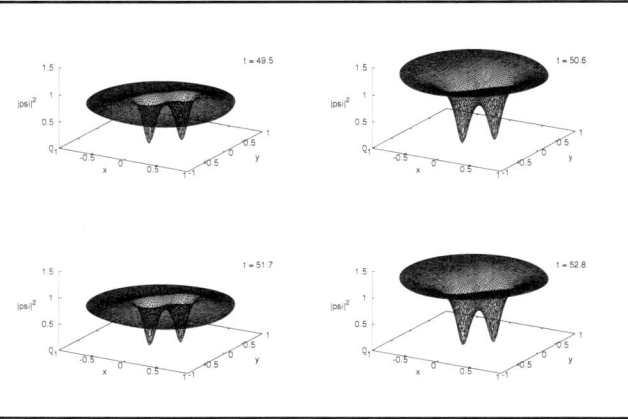

FIGURE 7.26: Evolution of the solution with $\Delta t = 1.1$: the semi-implicit scheme.

Advanced Topic III: Further Remarks

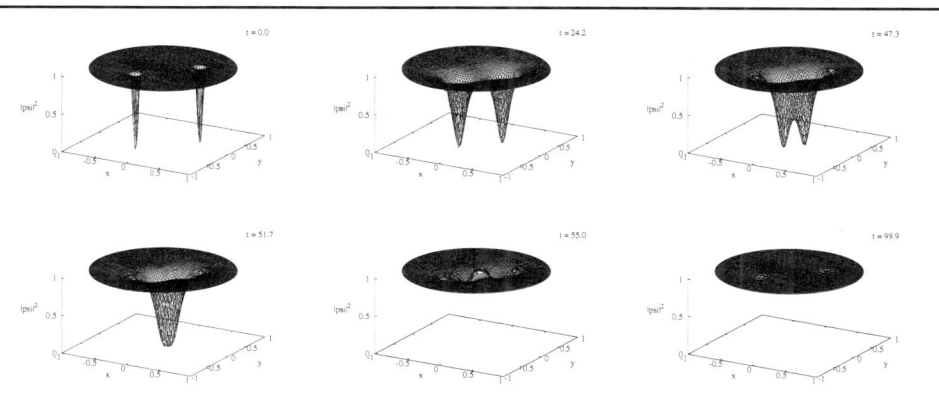

FIGURE 7.27: Evolution of the solution with $\Delta t = 1.1$: Scheme 7.10.

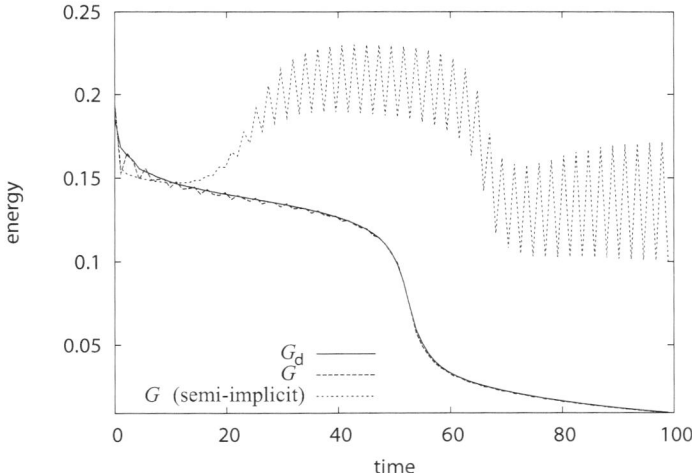

FIGURE 7.28: Evolution of the energies with $\Delta t = 1.1$: the semi-implicit scheme and Scheme 7.10 with $(a, b) = (2, -0.5)$.

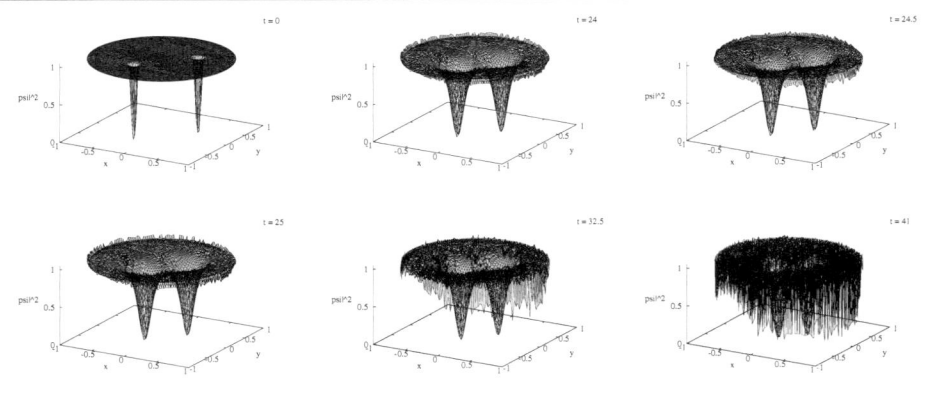

FIGURE 7.29: Evolution of the solution with $\Delta t = 0.5$: Scheme 7.10 with $(a, b) = (0.9, 0.5)$.

7.3 Extension to Non-Rectangular Meshes on 2D Region

In the previous section, we discussed the extension of the discrete variational derivative method to the Galerkin framework, with application to spatially two- or three-dimensional problems in mind. In this section, we explain a different approach for the same aim: the extension to non-uniform meshes on non-rectangular domains.

This challenge of "discrete variational calculus" on non-regular meshes starts by recalling the following fact:

> *The mathematical keystone of the discrete variational derivative method is the "summation-by-parts formula," on a given mesh.*

In other words, when we hope to generalize the method to general meshes, our main task should be to find the associated summation-by-parts formula on the designated mesh. Below we demonstrate such an example. As an example of flexible meshes where mesh points are arbitrarily set, we consider the Voronoi mesh. The following lemma is a summation-by-parts formula on Voronoi mesh. On a given mesh, we consider discrete functions, denoted by u and so on, which exhibit a value on each vertex.

LEMMA 7.1
Suppose a Voronoi mesh is given. Then for any discrete functions u and w

Advanced Topic III: Further Remarks

on the mesh, the following equality holds.

$$\sum_i \left\{ \sum_{j \in S_i} u_i \left(\frac{w_j - w_i}{l_{ij}} \right) \boldsymbol{s}_{ji} \Delta \Omega_{ij} \right\}$$
$$= -\sum_i \left\{ \sum_{j \in S_i} w_i \left(\frac{u_j - u_i}{l_{ij}} \right) \boldsymbol{s}_{ji} \Delta \Omega_{ij} \right\} + \sum_{i \in \partial \Omega_d} u_i w_i \boldsymbol{R}_i, \quad (7.104)$$

where $\boldsymbol{s}_{ji} \stackrel{\mathrm{d}}{=} (\boldsymbol{x}_j - \boldsymbol{x}_i)/l_{ij}$, S_i is an index set of neighbor points of $\boldsymbol{x}_i$, $\partial \Omega_d$ is the boundary surface, $\Delta \Omega_{ij} \stackrel{\mathrm{d}}{=} r_{ij} l_{ij}/4$, and $\boldsymbol{R}_i \stackrel{\mathrm{d}}{=} -\sum_{j \in S_i} r_{ij} \boldsymbol{s}_{ji}$. For the definitions of r_{ij} and l_{ij}, see Figure 7.30.

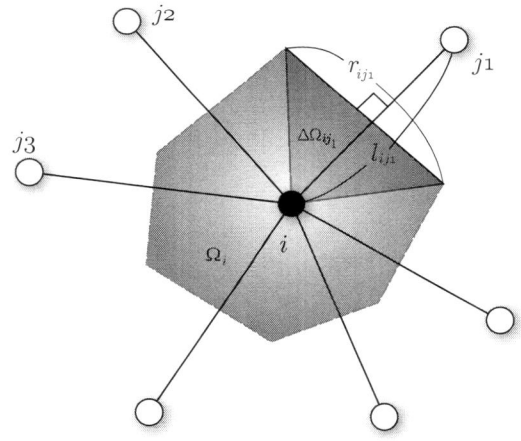

FIGURE 7.30: Voronoi mesh and finite difference points.

In a similar manner, we can deduce other formulas by which we can extend the whole framework of the discrete variational derivative method to higher-dimensional problems, whose domains are not necessarily rectangular with non-uniform meshes. We omit the detail in order to avoid exhaustive discussion. Instead, in Figure 7.31, we show numerical results to the linear diffusion equation under the Dirichlet boundary condition. The Voronoi mesh was generated with randomly distributed 2D points.

Another example is shown in Figure 7.32, where the Cahn–Hilliard equation was solved on a Voronoi mesh on a disk.

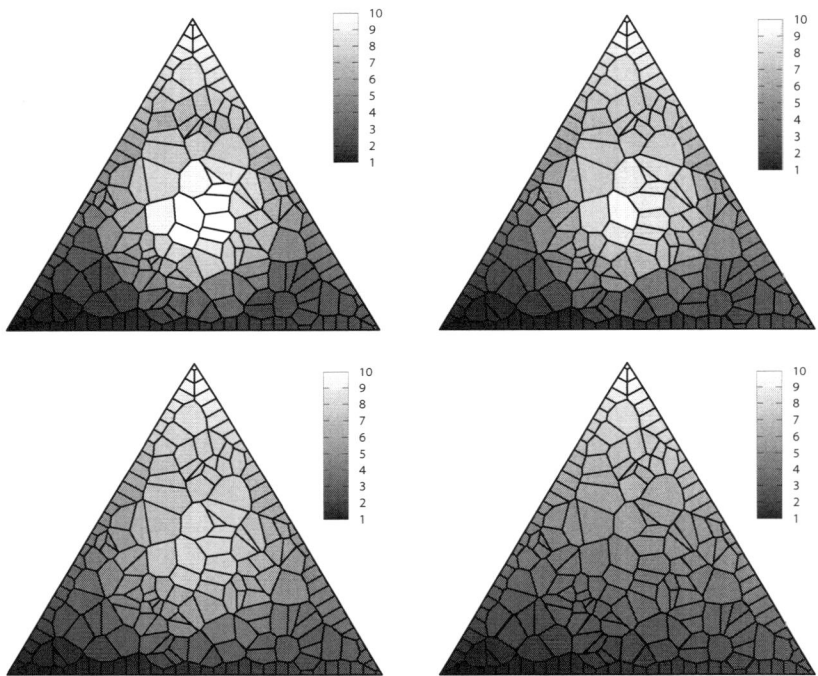

FIGURE 7.31: Numerical solutions of the linear diffusion equation under Dirichlet boundary condition by an extended scheme on a random Voronoi mesh with 200 points and $\Delta t = 5 \times 10^{-6}$. Top left: profile at time step $m = 0$, top right: at $m = 2000$, bottom left: at $m = 5000$, bottom right: at $m = 30000$.

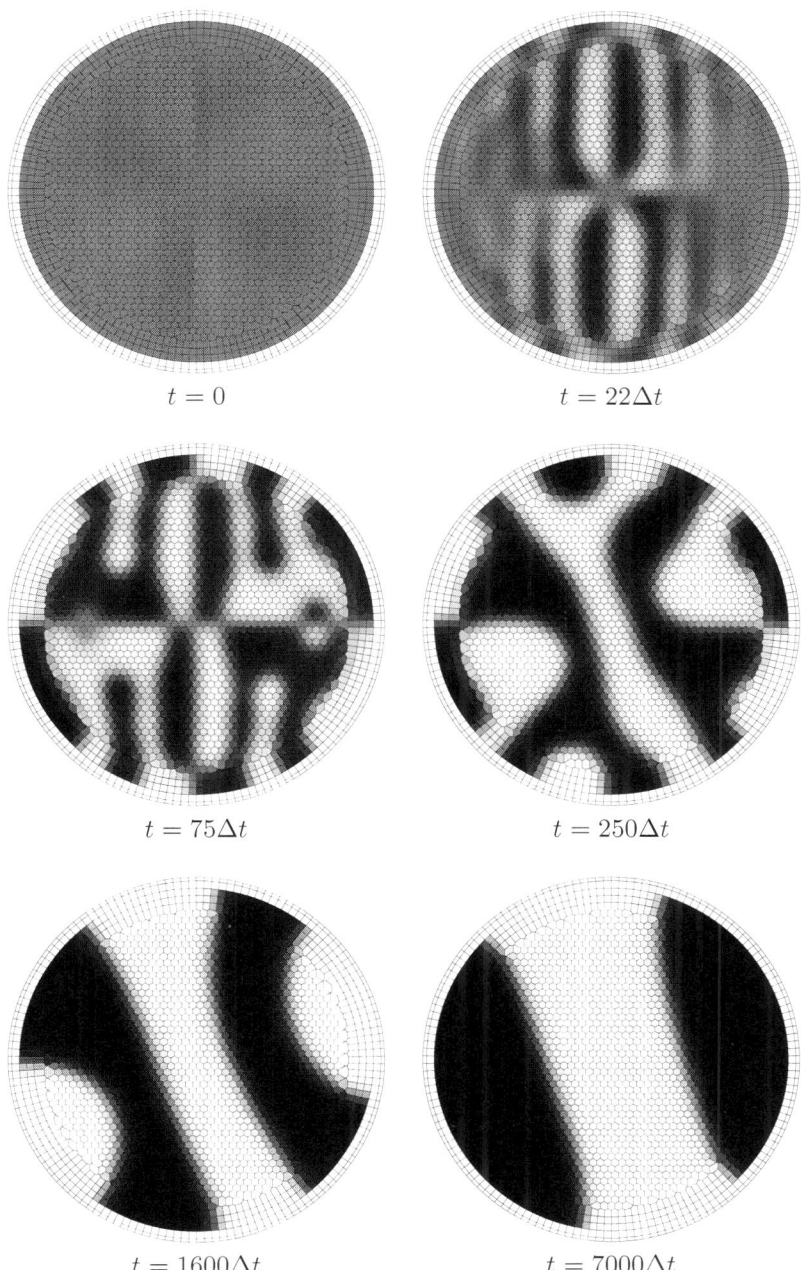

FIGURE 7.32: Numerical solutions of the Cahn–Hilliard equation on a unit disk with standard Neumann boundary conditions. With 2391 points and $\Delta t = 1 \times 10^{-5}$.

REMARK 7.10 As mentioned in the beginning of Section 7.2, quite recently (as of writing this book) several similar studies have started. For example, in Yaguchi–Matsuo–Sugihara [166], a generalization to non-uniform mesh has been proposed by a mapping technique. Related studies can be found in a research field called "compatible spatial discretization" or "mimetic schemes"; see, for example, [11, 153] and the references therein. ☐

Appendix A

Semi-Discrete Schemes in Space

In this appendix we show that

> for a given conservative or dissipative PDE, we can always appropriately discretize the space variable so that the resulting semi-discrete scheme still keeps some conservation or dissipation property.

It is surely possible for the target PDEs in this book, since if we consider the limit $\Delta t \to 0$ in the full discrete conservative or dissipative schemes presented in the preceding chapters, we surely obtain semi-discrete conservative or dissipative schemes. Below we present a more direct explanation. (See also McLachlan [129].)

Let us consider the first-order real-valued PDEs 1 and PDEs 2 for example. As in Chapter 2, we assume that the energy function $G(u, u_x)$ is of the form

$$G(u, u_x) = \sum_{l=1}^{\widetilde{M}} f_l(u) g_l(u_x), \tag{A.1}$$

and we consider its discrete version

$$G_{\mathrm{d},k}(\boldsymbol{U}) = \sum_{l=1}^{M} f_l(U_k) g_l^+(\delta_k^+ U_k) g_l^-(\delta_k^- U_k), \quad k = 0, \ldots, N, \tag{A.2}$$

where $U_k(t) \simeq u(k\Delta x, t)$ are approximate solutions. The associated global energy is then

$$\sum_{k=0}^{N} {}''G_{\mathrm{d},k}(\boldsymbol{U})\Delta x. \tag{A.3}$$

Note that this global energy is a function of the continuous variable t. Differentiating the global energy by t we obtain

$$\frac{\mathrm{d}}{\mathrm{d}t} \sum_{k=0}^{N} {}''G_{\mathrm{d},k}(\boldsymbol{U})\Delta x$$

$$= \sum_{k=0}^{N} {}'' \left\{ \sum_{l=1}^{M} \left(f_l' \dot{U}_k g_l^+ g_l^- + f_l(g_l^+)'(\delta_k^+ \dot{U}_k) g_l^- + f_l g_l^+ (g_l^-)'(\delta_k^- \dot{U}_k) \right) \right\}, \tag{A.4}$$

which corresponds to (3.26). We abbreviated $f_l(U_k)$ as f_l, $g_l^+(\delta_k^+ U_k)$ as g_l^+, and $g_l^-(\delta_k^- U_k)$ as g_l^-, and by $\dot{U}_k$ we mean $(\mathrm{d}/\mathrm{d}t) U_k(t)$. Let us introduce new notation

$$\frac{\partial G_\mathrm{d}}{\partial (\boldsymbol{U})_k} = \sum_{l=1}^M f_l'(U_k) g_l^+(\delta_k^+ U_k) g_l^-(\delta_k^- U_k), \tag{A.5a}$$

$$\frac{\partial G_\mathrm{d}}{\partial (\delta_k^- \boldsymbol{U})_k} = \sum_{l=1}^M f_l(U_k) g_l^+(\delta_k^+ U_k) (g_l^-(\delta_k^- U_k))', \tag{A.5b}$$

$$\frac{\partial G_\mathrm{d}}{\partial (\delta_k^+ \boldsymbol{U})_k} = \sum_{l=1}^M f_l(U_k) (g_l^+(\delta_k^+ U_k))' g_l^-(\delta_k^- U_k). \tag{A.5c}$$

The first one is an approximation to $\partial G/\partial u$, and the latter two are to $\partial G/\partial u_x$. The first symbol might look bizarre (one might feel that the brackets in the denominator should be deleted), but we like to use it here in order to be consistent with the full discrete symbols. With these symbols, the above formula can be simplified to

$$\frac{\mathrm{d}}{\mathrm{d}t} \sum_{k=0}^N {}'' G_{\mathrm{d},k}(\boldsymbol{U}) \Delta x = \sum_{k=0}^N {}'' \left\{ \frac{\partial G_\mathrm{d}}{\partial (\boldsymbol{U})_k} \dot{U}_k + \frac{\partial G_\mathrm{d}}{\partial (\delta_k^+ \boldsymbol{U})_k} (\delta_k^+ \dot{U}_k) + \frac{\partial G_\mathrm{d}}{\partial (\delta_k^- \boldsymbol{U})_k} (\delta_k^- \dot{U}_k) \right\}. \tag{A.6}$$

With the aid of the summation-by-parts formula (3.12a), we obtain

$$\frac{\mathrm{d}}{\mathrm{d}t} \sum_{k=0}^N {}'' G_{\mathrm{d},k}(\boldsymbol{U}) \Delta x$$

$$= \sum_{k=0}^N {}'' \left[\left\{ \frac{\partial G_\mathrm{d}}{\partial (\boldsymbol{U})_k} - \delta_k^- \left(\frac{\partial G_\mathrm{d}}{\partial (\delta_k^+ \boldsymbol{U})_k} \right) - \delta_k^+ \left(\frac{\partial G_\mathrm{d}}{\partial (\delta_k^- \boldsymbol{U})_k} \right) \right\} \dot{U}_k \right] \Delta x$$

$$+ \frac{1}{2} \left[\frac{\partial G_\mathrm{d}}{\partial (\delta_k^+ \boldsymbol{U})_k} (s_k^+ \dot{U}_k) + \left\{ s_k^- \left(\frac{\partial G_\mathrm{d}}{\partial (\delta_k^+ \boldsymbol{U})_k} \right) \right\} \dot{U}_k \right.$$

$$\left. + \frac{\partial G_\mathrm{d}}{\partial (\delta_k^- \boldsymbol{U})_k} (s_k^- \dot{U}_k) + \left\{ s_k^+ \left(\frac{\partial G_\mathrm{d}}{\partial (\delta_k^- \boldsymbol{U})_k} \right) \right\} \dot{U}_k \right]_0^N, \tag{A.7}$$

which again corresponds to (3.29). Then with the definitions

$$\frac{\delta G_\mathrm{d}}{\delta (\boldsymbol{U})_k} \stackrel{\mathrm{d}}{\equiv} \frac{\partial G_\mathrm{d}}{\partial (\boldsymbol{U})_k} - \delta_k^- \left(\frac{\partial G_\mathrm{d}}{\partial (\delta_k^+ \boldsymbol{U})_k} \right) - \delta_k^+ \left(\frac{\partial G_\mathrm{d}}{\partial (\delta_k^- \boldsymbol{U})_k} \right), \tag{A.8}$$

$$B_{\mathrm{r},3}(\boldsymbol{U}) \stackrel{\mathrm{d}}{\equiv} \frac{1}{2} \left[\frac{\partial G_\mathrm{d}}{\partial (\delta_k^+ \boldsymbol{U})_k} (s_k^+ \dot{U}_k) + \left\{ s_k^- \left(\frac{\partial G_\mathrm{d}}{\partial (\delta_k^+ \boldsymbol{U})_k} \right) \right\} \dot{U}_k \right.$$

$$\left. + \frac{\partial G_\mathrm{d}}{\partial (\delta_k^- \boldsymbol{U})_k} (s_k^- \dot{U}_k) + \left\{ s_k^+ \left(\frac{\partial G_\mathrm{d}}{\partial (\delta_k^- \boldsymbol{U})_k} \right) \right\} \dot{U}_k \right]_0^N, \tag{A.9}$$

we finally obtain the following expression:

$$\frac{\mathrm{d}}{\mathrm{d}t}\sum_{k=0}^{N}{''}G_{\mathrm{d},k}(\boldsymbol{U})\Delta x = \sum_{k=0}^{N}{''}\left[\frac{\delta G_{\mathrm{d}}}{\delta(\boldsymbol{U})}\right]_{k}\dot{U}_{k}\right]\Delta x + B_{\mathrm{r},3}(\boldsymbol{U}). \qquad (\text{A}.10)$$

Thus we find $\delta G_{\mathrm{d}}/\delta \boldsymbol{U}$, which is a semi-discrete approximation of $\delta G/\delta u$.

With the semi-discrete variational derivative, now we can define a semi-discrete conservative or dissipative scheme for PDEs 1 and PDEs 2. To avoid exhaustive discussions, let us limit ourselves to the special case $s = 0$.

Scheme A.1 (Semi-discrete scheme for the PDEs 1 ($s = 0$)) *Let a set of initial values $\boldsymbol{U}(0)$ be given. Then, a semi-discrete scheme for the PDEs 1 ($s = 0$) is given by*

$$\frac{\mathrm{d}}{\mathrm{d}t}U_{k}(t) = -\frac{\delta G_{\mathrm{d}}}{\delta(\boldsymbol{U})_{k}}, \qquad k = 0, \ldots, N. \qquad (\text{A}.11)$$

THEOREM A.1 Discrete dissipation property of Scheme A.1
Assume that a discrete boundary condition that satisfies the following condition is imposed:

$$B_{\mathrm{r},3}(\boldsymbol{U}(t)) = 0, \qquad t > 0. \qquad (\text{A}.12)$$

Then the scheme is dissipative in the sense that the inequality

$$\frac{\mathrm{d}}{\mathrm{d}t}\sum_{k=0}^{N}{''}G_{\mathrm{d}}(\boldsymbol{U}(t))\Delta x \leq 0. \qquad (\text{A}.13)$$

PROOF By the discrete variation equality (A.10), we have

$$\begin{aligned}\frac{\mathrm{d}}{\mathrm{d}t}\sum_{k=0}^{N}{''}G_{\mathrm{d},k}(\boldsymbol{U})\Delta x &= \sum_{k=0}^{N}{''}\left[\frac{\delta G_{\mathrm{d}}}{\delta(\boldsymbol{U})_{k}}\dot{U}_{k}\right]\Delta x + B_{\mathrm{r},3}(\boldsymbol{U})\\ &= \sum_{k=0}^{N}{''}\left[\frac{\delta G_{\mathrm{d}}}{\delta(\boldsymbol{U})_{k}}\left(-\frac{\delta G_{\mathrm{d}}}{\delta(\boldsymbol{U})_{k}}\right)\right]\Delta x\\ &\leq 0. \end{aligned} \qquad (\text{A}.14)$$

In the first equality the assumption on the discrete boundary condition is used. □

Scheme A.2 (Semi-discrete scheme for the PDEs 2 ($s = 0$)) *Let a set of initial values $\boldsymbol{U}(0)$ be given. Then, a semi-discrete scheme for the PDEs 2 ($s = 0$) is given by*

$$\frac{\mathrm{d}}{\mathrm{d}t}U_{k}(t) = \delta_{k}^{\langle 1 \rangle}\frac{\delta G_{\mathrm{d}}}{\delta(\boldsymbol{U})_{k}}, \qquad k = 0, \ldots, N. \qquad (\text{A}.15)$$

THEOREM A.2 Discrete conservation property of Scheme A.2
Assume that a discrete boundary condition that satisfies the following two condition is imposed:

(i) $B_{r,3}(\boldsymbol{U}(t)) = 0, \quad t > 0,$ *and*

(ii) $\left[\dfrac{\delta G_d}{\delta(\boldsymbol{U})_k} \cdot s_k^{\langle 1 \rangle} \left(\dfrac{\delta G_d}{\delta(\boldsymbol{U})_k} \right) \right]_0^N = 0.$

Then the scheme is conservative in the sense that the inequality

$$\frac{\mathrm{d}}{\mathrm{d}t} \sum_{k=0}^{N} {}'' G_d(\boldsymbol{U}(t)) \Delta x = 0. \tag{A.16}$$

PROOF By the discrete variation equality (A.10), we have

$$\begin{aligned}
\frac{\mathrm{d}}{\mathrm{d}t} \sum_{k=0}^{N} {}'' G_{d,k}(\boldsymbol{U}) \Delta x &= \sum_{k=0}^{N} {}'' \left[\frac{\delta G_d}{\delta(\boldsymbol{U})_k} \dot{\boldsymbol{U}}_k \right] \Delta x + B_{r,3}(\boldsymbol{U}) \\
&= \sum_{k=0}^{N} {}'' \left[\frac{\delta G_d}{\delta(\boldsymbol{U})_k} \cdot \delta_k^{\langle 1 \rangle} \left(\frac{\delta G_d}{\delta(\boldsymbol{U})_k} \right) \right] \Delta x \\
&= \frac{1}{2} \left[\frac{\delta G_d}{\delta(\boldsymbol{U})_k} \cdot s_k^{\langle 1 \rangle} \left(\frac{\delta G_d}{\delta(\boldsymbol{U})_k} \right) \right]_0^N \\
&= 0. \tag{A.17}
\end{aligned}$$

In the first and the third equalities, the assumption on the discrete boundary condition is used. In the second equality the summation-by-parts formula (3.12b) is used. □

Thus we obtained semi-discrete dissipative or conservative schemes for the PDEs 1 and PDEs 2, with $s = 0$. Extension to other PDEs in Chapter 2 is straightforward. Furthermore, although here we used the standard second-order approximation for spatial discretization, we can replace it with the high-order version described in Chapter 5.

The semi-discrete schemes are systems of ordinary differential equations (ODEs) with respect to $\boldsymbol{U}(t)$, whose dimension is the number of spatial grid points.

Appendix B

Proof of Proposition 3.4

In this section we prove Proposition 3.4. For a matrix (of vector) A in general its transpose is denoted by A^{T}.

LEMMA B.1

$$\delta_k^{\langle h \rangle} = e^{\mathrm{T}} D_k^h \, e \tag{B.1}$$

where

$$D_k \stackrel{\mathrm{d}}{\equiv} \begin{pmatrix} 0 & \delta_k^+ \\ \delta_k^- & 0 \end{pmatrix}, \tag{B.2}$$

$$e \stackrel{\mathrm{d}}{\equiv} \frac{1}{\sqrt{2}} \begin{pmatrix} 1 \\ 1 \end{pmatrix}. \tag{B.3}$$

PROOF Omitted since it is trivial. □

LEMMA B.2

$$\sum_{k=0}^{N}{}'' \left\{ a_k D_k a_k' + (D_k a_k^{\mathrm{T}})^{\mathrm{T}} a_k' \right\} \Delta x = \frac{1}{2} \left[a_k A_k a_k' + (A_k a_k^{\mathrm{T}})^{\mathrm{T}} a_k' \right]_{k=0}^{N} \tag{B.4}$$

where

$$a_k \stackrel{\mathrm{d}}{\equiv} \begin{pmatrix} \zeta_k & \eta_k \\ \theta_k & \xi_k \end{pmatrix}, \tag{B.5}$$

$$a_k' \stackrel{\mathrm{d}}{\equiv} \begin{pmatrix} \zeta_k' & \eta_k' \\ \theta_k' & \xi_k' \end{pmatrix}, \tag{B.6}$$

$$A_k \stackrel{\mathrm{d}}{\equiv} \begin{pmatrix} 0 & s_k^+ \\ s_k^- & 0 \end{pmatrix}. \tag{B.7}$$

PROOF Trivial from the first-order summation-by-parts formula (Prop. 3.2). □

Note that $(D_k a_k^{\mathrm{T}})^{\mathrm{T}} \neq a_k D_k^{\mathrm{T}}$ and $(A_k a_k^{\mathrm{T}})^{\mathrm{T}} \neq a_k A_k^{\mathrm{T}}$ since D_k and A_k are operator matrices.

LEMMA B.3

$$\sum_{k=0}^{N}{}''\left\{a_k D_k^h a_k'\right\}\Delta x$$
$$= (-1)^{h'}\sum_{k=0}^{N}{}''\left(D_k^{h'} a_k^{\mathsf{T}}\right)^{\mathsf{T}}\left(D_k^{h-h'} a_k'\right)\Delta x$$
$$+ \frac{1}{2}\left[\sum_{l=1}^{h'}(-1)^{l-1}\left\{\left(D_k^{l-1} a_k^{\mathsf{T}}\right)^{\mathsf{T}}\left(A_k D_k^{h-l} a_k'\right) + \left(A_k D_k^{l-1} a_k^{\mathsf{T}}\right)^{\mathsf{T}}\left(D_k^{h-l} a_k'\right)\right\}\right]_{k=0}^{N} \tag{B.8}$$

where $D_k^0 \stackrel{\mathrm{d}}{=} I$, $h \in \mathbb{N}^+$, $h' \in \mathbb{N}$ and $h' \leq h$.

PROOF By repeatedly using (B.4) on the left hand side of this equation, we see the claim. □

From this lemma and the lemma B.1 we obtain the following result.

$$\sum_{k=0}^{N}{}'' f_k \delta_k^{\langle h\rangle} f_k \,\Delta x$$
$$= (-1)^{h'}\sum_{k=0}^{N}{}''\left\{\boldsymbol{e}^{\mathsf{T}}\left(D_k^{h'} f_k\right)^{\mathsf{T}}\left(D_k^{h-h'} f_k\right)\boldsymbol{e}\right\}\Delta x$$
$$+ \frac{1}{2}\left[\sum_{l=1}^{h'}(-1)^{l-1}\left\{\boldsymbol{e}^{\mathsf{T}}\left(D_k^{l-1} f_k\right)^{\mathsf{T}}\left(A_k D_k^{h-l} f_k\right)\boldsymbol{e}\right\}\right.$$
$$\left.+ \sum_{l=1}^{h'}(-1)^{l-1}\left\{\boldsymbol{e}^{\mathsf{T}}\left(A_k D_k^{l-1} f_k\right)^{\mathsf{T}}\left(D_k^{h-l} f_k\right)\boldsymbol{e}\right\}\right]_{k=0}^{N} \tag{B.9}$$

where $h \in \mathbb{N}^+$, $h' \in \mathbb{N}$ and $h' \leq h$.

Substituting $h' = h/2$ into this equation we obtain Proposition 3.4 for even h. When h is odd, we obtain Proposition 3.4 by comparing this equation with $h' = (h-1)/2$ and the same equation with $h' = (h+1)/2$.

Bibliography

[1] Abdelgadir, A. A., Yao, Y., Fu Y. and Huang, P., A difference scheme for the Camassa–Holm equation, *Lecture Notes in Comput. Sci.*, **4682** (2007), 1287–1295. (Cited on p.197)

[2] Ablowitz, M. J., Kruskal, M. D. and Ladik, J. F., Solitary wave collisions, *SIAM J. Appl. Math.* **36** (1979), 428–437. (Cited on p.185)

[3] Ablowitz, M. J., Herbst, B. M. and Schober, C. M., Numerical simulation of quasi-periodic solutions of the sine-Gordon equation, *Physica D* **87**(1995), 37–47. (Cited on p.185)

[4] Ablowitz, M. J., Herbst, B. M. and Schober, C. M., On the numerical solution of the sine-Gordon equation — I. Integrable discretizations and homoclinic manifolds, *J. Comput. Phys.* **126** (1996), 299–314. (Cited on p.185)

[5] Ablowitz, M. J., Herbst, B. M. and Schober, C. M., On the numerical solution of the sine-Gordon equation — II. Performance of numerical schemes, *J. Comput. Phys.* **131** (1997), 354–367. (Cited on p.185)

[6] Aftalion, A. and Du, Q., Vortices in a rotating Bose–Einstein condensate: Critical angular velocities and energy diagrams in the Thomas–Fermi regime, *Phys. Rev. A*, **64** (2001), 063603. (Cited on p.182)

[7] Ahlfors, L. V., *Complex Analysis* (3rd. ed.), McGraw–Hill, New York, 1979. (Cited on p.50)

[8] Akrivis, G. D., Dougalis, V. A., and Karakashian, O. A., On fully discrete Galerkin methods of second-order temporal accuracy for the nonlinear Schrödinger equation, *Numer. Math.*, **59** (1991), 31–53. (Cited on p.10, 171, 315)

[9] Allen, S. M. and Cahn, J. W., A microscopic theory for antiphase boundary motion and its application to antiphase domain coarsening, *Acta Metallurgica*, **27** (1979), 1085–1095. (Cited on p.53)

[10] Ames, W. F., *Nonlinear Partial Differential Equations in Engineering*, Academic Press, New York, 1965. (Cited on p.54)

[11] Arnold, D. N., Bochev, P. B., Lehoucq, R. B., Nicolaides, R. A., and Shashkov, M. (Eds.), Compatible spatial discretizations, *IMA Vol. Math. Appl.*, **142**, Springer, New York, 2006. (Cited on p.352)

[12] Baillon, J.B. and Chadam, J.M., The Cauchy problem for the coupled Schrödinger-Klein-Gordon equations, in *Contemporary Developments in Continuum Mechanics and Partial Differential Equations*, edited by G.M. de la Penha and L.A. Medeiros, North Holland, Amsterdam, 1978. (Cited on p.64)

[13] Barzilai, J. and Borwein, J., Two-point step size gradient methods. *IMA J. Numer. Anal.*, **8** (1988), 141–148. (Cited on p.296)

[14] Benjamin, T. B., Bona, J. L. and Mahony, J. J., Model equations for long waves in nonlinear dispersive systems, *Phil. Trans. Roy. Soc. (London)*, **272** (1972), 47–78. (Cited on p.57, 212)

[15] Ben-Yu, G., Pascual, P. J., Rodriguez, M. J. and Vázquez, L., Numerical solution of the sine-Gordon equation, *Appl. Math. Comput.* **18** (1986), 1–14. (Cited on p.185, 186)

[16] Bullough, R. K. and Caudrey, P. H., The Multiple sine-Gordon equations in non-linear optics and in liquid ^{3}He, in F. Calogero ed., *Nonlinear Evolution equations Solvable by the Spectral Transform*, Pitman, London, 1978, 180–224. (Cited on p.185)

[17] Boling, G., The global solution of the system of equations for complex Schrödinger field coupled with Boussinesq type self-consistent field, *Acta Math. Sinica*, **26** (1983), 295–306. (Cited on p.64)

[18] Bourgain, J., *Global Solutions of Nonlinear Schrödinger Equations*, American Mathematical Society, Rhode Island, 1999. (Cited on p.60)

[19] Brezis, H., *Analyse Fonctionnelle: Théorie et Applications*, Masson, Paris, 1983. (Cited on p.58, 123, 315)

[20] Bratsos, A. G. and Twizell, E. H., A family of parametric finite-difference methods for the solution of the sine-Gordon equation, *Appl. Math. Comput.* **93** (1998), 117–137. (Cited on p.185)

[21] Brezzi, F. and Fortin, M., *Mixed and Hybrid Finite Element Methods*, Springer, New York, 1991. (Cited on p.305)

[22] Bridges, T. J., and Reich, S., Multi-symplectic integrators: numerical schemes for Hamiltonian PDEs that conserve symplecticity, *Phys. Lett. A*, **284** (2001), 184–193. (Cited on p.11)

[23] Budd, C., and Piggott, M. D., Geometric integration and its applications, in *Handbook of Numerical Analysis*, XI, North-Holland, Amsterdam, 2003, 35–139. (Cited on p.10, 11)

[24] Camassa, R. and Holm, D. D., An integrable shallow water equation with peaked solitons, *Phys. Rev. Lett.*, **71** (1993), 1661–1664. (Cited on p.57, 90, 195, 196, 320)

[25] Cahn, J.W. and Hilliard, J.E., Free energy of a non-uniform system. I. interfacial free energy, *J. Chem. Phys.*, **28** (1958), 258–267. (Cited on p.54)

[26] Celledoni, E., McLachlan, R. I., McLaren, D. I., Owren, B., Quispel, G. R. W., and Wright, W. M., Energy-preserving Runge-Kutta methods, *M2AN Math. Model. Numer. Anal.*, **43** (2009), 645–649. (Cited on p.11)

[27] Cohen, D., Owren, B., and Raynaud, X., Multi-symplectic integration of the Camassa-Holm equation, *J. Comput. Phys.*, **227** (2008), 5492–5512. (Cited on p.11)

[28] Constantin, A. and Escher, J., Global existence and blow-up for a shallow water equation, *Ann. Scuola Norm. Sup. Pisa Cl. Sci.*, **XXVI** (1998), 303–328. (Cited on p.316)

[29] Constantin, A. and Molinet, L., Global weak solutions for a shallow water equation, *Comm. Math. Phys.*, **211** (2000), 45–61. (Cited on p.316)

[30] Carrier, G. F., On the non-linear vibration problem of the elastic string, *Quart. Appl. Math.*, **3** (1945), 157–165. (Cited on p.67)

[31] Dağ, İ and Özer, M. N., Approximation of the RLW equation by the least square cubic B-spline finite element method, *Appl. Math. Model.*, **25** (2001), 221–231. (Cited on p.212)

[32] Dağ, İ, Saka, B. and Irk, D., Application of cubic B-splines for numerical solution of the RLW equation, *Appl. Math. Comput.*, **159** (2004), 373–389. (Cited on p.212)

[33] Dai, H. H., Model equations for nonlinear disperwive waves in a compressible Mooney–Rivlin rod, *Acta Mech.*, **127** (1998). (Cited on p.57)

[34] Dee, G. T. and W. van Saarloos, Bistable systems with propagating fronts leading to pattern formation, *Phys. Rev. Lett.*, **60** (1988), 2641–2644. (Cited on p.54)

[35] Delfour, M., Fortin, M., and Payre, G., Finite-difference solutions of a non-linear Schrödinger equation, *J. Comput. Phys.*, **44** (1981), 277–288. (Cited on p.10, 170)

[36] Dembo, R. S., Eisenstat, S. C., and Steihaug, T., Inexact newton methods, *SIAM J. Numer. Anal.*, **19** (1982), 400–408. (Cited on p.295)

[37] Deuflhard, P., *Newton Methods for Nonlinear Problems*, Springer, Heidelberg, 2004. (Cited on p.295)

[38] Djidjeli, K., Price, W. G. and Twizell, E. H., Numerical solutions of a damped sine-Gordon equation in two space variables, *J. Eng. Math.* **29** (1995), 347–369. (Cited on p.185)

[39] Du, Q. and Nicolaides, R.A., Numerical analysis of a continuum model of phase transition, *SIAM J. Numer. Anal.*, **28** (1991), 1310–1322. (Cited on p.10, 310)

[40] Du,Q., Global existence and uniqueness of solutions of the time-dependent Ginzburg-Landau model for superconductivity, *Appl. Anal.*, **53** (1994), 1–17. (Cited on p.65, 339)

[41] Du, Q., Finite element methods for the time-dependent Ginzburg–Landau model of superconductivity, *Comput. Math. Appl.*, **27** (1994), 119–133. (Cited on p.341)

[42] Duncan, D. B., Symplectic finite difference approximations of the nonlinear Klein–Gordon equation, *SIAM J. Numer. Anal.* **34** (1997), 1742–1760. (Cited on p.185)

[43] Durán, A. and López-Marcos, M. A., Conservative numerical methods for solitary wave interactions, *J. Phys. A: Math. Gen.*, **36** (2003), 7761–7770. (Cited on p.212)

[44] Ebihara, Y., Spherically symmetric solutions of some semilinear wave equations, in *Nonlinear Mathematical Problems in Industry II*, edited by H. Kawarada and N. Kenmochi and N. Yanagihara, Gakkotosho, Tokyo, 1992. (Cited on p.67)

[45] Eguchi, T., Oki, K., and Matsumura, S., Kinetics of ordering with phase separation, *Mat. Res. Soc. Symp. Proc.*, **21** (1984), 589–594. (Cited on p.65)

[46] Eilbeck, J. C. and McGuire, G. R., Numerical study of the regularized long-wave equation. I: numerical methods, *J. Comput. Phys.*, **19** (1975), 43–57. (Cited on p.212)

[47] Eilbeck, J. C. and McGuire, G. R., Numerical study of the regularized long-wave equation. II: interaction of solitary waves, *J. Comput. Phys.*, **23** (1977), 63–73. (Cited on p.212, 328)

[48] Eilbeck, J. C., Numerical studies of solitons, in A. R. Bishop and T. Schneider eds., *Solitons and Condensed Matter Physics*, Springer-Verlag, Berlin, 1978, 28–43. (Cited on p.185)

[49] Eisenstat, S. C. and Walker, H. F., Choosing the forcing terms in an inexact Newton method, Technical report CRPC-TR94463 in Center for Research on Parallel Computation, Rice University, 1994. (Cited on p.295)

[50] Elgamal, M. and Nakagiri, S., Weak solutions of sine-Gordon equation and their numerical analysis, *RIMS Kohkyuuroku*(Kyoto University), **984**, (1997), 123–137. (Cited on p.185)

[51] Evans, W. A. B., Cunha, M. D., Konotop, V. V. and Vázquez, L., Numerical study of various sine-Gordon "breathers", in L. Vazquez et al eds., *Nonlinear Klein–Gordon and Schrödinger Systems: Theory and Applications*, World Sci., Singapore, 1996, 293–302. (Cited on p.185)

[52] Faou, E., Hairer, E., and Pham, T.-L., Energy conservation with non-symplectic methods: examples and counter-examples, *BIT*, **44** (2004), 699–709. (Cited on p.10)

[53] Fei, Z. and Vázquez, L., Two Energy conserving numerical schemes for the sine-Gordon equation, *Appl. Math. Comput.*, **45** (1991), 17–30. (Cited on p.185, 186)

[54] Fei, Z., Pérez-García, V.M., and Vázquez, L., Numerical simulation of nonlinear Schrödinger systems: a new conservative scheme, *Appl. Math. Comput.*, **71** (1995), 165–177. (Cited on p.274)

[55] Feng, B., A regularized model equation for discrete breathers in nonlinear lattices, in the proceedings of Applied Mathematics Joint Conference (Kyoto, 2003), 205–210. (Cited on p.68, 222)

[56] Feng, B., Doi, Y. and Kawahara, T., Quasi-continuum approximation for discrete breathers in Fermi–Pasta–Ulam atomic chains, *J. Phys. Soc. Jpn.*, **73** (2004), 2100–2111. (Cited on p.68, 222)

[57] Fermi, E., Pasta, J. and Ulam, S., Studies of nonlinear problems I, *Los Alamos Sci. Lab. Rept.*, **LA-1490**, 1955. (Cited on p.66)

[58] Fletcher, R., On the Barzilai–Borwein method, in Optimization and Control with Applications, 235–256, *Appl. Optim.*, **96** (2005), Springer, New York. (Cited on p.296)

[59] Fock, V.A., Zur Schrödingerschen Wellenmechanik, *Z. Phys. A*, **38** (1926), 242–250. (Cited on p.67)

[60] Fock, V.A., Über die Invariante Form der Wellen- und der Bewegungsgleichungen für einen Geladenen Massenpunkt, *Z. Phys. A*, **39** (1926), 226–232. (Cited on p.67)

[61] Fornberg, B., *A Practical Guide to Pseudospectral Methods*, Cambridge University Press, New York, 1996. (Cited on p.229)

[62] Fuchssteiner, B. and Fokas, A. S., Symplectic structures, their Bäcklund transformations and hereditary symmetries, *Phys. D*, **4** (1981), 47–66. (Cited on p.57)

[63] Furihata, D., Discrete variational method for partial differential equation (in Japanese) PhD thesis, Dept. Applied Physics, Faculty of Engineering, Univ. of Tokyo, Tokyo, Japan, 1996 (in Japanese). (Cited on p.11)

[64] Furihata, D. and Mori, M., A stable finite difference scheme for the Cahn–Hilliard equation based on the Lyapunov functional, *ZAMM Z. angew. Math. Mech.*, **76** (1996), 405–406. (Cited on p.11)

[65] Furihata, D., Finite difference schemes for $\frac{\partial u}{\partial t} = \left(\frac{\partial}{\partial x}\right)^\alpha \frac{\delta G}{\delta u}$ that inherit energy conservation or dissipation property, *J. Comput. Phys.*, **156** (1999), 181–205. (Cited on p.11)

[66] Furihata, D., A stable and conservative finite difference scheme for the Cahn-Hilliard equation, *Numer. Math.*, **87** (2001), 675–699. (Cited on p.11, 135)

[67] Furihata, D., Finite-difference schemes for nonlinear wave equation that inherit energy conservation property, *J. Comput. Appl. Math.*, **134** (2001), 37–57. (Cited on p.11, 115)

[68] Furihata, D. and Matsuo, T., A stable, conservative, and linear finite difference scheme for the Cahn-Hilliard equation, *Japan J. Indust. Appl. Math.*, **20** (2003), 65–85. (Cited on p.11, 282)

[69] Furihata, D. and Mori, M., General derivation of finite difference schemes by means of a discrete variation (in Japanese) *Trans. Japan Soc. Indust. Appl. Math.*, **8** (1998), 317–340. (Cited on p.11)

[70] Furihata, D., Onda, T., and Mori, M., A numerical analysis of some phase separation problem, *Proceedings of the first China-Japan seminar of numerical mathematics*, 1992, 29–44. (Cited on p.3)

[71] Frutos, J. de, and Sanz-Serna, J. M., Accuracy and conservation properties in numerical integration: the case of the Korteweg-de Vries equation, *Numer. Math.*, **75** (1997), 421–445. (Cited on p.157)

[72] Gibbons, J., Thornhill, S.G., Wardrop, M.J., and Ter Haar, D., On the theory of Langmuir solitons, *J. Plasma Phys.*, **17** (1977), 153–170. (Cited on p.24, 61)

[73] Goda, K., On stability of some finite difference schemes for the Korteweg-de Vries equation, *J. Phys. Soc. Jpn.*, **39** (1975) 229–236. (Cited on p.157)

[74] Gonzalez, O., Time integration and discrete Hamiltonian systems, *Nonlinear Science*, **6** (1996), 449–467. (Cited on p.10, 11, 251)

[75] Gonzalez, O., Exact energy and momentum conserving algorithms for general models in nonlinear elasticity, *Comput. Methods Appl. Mech. Engrg.*, **190** (2000), 1763–1783. (Cited on p.11)

[76] Gordon, W., Der Comptoneffekt nach der Schrödingerschen Theorie, *Z. Phys. A*, **40** (1926), 117–133. (Cited on p.67)

[77] Greenspan, D., *Discrete Models*, Addison-Wesley, London, 1973. (Cited on p.10)

[78] Greig, I. S. and Morris, J. L., A Hopscotch method for the Korteweg-de Vries equation, *J. Comput. Phys.*, **20** (1976), 64–80. (Cited on p.157)

[79] Gross, E. P., Structure of quantized vortex in boson systems, *Nuovo Cimento*, **20** (1961), 454–477. (Cited on p.60, 180)

[80] Hanada, T., Ishimura, N. and Nakamura, M., Stable finite difference scheme for a model equation of phase separation, *Appl. Math. Comput.*, **151** (2004), 95–104. (Cited on p.65)

[81] Hairer, E., Long-time energy conservation of numerical integrators, foundations of computational mathematics, *London Math. Soc. Lecture Note Ser.*, **331** (2006), Cambridge Univ. Press, Cambridge, 162–180. (Cited on p.10)

[82] Hairer, E., Nørsett, S.P., and Wanner, G. *Solving Ordinary Differential Equations I — Nonstiff Problems*, Springer-Verlag, Heidelberg, 2000. (Cited on p.259)

[83] Hairer, E., Lubich, C., and Wanner, G., *Geometric Numerical Integration*, Springer-Verlag, Heidelberg, 2002. (Cited on p. x, 10, 11)

[84] Herman, R. L. and Knickerbocker, C. J., Numerically induced phase shift in the KdV soliton, *J. Comput. Phys.*, **104** (1993), 50–55. (Cited on p.157)

[85] Hirota, R., Nonlinear partial diffrence equations. I. A difference analogue of the Korteweg-de Vries equation, *J. Phys. Soc. Jpn.*, **43** (1977), 1424–1433. (Cited on p.11)

[86] Hirota, R., Nonlinear partial difference equations III; Discrete sine-Gordon equation, *J. Phys. Soc. Jpn.*, **43** (1977), 2079–2086. (Cited on p.11, 185)

[87] Hong, J., Jiang, S., and Li, C., Explicit multi-symplectic methods for Klein-Gordon-Schroedinger equations, *J. Comp. Phys.*, **228** (2009), 3517–3532. (Cited on p.11)

[88] Hong, J., Liu, H., and Sun, G., The multi-symplecticity of partitioned Runge-Kutta methods for Hamiltonian PDEs, *Math. Comp.*, **75** (2006), 167–181. (Cited on p.11)

[89] Hughes, T.J.R., Caughey, T.K., and Liu, W.K., Finite-element methods for nonlinear elastodynamics which conserve energy, *Trans. ASME*, **45** (1978), 366–370. (Cited on p.10)

[90] Ide, T., Error estimates for the implicit finite difference scheme for the Fujita problem by means of the discrete variational method, *Far East J. Appl. Math.*, **9** (2002), 137–155. (Cited on p.11, 55)

[91] Ide, T., Hirota, C. and Okada, M., Generalized energy integral for $\frac{\partial u}{\partial t} = \frac{\delta G}{\delta u}$, its finite difference schemes by means of the discrete variational

method and an application to Fujita problem, *Adv. Math. Sci. Appl.*, **12** (2002), 755–778. (Cited on p.11, 55)

[92] Jiménez, S., Derivation of the discrete conservation laws for a family of finite difference schemes, *Appl. Math. Comput.*, **64** (1994), 13–45. (Cited on p.11)

[93] John, F., *Lectures on Advanced Numerical Analysis*, Gordon and Breach, New York, 1967. (Cited on p.122)

[94] John, F., *Partial Differential Equations* (3rd ed.), Springer–Verlag, New York, 1978. (Cited on p.6)

[95] Keller, F. F. and Segel, L. A., Initiation of slime mold aggregation viewed as an instability, *J. Theor. Bio.*, **26** (1970), 399–415. (Cited on p.55)

[96] Klein, O., Quantentheorie und Fünfdimensionale Relativitätstheorie, *Z. Phys. A*, **37** (1926), 895–906. (Cited on p.67)

[97] Koide, S. and Furihata, D., Nonlinear and linear conservative finite difference schemes for regularized long wave equation, *Japan J. Indust. Appl. Math.*, **26** (2009), 15–40. (Cited on p.214, 221, 222)

[98] Korteweg, D.J. and G. de Vries, On the change of form of long waves advancing in a rectangular channel, and on a new type of long stationary waves, *Phil. Mag.*, **39** (1895), 422–443. (Cited on p.56)

[99] Kudryavtsev, A. E., Solitonlike solutions for a Higgs scalar field, *JETP Letter* **22** (1975), 82–83. (Cited on p.185)

[100] La Cruz, W., Martínez, J. M., and Raydan, M., Spectral residual method without gradient information for solving large-scale nonlinear systems of equations, *Math. Comp.*, **75** (2006), 1429–1448. (Cited on p.297, 298)

[101] La Cruz, W. and Raydan, M., Nonmonotone spectral methods for large-scale nonlinear analysis, *Opt. Meth. Soft.*, **18** (2003), 583–599. (Cited on p.296)

[102] Lees, M., Energy inequalities for the solution of differential equations, *Trans. Amer. Math. Soc.*, **94** (1960), 58–73. (Cited on p.126)

[103] Lee, I. J., Numerical solution for nonlinear Klein–Gordon equation by collocation Method with respect to spectral method, *J. Korean Math. Soc.* **32** (1995), 541–551. (Cited on p.185)

[104] Leimkuhler, B. and Reich, S., *Simulating Hamiltonian Dynamics*, Cambridge, London, 2005. (Cited on p.10, 11)

[105] Lenells, J., Conservation laws of the Camassa-Holm equation, *J. Phys. A: Math. Gen.*, **38** (2005), 869-880. (Cited on p.196)

[106] Levermore, C.D., and Oliver, M., The complex Ginzburg-Landau equation as a model problem, *Lectures in Appl. Math.*, **31** (1996), 141–190. (Cited on p.59)

[107] Li, S. and Vu-Quoc, L., Finite difference calculus invariant structure of a class of algorithms for the nonlinear Klein–Gordon equation, *SIAM J. Numer. Anal.*, **32** (1995), 1839–1875. (Cited on p.185, 186)

[108] Li, P. W., On the numerical study of the KdV equation by the semi-implicit and leap-frog method, *Comput. Phys. Comm.*, **88** (1995), 121–127. (Cited on p.157)

[109] Lord, G.J., Attractors and inertial manifolds for finite-difference approximations of the complex Ginzburg-Landau equation, *SIAM J. Numer. Anal.*, **34** (1997), 1483–1512. (Cited on p.339)

[110] Lubich, C., From Quantum to Classical Molecular Dynamics: Reduced Models and Numerical Analysis, European Math. Soc., Zurich, 2008. (Cited on p.11)

[111] Manoranjan, V.S., Ortega, T., and Sanz-Serna, J.M., Soliton and antisoliton interactions in the "good" Boussinesq equation, *J. Math. Phys.*, **29** (1988), 1964–1968. (Cited on p.63)

[112] Marsden, J. E., Patrick, G. W., and Shkoller, S., Multisymplectic geometry, variational integrators, and nonlinear PDEs, *Comm. Math. Phys.*, **199** (1998), 351–395. (Cited on p.11)

[113] Marsden, J. E. and West, M., Discrete mechanics and variational integrators, *Acta Numerica*, (2001), 357–514. (Cited on p.11)

[114] Matsuo, T., High-order schemes for conservative or dissipative systems, *J. Comput. Appl. Math.*, **152** (2003), 305–317. (Cited on p.40, 250, 261)

[115] Matsuo, T., A GBDF approach for designing conservative or dissipative schemes of any high-order, Proceedings of the Seventh China-Japan Seminar on Numerical Mathematics, Shi, Z.-C. and Okamoto, H. (eds.), Science Press, Beijing, 2006, 85-98. (Cited on p.250)

[116] Matsuo, T., New conservative schemes with discrete variational derivatives for nonlinear wave equations, *J. Comput. Appl. Math.*, **203** (2007), 32–56. (Cited on p.11)

[117] Matsuo, T., Dissipative/conservative Galerkin method using discrete partial derivatives for nonlinear evolution equations, *J. Comput. Appl. Math.*, **218** (2008), 506–521. (Cited on p.299)

[118] Matsuo, T., A Hamiltonian-conserving Galerkin scheme for the Camassa–Holm equation, *J. Comput. Appl. Math.*, **234** (2010), 1258–1266. (Cited on p.315)

[119] Matsuo, T. and Furihata, D., Dissipative or conservative finite difference schemes for complex-valued nonlinear partial differential equations, J. Comput. Phys., **171** (2001), 425–447. (Cited on p.11)

[120] Matsuo, T., Sugihara, M., Furihata, D. and Mori, M., Linearly implicit finite difference schemes derived by the discrete variational method, *Suriken Kokyuroku*, **1145** (2000), 121 – 129. (Cited on p.11)

[121] Matsuo, T., Sugihara, M., Furihata, D. and Mori, M., Spatially accurate dissipative or conservative finite difference schemes derived by the discrete variational method, *Japan J. Indust. Appl. Math.*, **19** (2002), 311–330. (Cited on p.11)

[122] Matsuo, T., Sugihara, M., and Mori, M., A derivation of a finite difference scheme for the nonlinear Schrödinger equation, advances in numerical mathematics; Proceedings of the fourth Japan-China joint seminar on numerical mathematics (Chiba, 1998), 243–250, GAKUTO Internat. Ser. Math. Sci. Appl., 12, Gakkotosho, Tokyo, 1999. (Cited on p.11)

[123] Matsuo, T. and Yamaguchi, H., An energy-conserving Galerkin scheme for a class of nonlinear dispersive equations, *J. Comput. Phys.*, **228** (2009), 4346–4358. (Cited on p.315)

[124] McLachlan, R. I., Symplectic integration of Hamiltonian wave equations, *Numer. Math.*, **66** (1994), 465–492. (Cited on p.11)

[125] McLachlan, R. I., On the numerical integration of ordinary differential equations by symmetric composition methods, *SIAM J. Sci. Comput.*, **16** (1995), 151–168. (Cited on p.248)

[126] McLachlan, R. I., Quispel, G. R. W., and Robidoux, N., Unified approach to Hamiltonian systems, Poisson systems, gradient systems, and systems with Lyapunov functions or first integrals, *Phys. Rev. Lett.*, **81** (1998), 2399–2403. (Cited on p.10, 11)

[127] McLachlan, R. I., Quispel, G. R. W., and Robidoux, N., Geometric integration using discrete gradients, *Phil. Trans. R. Soc. Lond. A*, **357** (1999), 1021–1045. (Cited on p.10, 11)

[128] Mclachlan, R. I., and Robidoux, N., Antisymmetry, pseudospectral methods, and conservative PDEs, in the proceedings of International Conference on Differential Equations (Berlin 1999), World Scientific Publishing, River Edge, 2000, 994–999. (Cited on p.11)

[129] McLachlan, R. I., Spatial discretization of partial differential equations with integrals, *IMA J. Numer. Anal.*, **23** (2003), 645–664. (Cited on p.11, 353)

[130] McLaren, D. I. and Quispel, G. R. W., Integral-preserving integrators, *J. Phys. A*, **39** (2004), L489–L495. (Cited on p.11)

[131] Montagne, R., Hernández-García, E., and San Miguel, M., Numerical study of a Lyapunov functional for the complex Ginzburg-Landau equation, *Physica D*, **96** (1996), 47–65. (Cited on p.339)

[132] Mori, S., Numerical schemes preserving the dissipation property of the Ginzburg–Landau equations (in Japanese), master's thesis, the University of Tokyo, 2008. (Cited on p.340, 341)

[133] Mu, M., A linearized Crank-Nicolson-Galerkin method for the Ginzburg-Landau model, *SIAM J. Sci. Comput.*, **18** (1997), 1028–1039. (Cited on p.339)

[134] Newell, A.C., and Whitehead, J.A., Finite bandwidth, finite amplitude convection, *J. Fluid Mech.*, **38** (1969), 279–303. (Cited on p.59)

[135] Olver, P. J., Euler operators and conservation laws of the BBM equation, *Math. Proc. Camb. Phil. Soc.*, **85** (1979), 143–160. (Cited on p.212)

[136] Olver, P. J., *Applications of Lie Groups to Differential Equations*, Springer-Verlag, New York, 1993. (Cited on p.196)

[137] Pen-Yu, K., and Sanz-Serna, J. M., Convergence of methods for the numerical solution of the Korteweg-de Vries equation, *IMA J. Numer. Anal.*, **1** (1981), 215–221. (Cited on p.157)

[138] Peregrine, D. H., Calculations of the development of an undular bore, *J. Fluid Mech.*, **25** (1966), 321–330. (Cited on p.57, 212)

[139] Perring, J. K. and Skyrme, T. H. R., A model unified field equation, *Nucl. Phys.*, **31** (1962), 550–555. (Cited on p.185)

[140] Pitaevskii, L. P., Vortex lines in an imperfect Bose gas, *Sov. Phys. JETP*, **13** (1961), 451–454. (Cited on p.60, 180)

[141] Powell, M. J. D., A hybrid method for nonlinear equations, in *Numerical Methods for Nonlinear Algebraic Equations*, P. Rabinowitz (Ed.), Gordon and Breach, London, 1970. (Cited on p.294)

[142] Qin, M.Z. and Zhao, P.F., Approximation for KdV equation as a Hamiltonian system, *Comput. Math. Appl.*, **39**,5–6(2000), 1–11. (Cited on p.157)

[143] Qin, M.Z. and Zhu, W.J., Construction of higher order symplectic schemes by composition, *Computing*, **47** (1992), 309–321. (Cited on p.247)

[144] Quispel, G. R. W. and Capel, H. W., Solving ODEs numerically while preserving a first integral, *Phys. Lett. A*, **218** (1996), 223–228. (Cited on p.11)

[145] Quispel, G. R. W. and Turner, G. S., Discrete gradient methods for solving ODEs numerically while preserving a first integral, *J. Phys. A*, **29** (1996), L341–L349. (Cited on p.11)

[146] Quispel, G. R. W. and McLaren, D. I., A new class of energy-preserving numerical integration methods, *J. Phys. A*, **41** (2008), 045206. (Cited on p.11)

[147] Richtmyer, R. D. and Morton, K. W., *Difference Methods for Initial-Value Problems*, John Wiley and Sons, New York, 1967. (Cited on p.83)

[148] Sakaguchi, H., Zigzag instability and reorientation of roll pattern in the Newell-Whitehead equation, *Prog. Theor. Phys.*, **86** (1991), 759–763. (Cited on p.286)

[149] Sanz-Serna, J. M., An explicit finite-difference scheme with exact conservation properties, *J. Comput. Phys.*, **47** (1982), 199–210. (Cited on p.157)

[150] Sanz-Serna, J. M., Methods for the numerical solution of the nonlinear Schroedinger equation, *Math. Comput.*, **43** (1984), 21–27. (Cited on p.10)

[151] Sanz-Serna, J. M. and Calvo, M. P., *Numerical Hamiltonian Problems*, Chapman and Hall, London, 1994. (Cited on p.10)

[152] Sekino, Y., On stability and convergence of nonlinear and linear implicit conservative finite difference schemes for the Gross–Pitaevskii equation, master's thesis, Osaka University, Osaka, 2008. (Cited on p.181)

[153] Shashkov, M., *Conservative Finite-Difference Methods on General Grids*, CRC Press, Florida, 1996. (Cited on p.352)

[154] Shimoji, S. and Kawai, T., Multi-valued solution to nonlinear wave equations (in Japanese), in the proceedings of Numerical Analysis Symposium (Nagano, 1997), 39–42. (Cited on p.67, 189)

[155] Strauss, W. and Vazquez, L, Numerical solution of a nonlinear Klein-Gordon equation, *J. Comput. Phys.*, **28** (1978), 271–278. (Cited on p.10, 185)

[156] Sulem, C., and Sulem, P., *The Nonlinear Schrödinger Equation*, Springer, New York, 1999. (Cited on p.60)

[157] Suzuki, M., Fractal decomposition of exponential operators with applications to many-body theories and Monte Carlo simulations, *Phys. Lett. A*, **146** (1990), 319–323. (Cited on p.38, 247)

[158] Swift, J. and Hohenberg, P. C., Hydrodynamic fluctuations at the convective instability, *Physical Review A*, **15** (1977), 319–328. (Cited on p.54)

[159] Taha, T.R. and Ablowitz, M.J., Analytical and numerical aspects of certain nonlinear evolution equations. II. numerical, nonlinear Schrödinger equation *J. Comput. Phys.*, **55** (1984), 203–230. (Cited on p.10)

[160] Taha, T.R. and Ablowitz, M.J., Analytical and numerical aspects of certain nonlinear evolution equations. III. numerical, KdV equation *J. Comput. Phys.*, **55** (1984), 231–253. (Cited on p.10, 157)

[161] Takeya, K., Conservative finite difference schemes for the Camassa–Holm equation (in Japanese), master's thesis, Osaka University, Osaka, 2007. (Cited on p.197)

[162] Takeya, K. and Furihata, D., Conservative finite difference schemes for the Camassa–Holm equation, in preparation. (Cited on p.197)

[163] Tanaka, H., and Nishi, T., Direct determination of the probability distribution function of concentration in polymer mixtures undergoing phase separation, *Phys. Rev. Lett.*, **59** (1987), 692–695. (Cited on p.2)

[164] Vu-Quoc, L. and Li, S., Invariant-conserving finite difference algorithms for the nonlinear Klein–Gordon equation, *Comput. Methods Appl. Mech. Eng.*, **107** (1993), 341–349. (Cited on p.185)

[165] Yaguchi, T., Matsuo, T., and Sugihara, M., An energy conservative numerical scheme on mixed meshes for the nonlinear Schrödinger equation, Proceedings of ICNAAM 2009, AIP Conference Proceedings, **1168** (2009), 892–894. (Cited on p.11, 299)

[166] Yaguchi, T., Matsuo, T., and Sugihara, M., An extension of the discrete variational method to nonuniform grids, *J. Comput. Phys.*, **229** (2010), 4382–4423. (Cited on p.11, 299, 352)

[167] Yaguchi, T., Matsuo, T., and Sugihara, M., Conservative numerical schemes for the Ostrovsky equation, *J. Comput. Appl. Math.*, **234** (2010), 1036–1048. (Cited on p.11)

[168] Yoshida, H., Construction of higher order symplectic integrators, *Phys. Lett. A*, **150** (1990), 262–268. (Cited on p.38, 247)

[169] Yoshinaga, T., Chaos in water waves due to the long-short wave interaction (in Japanese), *Butsuri*, **47** (1992), 300–303. (Cited on p.64)

[170] Zabusky, N. J. and Kruskal, M. D., Interaction of "solitons" in a collisionless plasma and the recurrence of initial states, *Phys. Rev. Lett.*, **15** (1965), 240–243. (Cited on p.157)

[171] Zakharov, V.E., Collapse of Langmuir waves, *Sov. Phys. JETP*, **35** (1972), 908–914. (Cited on p.61)

[172] Zakharov, V. E. and Kuznetsov, E. A., Three-dimensional solitons, *Sov. Phys. JETP*, **39** (1974), 285–286. (Cited on p.57)

[173] Zaki, S. I., Gardner, L. R. T. and Gardner, G. A., Numerical simulations of Klein–Gordon solitary-wave interactions, *Nuovo Cimento Soc. Ital. Fis. B(12)*, **112** (1997), 1027–1036. (Cited on p.185)

Index

a priori estimate, 6, 135, 171
accuracy
    global –, 227, 228
    local –, 228
    orders of –, 227, 234
adaptive integration, 3
Akrivis–Dougalis–Karakashian scheme, 315
Allen–Cahn equation, 53, 149, 151, 162

Bénard convection flow, ix, 59
BB method, 296
Ben-Yu scheme, 186
Benjamin–Bona–Mahony equation, 57, 212, 315, 328
bi-Hamiltonian
    – form, 195
    – structure, 196, 315
bilinear form, 11
bistable system, 55
blow up, 55, 60, 200, 209, 286, 336
Bose–Einstein condensation, 60, 180
Boussinesq equation
    good –, 63
Boussinesq–Schrödinger equation, 64
Brouwer's fixed-point theorem, 171, 172

Cahn–Hilliard equation, ix, 1, 54, 129, 280, 309, 349
Camassa–Holm equation, 57, 195, 212, 315
    limiting –, 315, 320, 329
Cauchy–Schwartz inequality, 121
composition method, 38, 247, 261

conservation property, 23, 27, 31, 34, 56, 66, 84, 99, 106, 114, 116, 236, 238, 259, 264, 267, 270, 278, 280, 304, 305, 308, 316–318, 356
contraction mapping theorem, 137, 139, 207
convection equation, 56, 85
convergence rate, 148, 207, 209
Crank–Nicolson scheme, 18, 23, 83, 86, 311, 321

Dai equation, 315, 325
Delfour–Fortin–Payre scheme, 170
diffusion equation, 13, 18, 36, 53, 82, 87, 92, 349
discrete chain rule, 300, 308, 319, 332, 333
discrete Fourier transform, 35
discrete functional analysis, 90, 119, 171
dissipation property, 7, 17, 37, 40, 46, 81, 97, 109, 234, 237, 256, 263, 267, 269, 277, 279, 301, 302, 341, 343, 355
domain coarsening, 54, 149
Du–Nicolaides scheme, 310

Ebihara equation, 67
Eguchi–Oki–Matsumura equation, 65
elliptic integral, 242
ergodic, 66
Euler scheme, 3, 335, 336
    backward –, 311, 312

implicit –, 321, 322, 324, 335, 336
Euler–Maclaurin summation formula, 137

Feng equation, 68, 222, 223
Fermi–Pasta–Ulam equation, 66
Fisher–Kolmogorov equation
  extended –, 54, 153
Fujita-type equation, 55

Gâteaux derivative, difference, 49, 76
Gagliardo–Nirenberg inequality
  discrete –, 123
Galerkin framework, scheme, 48, 58, 65, 293, 298
Gauss's truncation, 125
geometric numerical integration, x, 10
Ginzburg–Landau equation, 48, 59, 65, 283
  complex –, 23, 98, 164
  time-dependent –, 299, 339
Ginzburg–Landau theory, 59
Gonzalez's discrete gradient, 251
gradient-flow, 5
Green
  – function, 192, 315
  – operator, 192
Gronwall lemma
  discrete –, 126, 180
Gross–Pitaevskii equation, 60, 180–182

Hamiltonian
  – ODE, 10
  – PDE, 11, 62, 63
  – preserving, 261
  – structure, 60
  – system, x, 10
heat conduction equation, 54
Helmholtz operator, 191
Heun scheme, 199, 200, 207, 209

implicit, 42

linearly –, 40, 152, 198, 214, 271, 342
nonlinearly –, 238, 259, 271, 272
inner product, 119
integrable
  – equation, 57, 212
  completely –, 57, 60, 158, 329
integration-by-parts formula, 17, 52, 72
invariant, 168, 244, 248
inverse-scattering technique, 212

Jacobian, 294–297

Keller–Segel equation, 55, 191–193
Kepler problem, 261
Klein–Gordon equation, 10, 67, 185
Klein–Gordon–Schrödinger equation
  coupled –, 64
Korteweg–de Vries equation, ix, 10, 11, 19, 56, 57, 75, 157, 212, 239, 310

Lebesgue space, 119
Li scheme, 186
Lie group method, x
linearization technique, 151, 162, 181, 241, 293, 342
local truncation error, 228
long-wave interaction equation, 64
LU decomposition, 200
Lyapunov functional, 339, 340, 342, 344

mass conservation, 134, 149, 151, 158
mimetic, 299
multi-symplectic method, 11
multi-symplecticity, 11

negative-semidefinite, 62
Neumann boundary condition
  discrete –, 83, 91, 151
Newell–Whitehead equation, ix, x, 23, 59, 165, 285, 288

Newton method, 40, 87, 189, 216, 260, 293, 294, 298
  inexact –, 295, 296
  quasi- –, 295
non-rectangular domain, 348
nonuniform grid, mesh, 299, 348

operator
  averaging –, 70, 141
  backward difference –, 259
  central difference –, 3, 16, 76, 193, 229, 231
  difference –, 13, 35, 42, 69, 70, 141, 229, 249
  forward difference –, 111, 249
  high-order difference –, 35, 39, 232, 249, 259, 262
  second-order difference –, 250
  shift –, 69

partial derivative
  complex –, 308
  complex discrete –, 308
  discrete –, 300
pattern formation, 59
peakon, 199, 322
periodic boundary condition
  discrete –, 86, 99, 165, 167, 169, 196, 213, 223, 225, 229, 273
phase separation, ix, 8, 65, 149, 153
phase shift, 204, 220
phi-4 equation, 185
Poincaré–Wirtinger inequality
  discrete –, 122
Powell's hybrid algorithm, 294
predictor–corrector method, 298
prominence temperature equation, 54
pseudospectral, 239, 241, 244

regularized long wave equation, 57, 212
Runge–Kutta method, scheme, 188, 199, 216, 242, 286, 324

Schrödinger equation
  cubic nonlinear –, 240, 271
  nonlinear –, ix, 10, 24, 60, 100, 167, 180, 280, 313
  odd-order nonlinear –, 283
  relativistic –, 67
Schwartz inequality, 123
self-adjoint integrator, 247
shallow water wave, ix
Shimoji–Kawai equation, 67, 189, 190
short-wave interaction equation, 64
sine-Gordon equation, 67, 185, 187
  double –, 185
skew-symmetric, symmetry, 62, 116, 229, 251, 316, 330
Sobolev inequality, 6
  discrete –, 176
Sobolev lemma, 90
  discrete –, 121, 136, 206, 221, 225
Sobolev norm
  discrete –, 136
Sobolev space, 119
  discrete –, 120
Sobolev–Hilbert norm, 136
  discrete –, 206
soliton, 11, 57, 159, 162, 170, 185, 204, 244, 328, 335
  quasi- –, 57
spectral
  – differentiation, 35
  – residual method, 296
spinodal decomposition, 1, 54, 65
staggered grid, 117, 118
stationary breather, 68
Strauss scheme, 185, 188
string vibration equation, 67
structure-preserving, 10, 11, 13, 137
structure-preserving method, x
summation rule, 7, 70, 93
  rectangle –, 7
summation-by-parts formula, 16, 37, 92, 108, 230
  first-order –, 72

higher-order –, 73
second-order –, 73
superconductivity, 48, 59
Swift–Hohenberg equation, 54
symplectic method, x, 10, 11

Taylor expansion, 177
time-symmetric, 247
trapezoidal rule, 7, 70, 93, 230
traveling wave solution, 181

undular bore problem, 212

variational
   – form, 12, 61, 195, 329, 330
   – integrator, 11
   – structure, 11
variational derivative, 5, 20, 49, 50
   complex –, 20, 50
   complex discrete –, 21, 93
   discrete –, 13, 16, 29, 32, 79, 102
   high-order discrete –, 39, 231, 232, 251
   multiple-points discrete –, 46, 274
   semi-discrete –, 79, 355
   three-points discrete –, 29, 45, 46, 112, 114, 214, 273
Voronoi mesh, 298, 348

wave equation
   linear –, 28, 30, 66, 100, 185
   nonlinear –, 60
Wronskian form, 11

Zakharov equations, 24, 61, 63, 104, 183, 284
Zakharov–Kuznetsov equation, 57, 159
Zhang
   – explicit scheme, 186, 188
   – implicit scheme, 186, 188